CO-ALU-734

Neutrons in Biology

BASIC LIFE SCIENCES

Ernest H. Y. Chu, Series Editor
The University of Michigan Medical School
Ann Arbor, Michigan

Alexander Hollaender, Founding Editor

Recent Volumes in the series:

Volume 50	CLINICAL ASPECTS OF NEUTRON CAPTURE THERAPY Edited by Ralph G. Fairchild, Victor P. Bond, and Avril D. Woodhead
Volume 51	SYNCHROTRON RADIATION IN STRUCTURAL BIOLOGY Edited by Robert M. Sweet and Avril D. Woodhead
Volume 52	ANTIMUTAGENESIS AND ANTICARCINOGENESIS MECHANISMS II Edited by Yukiaki Kuroda, Delbert M. Shankel, and Michael D. Waters
Volume 53	DNA DAMAGE AND REPAIR IN HUMAN TISSUES Edited by Betsy M. Sutherland and Avril D. Woodhead
Volume 54	NEUTRON BEAM DESIGN, DEVELOPMENT, AND PERFORMANCE FOR NEUTRON CAPTURE THERAPY Edited by Otto K. Harling, John A. Bernard, and Robert G. Zamenhof
Volume 55	*IN VIVO* BODY COMPOSITION STUDIES: Recent Advances Edited by Seiichi Yasumura, Joan E. Harrison, Kenneth G. McNeill, Avril D. Woodhead, and F. Avraham Dilmanian
Volume 56	NMR APPLICATIONS IN BIOPOLYMERS Edited by John W. Finley, S. J. Schmidt, and A. S. Serianni
Volume 57	BOUNDARIES BETWEEN PROMOTION AND PROGRESSION DURING CARCINOGENESIS Edited by Oscar Sudilovsky, Henry C. Pitot, and Lance A. Liotta
Volume 58	PHYSICAL AND CHEMICAL MECHANISMS IN MOLECULAR RADIATION BIOLOGY Edited by William A. Glass and Matesh N. Varma
Volume 59	PLANT POLYPHENOLS: Synthesis, Properties, Significance Edited by Richard W. Hemingway and Peter E. Laks
Volume 60	HUMAN BODY COMPOSITION: *In Vivo* Methods, Models, and Assessment Edited by Kenneth J. Ellis and Jerry D. Eastman
Volume 61	ANTIMUTAGENESIS AND ANTICARCINOGENESIS MECHANISMS III Edited by G. Bronzetti, H. Hayatsu, S. DeFlora, M. D. Waters, and D. M. Shankel
Volume 62	BIOLOGY OF ADVENTITIOUS ROOT FORMATION Edited by Tim D. Davis and Bruce E. Haissig
Volume 63	COMPUTATIONAL APPROACHES IN MOLECULAR RADIATION BIOLOGY: Monte Carlo Methods Edited by Matesh N. Varma and Aloke Chatterjee
Volume 64	NEUTRONS IN BIOLOGY Edited by Benno P. Schoenborn and Robert B. Knott

---

A Continuation Order Plan is available for this series. A continuation order will bring delivery of each new volume immediately upon publication. Volumes are billed only upon actual shipment. For further information please contact the publisher.

Neutrons in Biology

Edited by

Benno P. Schoenborn

Los Alamos National Laboratory
Los Alamos, New Mexico

and

Robert B. Knott

Australian Nuclear Science and Technology Organisation
Menai, New South Wales, Australia

Plenum Press • New York and London

Library of Congress Cataloging-in-Publication Data

Neutrons in biology / edited by Benno P. Schoenborn and Robert B. Knott.
p. cm. -- (Basic life sciences ; v. 64)
"Proceedings of the Workshop on Neutrons in Biology, held October 24-28, 1994, in Santa Fe, New Mexico"--T.p. verso.
Includes bibliographical references and index.
ISBN 0-306-45368-1
1. Neutrons--Scattering--Congresses. 2. Biology--Methodology--Congresses. I. Schoenborn, Benno P. II. Knott, Robert B. III. Workshop on Neutrons in Biology (1994 : Santa Fe, N.M.) IV. Series.
QH324.9.N48N48 1996
574'.028--dc20

96-41800
CIP

QH
324
.9
.N48
N48
1996

Proceedings of the Workshop on Neutrons in Biology, held October 24–28, 1994, in Santa Fe, New Mexico

ISBN 0-306-45368-1

© 1996 Plenum Press, New York
A Division of Plenum Publishing Corporation
233 Spring Street, New York, N.Y. 10013

10 9 8 7 6 5 4 3 2 1

All rights reserved

No part of this book may be reproduced, stored in a retrieval system, or transmitted in any form or by any means, electronic, mechanical, photocopying, microfilming, recording, or otherwise, without written permission from the Publisher

Printed in the United States of America

PREFACE

This compendium presents some of the major applications of neutron scattering techniques to problems in biology. It is a record of the papers presented at the Neutrons in Biology Conference, the third in an occasional series held to highlight progress in the field and to provide a focus for future direction.

The strength of the neutron scattering technique remains principally in the manipulation of scattering density through hydrogen and deuterium atoms. The development of advanced detectors, innovative instrument and beamline components, and sophisticated data acquisition systems through the 1970s and early 1980s provided a sound foundation for the technique. With continued development, some of the exotic and expensive equipment has become affordable by the medium-sized facilities, thereby broadening the user base considerably.

Despite problems with the major neutron sources in the late 1980s and early 1990s, some spectacular results have been achieved. Whilst the high and medium flux beam reactors will continue to make a major impact in the field, the results from the first experiments, and the planned developments on spallation neutron sources, clearly indicate that the technique has enormous potential.

A theme for general discussion expressed succinctly by Steve White at the Conference dinner related to the continued growth of neutrons in biology. Competition for research funds has intensified over recent years and political reality dictates that unity within the neutron scattering community is essential. The priorities for types and locations of neutron sources and related facilities must be agreed to in the scientific forum and then aggressively promoted to funding authorities. With a well-coordinated strategy, neutron scattering in general, and neutrons in biology in particular, will play a major role in the understanding of structure and dynamics.

The Committee wishes to express their appreciation to the speakers and session chairpersons for their participation in the Conference, and to Jan Hull for conference planning. For the preparation of the papers for publication, we particularly wish to thank the referees, and Tricia Lewis and Cherylie Thorn for invaluable assistance.

The Third International Neutrons in Biology Conference was supported by the Los Alamos National Laboratory, the International Union of Crystallography and the US Department of Energy.

Conference Committee

Benno P. Schoenborn, *Chairman*
Robert B. Knott
Sax Mason
Andrew Miller
Nobuo Niimura
Charles C. Wilson

CONTENTS

Small Angle Scattering

Membrane Structures and Dynamics

Protein Structures

Fiber Diffraction

New Analysis and Experimental Techniques

NEUTRONS IN BIOLOGY

A Perspective

R. B. Knott[1] and B. P. Schoenborn[2]

[1] Australian Nuclear Science and Technology Organization
Private Mail Bag, Menai NSW 2234, Australia
[2] B. P. Schoenborn, Los Alamos National Laboratory,
Los Alamos New Mexico 87545

INTRODUCTION

After almost a decade of uncertainty, the field of neutrons in biology is set to embark on an era of stability and renewed vitality. As detailed in this volume, methodologies have been refined, new tools are now being added to the array, the two largest reactor sources have long term programs in place, and spallation sources are making an impact. By way of introduction, it is pertinent to reflect on the origins of the field and to highlight some aspects that have influenced the progress of the field. In an increasingly competitive environment, it is extremely important that the future capitalize on the substantial investment made over the last two to three decades.

The important similarities and differences that exist between X-ray and neutron scattering techniques are well documented. For X-rays, the scattering centers are the atomic electrons. Consequently, X-rays are scattered in proportion to the number of electrons thus leading to a strict correlation with the atomic position in the periodic table. For neutrons, on the other hand, the scattering centers are the atomic nuclei. Each nucleus has a characteristic interaction with a neutron. There is considerably less variation and the interaction can be different for different isotopes of the same element.

Indeed it was the neutron interaction with hydrogen and with deuterium that set neutron scattering apart. Hydrogenated or deuterated biomolecular components (atoms, molecules, molecular fragments and/or molecular aggregates) could be selected to address the particular structural issue under investigation, and the resultant contrast variation became a tool of immense potential value (Schoenborn & Nunes, 1972). The technique found application in three general areas: high resolution single crystal diffraction; low resolution diffraction from samples with long range order in one or two dimensions; and small angle scattering from randomly ordered structures.

Neutrons in Biology, edited by Schoenborn and Knott
Plenum Press, New York, 1996

In single crystal diffraction analysis using X-rays, the interaction with hydrogen is small, and too small to assign a hydrogen position with diffraction data of limited resolution. At best the analysis is guided by stereochemical data for the parent atom, which can often present more than one option for the hydrogen position, and cannot account for the natural variability in stereochemistry. The unique ability of high resolution neutron diffraction to image hydrogen and deuterium atoms was demonstrated initially in the case of many small biologically active molecules including vitamin B12 (Moore *et al.,* 1967). Whilst these studies provided a firm foundation, it still required a quantum leap in scientific methodology to investigate relevant problems in protein structure.

The early structural results on myoglobin and hemoglobin by Kendrew and Perutz in the 1960's, gave the first insight into the complex relationship between structure and function of proteins and laid the foundation of modern structural biology. Those pioneering efforts raised a myriad of questions about detailed molecular interactions involving basic structural motifs like hydrogen bonding, charge transfer and non bonding interactions generally referred to as van der Waals forces. The detailed analysis of these interactions by X-ray protein crystallography was impossible since proteins did not provide the high resolution data needed to localize hydrogen atoms - atoms that play a major role in protein structure and function. The oxygen carrying transport mechanism of hemoglobin proposed by Perutz (Bolton *et al.,* 1968) based on his work on deoxyhemoglobin involved the exact position of the iron atom in the heme and the presence or absence of a hydrogen atom on the distal histidine (Nobbs *et al.,* 1966). Similar questions, including the nature of non bonded interactions were posed by the observation that the inert gas xenon binds to myoglobin and hemoglobin (Nobbs *et al.,* 1966; Schoenborn, 1965; 1969a; Schoenborn *et al.,* 1965). Clearly there was a need for the exact location of hydrogen atoms and the resulting hydrogen bonds. Neutron diffraction was assessed to be the only way to determine hydrogen atom locations in proteins, and to localized solvent molecules. Given the fundamental and continued interest in the structure and function of myoglobin, it was clearly the protein to apply neutron diffraction methods (Schoenborn, 1969b). The basic motivation for the neutron diffraction study was to understand structural details of the heme environment. The strategy revolved around a comparison between myoglobin structure with CO and O_2 ligand binding. An important finding was in the case of CO binding. The imidazole Ne2 of the distal His-64(E7) was not protonated (Hanson & Schoenborn, 1981) in direct contrast to its state on O_2 binding (Phillips & Schoenborn, 1981). The myoglobin structure continues to provide important information on a range of structure-related issues.

Both neutron and X-ray diffraction pose the inherent phase problem. A single experiment measures only intensities and does not provide the phase information needed to calculate density maps. It is the density maps that image the molecular arrangement under investigation. Neutron diffraction, similar to the X-ray case does offer the possibility of determining phases experimentally through the use of heavy atom derivatives or anomalous dispersion effects using isotopes of gadolinium, cadmium or samarium etc. The use of these approaches in neutron diffraction studies is, however, time and neutron consuming and, usually, initial phases are derived from the structure obtained by X-ray diffraction. To show that such X-ray derived phases are sufficiently accurate to start refinement, a set of anomalous neutron phases for myoglobin was experimentally determined using the Cd isotope 113 (Schoenborn, 1975) and this demonstrated unequivocally that phases calculated from X-ray data are suitable to initiate neutron data refinement.

High resolution neutron diffraction investigations have consistently revealed hydrogen atom positions and, because of different scattering interactions, have also distinguished nitrogen from carbon or oxygen. This has been used to resolve ambiguities in X-ray crystallographic studies particularly the orientation of histidines, for example.

To reduce background scattering during data collection and enhance the localisation of exchangeable hydrogens, the H_2O solvent in myoglobin single crystals was exchanged with D_2O. This lead immediately to the realisation that H_2O/D_2O exchange can (i) provide some information on protein dynamics and (ii) enhance protein/solvent contrast in diffraction and small angle neutron scattering experiments.

Proteins are not static objects and any description of structure incorporates some element of dynamics. Indeed, the structure determined by X-ray diffraction techniques will explicitly include the thermal motion (and possible disorder) of the individual non-hydrogen atoms. A neutron diffraction analysis will also produce similar information with additional data on hydrogen atoms. Depending on the resolution of the data, the thermal motion of hydrogen atoms is included in the refinement. This leads to some interesting information on rotation of terminal methyl groups, for example. Spectroscopic studies had established that methyl groups exhibit rapid rotation but provided little information on preferred orientations. This information is directly available from the neutron diffraction structure. For example, 85% of the ordered methyl groups in trypsin are within 20 degrees of the staggered conformation. This does not mean that the groups are rigidly locked in position, but only that their rotation is quantized in 120 degree steps, about a position of highest stability (Kossiakoff, 1983).

Dynamics in another time domain are explored by the distribution of exchanged hydrogen atoms throughout the protein structure. This distribution will indicate the outcome of the exchange mechanism(s) on a scale dictated by the solvent-exchange time (from weeks to years). Clearly solvent accessibility plays a dominant role in the mechanism, and parts of the resultant exchange maps can be readily interpreted in terms of classical arguments. For example, the amide hydrogen atoms located on exposed turns of alpha-helical segments are readily exchanged (Schoenborn, 1972). However a number of exchange sites do present problems in interpretation. Therefore a number of exchange mechanisms have been proposed that involve cooperative motions including local breathing and global unfolding, but further investigation is required (Kossiakoff & Shteyn, 1984; Mason *et al.*, 1984).

These issues were further highlighted by the analysis of the structure of trypsin. The primary motivation for these studies was related to the mechanism of the serine protease enzyme activity (Kossiakoff & Shteyn, 1984; Kossiakoff & Spencer, 1980; 1981). The specific question was which of the two residues, Asp-102 or His-57, acted as the catalytic site. The neutron diffraction results demonstrated that the site of attachment of the proton is the imidazole of His-57 and not the carboxylate of Asp-102. In the process of addressing this specific mechanistic question, the study provided a wealth of information on more general aspects of H/D exchange mechanisms. There are 215 amide hydrogen atoms in trypsin. Of these, 67% are fully exchanged, 8% are partially exchanged and 25% are unexchanged after one year of solvent exchange. The locations of the unexchanged sites show a definite pattern. They are clustered in the regions occupied by the β-sheets which are characteristic of serine protease structure. Some of the β-sheets are located close to the protein surface; others are buried in the interior. The general lack of exchange within these structure elements appears to be independent of the distance to the solvent. This suggests that the type of structure can dominate over solvent accessibility. In summary, the degree of H/D exchange was found to be strongly correlated with the extent of hydrogen bonding of the site and those sites adjacent to it. Some correlation with solvent accessibility was found but is likely to be largely a consequence of the structure environment. On

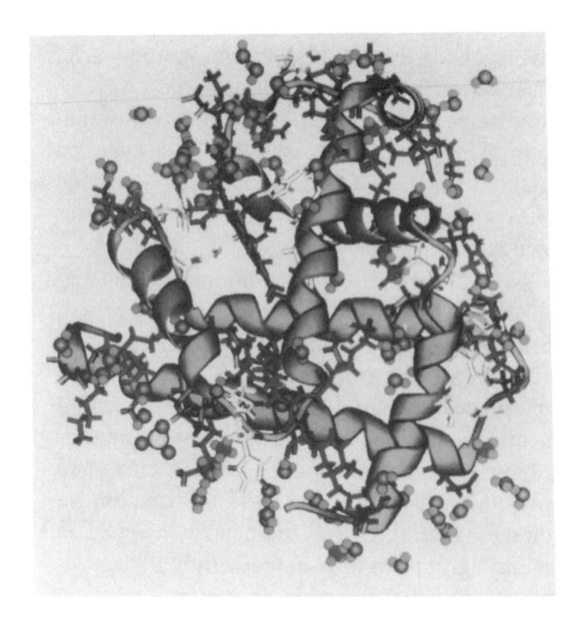

Figure 1. The myoglobin structure with water molecules as determined by neutron diffraction studies. Water is shown as dotted clouds on the surface of the protein (after Cheng & Schoenborn, 1990). Observed bound water molecules and 5 solvent ions identified in this study are depicted. Of the total of 89, 39 water molecules are bound to protein side chain atoms; 16 are bound to main chain atoms; and 12 are in bridges between protein atoms - 10 are intramolecular and 1 intermolecular. The 22 remaining water molecules are bound only to other water molecules. All water molecules bound to the protein are hydrogen bonded to polar or charged side chains that are depicted in the diagram. (A color version of this figure appears in the color insert following p. 214.)

the other hand, no significant correlation was observed with either the hydrophobicity or the thermal motion of the residue side chain.

A protein molecule is surrounded by layers of solvent that mediate its functional conformation as well as its biochemical characteristics. Information on this interface region has been obtained from high resolution neutron diffraction studies. Water constitutes between 40 - 60% of the volume in protein crystals. Water molecules that are hydrogen bonded to the protein surface can be directly visualized as integral components of the structure (Phillips & Schoenborn, 1981; Raghavan & Schoenborn, 1984; Teeter, 1984; Savage & Wlodawer, 1986). Of course the oxygen atoms of well ordered water molecules are assigned in X-ray structures. Such data usually correlate with the surface characteristics of the protein. Ordered water molecules have been located in hydrophilic regions, and small water clusters have been observed quasi-randomly distributed over the protein surface (Figure 1). Recently a formalism has been developed that models the solvent as a series of shells with spatial and physical characteristics (Schoenborn, 1988; Cheng & Schoenborn, 1990). Progressing outward from the protein surface, each shell is constructed of pseudoatoms arranged on a three-dimensional grid. Each pseudoatom is assigned coordinates and a global factor that represents the degree of order (or liquidity) within the shell. This information is included in the structure refinement. Convergence on a unique set of parameters can be expected with a data set that includes the low order diffraction data. It has been shown in studies of myoglobins and plastocyanin that such a solvent refinement enhances surface characteristics and even adjusts side chain locations. Recently it has been shown that such an approach is equally valid for X-ray data refinement (Jiang & Brünger, 1994; Shu, 1994).

The value of any experimental technique must be assessed in the light of comparable results from other techniques. Considerable progress is being made in constructing models that explain directly comparable data from X-ray scattering, NMR analysis, and/or computer simulation, with neutron scattering data. For example, molecular dynamics simulations of the 89 bound water molecules on carbonmonoxymyoglobin indicate that only four molecules are 'continuously' bound while the other 'bound' water molecules are labile and continuously break and make hydrogen bonds. Nevertheless at any selected time point 73 of the water sites are occupied (Gu & Schoenborn, 1995). These results explain in part, the apparent difference between NMR and diffraction experiments. In general, the residence time of a particular water molecule on the protein surface is too short to be observed by NMR techniques. However, the sites are nearly fully occupied on a time scale commensurate with data collection times usual in diffraction experiments.

Low resolution protein crystallography using contrast matching techniques are particularly effective in studying larger crystalline complexes. Examples include the nucleic acid in spherical viruses, the DNA in nucleosome core particles, the lipid in lipoproteins and detergent in crystals of detergent solubilized membrane proteins. Such neutron crystallographic techniques have been used to extend the X-ray studies of tomato bushy stunt virus by localising the RNA and missing protein enabling details of the protein-nucleic acid interactions to be elucidated (Timmins, 1988; Timmins *et al.*, 1994). Another example is the study of Lipovitellin, a lipoprotein found in the oocytes of egg-laying animals. Whilst the structure of the protein moiety was solved using X-ray crystallography, none of the bound lipid was located. Low resolution neutron diffraction methods located the lipid as a condensed phase in a cavity of the protein (Timmins *et al.*, 1992). There is some evidence that suggested the lipid is in the form of a bilayer.

Recognising the importance of ions and water to the hydration of DNA fibers, X-ray and neutron fiber diffraction data have been used to explore the location of water around the DNA double helix in both the D and A conformations (Fuller *et al.*, 1995; Langan *et al.*, 1993; Forsyth *et al.*, 1992; Forsyth *et al.*, 1995). Some of the results of these studies are presented in this volume, with the off-helix view of the water distribution in the major groove of the A form of DNA (Fuller *et al.*, 1995: Figure 9) providing a spectacular example. Computer calculations of two and three point correlation functions of polar solvents in water using a potential-of-mean force expansion, has accurately reproduced the hydration patterns of the A form of DNA (Garcia *et al.*, 1995). These experimental and theoretical studies are providing the basis for understanding the impact of changes in the concentration of water and ions in inducing conformational transitions in the DNA double helix.

The concept of contrast enhancement was also employed to study membrane structures. Contrast enhancement, particularly specific deuteration became a powerful tool in the one-dimensional diffraction analysis of membrane structures (Schoenborn *et al.*, 1970; Zaccai *et al.*, 1975; Yeager, 1976; Worcester, 1975; 1976; Blasie *et al.*, 1976; Schoenborn, 1976a; Chabre, 1975; Worcester & Franks, 1976; Pardon *et al.*, 1975). Neutron diffraction continues to make a useful contribution to membrane research (Blasie *et al.*, 1984; Herbette *et al.*, 1984; Herbette *et al.*, 1985; Bradshaw, 1995; Worcester *et al.*, 1995).

Early experiments with myoglobin and hemoglobin in solutions of various H_2O and D_2O mixtures showed that solvent contrast adjustment could be used to depict the hydration shell by analysing the observed radii of gyration as a function of protein/solvent contrast. Stuhrmann (1973; 1974) subsequently used these observations to develop the detailed analysis of proteins in solution resulting in the now commonly used Stuhrmann plot to analyse small angle scattering data. Such small angle solution scattering data were also frequently used to determine molecular weights using known proteins as calibration points (Wise *et al.*, 1979) or using water to calibrate the scattering curve (Jacrot & Zaccai, 1981). These scattering analyses have led to a better understanding of the tertiary structure of proteins and protein complexes in solution, as well as the protein/solvent interface (Stuhrmann, 1974; Ibel & Stuhrmann, 1975; Lehman & Zaccai, 1984).

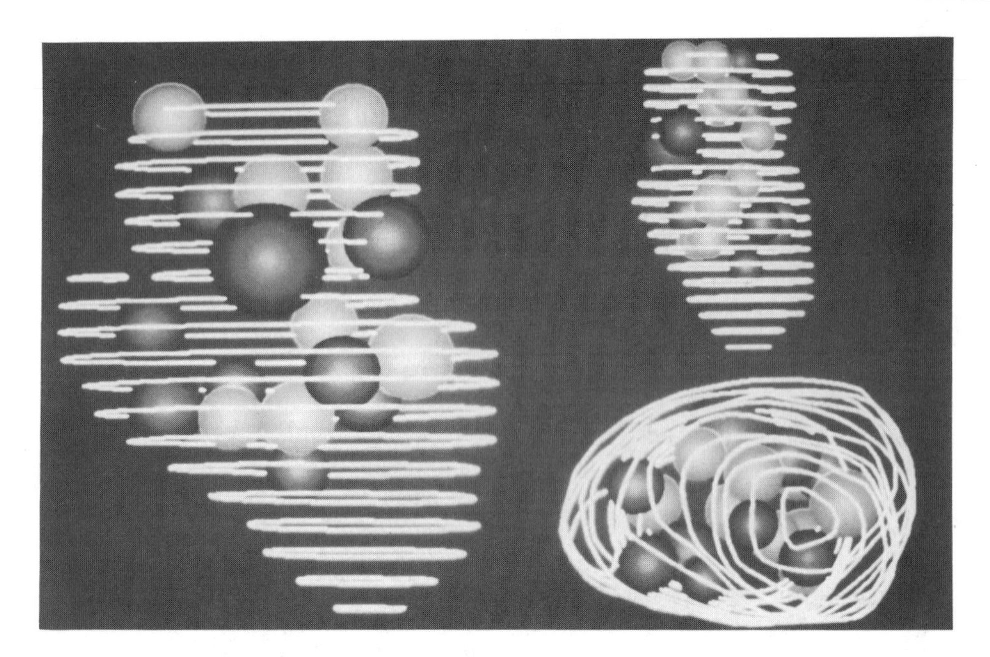

Figure 2. Three orthogonal views of the low resolution structure of the 30S ribosomal subunit of *E.coli* as determined by small angle neutron scattering techniques using specifically deuterated proteins. This structure was constructed by positioning the 21 proteins of known molecular weight by measuring the distance between each pair of proteins. The complete set of distances was determined by an extensive series of neutron scattering experiments. Literally hundreds of ribosome samples were required with one or two proteins deuterated according to a strict protocol (after Capel *et al.*, 1987). The maximum linear dimension of the array is about 190Å. The size of the spheres is proportional to their molecular weight. The numbering of the proteins follows the standard nomenclature for ribosomal proteins. The two views are front and back views of the model. (A color version of this figure appears in the color insert following p. 214.)

A notable expansion of contrast enhancement in the analysis of large protein complexes was proposed by Engelman and Moore (1972) which led to the eventual determination of the structural arrangement of the 21 proteins in the 30S subunit of the ribosome (Figure 2) (Capel *et al.*, 1987), and the proteins in the 50S subunit (May *et al.*, 1984). These were herculean tasks that integrated specific deuteration of protein subunits with advanced neutron scattering techniques.

In order to offset the inherent limitation of neutron source intensity, it was recognised from the outset that concerted strategies must be implemented to maximize the quantity and quality of experimental neutron scattering data. Consequently a number of major projects in the development of instrumentation and experimental protocols were initiated. The early uses of neutron scattering and diffraction for the analysis of biological structures were paralleled by extensive development of equipment that included the multilayer monochromator (Schoenborn *et al.*, 1974) and advanced detectors (Alberi, 1976; Cain *et al.*, 1976; Schoenborn *et al.*, 1978; Fischer *et al.*, 1983; Radeka *et al.*, 1995), as well as sophisticated data analysis techniques (Schoenborn, 1983a; 1983b).

Multilayer monochromators are extraordinarily efficient devices that can be designed for a specific application. High reflectivity (up to 98%), tuneable bandwidth and negligible harmonic contamination are unique features. In a field determined to maximize the neutron flux at the sample position, the multi-layer monochromator system has been used with some considerable success (Lynn *et al.,* 1976; Saxena & Schoenborn, 1976; 1977; 1988; Saxena & Majkrzak, 1984; Knott, 1995).

Major improvements in both small angle scattering research and in protein crystallography have been achieved by use of efficient position sensitive detectors with good resolution and positional stability (Glinka & Berk, 1983; Convert & Forsyth, 1983). The development of sophisticated detector technology has been very demanding but now detectors of incredible performance can be designed for each application (Radeka *et al.,* 1995).

Collecting diffraction data on a position sensitive detector with advanced computer technology leads to the possibility of new and more efficient data collection protocols. Not only is it possible to collect a number of reflections simultaneously, it is possible to collect the data in multi-dimensional space and integrate the data and its background (Schoenborn, 1983a; 1983b; 1992a). This leads to a significant improvement in the counting statistics for each reflection for no greater investment in neutron beamtime. Further innovation in experimental techniques continues - crystal growing techniques are improving, and structure as a function of temperature now offers the possibility of differentiating between disorder and thermal motion. Many of the major developments highlighted above have occurred at the large neutron facilities. With time many of the benefits are adopted by the medium flux reactors (Knott, 1995; Niimura, 1995a; 1995b). Careful selection of problems can lead to a worthwhile contribution from instruments with limited intensity. As outlined in more detail in this volume, the use of spallation neutron sources will add a new dimension to the field.

Many of the earlier developments in structural biology using neutrons were presented at a Brookhaven Symposium on Neutrons in Biology in 1975 (Schoenborn, 1976b). This symposium was held at the time when the capabilities of the new reactor at the Institute Laue-Langevin (Grenoble) started to have an impact. The sophisticated instrumentation, as exemplified by the small angle instrument D11 located on a cold source, immediately generated superb data that enabled the analysis of chromatin (Baldwin *et al.*, 1975; Hjelm *et al.,* 1977), and protein complexes (Stuhrmann, 1974; Marguerie & Stuhrmann, 1976). The development of structural biology at the ILL in Grenoble was greatly facilitated by the establishment of the EMBL outstation by Sir John Kendrew under the able scientific leadership of A. Miller.

The development of a dedicated small angle station for structural biology at the High Flux Beam Reactor (HFBR) at the Brookhaven National Laboratory (BNL) improved the availability of neutrons for structural biology research in the US (Schoenborn *et al.,* 1978). With the completion of the cold source, the utility of that instrument was greatly enhanced. The small angle scattering instrument was soon followed by the establishment of a dedicated protein crystallography station at BNL, followed by the installation of a banana detector for protein crystallography at the ILL. These developments were rewarded with a number of protein crystallographic studies of proteins including myoglobin (Hanson & Schoenborn, 1981; Phillips & Schoenborn, 1981; Cheng, 1990; Cheng & Schoenborn, 1990), trypsin (Kossiakoff & Shteyn, 1984; Kossiakoff & Spencer, 1980; 1981), lysozyme (Mason *et al.,* 1984; Bentley & Mason, 1981; Bentley *et al.,* 1979; 1983), ribonuclease (Wlodawer & Sjölin, 1984), RNase Wlodawer *et al.,* 1983), trypsin inhibitor (Wlodawer *et al.,* 1984), crambin (Teeter, 1984; Teeter & Kossiakoff, 1984) and later by plastocyanin (Church, 1992), concanavalin (Gilboa & Yariv, unpublished) and fatty acid binding protein (Sacchetini & Scapin, unpublished). Such advances in protein crystallography were paralleled with an explosion in the study of oriented but non-crystalline systems such as retinal rods (Yeager, 1976; Yeager *et al.,* 1980a; 1980b), lipid bilayers (King *et al.*, 1984; White & Weiner, 1995; White *et al.*, 1981; Wiener & White, 1991a; 1991b), lipoproteins (Timmins *et al.*, 1992; Timmins & Pebay-Peyroula, 1995; Trewhella *et al.*, 1994; Burks & Engelman, 1981), bacteriorhodopsin (King *et al.*, 1979; King & Schoenborn, 1985), spherical viruses (Timmins, 1995), oriented helical viruses (Nambudripad *et al.*, 1991; Stark *et al.*, 1988) and muscle (Curmi *et al.*, 1988; Curmi *et al.*, 1991). Another study of major significance is the continuation of the chromatin structure investigation - the complex between DNA and histones of chromosomes (Graziano *et al.*, 1995; Ramakrishnan *et al.*, 1993). Much of the earlier progress in this period was presented at the second Neutrons in Biology Conference held in 1981 (Schoenborn, 1984).

Inspection of the Brookhaven Protein Data Bank clearly indicates that at least in terms of quantity, neutron protein crystallography has not mirrored the enormous growth of X-ray protein crystallography. With the low neutron flux as compared to X-ray sources, data collection proved to be very demanding. Even with very large crystals (10 to 30mm^3) a high resolution data set would take a few months to collect. This posed a number of logistical difficulties. Staff numbers at the large neutron sources were too few to collect data for external users and external users frequently could not spend the lengthy periods at reactors or pay for the extended travel costs. Efforts to establish local crystallographic groups of sufficient size to collect and analyse data were not entirely successful. The Center for Structural Biology was established at BNL in 1983 but did not reach an efficient size to properly develop instruments, attend to users and expand the frontiers of structural neutron science. Unfortunately the leadership of that center has been vacant since 1991.

With shortages in personnel and funding, technical developments were slow and data collection was hampered by data acquisition systems with limited computing power and software; minimal scientific and technical support and frequent shutdowns of the cold moderator and the HFBR itself. With these problems and the uncertainties of reactor availability particularly at the HFBR, more and more scientists were reluctant to commit resources to problems requiring neutrons. The operation of the small angle station at the HFBR serves as an example with an average operational time of 66% for the period from 1982 to 1987. The range was from a low of 44 days per year to full operation of 234 days. From 1988 through 1994 the record was worse with an average reactor uptime of 45%. This ranged from a virtual complete shutdown from 1989 through 1991, with full operation only achieved in 1988. The development of neutron scattering facilities in the US was further impacted by the lack of funds to extend the Brookhaven's HFBR facility with a guide hall and perceived safety standards that led to a reduction in reactor power. Fortunately the guide hall and cold moderator project at NIST was funded by the Department of Commerce resulting in vastly improved US small angle scattering instrumentation since it came on line in 1993.

In addition to the reactor facilities, a small angle instrument for structural biology research became available in 1988 at the pulsed spallation source LANSCE (Los Alamos Neutron Scattering Center) (Seeger & Hjelm, 1991). The facility was parasitic to the nuclear science program of the Los Alamos Meson Physics Facility (LAMPF) and beam time was difficult to schedule within the operational time that varied from 2 to 4 months per year. There were frequent unscheduled interruptions. However, as a consequence of major refocussing of research efforts, the outlook for the LANSCE facility (see Schoenborn & Pitcher,1995: Figure 1a) looks much better. The present upgrade of the accelerator, LAMPF, and dedicated beam time at LANSCE of 8 month per year by 1998 (Pynn, 1995) will permit proper scheduling of experiments. The conceptual design of a protein crystallography station for LANSCE is presented in this volume (Schoenborn & Pitcher, 1995: Figures 3 and 4). Funding for diffraction instruments for structural biology by the

Department of Energy (that supports major facilities) is not yet available thereby delaying the installation of such instruments by at least another 2 years.

Apart from the recent lengthy shutdowns at the ILL for a planned upgrade and an unplanned vessel repair, the development of neutron science fared much better in Europe. These developments include the Orphee reactor in Paris; the expansion of the already superb facilities at ILL with the addition of a second guide hall with new instrumentation; the HMI reactor in Berlin; the completion of the spallation source ISIS at the Rutherford Appleton Laboratory; the construction of a continuous spallation source SINQ in Switzerland; and the planned new reactor in Germany. Instrumentation for structural biology studies at the ISIS neutron source are still incomplete but recent experiments on ordered DNA (Langan *et al.,* 1993; Forsyth *et al.,* 1992; Forsyth *et al.,* 1995) showed the utility of spallation neutrons for the investigation of biological systems. A prototype protein crystallography facility has been evaluated and plans for a major upgrade have been formulated (Wilson, 1995). Plans to develop a structural biology program at SINQ are unclear at present. Many of these major advances are presented in this volume.

The recent decision by the US Department of Energy and Congress to halt the development of the Advanced Neutron Source in Oakridge National Laboratory is a serious blow to the neutron community. However, with appropriate planning this decision will allow the extension of present neutron sources and particularly encourage the development of advanced spallation sources with long and short pulse characteristics.

Details of the principles of pulsed neutron sources are presented in this volume (Schoenborn & Pitcher, 1995; Wilson, 1995). In general, the production of neutrons by the spallation process leads to new opportunities for neutron scattering techniques. Pulsed neutron sources use high energy protons from an accelerator (10 to 120Hz) to bombard a heavy metal target. The resultant pulse of high energy neutrons is moderated by interaction with a suitable material (water, hydrogen etc) to reduce the neutron energy to the thermal range (1meV to 100meV, or 10Å to 1Å). Since the velocity of neutrons is energy dependent, the moderated neutrons will have an energy-dependent arrival time at a distant point in space. With appropriate geometry, diffraction experiments are carried out as a function of time and use a large part of the total neutron energy spectrum. This is particularly advantageous for protein crystallography. Essentially the data are collected in the Laue fashion with each diffraction peak having an associated wavelength thus eliminating physical spot overlaps - in essence, the best of a monochromatic and a white radiation experiment are retained (Schoenborn, 1992b).

The present pulsed neutron sources rely on proton pulses of less than one microsecond duration yielding very high energy-resolved neutrons (Schoenborn & Pitcher, 1995: Figure 2). Such resolution is, however, not required for protein crystallography. The use of long pulse sources for diffraction studies has been suggested by Bowman and described by Mezei (1994). These proposed devices would have a proton current increased by an order of magnitude with equivalent increases in spallation neutrons. The existing linear accelerator at Los Alamos (LAMPF) can produce proton pulses with 1mA current and a width of 1msec at 60Hz. This would lead immediately to a 10 fold increase in neutron flux. In addition, the design of moderators tailored for protein crystallography can produce further increases in flux, enhancing the most useful wavelength range (1 to 5Å). The use of such long pulse spallation neutron sources (LPSS) with advanced large, efficient position sensitive detectors and focusing optics should enable the collection of high resolution protein data within days instead of months.

The development of the present LAMPF as an LPSS requires only the addition of a neutron target complex and associated neutron beam facilities. Such a facility would provide a major new neutron source surpassing the capabilities of the ILL for many experiments particularly protein, virus or membrane crystallography.

In a technical sense, neutron scattering methods are now mature. The application of the neutron diffraction technique to assign hydrogen atom positions in proteins and to differentiate between hydrogen and deuterium atoms has focused on structural issues in protein reaction mechanisms, protein dynamics, and protein-water interactions. Results are, however, available only on a limited number of systems. Neutron scattering techniques will remain powerful techniques for the solution of well formulated problems in structural biology. The issues that motivated the establishment of a neutron scattering capability are still valid. Clearly other dimensions have been added to the original concept. High resolution single crystal studies remain of utmost importance, with lower resolution single crystal studies for solvent, lipid, detergent etc interactions becoming increasingly important. New information derived thus far includes direct observation of the protonation states of histidine in myoglobin, trypsin and ribonuclease; deamination states of amide side-chains in trypsin; detailed conformation of methyl rotors in trypsin, crambin, BPTI and myoglobin; extensive H/D exchange data on a range of proteins; information on protein hydration and bulk solvent at ever increasing resolution; and the location of lipids, detergents and RNA in complexes with protein. Mirroring similar trends in X-ray scattering studies where spectroscopic information is of increasing interest, studies of dynamic behaviour of biomolecules using inelastic neutron scattering are progressing but are still in their infancy (Cusack, 1984; Engelman *et al.*, 1984; Middendorf, 1984; Middendorf *et al.*, 1984). Major contributions can be expected from these techniques in the next decade.

Significant advances are still needed in the performance of instruments and data processing systems particularly for spallation neutron studies. Many of these issues have been identified in detail and funding is being actively sought. This is by no means an exhaustive review of the use of neutrons for structural biology research but highlights some of the milestones that have established the technique and could shape future directions.

ACKNOWLEDGMENTS

The authors wish to acknowledge a critical review of this paper by Hans Frauenfelder. The authors also wish to thank X. Cheng and D. Engelman for providing Figures 1 and 2 respectively.

REFERENCES

Alberi, J., (1976). Development of large-area, position-sensitive neutron detectors. In *Neutron Scattering for the Analysis of Biological Structures*. (B.P. Schoenborn, editor) VIII24-VIII42. (Nat. Technical Infor. Serv.; U.S. Dept. of Commerce, Springfield VA).

Baldwin, J.P., Boseley, P.G., Bradbury, E.M., & Ibel, K., (1975). The subunit structure of the eukaryotic chromosome. *Nature*, 253:245–249.

Bentley, G.A., & Mason, S.A., (1981). In *Structural Studies on Molecules of Biological Interest* (G. Dodson, J.P. Glusker and D. Sayre, editors) pp246–255. (Clarendon Press, Oxford).

Bentley, G.A., Duee, E.D., Mason, S.A., & Nunes, A.C.J., (1979). Protein structure determination by neutron diffraction: lysozyme. *Chim. Phys.*, 76:817–821.

Bentley, G.A., Delepierre, M., Dobson, C.M., Wedin, R.E., Mason, S.A., & Poulsen, F.M.J., (1983). Exchange of individual hydrogens for a protein in a crystal and in solution. *J. Mol. Biol.*, 170:243–247.

Blasie, J.K., Pachence, J.M., & Herbette, L.G., (1984). Neutron diffraction and the decomposition of membrane scattering profiles into the scattering profiles of their molecular components. In *Neutrons in Biology* (B.P. Schoenborn, editor) pp201–210. (Plenum Publishing Corporation, New York).

Blasie, J.K., Schoenborn, B.P., & Zaccai, G., (1976). Direct methods for the analysis of lamellar neutron diffraction from oriented multilayers: A difference patterson deconvolution approach. In *Neutron Scattering for the Analysis of Biological Structures* (B.P. Schoenborn, editor) ppIII58-III67. (Nat. Technical Infor. Serv.; U.S. Dept. of Commerce, Springfield, VA).

Bolton, W., Cox, J.M., & Perutz, M.F., (1968). Structure and function of haemoglobin IV. A three dimensional Fourier synthesis of horse deoxyhaemoglobin at 5.5Å resolution. *J. Mol. Biol.*, 33:283–297.

Bradshaw, J., (1995). Neutron diffraction studies of amphipathic helices in phospholipid bilayers. In *Neutrons in Biology* (B.P. Schoenborn and R. Knott, editors) (Plenum Publishing Corporation, New York).

Burks, C., & Engelman, D.M., (1981). Cholesteryl myristate conformation in liquid crystalline mesophases determined by neutron scattering. *Proc. Natl. Acad. Sci. USA*, 78:6863–6867.

Cain, J.E., Norvell, J.C., & Schoenborn, B.P., (1976). Linear position-sensitive counter system for protein crystallography. In *Neutron Scattering for the Analysis of Biological Structures.* (B.P. Schoenborn, editor) ppVIII43-VIII50. (Nat. Technical Inform. Serv.; U.S. Dept. of Commerce, Springfield, VA).

Capel, M., Moore, P.B., Engelman, D.M., Schneider, D., Schoenborn, B.P., Kjeldgaard, M., Langer, J., Ramakrishnan, V., Sillers, I.-Y., & Yabuki, S., (1987). Complete mapping of the proteins in the small ribosomal subunits of *E.coli. Science,* 238:1403–1406.

Chabre, M., (1975). X-ray diffraction studies of retinal rods I. Structure of the disc membrane, effect of illumination. *Biochim. Biophys. Acta,* 382:322–355.

Cheng, X., (1990). *Hydration and solvent structure in proteins and solvent effect on protein refinement: a neutron diffraction study of myoglobin crystals.* SUNY Stonybrook. Ph.D. Thesis.

Cheng, X., & Schoenborn, B.P., (1990). Hydration in protein crystals. A neutron diffraction analysis of carbonmonoxymyoglobin. *Acta Cryst.,* B46:195–208.

Church, B., (1992). *X-ray and Neutron Crystallography of Plastocyanin.* University of Sydney NSW Australia. PhD Thesis.

Convert, P., & Forsyth, J.B., (1983). *Position-Sensitive Detection of Thermal Neutrons.* (Academic Press, New York).

Curmi, P.M.G., Stone, D.B., Schneider, D.K., & Mendelson, R.A., (1991). Mechanism of force generation studied by neutron scattering. *Adv. Biophys.,* 27:131–141.

Curmi, P.M.G., Stone, D.B., Schneider, D.K., Spudich, J.A., & Mendelson, R.A., (1988). Comparison of the structure of myosin subfragment 1 bound to actin and free in solution. A neutron scattering study using actin made 'invisible' by deuteration. *J. Mol. Biol.,* 203:781–798.

Cusack, S., (1984). Neutron scattering studies of virus structure. In *Neutrons in Biology* (B.P. Schoenborn, editor), pp173–188. (Plenum Publishing Corporation, New York).

Engelman, D.M., & Moore, P.B., (1972). A new method for the determination of biological quarternary structure by neutron scattering. *Proc. Natl. Acad. Sci. USA,* 69:1997–1999.

Engelman, D.M., Dianoux, A.J., Cusack, S., & Jacrot, B., (1984). Inelastic neutron scattering studies of hexokinase in solution. In *Neutrons in Biology* (B.P. Schoenborn, editor) pp365–380. (Plenum Publishing Corporation, New York).

Fischer, J., Radeka, V., & Boie, R.A., (1983). High position resolution and accuracy in ^{3}He two-dimensional thermal neutron PSDs. In *Position-Sensitive Detection of Thermal Neutrons* (P. Convert and J.B. Forsyth, editors) pp129–140. (Academic Press, London).

Forsyth, V.T., Langan, P., Mahendrasingam, A., Fuller, W., & Mason, S.A., (1992). High-angle neutron fiber diffraction studies of DNA. *Neutron News,* 3(4):21–24.

Forsyth, V.T., Langan, P., Whalley, M.A., Mahendrasingam, A., Wilson, C.C., Giesen, U., Dauvergne, M.T., Mason, S.A., & Fuller, W., (1995). Time of flight Laue fiber diffraction studies of perdeuterated DNA. In *Neutrons in Biology* (B.P. Schoenborn and R. Knott, editors) (Plenum Publishing Corporation, New York).

Fuller, W., Forsyth, V. T., Mahendrasingam, A., Langan, P., & Pigram, W. J., (1995). DNA hydration studied by neutron fiber diffraction. In *Neutrons in Biology* (B.P. Schoenborn and R. Knott, editors) (Plenum Publishing Corporation, New York).

Garcia, A.E., Hummer, G., & Soumpasis, D.M., (1995). Theoretical description of biomolecular hydration - application to A-DNA. In *Neutrons in Biology* (B.P. Schoenborn and R. Knott, editors) (Plenum Publishing Corporation, New York).

Glinka, C.J., & Berk, N.F., (1983). The two-dimensional PSD at the National Bureau of Standards' Small Angle Neutron Scattering Facility. In *Position-Sensitive Detection of Thermal Neutrons.* (P. Convert and J.B. Forsyth, editors). pp141–148. (Academic Press, London).

Graziano, V., Gerchman, S.E., Schneider, D.K., & Ramakrishnan, V.R., (1995). Neutron scattering studies on chromatin higher-order structure. In *Neutrons in Biology* (B.P. Schoenborn and R. Knott, editors) (Plenum Publishing Corporation, New York).

Gu, W., & Schoenborn, B.P., (1995). Molecular dynamics simulation of hydration in myoglobin. *Proteins,* 22:20–26.

Hanson, J.C., & Schoenborn, B.P., (1981). Real space refinement of neutron diffraction data from sperm whale carbonmonoxymyoglobin. *J. Mol. Biol.,* 153:117–146.

Herbette, L., Napolitano, C.A., & McDaniel, R.V., (1984). Direct determination of the calcium profile structure for dipalmitoyllecithin multilayers using neutron diffraction. *Biophys. J.,* 46:677–685.

Herbette, L., DeFoor, P., Fleischer, S., Pascolini, D., Scarpa, A., & Blasie, J.K., (1985). The separate profile structures of the functional calcium pump protein and the phospholipid bilayer within isolated sarcoplasmic reticulum membranes determined by X-ray and neutron diffraction. *Biochim. Biophys. Acta,* 817:103–122.

Hjelm, R.P., Kneale, G.G., Suau, P., Baldwin, J.P., Bradbury, E.M., & Ibel, K., (1977). Small angle neutron scattering studies of chromatin subunits in solution. *Cell,* 10:139–151.

Ibel, K., & Stuhrmann, H.B., (1975). Comparison of neutron and X-ray scattering of dilute myoglobin solutions. *J. Mol. Biol.,* 93:255–265.

Jacrot, B., & Zaccai, G., (1981). Determination of molecular weight by neutron scattering. *Biopolymers,* 20:2413–2426.

Jiang, J., & Brünger, A.T., (1994). Protein hydration observed by X-ray diffraction. Solvation properties of penicillopepsin and neuraminidase crystal structures. *J. Mol. Biol.,* 243:100–115.

King, G.I., & Schoenborn, B.P., (1985). Neutron scattering of bacteriorhodopsin. *Methods Enzymol.,* 88:241–248.

King, G.I., Stoeckenius, W., Crespi, H.L., & Schoenborn, B.P., (1979). The location of low retinal in the purple membrane profile by neutron diffraction. *J. Mol. Biol.,* 130:395–404.

King, G.I., Chao, N.-M., & White, S.H., (1984). Neutron diffraction studies on incorporation of hexane into oriented lipid bilayers. In *Neutrons in Biology* (B.P. Schoenborn, editor) pp159–172. (Plenum Publishing Corporation, New York).

Knott, R. (1995). Neutron scattering in Australia. In *Neutrons in Biology* (B.P. Schoenborn and R. Knott, editors) (Plenum Publishing Corporation, New York).

Kossiakoff, A.A., (1983). Protein dynamics investigated by the neutron diffraction-hydrogen exchange technique. *Nature,* 296:713–721.

Kossiakoff, A.A., & Shteyn, S., (1984). Effect of protein packing structure on side-chain methyl rotor conformations. *Nature,* 311:582–583.

Kossiakoff, A.A., & Spencer, S.A., (1980). Neutron diffraction identifies His57 as the catalytic base in trypsin. *Nature,* 288:414–416.

Kossiakoff, A.A., & Spencer, S.A., (1981). Direct determination of the protonation states of aspartic acid-102 and histidine-57 in the tetrahedral intermediate of the serine proteases. *Biochem.,* 20:6462–6474.

Langan, P., Forsyth, V.T., Mahendrasingam, A., Alexeev, D., Mason, S.A., & Fuller, W., (1993). In *Water-Biomolecule Interactions.* Conference Proceedings (M.U. Palma, M.B. Palma-Vittorelli and F. Parak, editors) pp235–238. (SIF, Bologna).

Lehman, M.S., & Zaccai, G., (1984). Neutron small-angle scattering studies of ribonuclease in mixed aqueous solutions and determination of the preferentially bound water. *Biochem.,* 23:1939–1942.

Lynn, J.W., Kjems, J.K., Passell, L., Saxena, A.M., & Schoenborn, B.P., (1976). Iron-germanium multilayer neutron polarizing monochromators. *J. Appl. Cryst.,* 9:454- 459.

Marguerie, G., & Stuhrmann, H.B., (1976). A neutron small-angle scattering study of bovine fibrinogen. *J. Mol. Biol.,* 102:143–156.

Mason, S.A., Bentley, G.A., & McIntyre, G.J., (1984). Deuterium exchange in lysozyme at 1.4Å resolution. In *Neutrons in Biology* (B.P. Schoenborn, editor) pp323–334. (Plenum Publishing Corporation, New York).

May, R.P., Stuhrmann, H.B., & Nierhaus, K.H., (1984). Structural elements of the 50S subunit of *E.coli* ribosomes. In *Neutrons in Biology* (B.P. Schoenborn, editor) pp25–46. (Plenum Publishing Corporation, New York).

Mezei, F., (1994). On the comparison of continuous and pulsed sources. *Neutron News,* 5:2–3.

Middendorf, H.D., (1984). Inelastic scattering from biomolecules: Principles and prospects. In *Neutrons in Biology* (B.P. Schoenborn, editor) pp401–437. (Plenum Publishing Corporation, New York).

Middendorf, H.D., Randell, J.T., & Crespi, H., (1984). Neutron spectroscopy of hydrogeneous and biosynthetically deuterated proteins. In *Neutrons in Biology* (B.P. Schoenborn, editor) pp381–400. (Plenum Publishing Corporation, New York).

Moore, F.M., Willis, B.T.M., & Crawfoot-Hodgkin, D., (1967). Crystal and molecular structure from neutron diffraction analysis. *Nature,* 214:130–133.

Nambudripad, R., Stark, W., Opella, S.J., & Makowski, L., (1991). Membrane-mediated assembly of filamentous bacteriophage Pf1 coat protein. *Science,* 252:1305–1308.

Niimura, N., (1995a). Neutron scattering and diffraction instrumentation for structural biology in Japan. In *Neutrons in Biology*. (B.P. Schoenborn and R. Knott, editors) (Plenum Publishing Corporation, New York).

Niimura, N., (1995b). Neutron diffractometer for bio-crystallography (BIX) with an imaging plate neutron detector. In *Neutrons in Biology*. (B.P. Schoenborn and R. Knott, editors) (Plenum Publishing Corporation, New York).

Nobbs, C.L., Watson, H.C., & Kendrew, J.C., (1966). Structure of deoxymyoglobin: A crystallographic study. *Nature*, 209:339–341.

Pardon, J.F., Worcester, D.L., Wooley, J.C., Tatchell, K., van Holde, K.E., & Richards, B.M., (1975). Low angle neutron scattering from chromatin subunit particles. *Nucleic Acids Res.*, 2:2163–2176.

Phillips, S.E.V., & Schoenborn, B.P., (1981). Neutron diffraction reveals oxygen-histidine hydrogen bond in oxymyoglobin. *Nature*, 292:81–82.

Pynn, R., (1995). LANSCE expands future operations. *Neutron News*, 6(1):30.

Radeka, V., Schaknowski, N.A., Smith, G.C., & Yu, B., (1995). High precision thermal neutron detectors. In *Neutrons in Biology* (B.P. Schoenborn and R. Knott, editors) (Plenum Publishing Corporation, New York).

Raghavan, N.V., & Schoenborn, B.P., (1984). The structure of bound water and refinement of acid metmyoglobin. In *Neutrons in Biology* (B.P. Schoenborn, editor) pp247–259. (Plenum Publishing Corporation, New York).

Ramakrishnan, V.R., Finch, J.T., Graziano, V., Lee, P.L., & Sweet, R.M., (1993). Crystal structure of globular domain of histone H5 and its implications for nucleosome binding. *Nature*, 362:219–223.

Savage, H.F.J., & Wlodawer, A., (1986). Determination of water structure around biomolecules using X-ray and neutron diffraction methods. *Methods Enzymol.*, 127:162–183.

Saxena, A.M., & Majkrzak, C.F., (1984). Neutron optics with multilayer monochromators. In *Neutrons in Biology*. (B.P. Schoenborn, editor) pp143–158. (Plenum Publishing Corporation, New York).

Saxena, A.M., & Schoenborn, B.P., (1976). Multilayer monochromators for neutron scattering. In *Neutron Scattering for the Analysis of Biological Structures* (B.P. Schoenborn, editor) ppVII30-VIII48. (Nat. Technical Infor. Serv.; U.S. Dept. of Commerce, Springfield, VA).

Saxena, A.M., & Schoenborn, B.P., (1977). Multilayer neutron monochromators. *Acta Cryst.*, 833:805–813.

Saxena, A.M., & Schoenborn, B.P., (1988). Multilayer monochromators for neutron spectrometers. *Material Science Forum*, 27/28:313–318.

Schoenborn, B.P., (1965). Binding of xenon to haemoglobin. *Nature*, 208:760–762.

Schoenborn, B.P., (1969a). Neutron diffraction analysis of myoglobin. *Nature*, 224:143–146.

Schoenborn, B.P., (1969b). Structure of alkaline metmyoglobin-xenon complex. *J. Mol. Biol.*, 45:279–303.

Schoenborn, B.P., (1972). A neutron diffraction analysis of myoglobin II. Hydrogen-deuterium bonding in the main chain. In *Structure and Function of Oxidation Reduction Enzymes* (A. Akeson and A. Ehrenberg, editors) pp109–116. (Pergamon Press, Oxford).

Schoenborn, B.P., (1975). In *Anomalous Scattering* (S. Ramaseshan and S.C. Abrahams, editors) pp407–421. (Munksgaard, Copenhagen).

Schoenborn, B.P., (1976a). Neutron scattering for the analysis of membranes. *Biochim. Biophys. Acta*, 457:41–55.

Schoenborn, B.P., (1976b). *Neutron Scattering for the Analysis of Biological Structures*. (Nat. Technical Infor. Serv.; U.S. Dept. of Commerce, Springfield, VA).

Schoenborn, B.P., (1983a). Peak shape analysis for protein neutron crystallography with position-sensitive detectors. *Acta Cryst.*, A39:315–321.

Schoenborn, B.P., (1983b). Data processing in neutron protein crystallography using position-sensitive detectors. In *Position-Sensitive Detection of Thermal Neutrons* (P. Convert and J.B. Forsyth, editors) pp321–331. (Academic Press, London).

Schoenborn, B.P., (1984). *Neutrons in Biology*. (Plenum Publishing Corporation, New York).

Schoenborn, B.P., (1988). The solvent effect in protein crystals. A neutron diffraction analysis of solvent and ion density. *J. Mol. Biol.*, 201:741–749.

Schoenborn, B.P., (1992a). Area detectors for neutron protein crystallography. *SPIE*, 1737:235–243.

Schoenborn, B.P., (1992b). Multilayer monochromators and super mirrors for neutron protein crystallography using a quasi Laue technique. *SPIE*, 1738:192–199.

Schoenborn, B.P., & Nunes, A.C., (1972). Neutron scattering. *Ann. Rev. Biophys. Bioengineer.*, 1:529–552.

Schoenborn, B.P., & Pitcher, E., (1995). Neutron diffractometers for structural biology at spallation neutron sources. In *Neutrons in Biology* (B.P. Schoenborn and R. Knott, editors) (Plenum Publishing Corporation, New York).

Schoenborn, B.P., Caspar, D.L.D., & Kammerer, P.F.J., (1974). A novel neutron monochromator. *J. Appl. Cryst.*, 7:508–510.

Schoenborn, B.P., Nunes, A.C., & Nathans, R., (1970). Neutron diffraction analysis of biological structures. *Berichte Bunsengessellschaft fur Physical Chemistry*, 74:1202–1207.

Schoenborn, B.P., Watson, H.C., & Kendrew, J.C., (1965). Binding of xenon to sperm whale myoglobin. *Nature*, 207:28–30.

Schoenborn, B.P., Alberi, J., Saxena, A.M., & Fischer, J.J., (1978). A low angle neutron data acquisition system for molecular biology. *J. Appl. Cryst.*, 11:455–460.

Seeger, P.A., & Hjelm, R.P., (1991). Small-angle neutron scattering at pulsed spallation sources. *J. Appl. Cryst.*, 24:467–478.

Shu, F., (1994). *The Structure of Solvent Molecules Bound to Per-Deuterated, Recombinant Sperm-Whale Myoglobin*. SUNY Stonybrook. Ph.D. Thesis.

Stark, W., Glucksman, M.J., & Makowski, L., (1988). Conformation of the coat protein of filamentous bacteriophage Pf1 determined by neutron diffraction from magnetically oriented gels of specifically deuterated virons. *J. Mol. Biol.*, 199:171–182.

Stuhrmann, H.B., (1973). Comparison of the three basic scattering functions of myoglobin in solution with those from the known structure in crystalline state. *J. Mol. Biol.*, 77:363–369.

Stuhrmann, H.B., (1974). Neutron small-angle scattering of biological macromolecules in solution. *J. Appl. Cryst.*, 7:173–178.

Teeter, M.M., (1984). Water structure of a hydrophobic protein at atomic resolution: Pentagon rings of water molecules in crystals of crambin. *Proc. Natl. Acad. Sci. USA*, 81:6014–6018.

Teeter, M.M., & Kossiakoff, A.A., (1984). The neutron structure of the hydrophobic plant protein crambin. In *Neutrons in Biology* (B.P. Schoenborn, editor) pp335–348. (Plenum Publishing Corporation, New York).

Timmins, P.A., (1988). Neutron scattering studies of the structure and assembly of spherical viruses. *Makromol. Chem. Makromol. Symp.*, 15:311–321.

Timmins, P.A., (1995). Low resolution neutron crystallography of large biological macromolecular assemblies. *Neutron News*, 6(1):13–18.

Timmins, P.A., & Pebay-Peyroula, E., (1995). Protein-detergent interactions in single crystals of membrane proteins studied by neutron crystallography. In *Neutrons in Biology* (B.P. Schoenborn and R. Knott, editors) (Plenum Publishing Corporation, New York).

Timmins, P.A., Wild, J., & Witz, J., (1994). The three-dimensional distribution of RNA and protein in the interior of tomato bushy stunt virus: A neutron low-resolution single-crystal diffraction study. *Structure*, 2:1191–1201.

Timmins, P.A., Poliks, B., & Banaszak, L.J., (1992). The location of bound lipid in the lipovitellin complex. *Science*, 257:652–655.

Trewhella, J., Gogel, E., Zaccai, G., & Engelman, D.M., (1984). Neutron diffraction studies of bacteriorhodopsin structure. In *Neutrons in Biology* (B.P. Schoenborn, editor) pp227–246 (Plenum Publishing Corporation, New York).

White, S.H., & Wiener, M.C., (1995). Fluid bilayer structure determination: joint refinement in 'composition space' using X-ray and neutron diffraction data. In *Neutrons in Biology* (B.P. Schoenborn and R. Knott, editors) (Plenum Publishing Corporation, New York).

White, S.H., King, G.I., & Cain, J.E., (1981). Location of hexane in lipid bilayers determined by neutron diffraction. *Nature*, 290:161–163.

Wiener, M.C., & White, S.H., (1991a). Fluid bilayer structure determination by the combined use of X-ray and neutron diffraction. I. Fluid bilayer models and the limit of resolution. *Biophys. J.*, 59:162–173.

Wiener, M.C., & White, S.H., (1991b). Fluid bilayer structure determination by the combined use of X-ray and neutron diffraction. II. 'Composition-space' refinement method. *Biophys. J.*, 59:174–185.

Wilson, C.C., (1995). The potential for biological structure determination with pulsed neutrons. In *Neutrons in Biology* (B.P. Schoenborn and R. Knott, editors) (Plenum Publishing Corporation, New York).

Wise, D.S., Karlin, A., & Schoenborn, B.P., (1979). An analysis by low-angle neutron scattering of the structure of the acetylcholine receptor from *torpedo californica* in detergent solution. *Biophys. J.*, 28:473–496.

Wlodawer, A., & Sjölin, L., (1984). Application of joint neutron and X-ray refinement to the investigation of the structure of ribonuclease A at 2.0Å resolution. In *Neutrons in Biology* (B.P. Schoenborn, editor) pp349–364. (Plenum Publishing Corporation, New York).

Wlodawer, A., Miller, M., & Sjölin, L., (1983). Active site of RNase: Neutron diffraction study of a complex with uridine vanadate, a transition-state analog. *Proc. Natl. Acad. Sci. USA*, 80:3628–3631.

Wlodawer, A., Walter, J., Huber, R., & Sjölin, L., (1984). Structure of bovine pancreatic trypsin inhibitor. Results of joint neutron and X-ray refinement of crystal form II. *J. Mol. Biol.*, 180:301–329.

Worcester, D.L., (1975). In *Biological Membranes* (D. Chapman and D.F.H. Wallach, editors) pp1–48. (Academic Press, London).

Worcester, D.L., (1976). Neutron diffraction studies of biological membranes and membrane components. In *Neutron Scattering for the Analysis of Biological Structures* (B.P. Schoenborn, editor) ppIII37-III57. (Nat. Technical Infor. Serv.; U.S. Dept. of Commerce, Springfield, VA).

Worcester, D.L., & Franks, N.P., (1976). Structural analysis of hydrated egg lecithin and cholesterol bilayers. II. Neutron diffraction. *J. Mol. Biol.,* 100:359–378.

Worcester, D.L., Hamacher, K., Kaiser, H., Kulasekere, R., & Torbet, J., (1995). Intercalation of small hydrophobic molecules in lipid bilayers containing cholesterol. In *Neutrons in Biology* (B.P. Schoenborn and R. Knott, editors) (Plenum Publishing Corporation, New York).

Yeager, M., (1976). Neutron diffraction analysis of the structure of retinal photoreceptor membranes and rhodopsin. In *Neutron Scattering for the Analysis of Biological Structures.* (B.P. Schoenborn, editor) ppIII3-III36. (Nat. Technical Infor. Serv.; U.S. Dept. of Commerce, Springfield, VA).

Yeager, M., Schoenborn, B.P., Engelman, D.M., Moore, P.B., & Stryer, L., (1980a). Neutron diffraction of intact retinas. *J. Mol. Biol.,* 137:1–34.

Yeager, M., Schoenborn, B.P., Engelman, D.M., Moore, P.B., & Stryer, L., (1980b). Neutron diffraction analysis of the structure of rod photoreceptor membranes in intact retinas. *J. Mol. Biol.,* 137:315–348.

Zaccai, G., Blasie, J.K., & Schoenborn, B.P., (1975). Neutron diffraction studies on the location of water in lecithin bilayers. *Proc. Natl. Acad. Sci. USA,* 72:376–380.

1

NEUTRON ANATOMY

G. E. Bacon*

University of Sheffield
Sheffield, England

ABSTRACT

The familiar extremes of crystalline material are single-crystals and random powders. In between these two extremes are polycrystalline aggregates, not randomly arranged but possessing some preferred orientation and this is the form taken by constructional materials, be they steel girders or the bones of a human or animal skeleton. The details of the preferred orientation determine the ability of the material to withstand stress in any direction. In the case of bone the crucial factor is the orientation of the c-axes of the mineral content - the crystals of the hexagonal hydroxyapatite - and this can readily be determined by neutron diffraction. In particular it can be measured over the volume of a piece of bone, utilising distances ranging from lmm to 10mm. The major practical problem is to avoid the intense incoherent scattering from the hydrogen in the accompanying collagen; this can best be achieved by heat-treatment and it is demonstrated that this does not affect the underlying apatite. These studies of bone give leading anatomical information on the life and activities of humans and animals - including, for example, the life history of the human femur, the locomotion of sheep, the fracture of the legs of racehorses and the life-styles of Neolithic tribes. We conclude that the material is placed economically in the bone to withstand the expected stresses of life and the environment.

The experimental results are presented in terms of the magnitude of the 0002 apatite reflection. It so happens that for a random powder the 0002, 1121 reflections, which are neighbouring lines in the powder pattern, are approximately equal in intensity. The latter reflection, being of manifold multiplicity, is scarcely affected by preferred orientation so that the numerical value of the 0002/1121 ratio serves quite accurately as a quantitative measure of the degree of orientation of the c-axes in any chosen direction for a sample of bone.

* Address for correspondence: Windrush Way, Guiting Power, Cheltenham GLOS GL54 5US, England.

Neutrons in Biology, edited by Schoenborn and Knott
Plenum Press. New York. 1996

INTRODUCTION

Neutron diffraction has advantages which make it a powerful technique for carrying out quantitative study of the anatomy of human and animal skeletons. In the past some attempts have been made to probe the architecture of bones by X-ray diffraction but these are very restricted by the relatively poor penetrating power of X-rays of wavelength 1 or 2Å (0.1 - 0.2nm); this means that conclusions are restricted to the regions of the bones near the surface. For neutrons the penetrating power is much greater and it is possible to examine thicknesses of material ranging from a millimetre to a centimetre, thus yielding integrated data for the whole thickness of a bone and offering immense possibilities for the study of the anatomy of humans and animals.

Bone is a composite of collagen and a mineral component which is basically hydroxyapatite, $Ca_5(PO_4)_3OH$. The two components are intertwined in a manner which is only partially known and far from being well understood. Essentially, a bone is an oriented assembly, being neither a single-crystal nor a random powder, and this description applies whether one is concentrating thought on the collagen or the apatite. With neutrons the diffraction pattern readily produces details of the orientation of the apatite crystals, which have the symmetry of the space group $P6_3/m$. What we shall find is that the c axes of each bone have a direction preference which is determined by the stresses on the individual bone as it plays its part in the functioning of the living animal or human body. Bone is a living material and the form and constitution of a bone change during its lifetime. In 1870, long before it was possible to measure experimentally the atomic structure of materials, Wolff propounded a law which stated that 'every change in the function of bone is followed by certain changes in its internal architecture and its external conformation' (Wolff, 1870; 1899). We shall seek to demonstrate the truth of this law in terms of the atomic structure of the bone and to indicate how it is the stress which the bone has to withstand that determines the preferential orientation of the crystals. It will remain a moot point how, maybe in some piezoelectric fashion, the physical and biochemical development of the bone can take account of the stress as the bone grows. It may also be speculated whether it is averaged stress or a peak stress over some finite period which is important.

Typical sections of bone, such as are shown in Figure 1 (Bacon *et al.*, 1984) for the bones of the foot (the tibia, talus and calcaneus) reveal the two types of bone, namely the solid outside layer of compact bone and the fibrous trabecular bone which links it within. It can be shown by X-ray diffraction that the c-axes of the apatite crystals in the trabecular fibers are preferentially aligned in the fibre direction; with neutrons it is possible to assess the orientation of the solid and trabecular material as a whole, and notably for the substantial solid shell of long bones where the preferential orientation of the c-axes may be very considerable.

The principle of our method, which has been described in several earlier papers (Bacon *et al.*, 1979a; 1984), is to mount the bone sample on the axis of a neutron diffractometer and record the diffraction pattern for a variety of inclinations of the sample, and hence of the reflecting planes, in relation to the neutron counter or position sensitive detector lying in the horizontal plane. Before discussing the experimental problems it will be helpful to show a typical result. Figure 2 shows two patterns for the shaft of an ox femur recorded in 20 minutes with neutrons of wavelength 2.4Å on the instrument D1B at the ILL Grenoble (Bacon *et al.*, 1977). In the upper diagram the shaft of the bone is vertical; in the lower diagram it is horizontal. The essential difference between the two patterns is that in the upper one the intensities from the prism planes, such as 110 and 300, are substantial and

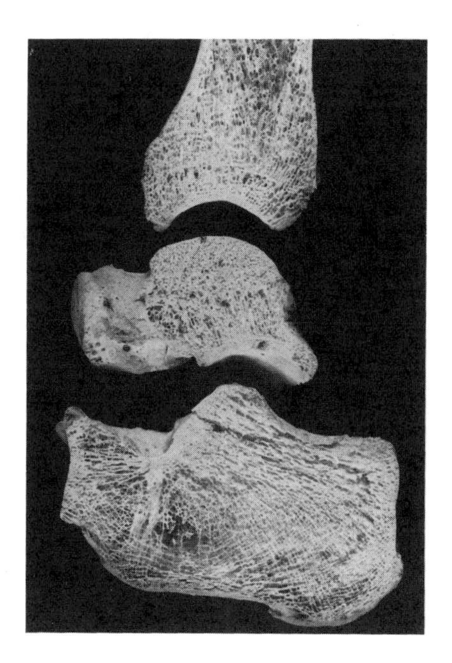

Figure 1. Sections in the sagittal plane of the three foot bones (tibia, talus and calcaneus), after heat treatment, showing the compact cortical bone and the trabeculae. The bones have been slightly separated vertically for clarity in the photograph (from Bacon *et al.*, 1984; Figure 3).

those of the basal planes, such as 002 and 004, are small. In the lower pattern the converse is true. [In our nomenclature for hexagonal reflections we omit the redundant third hexagonal index.] These features of the two patterns indicate that there is substantial preferential orientation of the c-axes along the shaft of the bone and this can be evaluated quantitatively. It is to be noted that the 111 reflection, which comes from a plane of manifold multiplicity, has approximately the same intensity in the two patterns. Moreover it so happens that the atomic structure of hydroxyapatite is such that for a random powder the intensities of the 111 and 002 reflections are roughly equal (more exactly the 002/111 ratio is 0.92 for the powder). This means that for an oriented sample the numerical value of the 002/111 ratio may be taken as the approximate value of the preferential orientation of the apatite c axes in the appropriate direction. In Figure 2 its value is 3.0 along the shaft of the femur.

There is one troublesome difficulty in carrying out the neutron experiment which we have just described. The collagen in the bone contains much hydrogen, which has an incoherent neutron scattering cross-section of 80 barns, and this will contribute a very large background to the diffraction pattern, against which any weak diffraction lines may be indistinguishable. Consequently, in order to achieve any high accuracy of intensity measurement the collagen must be largely removed and we have described elsewhere (Bacon *et al.*, 1979b) the various treatments which were explored for adequately reducing the background. We concluded that the most satisfactory treatment was to heat the bone sample at 620°C for 20 minutes. The paper just mentioned also describes the subsidiary measurements which were made with both X-rays and neutrons to establish that the apatite texture was not altered significantly, either as regards crystallinity or preferential orientation, by this heat treatment which reduced the background by about nine times. It may be noted that the bones in the photograph of Figure 1 had already been heat-treated and it is perhaps remarkable how little the physical appearance of even the trabeculae has been altered by the extraction of the collagen.

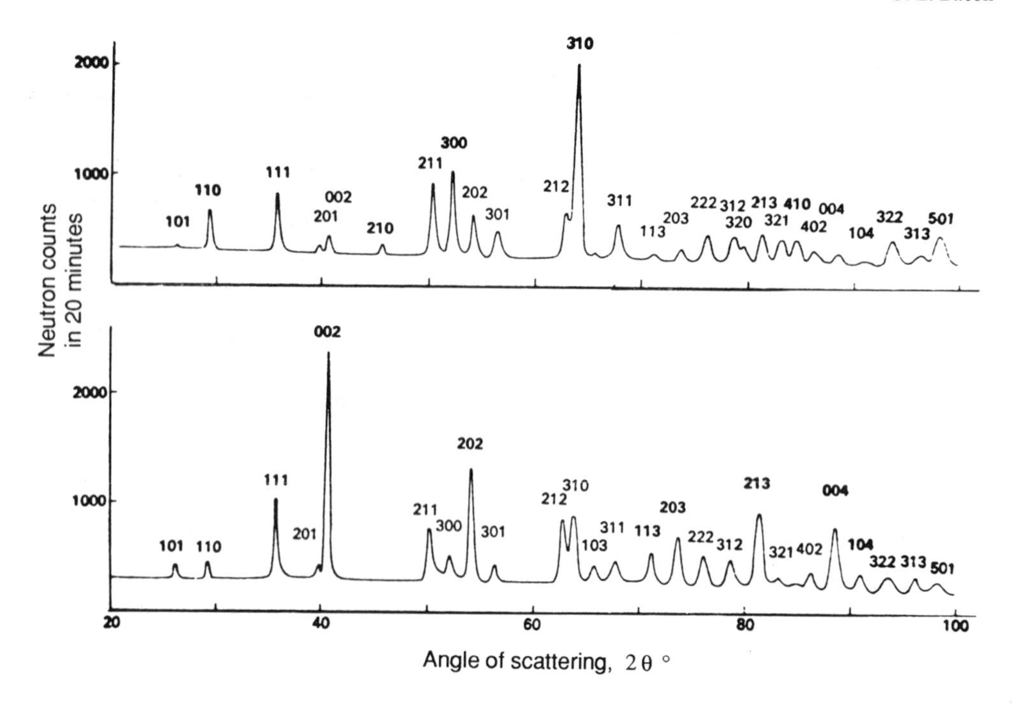

Figure 2. Neutron diffraction patterns for a section of the shaft of an ox femur for a wavelength of 2.4Å. For the upper curve the bone shaft is vertical; in the lower curve the shaft is horizontal. The indices identify the reflections of the hexagonal hydroxyapatite crystals; the redundant third Miller index is omitted (from Bacon *et al.*, 1977; Figure 4).

ANATOMICAL STUDIES

Having established a suitable experimental procedure for measuring preferred orientation we first studied some anatomical examples in which the nature of the stresses in the bones might be regarded as fairly self-evident. As a start we looked at the scapula (Bacon *et al.*, 1979a), the shoulder bone, which is mechanically controled in a very evident way by the attached muscles. Figure 3 is a diagram of the scapula showing the seven sites, A - G, at which measurements were made with a neutron beam 7mm in diameter. At each position the polar diagram for the distribution of the apatite c axes is shown, as determined from the diffraction patterns measured at successive rotations of the bone by 20°. In Figure 4 the attached muscles have been superimposed on the polar diagram. At sites A, C and D we have a sharply defined polar diagram pointing along the edge-pulling muscles. At E, F the c axes have a noticeably multiple distribution, giving a wider diagram in which the main peak due to the vertical edge-pulling muscles is accompanied by a secondary peak in the direction of the infra-spinatus fibers. At site G, and even more so at B, the infra-spinatus predominates.

Another example where the directions of stress are fairly obvious is provided by the bones of the foot (Bacon *et al.*, 1984), for which Figure 5 shows how the forces from the tibia bone of the leg pass down through the ankle bone (talus) to the toes and the calcaneus heel-bone. The drawing of the highly oriented first metatarsal bone, drawn in the bottom left hand corner, should be displaced substantially to the left of the diagram. There are some noticeable features in this assembly of bones. There are no muscles along the top

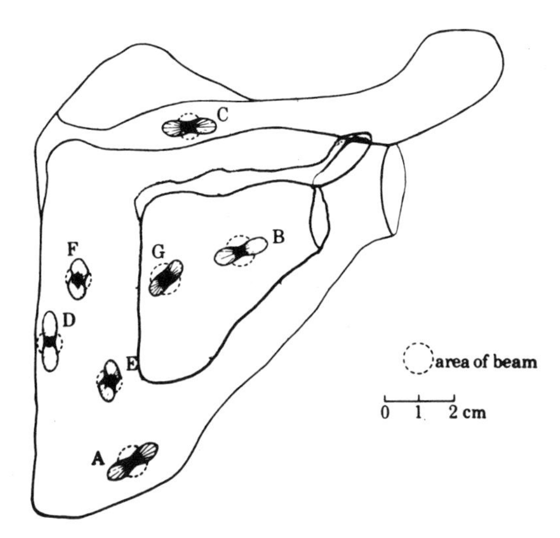

Figure 3. The polar diagrams at seven sites, A - G, on the scapula, showing the amount of apatite with its c axis in each direction. For example the sharp orientation at A is to be contrasted with the broader, double-peaked curve at F (from Bacon *et al.*, 1979a; Figure 4).

edge of the heel, which is controled by the anti-clockwise pull of the Achilles tendon: accordingly the edge of the heel from C towards F functions as a rigid beam, with a well-defined c axis direction. Along the bottom of the heel the spread of the axes has a dual character, corresponding to both the pull of the roughly horizontal muscles and to the weight bearing down from the joint with the talus. The centre of the talus acts like a junction-box, receiving vertical weight from the tibia and distributing it to the toes and the heel.

On a more ambitious level we have examined the vertebrae of the human spine and compared the results with corresponding data for the rhesus monkey. The examination of no more than two samples of each of 20 vertebrae makes this quite a substantial project

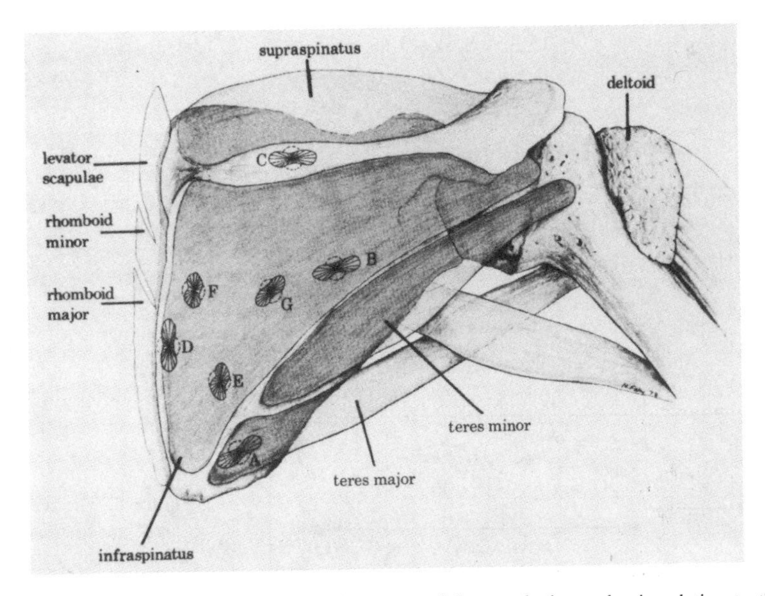

Figure 4. A perspective view of the scapula, showing some of the attached muscles in relation to the sites of measurement and their polar diagrams (from Bacon *et al.*, 1979a, Figure 5).

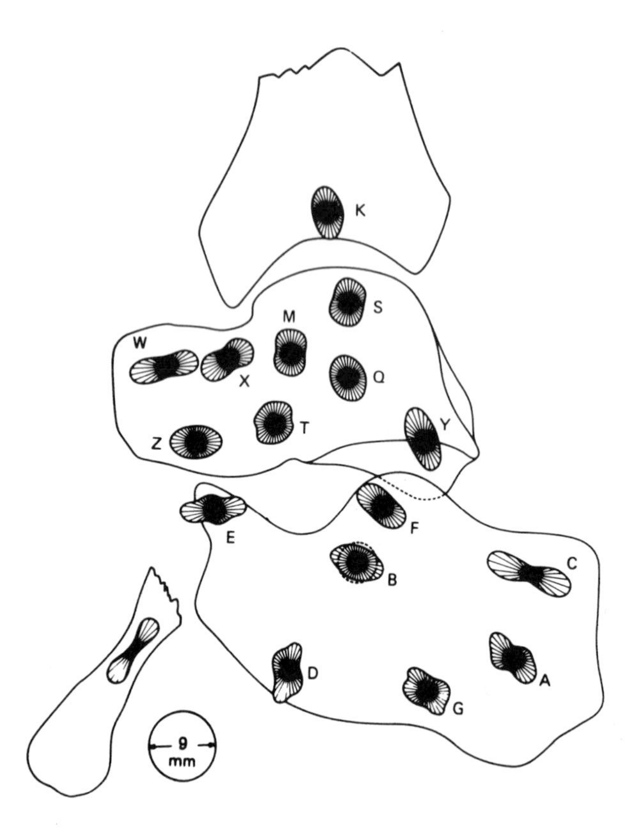

Figure 5. A section through the bones of the foot indicating the sites of the neutron measurements and their polar diagrams. In the bottom left-hand corner, but not in location, is a similar diagram showing the very high orientation of the first metatarsal bone. The 9mm circle indicates the size of the neutron beam (from Bacon *et al.*, 1984; Figure 5).

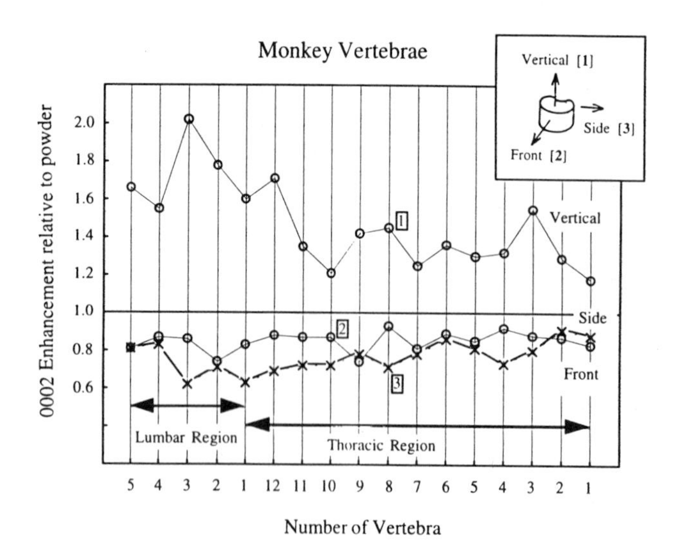

Figure 6. The preferential orientation of the apatite c axes in the spinal vertebrae of a rhesus monkey. The three curves denote (1) the vertical direction (2) horizontally forwards and (3) horizontally sideways.

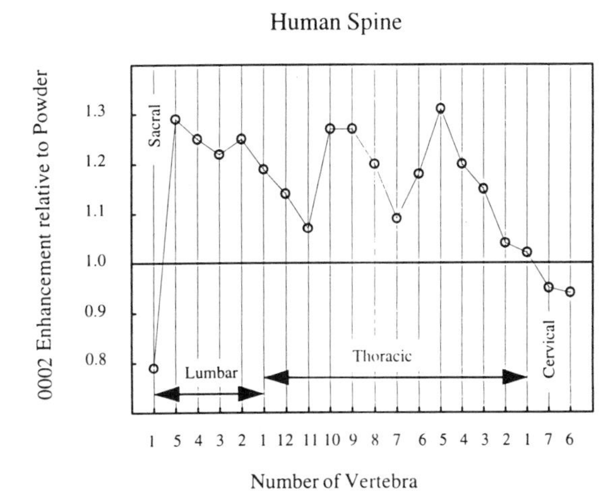

Figure 7. The preferential orientation in the vertical direction of the c axes for the vertebrae of a human spine. Note the negative preference of the first sacral vertebra.

and it soon became evident that in order to draw detailed conclusions it would be necessary to examine many more samples in order to take account of biological variations between the bones of different individuals. Nevertheless the observations gave an idea of the information which would be forthcoming from further work and we are able to draw some preliminary conclusions. Figure 6 for the monkey shows that the vertical orientation is much larger at the base of the spine where weight-bearing predominates; the orientation falls off quickly further up the spine where the body is much more flexible and more movement takes place. It is also evident from the curves marked 'side' and 'front' in Figure 6 that the orientation is less in the front-back plane than for the side-side, indicating the larger backwards and forwards bending which takes place. Figure 7 contrasts the situation for the human spine. Here the vertical orientation remains much more uniform as we go up the spine until the shoulders and neck are reached, and this accords with the more upright human stance. It is to be expected that these details of the spinal textures displayed in Figures 6 and 7 can be correlated with the influence of the ribs and the muscles attachments, but many more samples would need to be studied before speculating on more precise conclusions.

CHANGES DURING LIFE

So far we have thought of the human and animal skeletons as rather like a mechanical construction, designed to withstand the expected forces and incorporating some factor of safety. However we have to take account of the fact that the bones renew as humans and animals grow, or, recalling Wolff's Law, as changes in the function of the bones take place. How does the texture of a bone, as indicated by the preferred orientation of the c axes of the apatite crystals, vary with age or, perhaps with the lifestyle of the individual? We sought to answer this question for the human femur by examining 75 femurs distributed fairly evenly in age between still-born babies and adults up to an age of 82 years (Bacon & Griffiths, 1985). The vertical orientation, represented as usual by the ratio of the 002/111 intensities, is plotted as a function of age in Figure 8. The orientation is low at birth, where it is just over 2, and then steadily increases as the child grows and ceases to spend all its time lying down, reaching a maximum at an age of about 3 years when the

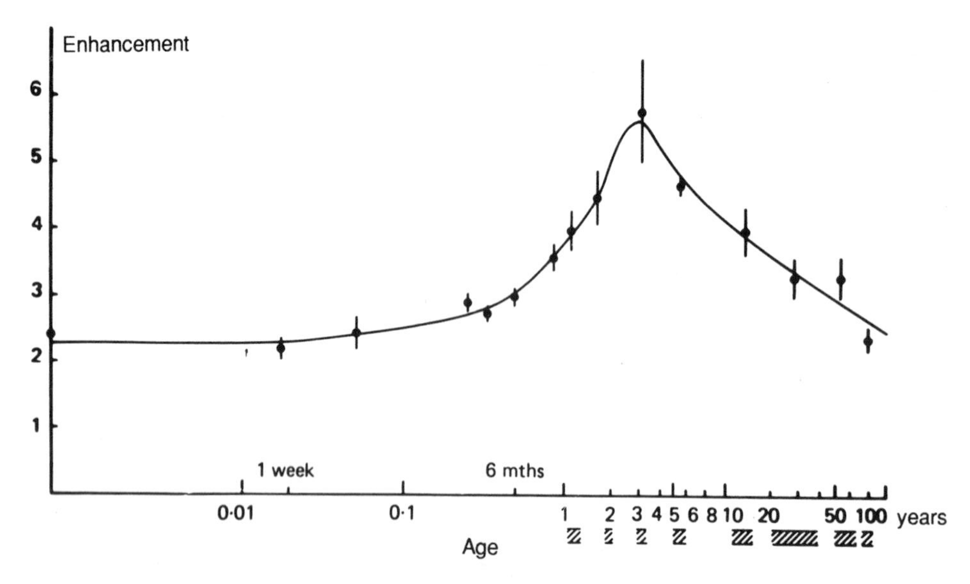

Figure 8. The enhancement of the 002 reflection, indicating the preferential vertical orientation of the apatite crystals in the human femur as a function of age on a logarithmic scale. The hatched areas beneath the age-scale give the age bands covered by the data points (from Bacon & Griffiths, 1985; Figure 2).

child has become fully mobile. With further growth the orientation falls as more of life is spent sitting down.

We can contrast the pattern of Figure 8 with corresponding data for the radius bone of the sheep (Bacon & Goodship, 1991), an animal which can stand up and run as soon as it is born. Figure 9 summarises the information for 96 bones, again plotting the 002/111 ratio as a function of age. We see that the orientation has already reached a value of 4 at the birth of the sheep and that it has been increasing steadily from a value of about 1.7 at 80 days after conception, when the bone had become large enough to handle and measure with neutrons. After an age of about 1 year the orientation falls off, but not so quickly as for a human. Maybe the slower fall reflects the fact that older sheep still have to stand up to eat and, losing their teeth, have to spend a large proportion of their time in this occupation.

In the hey-day of X-ray powder photographs it was often said that the powder photograph was the finger-print of a chemical substance. We can now suggest that the orientation of the c-axes of the apatite crystals in bones might be regarded as an imprint of the activities and lifestyle of an individual. It is likely that we should find distinctive features in bones of, say, tennis players or acrobats. Until we have a method of making the neutron measurements on living individuals we are thwarted in carrying out definitive experiments on this topic by the need for an adequate number of nominally similar samples. However it is possible to offer a demonstration of the effect of life-style on bone orientation by going back to Neolithic times.

Ten years ago a colleague at Sheffield, Judson Chesterman, who was an anatomist and an archaeologist, told me about two tribes living in the Orkney islands whose bones were found in chambered tombs dating from 3000BC. The two tribes appeared to have developed quite independently and there was no suggestion of any family relation between them. One tribe from Isbister lived on very hilly ground and the lower front edge of their tibiae showed the so-called 'squatting facets', usually associated with hard work, such as

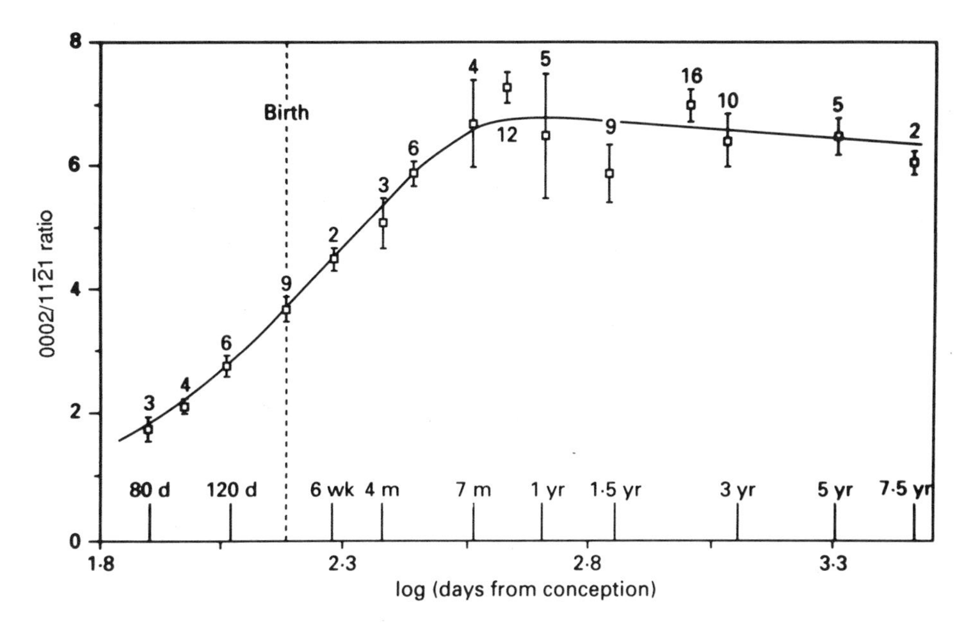

Figure 9. The variation with age, on a logarithmic scale, of the preferential vertical orientation of the c axes in the radius bone of the sheep. The vertical lines through the recorded points give the standard deviation of the mean value; the numerals above the points give the number of samples (from Bacon & Goodship, 1991; Figure 2).

carrying, pushing and pulling loads uphill. The site where they lived was fringed with cliffs 100ft high and these had to be climbed to secure seabirds and their eggs for food and also to reach the beach. A second tribe lived on level ground, about 20 miles away at Quanterness, and their bones did not show the squatting facets. Chesterman suggested that the crystal orientations of the ankle bones of the two tribes would be different, so we examined four samples of the tibiae of each tribe and also four present-day tibiae (Bacon, 1990). Figure 10 shows the twelve individual polar diagrams. The samples from Isbister

Isbister Quanterness Modern

Figure 10. Polar diagrams indicating the preferred orientation of the apatite crystals at the lower front edge of the tibia for sets of four tibiae from Isbister, Quanterness and Sheffield (modern) respectively. In each diagram the radius vector in any direction is proportional to the number of c axes in that direction. The vertical direction on the page is the vertical line of the leg (from Bacon, 1990; Figure 2).

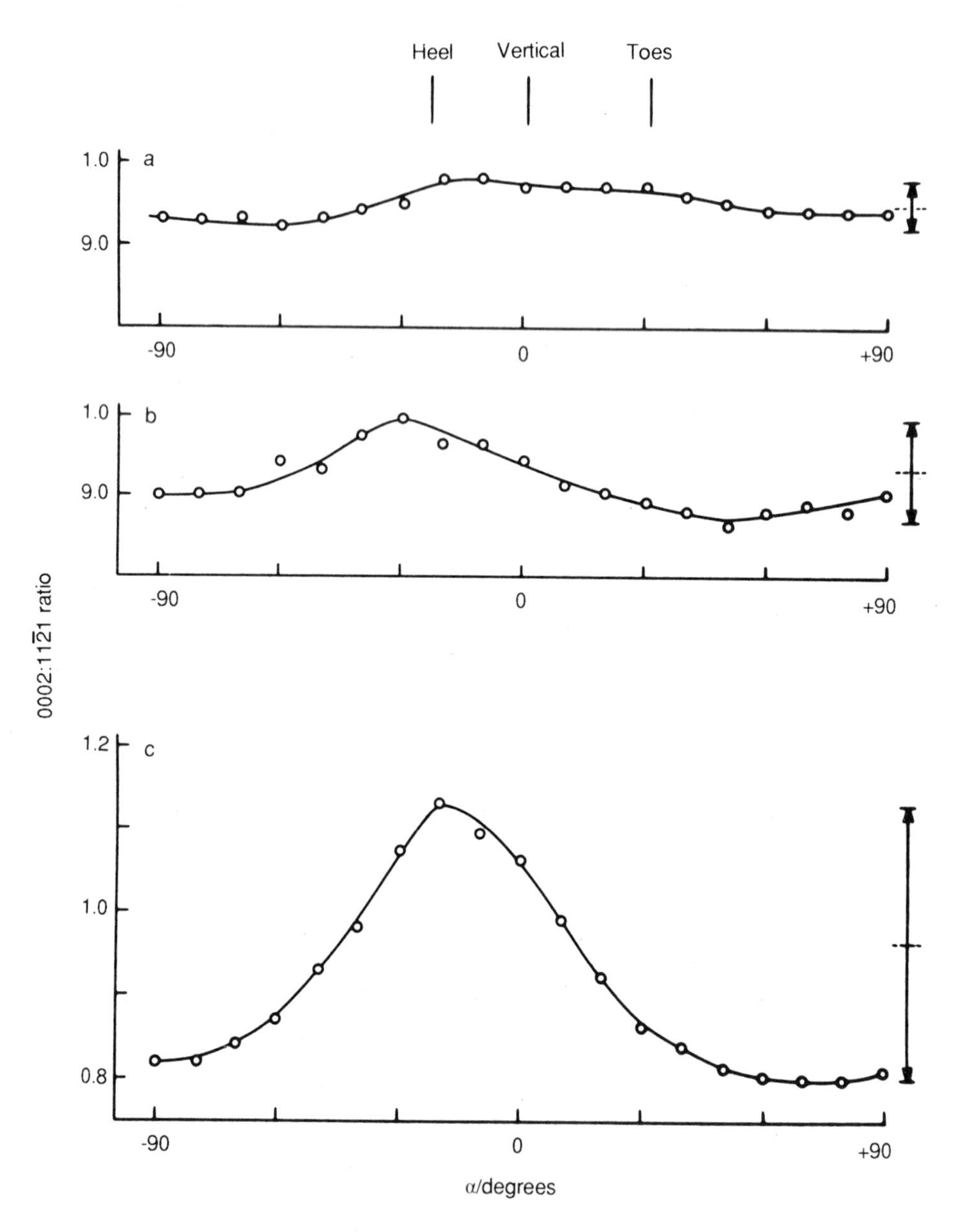

Figure 11. Cartesian plots expressing the mean curve of preferred orientation for each group of four tibiae, as a function of direction in the tibia. The distances between the arrowheads at the right-hand side of the diagram indicate the amount of anisotropy ie the extent of departure from uniform circularity, for the (a) Isbister (b) Quanterness and (c) modern tibiae respectively. The centre-point marked between arrowheads is close to the value of 0.92 expected for a randomly-oriented powder (from Bacon, 1990; Figure 3).

are almost circular, showing that these tibiae act almost like a universal joint in the front-to-back (sagittal) plane. The diagrams for Quanterness are more anisotropic and in today's diagrams, for elderly people, the predominant effect of an upright stance is clearly evident. The mean diagrams for the three groups of bones are shown as Cartesian plots in Figure 11 which average out the individual biological variations. The differences between the maximum and minimum values of the 002/111 ratio in the three curves in the Figure are 0.06, 0.12 and 0.33. We can take the reciprocals of these values to define what we may call an 'index of manoeuvrability' for the ankle, obtaining values of 16, 8 and 3 respectively.

Among other topics which we have explored in a preliminary way are a possible relation between the texture of the cannon bones of horses and their vulnerability to fracture, the function of the enormous antlers of the extinct Irish elk (Kitchener *et al.*, 1994) - for combat or only display - and the stresses at the human hip-joint where the head of the femur pivots in the acetabulum.

CONCLUSION

There is no limit to the number of anatomical problems which can profitably be studied. At the same time there is scope for experiments designed to further our knowledge of the linkage between hydroxyapatite and collagen which produces the composite of bone itself.

REFERENCES

Bacon, G.E., (1990). The dependence of human bone texture on life style. *Proc. R. Soc. Lond.*, B240:363–370.

Bacon, G.E., & Goodship, A.E., (1991). The orientation of the mineral crystals in the radius and tibia of the sheep, and its variation with age. *J. Anat.*, 179:15–22.

Bacon, G.E., & Griffiths, R.K., (1985). Texture, stress and age in the human femur. *J. Anat.*, 143:97–101.

Bacon, G.E., Bacon, P.J., & Griffiths, R.K., (1977). The study of the bones by neutron diffraction. *J. Appl. Cryst.*, 10:124–126.

Bacon, G.E., Bacon, P.J., & Griffiths, R.K., (1979a). Stress distribution in the scapula studied by neutron diffraction. *Proc. R. Soc. Lond.*, *B*204:355–362.

Bacon, G.E., Bacon, P.J., & Griffiths, R.K., (1979b). The orientation of apatite crystals in bone. *J. Appl. Cryst.*, 12:99–103.

Bacon, G.E., Bacon, P.J., & Griffiths, R.K., (1984). A neutron diffraction study of bones of the foot. *J. Anat.*, 139:265–273.

Kitchener, A.C., Bacon, G.E., & Vincent, J.F.V., (1994). The Irish elk used its antlers for fighting. *Biomimetics*, 2:297–307.

Wolff, J., (1870). Die innere Architecktur der Knocken. *Virchows Archiv fur pathologische Anatomie und Physiologie.*, 50:389–345.

Wolff, J., (1899). Die Lehre von der functionellen Knockengestallt. *Virchows Archiv fur pathologische Anatomie und Physiologie.*, 155:256.

2

NEUTRON SCATTERING FACILITIES

History, Performance, and Prospects

D. L. Price

Argonne National Laboratory
Argonne Illinois 60439

ABSTRACT

The past history, present performance and future prospects for neutron scattering facilities will be discussed. Special features of neutron scattering techniques applicable to biological problems will be reviewed, emphasizing the relation between structure, dynamics and function. The use of isotope labeling and anomalous neutron scattering to identify the environments of specific atoms of structural elements will be highlighted.

Work supported by U.S. DOE Contract #W-3 1-1 092ENG-38.

3

THE NIST NBSR AND COLD NEUTRON RESEARCH FACILITY

J. J. Rush

Institute for Materials Science and Engineering
National Institute of Standards and Technology
Gaithersburg Maryland 20899

ABSTRACT

The 20 MW Neutron Beam Split-Core Reactor (NBSR) has nine radial thermal beam tubes, and a large, highly accessible (35 cm) cold source serving an extensive network of eight guide tubes. In operation or under construction are twenty-five neutron beam instruments (20 for neutron scattering) and about a dozen other facilities for neutron trace analysis, dosimetry and irradiation. The 6×15 cm^2 cold neutron guides are coated with ^{58}Ni, and the last three being installed this fall are coated top and bottom with supermirrors for further increases in intensity. The new semi-spherical liquid hydrogen source will be described, along with the eight scattering instruments (reflectometry, SANS and high-resolution spectroscopy) which have, or will have, an extensive use in biological research. These instruments will likely provide the best overall capability in the U.S. for the next decade for a number of applications in biomolecular structure and dynamics.

4

NEUTRON SCATTERING INSTRUMENTATION FOR BIOLOGY AT SPALLATION NEUTRON SOURCES

Roger Pynn

Los Alamos National Laboratory
Los Alamos New Mexico 87545

ABSTRACT

Conventional wisdom holds that since biological entities are large, they must be studied with cold neutrons, a domain in which reactor sources of neutrons are often supposed to be pre-eminent. In fact, the current generation of pulsed spallation neutron sources, such as LANSCE at Los Alamos and ISIS in the United Kingdom, has demonstrated a capability for small angle scattering (SANS) - a typical cold-neutron application - that was not anticipated five years ago. Although no one has yet built a Laue diffractometer at a pulsed spallation source, calculations show that such an instrument would provide an exceptional capability for protein crystallography at one of the existing high-power spallation sources. Even more exciting is the prospect of installing such spectrometers either at a next-generation, short-pulse spallation source or at a long-pulse spallation source. A recent Los Alamos study has shown that a one-megawatt, short-pulse source, which is an order of magnitude more powerful than LANSCE, could be built with today's technology. In Europe, a preconceptual design study for a five-megawatt source is under way. Although such short-pulse sources are likely to be the wave of the future, they may not be necessary for some applications - such as Laue diffraction - which can be performed very well at a long-pulse spallation source. Recently, it has been argued by Mezei that a facility that combines a short-pulse spallation source similar to LANSCE, with a one-megawatt, long-pulse spallation source would provide a cost-effective solution to the global shortage of neutrons for research. The basis for this assertion as well as the performance of some existing neutron spectrometers at short-pulse sources will be examined in this presentation.

5

THE POTENTIAL FOR BIOLOGICAL STRUCTURE DETERMINATION WITH PULSED NEUTRONS

C. C. Wilson

ISIS Facility
CLRC Rutherford Appleton Laboratory
Chilton, Didcot, Oxon OX11 0QX, United Kingdom

ABSTRACT

The potential of pulsed neutron diffraction in structural determination of biological materials is discussed. The problems and potential solutions in this area are outlined, with reference to both current and future sources and instrumentation. The importance of developing instrumentation on pulsed sources in emphasised, with reference to the likelihood of future expansion in this area. The possibilities and limitations of single crystal, fiber and powder diffraction in this area are assessed.

THE ADVANTAGES OF NEUTRONS IN THE STUDY OF BIOLOGICAL STRUCTURES

Much of the progress in studying biological materials with neutrons has been made at the high flux reactor sources at the ILL, Grenoble and at Brookhaven National Laboratory in the USA. Neutron scattering results from these and other facilities have proven a valuable complement to X-ray and other measurements on biological systems. Exploiting the usual benefits of neutrons in the location of light atoms; in determining the vibrations of hydrogen atoms in often complex molecular environments; and in probing even delicate samples non-destructively, workers in this area have had considerable success in fine tuning the detail already known from other techniques. Often this fine tuning, of course, provides information critical to understanding the system under study.

There is one area, however, in which neutrons have a profound and unique advantage in the study of biological systems. By exploiting the very large difference in scattering power between hydrogen and its heavier isotope deuterium, the method of contrast variation can be used to highlight particular components of a system. This can be used to

Neutrons in Biology, edited by Schoenborn and Knott
Plenum Press, New York, 1996

allow large and complex structures to be effectively broken into smaller components (*in situ* and non-destructively), allowing more detailed analysis. Combining different H/D substitution schemes thus allows detail to be built up from a series of experiments where a single scattering experiment would yield more limited information.

Since H_2O (often easily substituted by D_2O simply by soaking) forms such a crucial part of many biological systems—for example often determining the secondary or higher order structure—it follows that neutron scattering can have a substantial impact in this area. Appropriate experimental methodology can yield a high degree of structural detail, allowing neutron scattering to play a full role, often in combination with other methods. For example, combining neutron scattering with NMR yields an extremely powerful method for allowing the determination of both the chemical (NMR) and structural (neutron scattering) environment of hydrogen atoms.

In addition to the obvious possibilities engendered by H_2O/D_2O substitution, however, chemical isotopic enrichment or substitution is becoming more widespread as preparation techniques are refined. Allowing particular sites, in small molecules; particular residues, in larger molecules; or particular molecules, in aggregates, to be highlighted again accrues the benefits mentioned above. Not only chemical but biological preparation techniques are being enhanced and culturing from deuterated media could lead to still more H/D contrast possibilities in larger systems.

This paper aims to review the current and future potential for pulsed neutron sources to continue and expand upon the successful work already performed in these areas.

THE ISIS PULSED NEUTRON SOURCE

ISIS, the UK Spallation Neutron Source sited at the Rutherford Appleton Laboratory near Oxford (Figure 1; Wilson, 1995a), is the world's most intense pulsed neutron source. The 800MeV proton accelerator produces a current of >190μA of protons (designed to reach 200μA), pulsed at 50Hz, impacting on a heavy metal target (typically U or Ta). The resulting neutrons, produced by the spallation process, are slowed down for condensed matter studies in an array of hydrogenous moderators surrounding the target. These are of three types: H_2O at 316K; CH_4 at 95K and H_2 at 20K, each producing a distinctive spectrum of neutrons with a high epithermal flux at short wavelengths and a peak Maxwellian flux at a wavelength dependent on the moderator temperature. The pulse structure of the neutrons produced at ISIS is maintained by using small moderators of cross-section area $10 \times 10\ cm^2$ and depth 5cm. The white neutron beams are sorted using the time-of-flight technique in which the arrival time of each neutron at the detector allows the determination of its wavelength. The useful wavelength range in these white beams is around 0.2–20Å.

Surrounding the ISIS target station is an array of 15 neutron and muon instruments for condensed matter research (Figure 2). These allow studies to be made of structure—from crystals, powders, liquids, amorphous, polymeric materials, surfaces etc—and of dynamics—atomic excitations, vibrations, rotations, diffusive processes, quantum tunnelling etc.

Until the advent of ISIS, the flux available at pulsed neutron sources was wholly inadequate for the study of biological materials. Indeed ISIS, with a high flux of short wavelength neutrons, is easy to perceive as a high resolution source, ideal for examining the harder end of condensed matter science: physics, materials and small molecule chemistry.

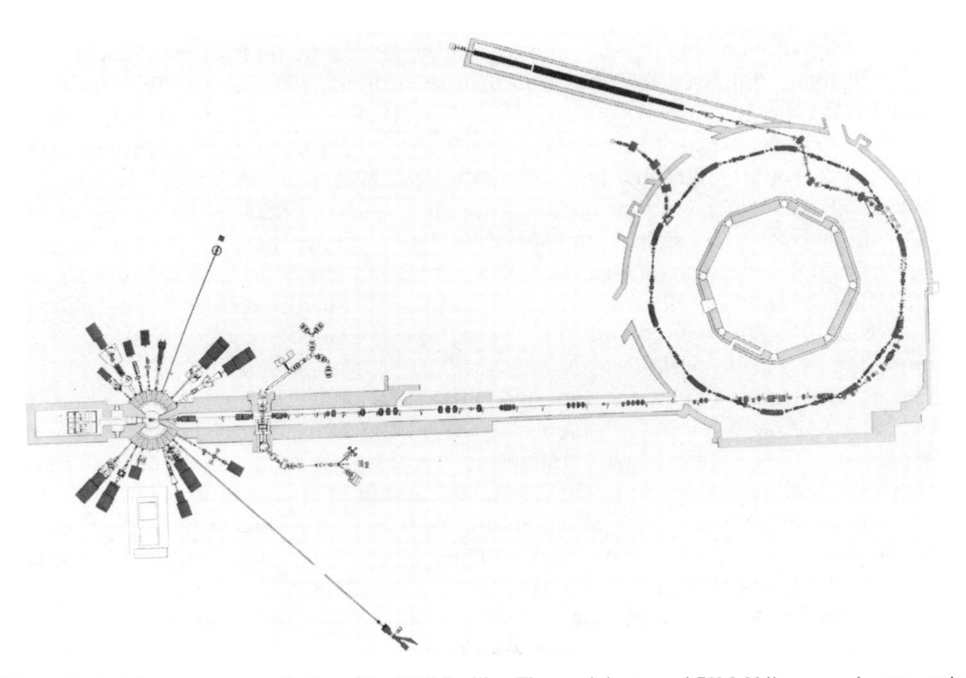

Figure 1. Aerial view and overall plan of the ISIS Facility. The pre-injector and 70MeV linear accelerator are in the small building towards the top, the 800MeV synchrotron under the grass mound at the top right, and the Target Station and Experimental Hall in the large building in the center of this photograph.

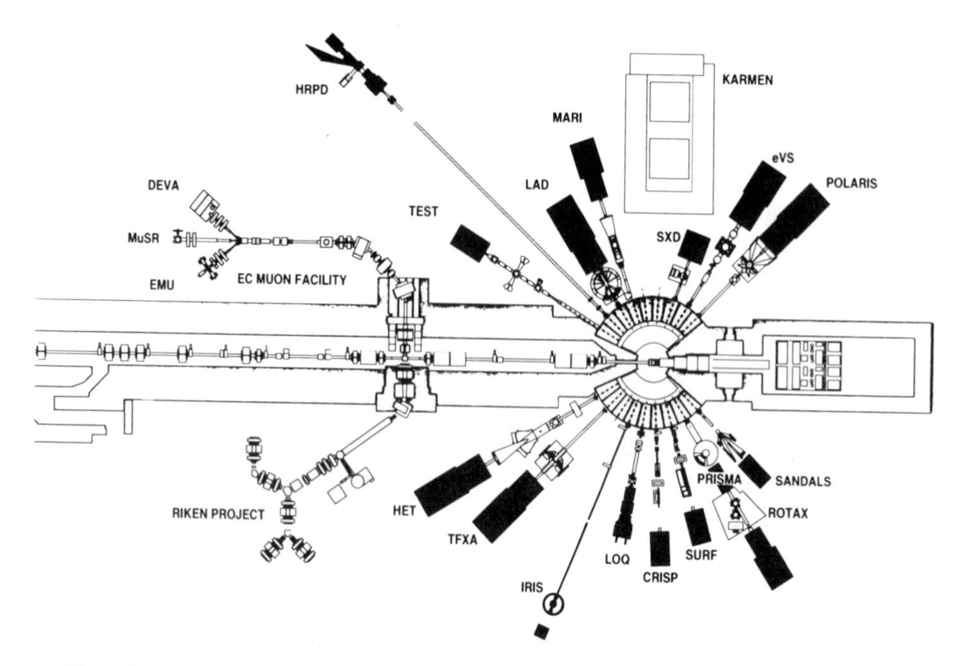

Figure 2. Layout of the neutron (and muon) scattering instruments in the ISIS Experimental Hall.

However, a pulsed neutron source also has a rich flux of the colder neutrons required for biology, and ISIS has been the first such source at which this has been realised and exploited. Although the present cold 'source' at ISIS, a liquid hydrogen moderator at 25K is still more optimised for resolution than flux, the number of cold neutrons is impressive.

One of the first major successes in the cold neutron area at ISIS was the establishment of a routine, high quality, rapid throughput neutron reflectometry programme. Allowing the study of molecular conformation at surfaces, reflectometry is one area in which contrast variation has had major success. For example, the conformation of polymers, surfactants and membrane-simulating phospholipid layers have been studied in detail (Penfold & Thomas, 1990; Naumann *et al.*, 1994). Small angle scattering has also been successful in studying colloids, microemulsions, aggregates and membranes including bio-systems (Eastoe *et al.*, 1994; Winter *et al.*, 1991; Meyer *et al.*, 1995).

The combination of cold neutrons and high resolution has been exploited in spectroscopic studies of the dynamics of biopolymer gels, water transport in bio-systems and the examination of large scale, slow librations and translations of macromolecules (Deriu, 1993; Deriu *et al.*, 1993; Konig *et al.*, 1994). In addition, the possibilities of long wavelength powder diffraction studies of larger molecules can be evaluated, with every promise of success in the quantitative analysis of the structure of drugs and pharmaceutical preparations and possible extension to larger systems (see below).

The warmer neutrons available elsewhere at ISIS can also be exploited in biology related areas but mostly in the examination of model systems. Elucidating detail on components, be it structural or vibrational, can be of great use when extrapolated to macromolecular systems, but it must be combined with work on the derived macromolecule to allow relevant deductions to be made. The biology studied at ISIS to date is summarised briefly in Figure 3.

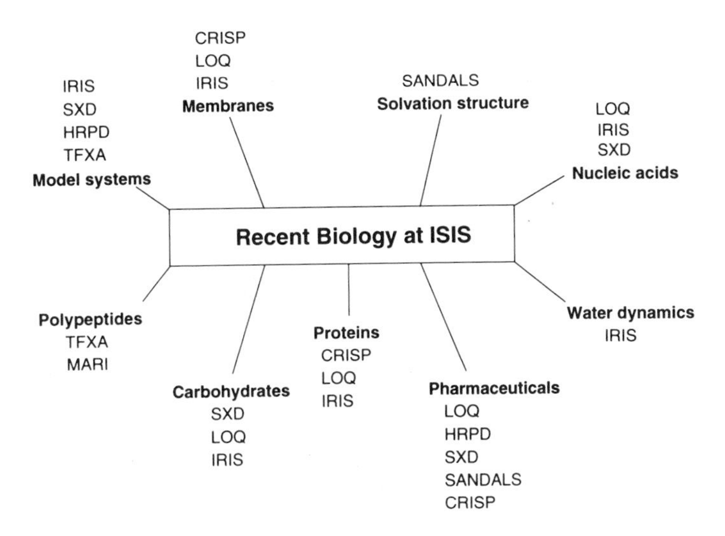

Figure 3. A brief schematic summary of the biology studied at ISIS by subject area and instrument. Key to instruments: IRIS—Long wavelength spectrometer and diffractometer; SXD—Time-of-flight Laue single crystal diffractometer; HRPD—High resolution powder diffractometer; TFXA—Wide energy range spectrometer; CRISP—Surface reflectometer; SANDALS—Liquids diffractometer; LOQ—Small angle scattering spectrometer.

THE AIM—BIOLOGICAL NEUTRON CRYSTALLOGRAPHY

The power of neutrons in determining solvent water structure, hydrogen bonding schemes and precise active site geometry in proteins is proven (Schoenborn, 1984; and references therein) but applications to date, especially at pulsed neutron sources, have been limited in part by the lack of appropriate instrumentation. Much of large molecule organic and modest sized biological single crystal work should ideally be accessible in data collection times of around a week even from the small single crystals which are likely to be available. Higher flux, improved dedicated instrumentation, more cold neutrons and more detectors are the keys to success in this area.

A typical specification for such work with neutrons would be to tackle unit cell edges of the order of 100Å, cell volumes of > 10^5 $Å^3$, to a d_{min} of around 1.2–1.5Å in favorable cases, depending on the crystal diffraction characteristics. An essential prerequisite for successful work in this area would be to have reasonable data collection times (days or weeks rather than months) but the data collection is always likely to be limited by small sample size which for protein single crystals is rarely greater than $1mm^3$ in volume and is frequently considerably less than this.

SXD—The ISIS Time-of-flight Laue Diffractometer

Time-of-flight (tof) Laue diffraction remains a relatively novel technique for the determination of crystal structures. The method exploits the time-sorted white beams available from a pulsed source such as ISIS, along with large area position-sensitive detectors, to allow the simultaneous measurement of fully-resolved three dimensional volumes in reciprocal space. One of the major requirements of a tof Laue diffractometer such as the SXD at ISIS (Wilson, 1990; Figure 4), is the availability of reliable, large area position-sensitive detectors. There has been a major development effort at ISIS in this area over the

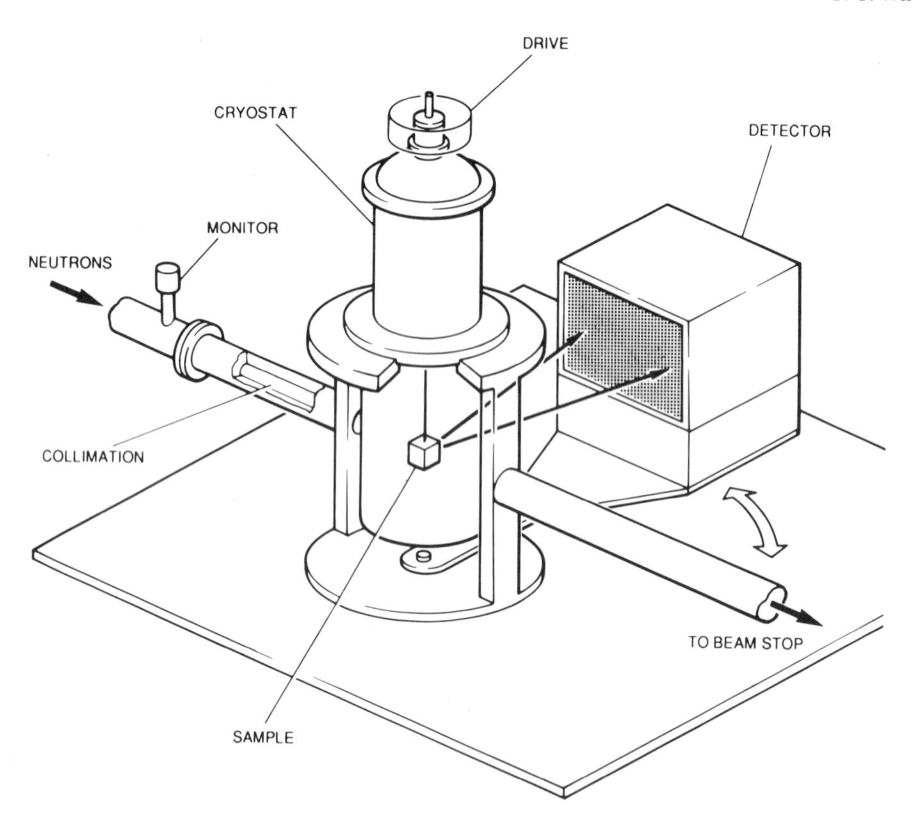

Figure 4. Schematic layout of the time-of-fight Laue diffractometer SXD at ISIS. The white beam of neutrons, wavelength sorted by the time-of-flight technique, is incident on the sample from the left of the figure and the scattered neutrons are detected by a large area position-sensitive detector.

past few years, the outcome of which has recently been summarised (Wilson, 1993). The detectors currently in use (Figure 5) are based on fiber optic coupling of pixel elements to photomultiplier tubes (PMTs). Four fibers extend from each pixel element to a unique set of four PMTs. Use of a quad coincidence method then allows the 64 × 64 pixel elements to be encoded onto just 32 tubes. The scintillator used at present in 0.4mm thick ZnS doped plastic, which has relatively low efficiency (around 20% at 1Å) but is virtually insensitive to γ radiation, leading to good signal-to-noise ratios in the detector. The detector complement on SXD has recently been enhanced by the construction of a second module. The typical layout of the instrument at present is shown in Figure 6.

The ability of an instrument such as SXD to sample reciprocal space volumes renders the technique particularly effective in reciprocal space surveying applications, for example in rapid, quantitative diffuse scattering measurements. However, by appropriate use of wavelength dependent corrections, the technique can also be used in 'standard' crystallography—structural refinement.

Applications in Chemical Crystallography

In recent years, once stable large area detectors had been developed (Wilson, 1993), SXD has been successfully used for applications in small molecule crystallography in the

Figure 5. View of the construction of the current position-sensitive detectors used on SXD. The ZnS scintillator sits on the front of the pixelated grid, and the 16384 optical fibers are coupled to 32 photomultiplier tubes (not shown), allowing 4096 elements to be coded by just 32 PMTs.

traditional areas of hydrogen atom location, as a complement to high resolution X-ray structure determination in charge density studies and in related areas more explicitly using the unique characteristics of time-of-flight Laue diffraction (Wilson, 1995b).

Single crystal neutron diffraction is an extremely powerful technique for accurate chemical crystallography. Good neutron single crystal data will:

- yield high precision atomic and vibrational parameters, including higher order thermal effects;

Figure 6. The current set-up on SXD, showing two detectors in position. Geometric constraints at present restrict the detector array to a 'low angle-high angle' pair but the optimised construction of the new modules (on the left) will allow more efficiently stacking of multiple detectors.

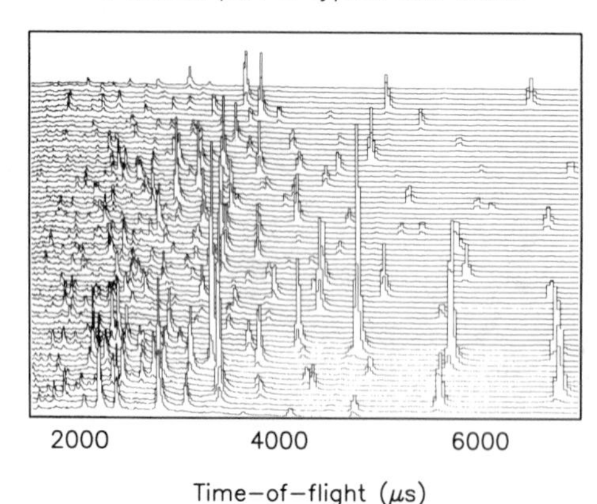

Figure 7. Many reflections are accessed in a single measurements in time-of-flight Laue diffraction. The figure shows a typical equatorial plot across the detector from low (bottom) to high (top) 2θ values. In this plot of scattering from the neurotransmitter acetylcholine, the diffraction pattern has been condensed vertically on the detector and plotted in approximately constant 2θ strips.

- permit detailed analysis of conformations, molecular energetics etc;
- allow detailed charge density work to be carried out in combination with high resolution X-ray studies ($\sin\theta/\lambda > 1\text{Å}^{-1}$) using X-n methods. Such studies yield further information on the nature of the chemical bonds in a molecule;
- allow very accurate definition of hydrogen bonding, further enhancing the potential for reliable molecular design, for example.

Some recent applications of SXD in chemical crystallography are given in Appendix I.

It is clear that there are several high profile areas in which such neutron structure determination can have an impact. For example:

- Pharmaceuticals, where many drug molecules are in the accessible cell range up to 10^4Å^3. Detailed neutron data can be vital to the understanding of molecular conformation, especially with regard to the often very small energy differences between active and inactive polymorphs. Neutrons also sample the bulk of such materials, again vital in the study of polymorphism in relation to production processes;
- Organometallic materials, widely used as catalysts (for example H_2–2H behavior in homogeneous catalysis involving C-H activation) and fertilisers. Such materials frequently have important hydrogen atoms located close to a heavy metal atom, rendering neutron diffraction the only way of adequately determining the structure;
- Organic structures, where neutrons have a vital role to play in the study of basic bonding, charge density studies, hydrogen bonding and in non-destructive phase transitions, particularly those involving hydrogen atom shifts;

The advantages of tof Laue diffraction in structure determination are that many Bragg reflections are collected simultaneously (Figure 7), allowing the instrumentation to be optimised fairly easily for each data collection. Tof Laue is the optimal white beam method, allowing the separation of Laue orders and also separating the diffraction pattern generally into three dimensions. This both avoids accidental reflection overlaps (frequently as much or more of a problem than the precise overlap of Laue orders) and stretches the 'Laue background' resulting from the incoherent scattering from the hydrogen content of the samples. Thus the whole pulse (all wavelengths) can be used with impunity.

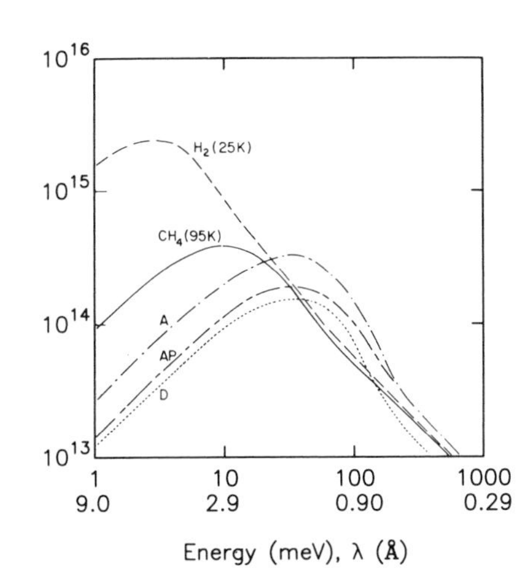

Figure 8. A comparison of the flux from the ISIS poisoned ambient water moderator (as viewed by SXD), the liquid methane moderator at 95K and the liquid hydrogen at 25K.

The disadvantages, however, especially in relation to large molecule crystallography, are the lack of time-averaged flux in current pulsed sources; the fact that the present SXD is sited on the 'hot' water moderator which produces a peak neutron flux at 1.14Å with a rapid fall off at longer wavelengths (see Figure 8; Taylor, 1984); and that the technique itself remains relatively novel and is still being established in structure determination.

TIME-OF-FLIGHT LAUE AND FIBER DIFFRACTION

The application of SXD in fiber diffraction of biological molecules is described elsewhere in this volume (Forsyth *et al.*, 1995) but can be briefly summarised here in relation to the technical aspects of data collection and analysis. Fiber diffraction involves a sample with one unique, ordered axis. The technique normally uses an area detector with monochromatic radiation to produce a two dimensional diffraction pattern and can reveal fairly detailed structural information given a good starting model. In favorable cases, quasi-atomic resolution can be obtained.

In tof Laue fiber diffraction, each time (wavelength) slice is a more or less complete 'fiber pattern' but in this narrow slice, the pattern is measured with very low statistics as the point-by-point flux is low. These wavelength slices must therefore be focused to give adequate patterns, involving some fairly complex focusing algorithms and careful consideration of matters such as flux profile normalisation and the variation of instrumental resolution with angle. In practice, the initial processing of such patterns uses the time, rather than the spatial distribution, with various methods employed to normalise each spatial element to take into account incident beam flux, absorption and detector response (Figure 10 below is an example of such a normalised spectrum). These normalised spectra are subsequently focused in reciprocal space to reconstruct the overall fiber pattern. In spite of the relative complexity of these procedures, this can be achieved fairly successfully and quantitative fiber patterns obtained which can be analysed in the same way as those collected in monochromatic experiments.

Applications to date have principally involved studies of the water structure and conformation of DNA (Forsyth *et al.*, Fuller *et al.*, this volume); studies of cellulose II fi-

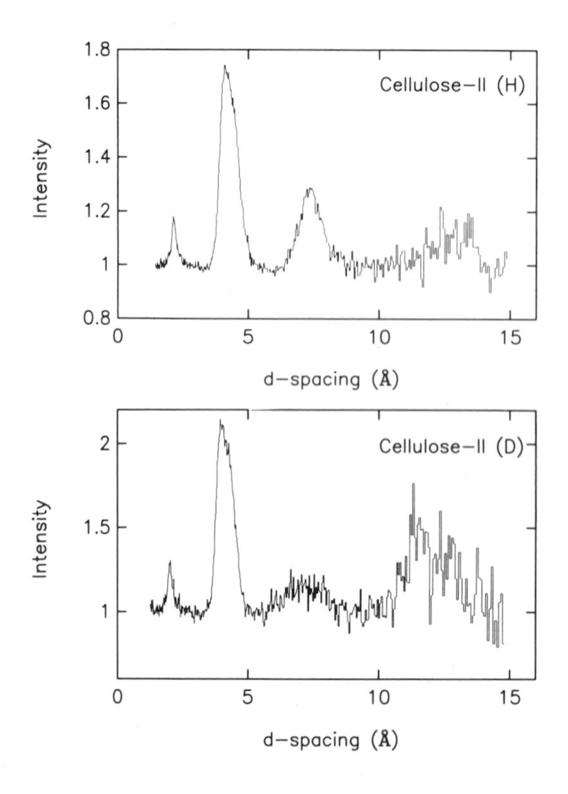

Figure 9. The zeroth layer fiber pattern from cellulose-II. (top) Fully hydrogenous fiber, (bottom) with the exchangeable hydrogens replaced with deuterium. The structural information available from these is currently under study.

bers utilising contrast variation from the exchangeable hydrogen atoms (Kroon-Batenberg *et al.*, 1994; Figure 9); and initial measurements from hyaluronate fibers (Deriu *et al.*, 1995). While the fiber patterns generally have similar appearance to those measured on a reactor source, for example on D19 (ILL Grenoble), the overall statistics are rather low. However, in certain areas of the pattern, the characteristics of a pulsed source can be extremely favorable in measurements of this type. For example, the high flux and resolution at short wavelengths can enable higher resolution portions of the pattern to be successfully measured (Whalley *et al.*, in preparation; Figure 10), and the nature of the tof Laue dif-

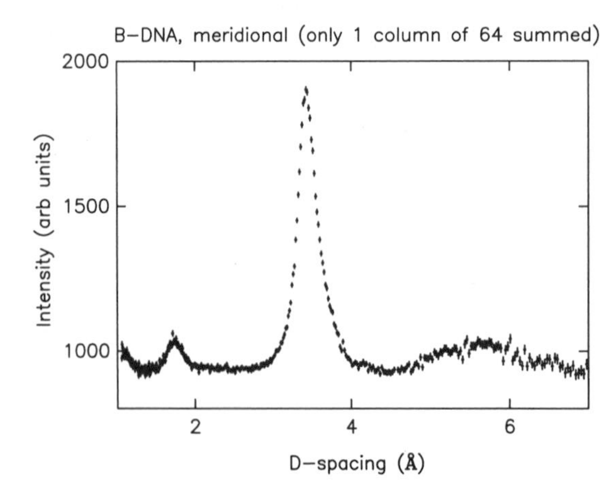

Figure 10. The meridional diffraction from a fiber of B-DNA measured on SXD. In addition to the well known strong diffraction feature at 3.4Å, there is clear evidence for the presence of the second order reflection at 1.7Å. The availability of such high resolution information should allow better structural models to be constructed from the fiber diffraction pattern.

fraction technique allows such regions of the diffraction pattern to be measured without penalty in terms of data collection time due to the variable wavelength, variable angle nature of the method. There is an obvious complementarity evident here, where tof Laue fiber diffraction can focus on important regions of a pattern while the whole pattern is more efficiently measured using steady state methods.

The problems associated with tof fiber diffraction have largely been alluded to in the above—the need for wavelength and angle dependent corrections, the non-trivial nature of the focusing, the need for large sample volumes, preferably of various H/D exchanged materials to maximise information content and the relatively low integrated flux on present pulsed sources. These problems can all be tackled, however, with improved data reduction and analysis software resulting from the experience gained in collecting and analysing data from such materials; and with improved preparation methods—the sample volume problem is not as serious for fibers as for crystals. One of the major problems, however, the fact that the peak flux on SXD is optimised for short wavelengths, can be tackled by performing fiber diffraction experiments at ISIS on a colder moderator.

IMPROVING THE PROSPECTS FOR NEUTRON DIFFRACTION STUDIES OF BIOLOGICAL STRUCTURE

The problems associated with studying large and weakly scattering biological systems provide a challenging problem to the designers of neutron sources and instruments. The basics of the problem are simple to state:

- in terms of sources the highest possible flux must be obtained consistent with carrying out the scattering experiment—traditionally with reactors but more recently with new possibilities from pulsed sources;
- in terms of instrumentation large amounts of data must be recorded and analysed as efficiently as possible.

Stark simplicity then, but of course hiding complex problems.

Improvements in Instrumentation

Since reactor sources have been and remain at present the most appropriate for neutron biological crystallography, it is important that developments of appropriate instruments continue at such sources. For example, the quasi-Laue diffractometer LADI, newly in operation at the ILL, exploits a relatively wide wavelength band cold neutron beam with a large area of high resolution detectors close to the sample to optimise the count rate possibilities from protein crystals. With the exploitation of the rapidly developing technology of image plates (Cipriani et al., this volume) to surround the sample with detectors, LADI allows the use of simple sample geometry and well established Laue techniques, and promises to be a powerful instrument for protein crystallography.

For pulsed sources, it is similarly vital to have a large detector array, allowing the largest possible number of reflections to be measured simultaneously. In principle, given the wide wavelength range which the tof technique allows one to use, the diffraction data could be accumulated in very few crystal orientations, assuming that different wavelength ranges can be used for detectors at different angles. The basic detector technology and the associated electronics are now proven from developments at ISIS and elsewhere (Rhodes *et al.*, 1994; Schoenborn & Pitcher, this volume). It should be noted that image plates are not at present an option for tof studies as the wavelength discrimination is lost using current methods of signal storing and reading.

As stated above, the tof technique allows the Laue method to be used without penalty and the continuous nature of the data can be used to optimise intensity extraction, perhaps in a manner analogous to Pawley refinement of powder patterns. Such a method has recently been implemented for the first time on SXD in small molecule studies (Wilson, 1996) with significant improvements in intensity extraction, especially for weaker reflections and those at higher $\sin\theta/\lambda$ values where there is more likelihood of overlap.

The broad outline of an instrument for large molecule crystallography (LMX) at a pulsed neutron source is very straightforward (Lehmann & David, 1992; Appendix II). The instrument would be sited on a cold moderator, optimising the flux of neutrons in the 2–6Å range, for example. An array of 2D PSD's are placed close to the sample, which is held within a very simple geometry. The beam at the sample should ideally be focused in some manner to improve the flux on the small sample volume. Even focusing in one dimension would be of great assistance and the resolution costs in the time-of-flight technique are not too serious. The improved intensity extraction software and rapid data processing requirements are also tractable problems. The aim of such an LMX instrument would be to allow 1–2 week throughput for very large organic and small protein structures.

The immediate prospects for such an instrument at ISIS involve installing one or more large area PSD's on a cold neutron beam, possibly as part of the funded OSIRIS beam line development (Ross, 1993). Such a prototype LMX would allow the design parameters for a future instrument to be optimised. This arrangement may also allow some initial experiments on large molecule organic materials and very small proteins using the SXD-style of data collection.

Improvements in Sources

However, it may well be that a step function increase in studies in this area will only happen when the neutron sources themselves are improved, yielding the increases in flux which will help to overcome the intrinsic problems of studying biological materials. On reactors, the prospect for improvements are somewhat limited, given that the power density in the reactor core limits the potential flux increase to around five times that of the present ILL at best. However, it is simpler to construct moderators for cold neutrons in reactors and there are no pulse length considerations. It is also easy to select, guide and detect the longer wavelength neutrons required for biology and well established steady state methods can be exploited to the full in extending the range of such studies.

For pulsed sources, still relatively in their infancy, there is more scope for improvement. On both the accelerator and target station side, advancing technology allied with increasing experience of operating such sources promise a rich future.

The proposed ISIS second target station promises a source more optimised for biological studies than the present high resolution set-up. This project, proposed for the late 1990s, would involve construction of a target station optimised for cold neutrons operating with one pulse in five from the current ISIS accelerator—a 10 Hz source (Figure 11). There are many moderator options for optimising wavelength, flux and pulse width characteristics (Figure 12). For example, it is possible to obtain ultra-high neutron flux from purpose built coupled liquid hydrogen moderators. However, the resulting wide pulse from such a moderator would provide extra challenges to the optimisation of intensity extraction software.

On the accelerator side, the present design study for the ESS, the European Spallation Source (Taylor, 1992), aims to provide a source with a peak flux of some 30 times

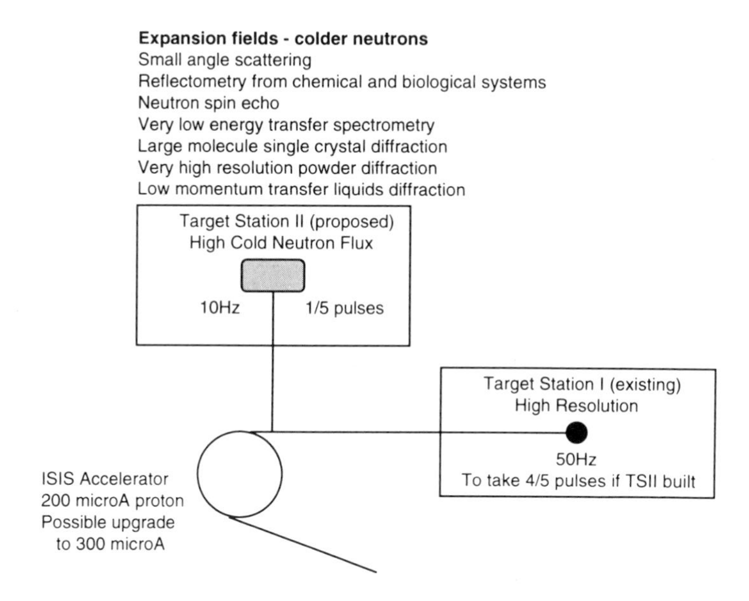

Figure 11. A second Target Station for ISIS.

that of the present ISIS, with a time averaged flux equivalent to that of the ILL, the most intense reactor neutron source in the world. The use of time-sorted white beams from such a source means that full advantage can be taken of this flux increase. Combining such a source with appropriately optimised moderators, as for ISIS Target Station II, should allow for the provision of extremely powerful instrumentation for large molecule crystallography, yielding rapid throughput of moderately large protein structures.

POSSIBILITIES IN NEUTRON POWDER DIFFRACTION

The role of powder diffraction in structure determination of organic structures has recently undergone a revolution as improved resolution instrumentation has allowed the refinement of more complex materials. At present, the structures of deuterated organic

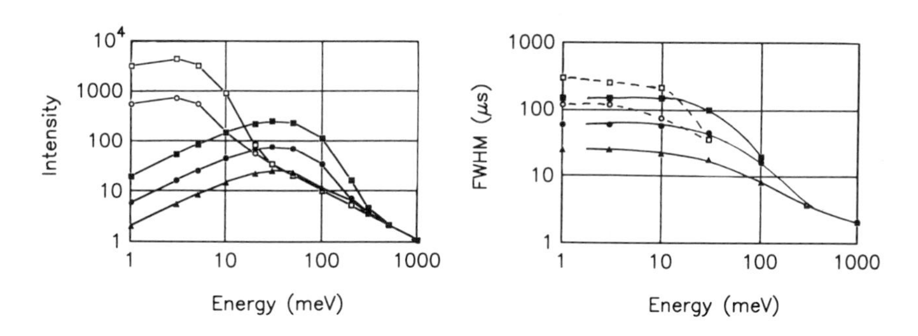

Figure 12. Moderator options for ISIS Target Station II and ESS, the European Spallation Source in terms of flux (left) and pulse width (right). For a 5MW spallation source (ESS), the flux is expressed in units of 2.2×10^{14} n/(eV.sr.100cm^2.s).

molecules with unit cell volumes of > ~2000Å^3 can be studied and refined accurately, with care, using sub-Å data (Ibberson & Prager, 1995; Lehmann *et al.*, 1994; Ibberson *et al.*, 1992). Such materials are useful as model systems. However, the extension to larger systems will certainly require increased use of detailed constraints in stabilising and making tractable refinements of such molecules.

Figure 13 illustrates the scale of the problem when trying to extend powder diffraction measurements to larger unit cell systems. The Figure shows simulations of the diffraction peaks for various unit cell volumes on a high resolution powder diffractometer on a cold moderator at a pulsed source. These simulations make two very favorable (and idealistic) assumptions. First, the overall temperature factor (B) is assumed to be 1Å^2, which is approximately correct for a small organic molecule at low temperature, but is a severe underestimate for larger systems. With a higher temperature factor the short d-spacing end of the diffraction patterns would be further suppressed. Secondly, there is no background included in these simulations; in particular the large background which would be present from incoherent scattering from any hydrogen atoms present in the material has been ignored. The apparent 'background' evident on plots (b) and (c) is in fact due entirely to the build-up of peak overlap. It is clear from the extreme peak overlap in even these idealised calculations, that the problems of trying to refine structures from powder diffraction patterns of larger unit cell materials multiply rapidly, emphasising the need for the extensive use of constraints to try to extract useful structural information in such cases.

The limits in this area are difficult to estimate at present, with developments in instrumentation and software offering improved prospects. For example, it may eventually be possible to refine fairly large structures to 2–3Å resolution from powder data by learning from the techniques used to constrain single crystal protein structure refinements. In addition, the recent increased exploitation of the surprisingly rich flux of cold neutrons from ISIS has enhanced capabilities in this area, and even hydrogenous small organic molecules are now becoming amenable to study. There is a clear role for neutron powder diffraction in the study of the larger model systems and the moderately large organic molecules concerned with biological function, even though protein structures, for example, are likely to remain well out of reach in the foreseeable future.

CONCLUSIONS

Provided the production of cold neutrons is optimised in future sources, and provided the whole pulse is exploited using the tof technique, pulsed neutrons have the potential to make major contributions to the study of biological structure. For example, an optimised neutron single crystal diffractometer such as LMX could yield useful high resolution information in a wide range of biological structural problems.

Developing the study of biology on pulsed neutron sources is important as there are many important problems for neutron scattering to tackle in biology: more data on more structures to better resolution; assisting molecular dynamics calculations which involve solvent structure as well as the molecule; biological phase and conformational transitions where neutrons can help resolve the possible role of water; low resolution studies of more complex, less ordered systems, eg, the outline of structures of assemblies of complexes, where individual components are known from high resolution studies.

The exploitation of neutron scattering in the areas of biological and macromolecular chemistry is not as widespread as one might expect given the current high interest in such areas. For example, the contrast with the recent explosion in the use of X-rays in protein

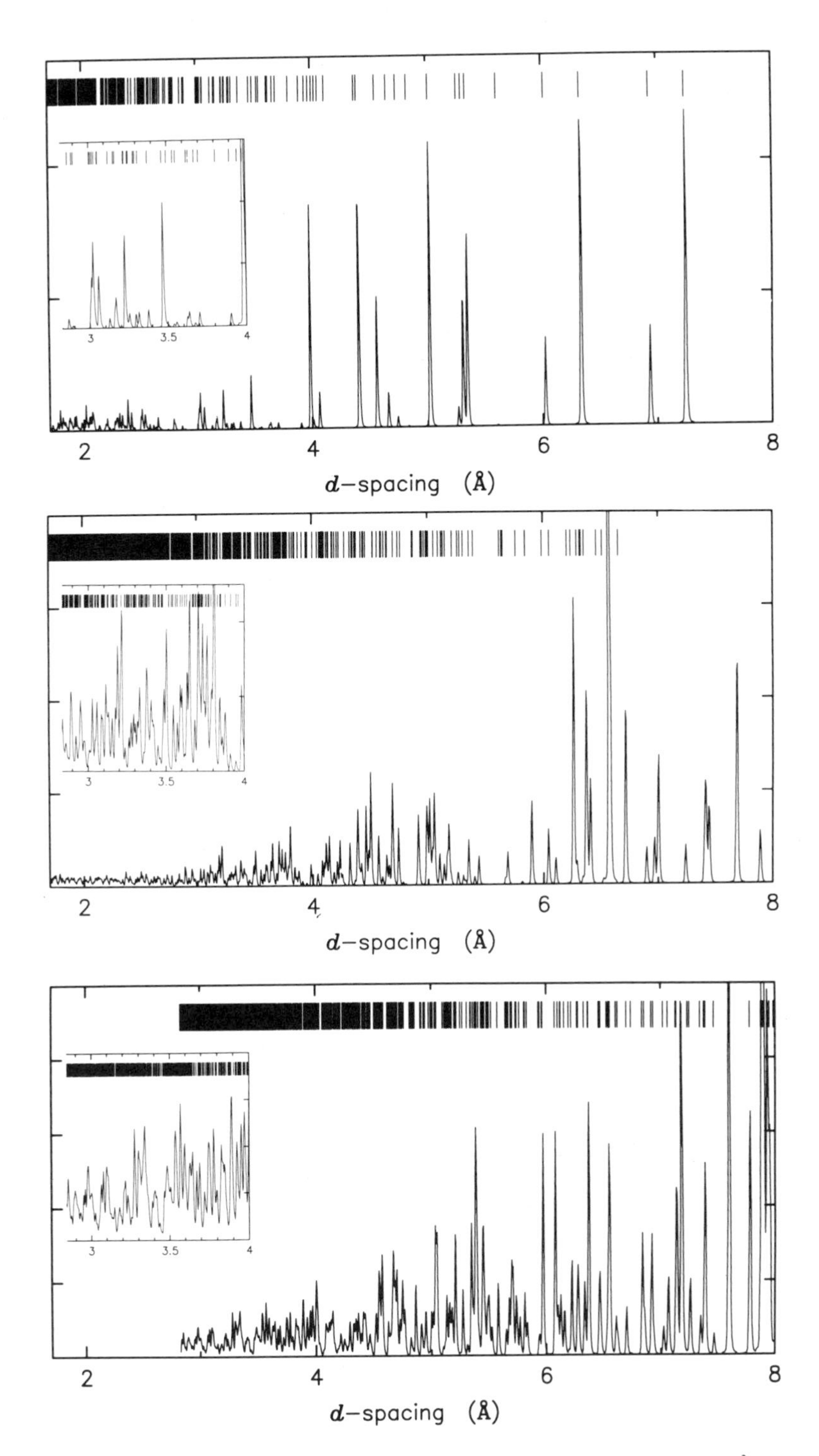

Figure 13. Simulated powder diffraction patterns (at Δd/d resolution of approximately 2×10^{-3}) for various unit cell volumes: (top) ~1230Å^3, $P2_1/c$ (typical of a small organic; (centre) ~9000Å^3, $P2_12_12_1$ (a fairly large organic); (bottom) ~31000Å^3, C2 (a large organic or very small protein). These simulations make the very favorable assumptions that the overall temperature factor (B) is around 1Å^2 and the incoherent background from any hydrogen atoms present has been ignored. The tag marks at the top of each pattern represent the positions of individual peaks.

crystallography is marked. This is due in large part to the intrinsically weak scattering of most samples of biological interest, rendering neutron scattering experiments in this area particularly challenging. It is therefore vital to lay the groundwork for future instruments and sources now, even though the current pulsed sources and instrumentation are less favorable, if we are properly to exploit neutron biological crystallography in the future.

Given current political tides, substantial enhancement of neutron production over the next few decades is likely only to come from development and construction of new pulsed sources. Existing steady state sources will continue to contribute substantially to the science programme but are unlikely to be augmented by significant upgrades or by new reactor sources. Neutron scattering in structural biology must therefore place itself in a strong position to benefit from current plans regarding the next generation of pulsed sources, or we may find that a golden opportunity has passed us by.

REFERENCES

Deriu, A., (1993). The power of quasielastic neutron scattering to probe biophysical systems. *Physica B*, 183:331–342.

Deriu, A., Cavatorta, F., Cabrini, D., Carlile, C.J., & Middendorf, H.D., (1993). Water dynamics in biopolymer gels by quasielastic neutron scattering. *Europhysics Letts.*, 24:351–357.

Deriu, A., Cavatorta, F., Moze, O., & diCola, D., (1995). Distribution of water in a highly hydrated fibre of K hyaluronate. *ISIS Annual Report* 1995, Vol. II, p A461.

Eastoe, J., Steytler, D.C., Robinson, B.H., & Heenan, R.K., (1994). Pressure induced structural changes in water-in-propane microemulsions. *J. Chem. Soc. Faraday Trans.*, 90:121–3127.

Ibberson, R.M., & Prager, M., (1995). The *ab initio* structure determination of dimethylacetylene using high resolution neutron powder diffraction. *Acta Cryst.*, B51:71–76.

Ibberson, R.M., David, W.I.F., & Prager, M., (1992). Accurate determination of hydrogen atom positions in toluene by neutron powder diffraction. *J. Chem. Soc. Chem. Comm.*, 1992:1438–1439.

Konig, S., Sackmann, E., Richter, D., Zorn, R., Carlile, C.J., & Bayerl, T.M., (1994). Molecular dynamics of water in oriented DPPC multilayers studied by quasielastic neutron scattering and deuterium NMR relaxation. *J. Chem. Phys.*, 100:3307–3316.

Kroon-Batenberg, L., Schreurs, A.M.M., & Wilson, C.C., (1994). Structure determination of cellulose II by fibre diffraction. *ISIS Annual Report 1994*, Vol. II, p A434.

Lehmann, A., Luger, P., Lehmann, C.W., & Ibberson, R.M., (1994). Oxalic acid dihydrate—an accurate low temperature structural study using high resolution neutron powder diffraction. *Acta Cryst.*, B50:344–348.

Lehmann, M.S., & David, W.I.F., (1992). Report of the Crystallography Working Group. In *Instrumentation and Techniques for the European Spallation Source*, p. CRYST-22, editor A. D. Taylor, *Rutherford Appleton Laboratory Report*, RAL-92–040.

Meyer, D.F., Nealis, A.S., Bruckdorfer, K.R., & Perkins, S.J., (1995). Characterisation of the structure of polydisperse human low density lipoprotein by neutron scattering. *Biochem. J.*, 310:407–415.

Naumann, C., Dietrich, C., Lu, J.R., Thomas, R.K., Rennie, A.R., Penfold, J., & Bayerl, T.M., (1994). Structure of mixed monolayers of dipalmitoylglycerophosphocholine and polyethylene-glycol monododecyl ether at the air-water interface. *Langmuir*, 10:1919–1925.

Penfold, J., & Thomas, R.K., (1990). The application of the specular reflection of neutrons to the study of surfaces and interfaces. *J. Phys. Condensed Matter*, 2:1369–1412.

Rhodes, N.J., Boram, A.J., Johnson, M.W., Mott, E.M., & Wardle, A.G., (1994). ISIS neutron detectors. In ICANS-XII, *Rutherford Appleton Laboratory Report*, RAL-94–025, Vol. I, pp. 185–199.

Ross, D.K., (1993). OSIRIS—Cold Neutron Spectrometer and Diffractometer. Proposal of a UK Collaborative Research Group. University of Salford.

Schoenborn, B.P., (editor) (1984). *Neutrons in Biology* (Basic Life Sciences Vol. 27, A. Hollaender, general editor). Plenum, New York.

Taylor, A.D., (1984). SNS moderator performance predictions. *Rutherford Appleton Laboratory Report*, RAL-84–120.

Wilson, C.C., (1990). The data analysis of reciprocal space volumes. In *Neutron Scattering Data Analysis 1990*, pp. 145–163, editor M.W. Johnson, IOP Conference Series 107. Adam Hilger, Bristol.

Wilson, C.C., (1993). The development and performance of position-sensitive detectors at ISIS. In *Neutrons, X-rays and Gamma Rays: Imaging Detectors, Material Characterisation Techniques and Applications*, pp. 226–234, editors J.M. Carpenter and D.F.R. Mildner, SPIE Conference Proceedings 1737, San Diego.
Wilson, C.C., (1995a). A guided tour of ISIS—the UK spallation neutron source. *Neutron News*, 6:27–34.
Wilson, C.C., (1995b). Monitoring phase transitions using a single data frame in neutron time-of-flight Laue diffraction. *J. Appl. Cryst.*, 28:7–13.
Wilson, C.C., (1996). Improving the extraction of reflection intensities in single crystal time- of-flight diffraction using a peak profile fitting procedure. *J Appl. Cryst.*, submitted.
Winter, R., Christmann, M.-H., Böttner, M., Thiyagarajan, P., & Heenan, R.K., (1991). The influence of the local anaesthetic tetracaine on the temperature and pressure dependant phase behaviour of model biomembranes. *Ber. Bunsenges. Phys. Chem.*, 95:811–820.

APPENDIX I

Examples of Recent Applications of SXD in Chemical Crystallography

Due to the recent developments in both hardware and software on SXD, the most exciting applications in chemical crystallography are necessarily also recent. The following is a representative selection of some of this work.

i. An accurate neutron determination of the structure of the energy rich compound creatine monohydrate was carried out on SXD to complement ultra-high resolution low temperature X-ray studies. In spite of the high quality of the X-ray data, the neutron study has added valuable information for use in charge density studies, allowing the decoupling of thermal parameters from multipole refinements, especially for the hydrogen atoms (Frampton *et al.*, *Acta Cryst. B*, to be published);
ii. The weak hydrogen bonded system 2-ethynyl-2-adamantanol was examined to verify infrared work which had suggested a complex 'weak' hydrogen bond scheme in this material. The precise molecular geometry was determined on SXD (Allen *et al.*, *J. Amer. Chem. Soc.*, submitted), with a full anisotropic refinement involving 525 parameters (Figure AI.1). The crystal packing from the neutron data does indeed reveal a hydrogen bonded network, with the O-H..π bonded system, along with the C-H..O bond, unusually short and linear examples of their type.
iii. The weak intermolecular contacts in acetylcholine bromide are of interest as this molecule is an important neurotransmitter involved in the modulation of many

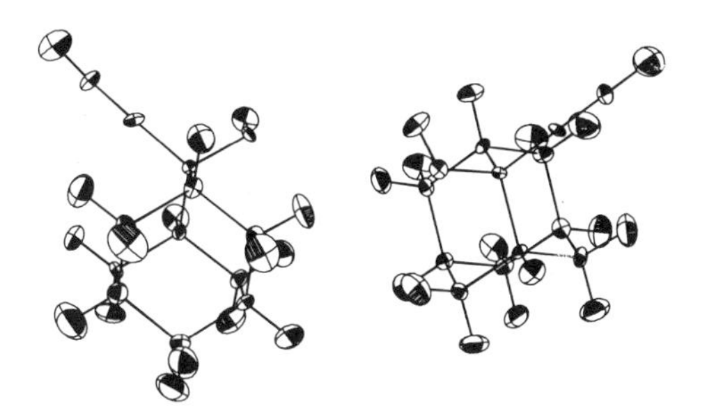

Figure AI.1. ORTEP view of the two independent molecules in the structure of 2-ethynyl-2- adamantanol.

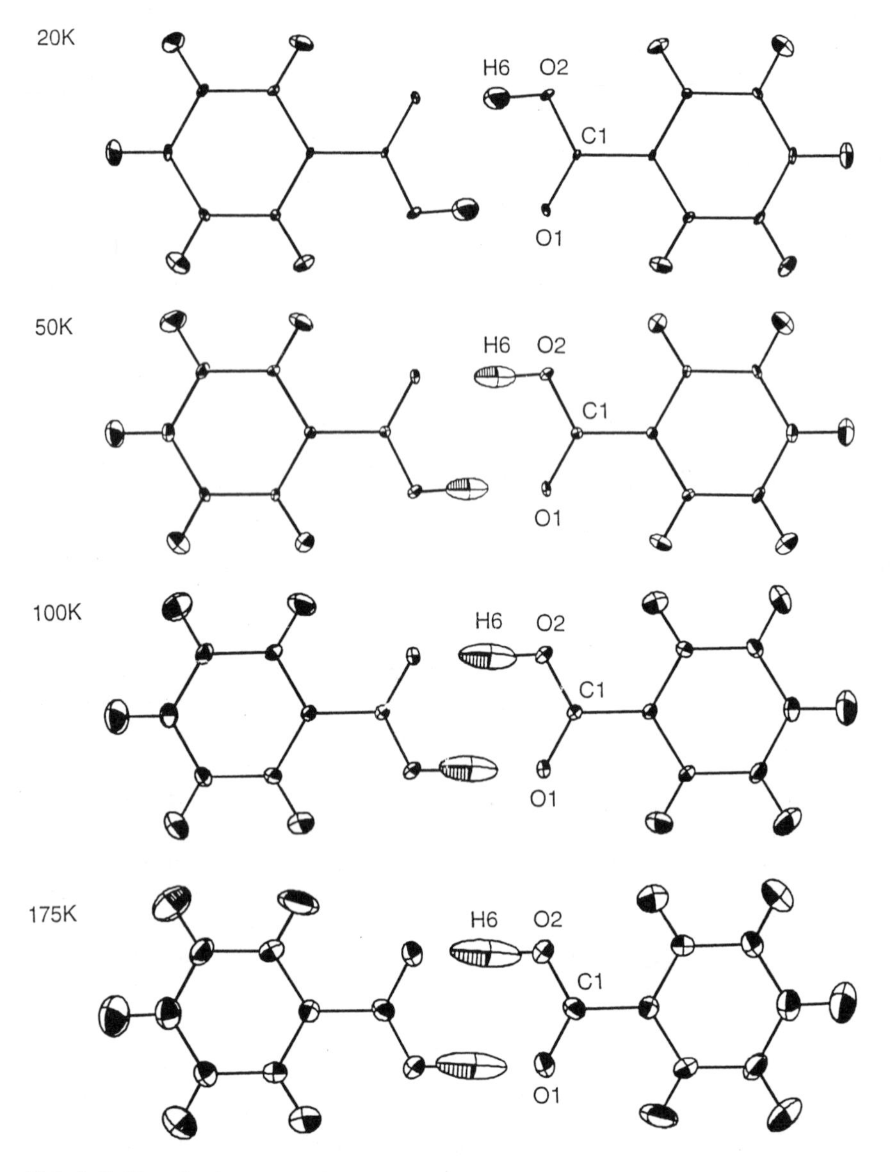

Figure AI.2. (left) The refined structure of benzoic acid assuming one proton site for the acidic proton. The evidence for disorder at higher temperatures is clear and indeed this disorder can be refined accurately, revealing a split proton site with occupancies varying smoothly with temperature (right).

functions in the human body. The SXD data (see Figure 7 in main text) revealed a previously unexplored network of significant non-bonded contacts involving the hydrogen atoms, helping gain an understanding of the binding site behavior of this molecule (Shankland *et al.*, *Nature*, submitted).

iv. Temperature dependent single crystal neutron diffraction of organic materials remains relatively novel, and the rapid data collection methods possible on SXD have recently begun to produce results in this area, both for limited and full data sets. For the latter, the temperature dependence of the hydrogen bonding system

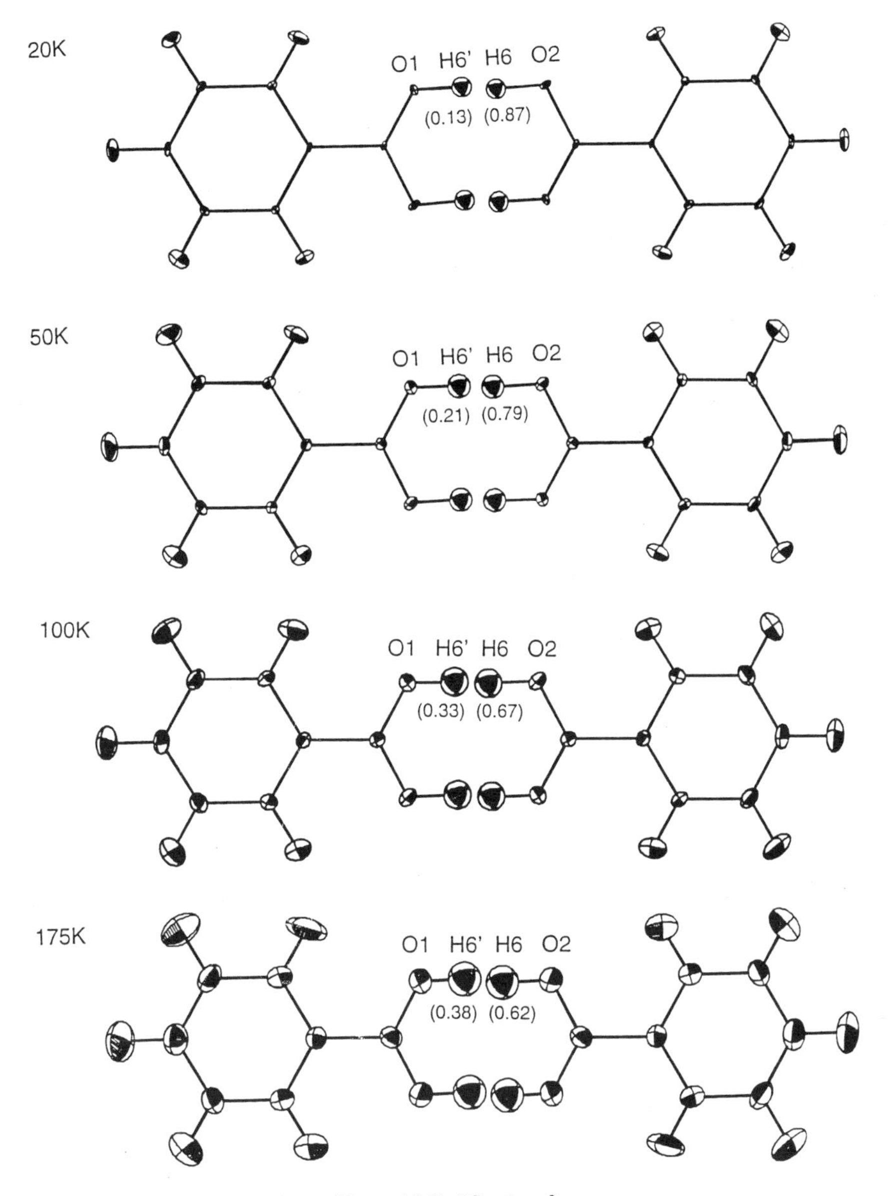

Figure AI.II. (*Continued*)

in benzoic acid (Wilson *et al.*, *Chem. Phys. Letts.*, submitted; *Acta Cryst. B*, submitted), an archetypal carboxylic acid dimer, was studied with a data collection time of under two days per temperature. The refined structure at four temperatures (20K, 50K, 100K & 175K) is shown in Figure AI.2. Clearly one hydrogen (H6), that involved in the dimeric hydrogen bond, is disordered at the higher temperatures. This disorder was successfully followed as a function of temperature, showing the potential of tof Laue diffraction in temperature dependent single crystal studies to high resolution in small molecular systems.

APPENDIX II

LMX—Instrument for Medium to Large Molecule Crystallography

At present the scope for medium/large molecule crystallography at pulsed neutron sources is limited. Current single crystal diffractometers such as SXD at ISIS are limited to modest sized cells and hence modest sized molecules. This is primarily because of the flux profile of the instrument, which views a liquid water moderator giving a Maxwellian peak in the region of 1Å. In spite of the recent successes of powder diffraction in molecular studies and in demonstrating the improved resolution available at long d-spacings on a cold source, single crystal diffraction remains the first choice for accurate molecular crystallography. The proposed Large Molecule Crystallography instrument LMX (Figure AII.1) would be a first step on the road to allowing time-of-flight Laue diffraction studies of biological systems. While the scope of the instrument itself would remain in large molecule organic and supramolecular chemistry and the very small unit cell end of protein crystallography, it would allow the potential for future expansion to larger systems to be explored.

In order to access these vitally important areas of chemistry and small biomolecules, we require a single crystal diffractometer on a cold neutron beam. At ISIS, this large molecule crystallography instrument, LMX, would be situated either on a liquid CH_4 or a liquid H_2 beamline, possibly in a break in the guide leading to an established instrument. Equipped initially with one, but ultimately a whole array, of large area ZnS PSD's, LMX would exploit the high flux of supra-Å neutrons on such a beam line to perform single crystal studies on materials of medium to large cell volumes in reasonable time scales (initially typically one month, reducing to several days with increased detector array). For all but the very largest cells in this range, the first detectors would be placed at high angles (typically 130–150), allowing good resolution to be obtained.

For large molecule organic chemistry, LMX would be sited at approximately 11m (L1) from the cold moderator. Ideally for biological studies this would be reduced to around 7m to increase flux, but the requirements of resolution may demand a longer beamline, especially if this moderator is coupled on some future Target Station or source to further increase flux at the expense of resolution. There is a current study commissioned by the ESS project (*Lehmann et al.*, in preparation) aimed at resolving the conflicting requirements of flux and resolution in a large unit cell pulsed source instrument. The detectors are placed close to the sample (short L2) to maximise the number of reflections accessed in a single measurement. This would require a PSD of at least the performance of the present SXD module (3mm resolution, $192 \times 192mm^2$—and in fact smaller pixel sizes of the order of 1–2mm would be more favorable) and continued provision of improved intensity extraction software. Choppers upstream and downstream of the instrument would allow selection of different wavelength ranges for LMX and any downstream instrument. LMX itself would typically use neutrons in the 1–10Å region, sampled in approximately 4Å windows, eg 2–6Å might be typical for the study of a small biomolecule to good resolution.

Requirements for the instrument

- Cold neutron guide with breaks at ~7m, ~11m and ~15 m (1–2m break in the guide);
- Upstream and downstream choppers for wavelength selection;

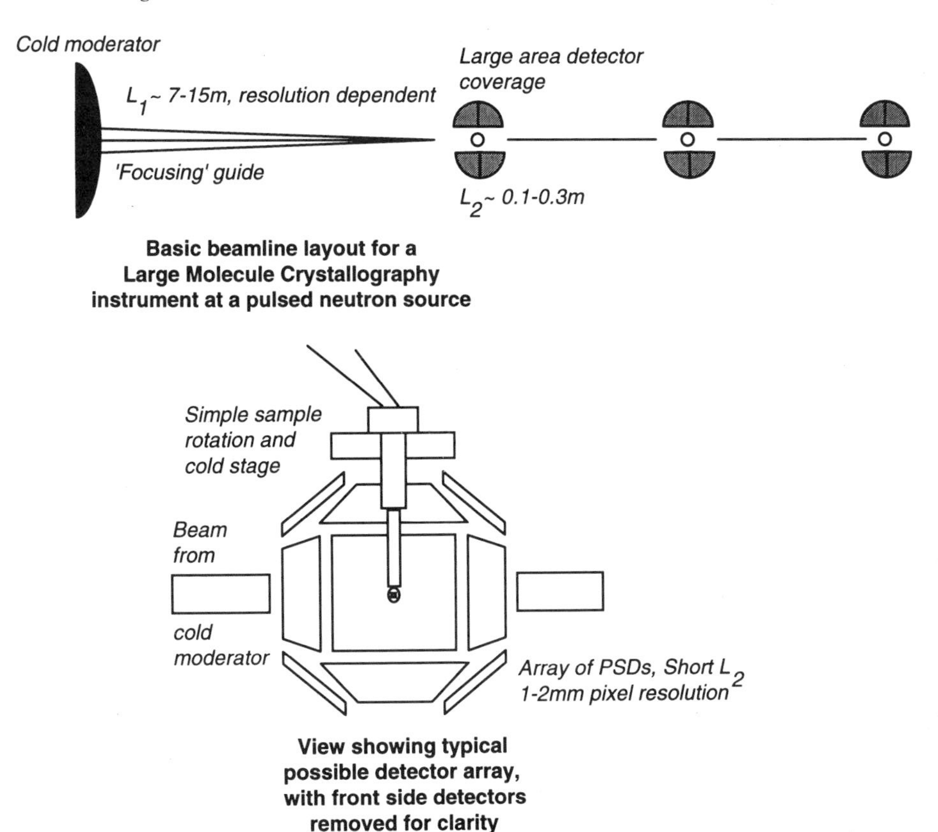

Figure AII.1. General layout and possible detector arrangement for a Large Molecule Crystallography instrument (LMX) at a pulsed neutron source.

- ZnS PSD detector (initially one of these). At least 3mm resolution, at least 200 × 200mm^2. Larger better, and smaller pixel size also better;
- Simple orientation device for Eulerian geometry in CCR—these materials do not generally require extreme sample environment. The simple specimen environment conditions make the instrument itself rather compact, making the possibility of multiple sample positions a realistic one;
- Novel intensity extraction software, exploiting the continuous nature of the reciprocal space volume sampled to allow overlapping peaks to be separated by profile fitting methods.

LMX would allow single crystal diffraction at ISIS to access medium molecule crystallography at fairly high resolution; large molecule crystallography at good resolution; and very large molecule crystallography, extending into the lower end of biological molecule crystallography, at medium resolution.

6

HIGH PRECISION THERMAL NEUTRON DETECTORS

V. Radeka, N. A. Schaknowski, G. C. Smith, and B. Yu

Brookhaven National Laboratory
Upton, New York 11973-5000

ABSTRACT

Two-dimensional position sensitive detectors are indispensable in neutron diffraction experiments for determination of molecular and crystal structures in biology, solid-state physics and polymer chemistry. Some performance characteristics of these detectors are elementary and obvious, such as the position resolution, number of resolution elements, neutron detection efficiency, counting rate and sensitivity to gamma-ray background. High performance detectors are distinguished by more subtle characteristics such as the stability of the response (efficiency) versus position, stability of the recorded neutron positions, dynamic range, blooming or halo effects. While relatively few of them are needed around the world, these high performance devices are sophisticated and fairly complex; their development requires very specialized efforts. In this context, we describe here a program of detector development, based on ^{3}He filled proportional chambers, which has been underway for some years at the Brookhaven National Laboratory. Fundamental approaches and practical considerations are outlined that have resulted in a series of high performance detectors with the best known position resolution, position stability, uniformity of response and reliability over time, for devices of this type.

INTRODUCTION

An important role is played by position-sensitive neutron detectors in experiments which use thermal neutrons from reactors or spallation sources, particularly neutron scattering studies of biological, physical and chemical samples. The main nuclei whose neutron cross-sections are large enough to make them suitable choices for neutron detectors are: ^{3}He, used solely in gas proportional detectors, ^{6}Li, used as a thin converter foil or in scintillators such as LiF and LiI, ^{10}B, used as a thin converter foil or in the form of BF_3 in gas proportional detectors, and ^{155}Gd, ^{157}Gd or natural Gd, used as a converter foil or as a partial constituent of a phosphor. Thermal neutron detection is based on nuclear reactions, whose products determine the ultimate position resolution of the detector, but the choice of detector technology for a particular application is determined by the experimenter's requirements with respect to additional detector characteristics, such as efficiency, counting rate capability, required size, whether or not dynamic studies are to be performed, sensitivity to background radiation such as gamma-rays, and position stability as a function of time.

A program of detector development for thermal neutron scattering experiments in the study of biological structures has been underway for some years at Brookhaven. The technology of choice has been the gas proportional detector, with ^{3}He as the neutron absorbing gas, because of several beneficial factors. These detectors offer superior efficiency to that of all other detectors, their maximum counting rate capability is adequate for that required in most reactor and spallation source applications, they can be built in a wide range of sizes, they are very insensitive to gamma radiation with appropriate choice of gas mixture, and their position stability is better than most other detectors. In single neutron counting mode, they offer infinite dynamic range for static experiments, while also offering the capability of performing dynamic studies with time frames below one second. We now describe some of the basic characteristics of these gas proportional detectors.

POSITION RESOLUTION AND EFFICIENCY

Position Resolution

The reaction by which neutrons are stopped in ^{3}He is:

$$^3\text{He} + \text{n} \rightarrow {}^3\text{H} + \text{p} + 0.764\text{MeV}$$

The energetic reaction products, a 191keV triton and a 573keV proton, are emitted in opposite directions, as shown schematically in Figure 1. The ionization centroid of the two particle tracks is displaced significantly from the neutron interaction position because the proton is more heavily ionizing than the triton, and also has a larger range. It is the displaced centroid which the position-sensitive cathodes of the detector measure, independent of the electronic encoding principle. The loci of centroids from many events describe a sphere; when projected in one dimension, these loci describe a a rectangular

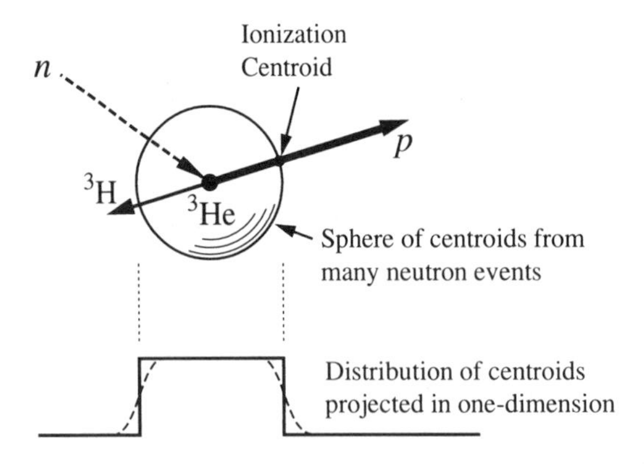

Figure 1. Representation of thermal neutron interaction with ^{3}He nucleus. The reaction products, a triton with kinetic energy 191keV and a proton with kinetic energy 573keV, are emitted in opposite directions. The greater range and ionizing power of the proton results in a displacement of the ionization centroid from the interaction point. Successive events form a sphere of centroids, whose projection in one dimension is a rectangle.

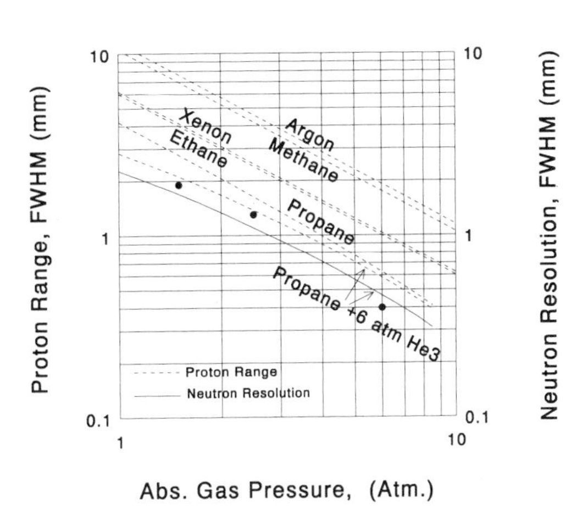

Figure 2. Calculated values for range (dashed lines) of 573keV proton in some suitable gases for admixture with ^{3}He. Curve for propane decreases more rapidly than 1/pressure because of non-ideal gas behavior. Lower curve (solid line) is predicted resolution for a mixture of 6atmos ^{3}He and propane over a range of propane pressures. Solid circles show experimental resolution measurements from operational BNL detectors.

distribution whose width is equal to the diameter of the sphere. In practice, range straggling of the triton and proton, and electronic noise, introduce a small broadening term, and the resulting distribution is shown by the dashed lines in Figure 1, but the FWHM remains that of the rectangular distribution. It is possible to show that (Fischer *et al.,* 1983), to a good approximation, the displacement of the total ionization centroid is $0.4R_p$, where R_p is the proton range; thus, the FWHM position resolution is $0.8R_p$.

Proton range, which varies inversely with the density of a particular gas, is several centimeters in 1atmos He. To achieve resolution in the millimeter range, it is necessary to use an additional gas for the purpose of providing stopping power for the proton and triton. Different gases have been used by various workers to accomplish this task, and the upper five curves in Figure 2 show the expected proton range in some suitable gases (Anderson & Ziegler, 1977). From these curves it is clear that propane is an appropriate choice as an additive to ^{3}He for the proportional chamber gas. Although the higher hydrocarbons have even greater stopping power, there are several reasons that render them unsuitable in this application; they are more likely to form anode wire deposits under intense irradiation, and they are incompatible with most forms of gas purifier (see later). Propane has an additional advantage that its density increases at a slightly greater rate than its pressure (Goodwin & Haynes, 1982). At an absolute pressure of 8.5atmos, its density is about 15% greater than if it were an ideal gas (above 8.5atmos propane condenses); this phenomenon gives rise to the advantageous deviation from 1/pressure of the proton range curve for propane in Figure 2.

Efficiency and Gamma Sensitivity

A thermal neutron chamber is, typically, about 1.5cm deep, and calculations have been carried out to determine the detection efficiency in this gas depth over a range of operating gas pressures. The curves are shown in Figure 3, where the lower and upper abscissa have units of Å and meV, respectively, using published cross-section data for ^{3}He (Garber & Kinsey, 1976). Very high detection efficiencies can be achieved for cold neutrons, around 9Å, with just 1 to 2atmos of ^{3}He; as wavelength decreases, efficiency falls, but even at 1Å an efficiency of about 50% can be achieved with 6atmos of ^{3}He. The detection efficiency of this class of detector is significantly greater than that of other detecting media.

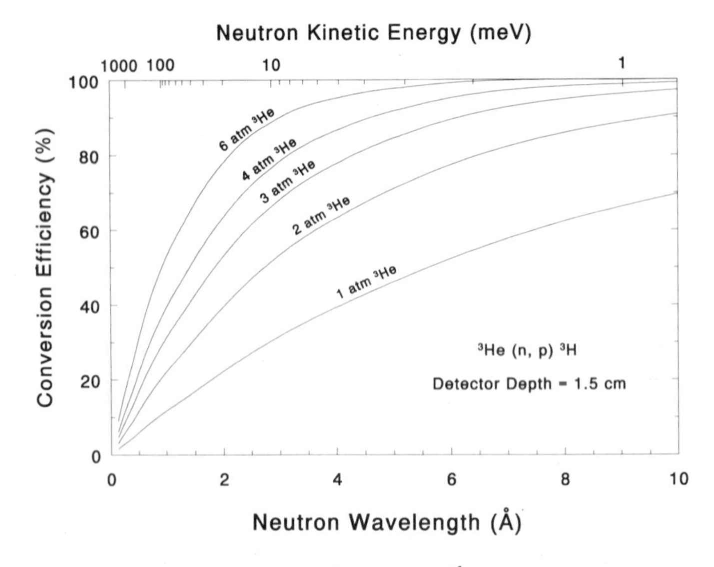

Figure 3. Calculated neutron conversion efficiency for a range of ^{3}He partial pressures, for a detector with gas depth of 1.5cm. Note that for wavelengths above 1Å, efficiencies of >50% are readily achievable; cold neutron conversion efficiencies are in excess of 80%.

As an admixture with ^{3}He, propane has two advantages in addition to its proton/triton stopping power. First, it is a good quench gas for the proportional avalanche. Second, it has a relatively low photon absorption cross-section. Position sensitive detectors used in neutron scattering studies usually operate in an environment with a high photon background; therefore, the gas added to reduce proton range should have a low sensitivity to X-rays and gamma-rays. Propane has a lower cross-section for gamma-rays than the other four proton stopping gases in Figure 2 (Fischer *et al.,* 1983). Carbon tetrafluoride, not shown in Figure 2, has similar stopping power to propane, and has been considered by other groups (Kopp *et al.*, 1982); however, CF_4 has a somewhat higher photon cross-section, and also is not properly suited to the gas purifier.

BASIC OPERATING PRINCIPLES

A cross-section of the basic proportional chamber structure is shown schematically in Figure 4. Neutrons enter through a window, usually aluminum, and most are stopped in the absorption and drift region. The primary ionization created by the proton and triton, about 30,000 electrons, then drifts through the upper wire cathode and an avalanche takes place on the nearest anode wire, or wires. The upper cathode wires and anode wires, though not necessarily of the same pitch, normally run in the same direction. The lower cathode is usually fabricated on a large glass plate; it has vacuum deposited copper strips, running at right angles to the anode wires. The anode avalanche induces positive charge on both the upper and lower cathodes, with a quasi-Gaussian profile that has a FWHM of approximately 1.5 times the anode-cathode spacing. The sampling of the cathode induced charge with wires or strips yields the center of gravity of the anode avalanche with high precision when appropriate design criteria are followed (Gatti *et al.,* 1976; Mathieson & Smith, 1989). For convenience, the lower cathode is designated the X-axis and the upper

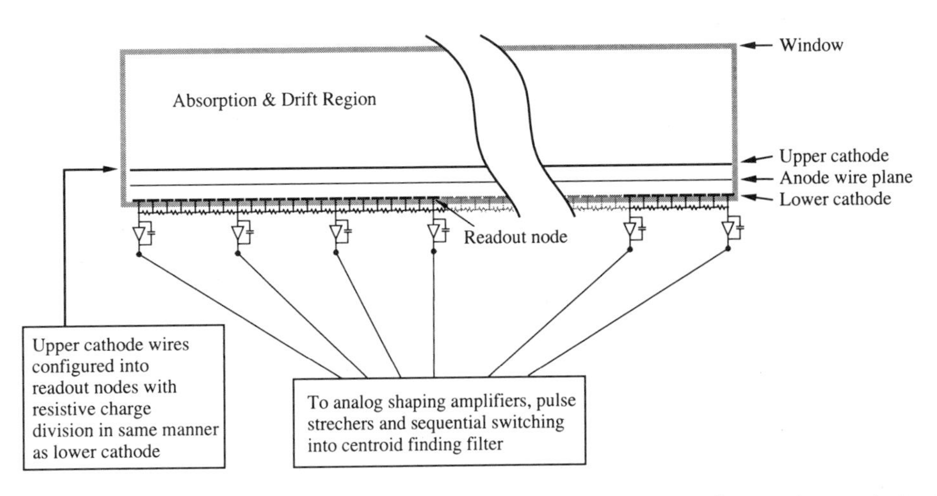

Figure 4. Schematic diagram of proportional chamber geometry. A neutron, entering from top, interacts in the absorption and drift region, secondary ionization then drifts down to the anode. Induced signal charge is collected by two, or three, nodes on each cathode, after resistive charge division; sampled signals are switched to centroid finding electronics.

cathode the Y-axis; the anode wires, which are all connected together, yield the total energy signal which is used for gating the position encoding electronics to further eliminate gamma-rays.

Each cathode has a number of readout nodes, connected to charge sensitive preamplifiers. Charge division between nodes is carried out resistively. For combined optimization of position linearity and position resolution, the RC time constant of one subdivision is approximately equal to the shaping amplifier time constant. The anode avalanche from each detected neutron gives rise to collected charge on two, or three, adjacent cathode nodes; these signals are then switched sequentially into a centroid finding filter which yields a timing signal commensurate with the linear position along the respective axis that the event occurred (Radeka & Boie, 1980; Boie *et al.*, 1982). The method achieves a high absolute position accuracy and a high position resolution with small avalanche size because of the number of signal outputs from the detector. The position resolution limit due to electronic noise is given approximately by:

$$\mathrm{FWHM} \approx 7(kTC_D)^{1/2}(l/N^{3/2})(1/Q_s)$$

where k is Boltzmann's constant, T is absolute temperature, C_D is the total readout electrode capacitance, l the detector length, N the number of subdivisions and Q_s the charge induced on the readout cathode. It is important to note that larger area detectors can be fabricated with the same electronic noise as smaller ones just by using additional nodes and keeping the node spacing, l/N, constant; this major attribute is not possible in global RC encoding methods with only two outputs per axis.

DESIGN AND FABRICATION DETAILS

Detectors with sensitive areas ranging from 5cm × 5cm to 50cm × 50cm have been designed and fabricated in our detector development program. One device with a sensing

Figure 5. Array of three detectors, each with 20cm × 20cm sensitive area, built to upgrade HFBR protein crystallography spectrometer, which previously had just a single detector. This triples the spectrometer's acceptance range and significantly decreases data collection time for structure determination.

area of 20cm × 20cm, for example, is housed in a robust aluminum enclosure, consisting of a 38cm diameter base plate, 3cm thick, upon which the electrodes comprising the two cathode and anode planes are mounted. A front cover is bolted to the base, and a double O-ring seal between these two halves serves to provide gas containment. The 20cm × 20cm window, machined into the front cover, has a thickness of about 9mm. All electrical feedthroughs are positioned on the thick base plate and are doubly sealed. The enclosure has a maximum allowable operating pressure in excess of 10atmos absolute. Figure 5 shows a front view of an array of three of these detectors, recently completed for installation in the macromolecular neutron crystallography beam line of Brookhaven's High Flux Beam Reactor (HFBR).

Inside each 20cm × 20cm detector, the X-cathode consists of 134 copper strips on a pitch of 1.6mm. Every seventh strip forms a readout node, connecting to a charge preamplifier, as shown in Figure 4. There are 20 nodes on this axis. The Y-cathode consists of wires with a pitch of 0.8mm, electrically connected in pairs; every seventh pair forms a readout node, of which there are 17 on this axis. Figure 6 shows a section along the edge of the X-axis, each node being fed to its corresponding preamplifier (on the outside of the housing) via the glass-to-metal feedthroughs. Chip resistors for the inter-node charge division are positioned between each cathode strip. The anode plane consists of 112 Au plated tungsten wires, with a diameter of 12μm and a pitch of 1.6mm. There are two guard wires of larger diameter at each side of the anode plane to reduce the electric field.

In the set of detectors constructed in our program, two additional sizes have been fabricated. Most recently, a very high resolution device with a sensitive area 5cm × 5cm was completed for an HFBR experiment investigating Rayleigh-Bénard convection, which required position resolution of less than half a millimeter. This was achieved primarily by a combination of the required high pressure to limit the proton/triton range (8atmos of ^{3}He and 6atmos of propane) and reduced node spacing on each position sensing axis. It was also necessary to employ electrode separations and wire spacings of less than 1mm. A picture of the completed detector, taken from the rear, is shown in Figure 7; the array of hybrid preamplifier cards for readout of the cathodes can be seen in the electronics housing. A very large area device, 50cm × 50cm, has been in operation for several years at the

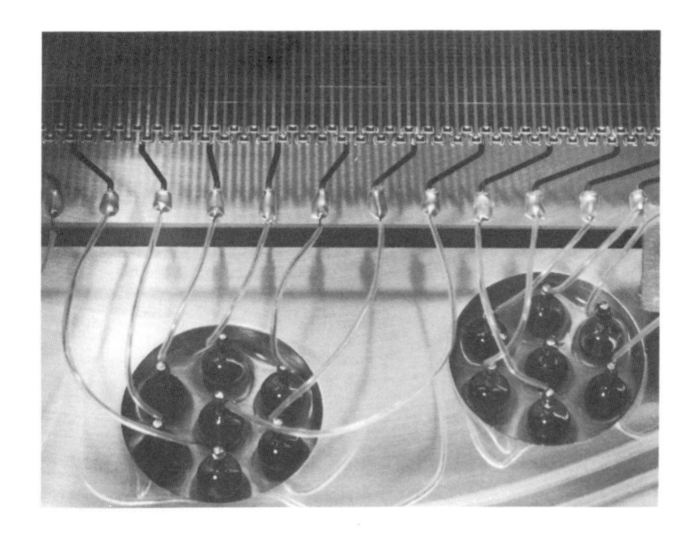

Figure 6. Close-up photograph along the edge of lower cathode of 20cm × 20cm neutron detector. In upper part, running left to right, are last few wires of upper cathode and anode planes, below which are copper strips of lower cathode running at right angles. Chip resistors between ends of strips form RC network for charge division. Cathode nodes are connected to preamplifiers (not shown) on outside of detector housing, via the glass-to-metal feedthroughs.

small angle scattering beam line of the HFBR. This device has a more modest resolution requirement, about 3mm FWHM.

Detectors of this nature, which are fabricated with small quantities of organic materials, will operate over long periods of time only if measures are taken to keep the gas mixture free from impurities. A closed loop gas purification and circulation system is a sensible solution to this problem, bearing in mind the expense of ^{3}He, about $US150 per STP liter. A small pump, enclosed by a pair of stainless steel UHV flanges, circulates the gas through the detector and a purifier at a rate of approximately 1–2 liters/min. The purifier consists of a stainless steel cylinder with one half length filled with Ridox (Fisher Scientific Co., Fair Lawn, NJ 07410), an oxygen absorber, and the other half length filled with a 3Å pore size molecular sieve material, mainly for water absorption. The pump and

Figure 7. Photograph of rear of the very high resolution detector. With a gas filling of 8atmos ^{3}He and 6atmos propane, this detector has achieved the best resolution ever measured in a gas-filled thermal neutron detector, less than 400μm FWHM. The gas circulation pump is contained in the UHV flange assembly; above this, the vertical cylinder contains the gas purifier.

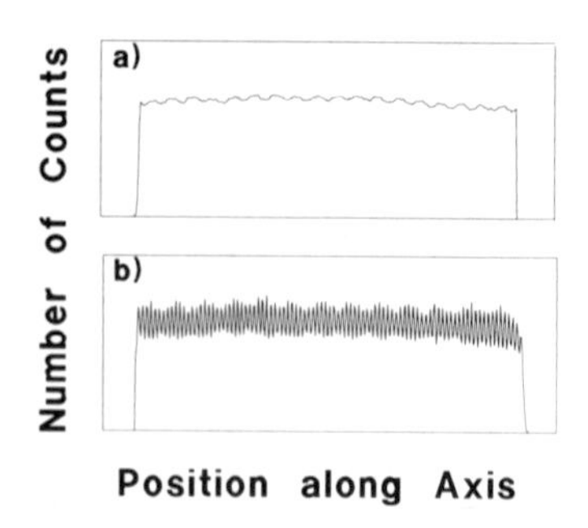

Figure 8. Uniform irradiation responses from 20cm × 20cm detector. a) the lower cathode, X-axis response, and b) the upper cathode, Y-axis response.

purifier can be seen on the detector assembly in Figure 7. This arrangement permits limited use of chamber construction materials such as G10 fiberglass. The purifier has been very effective in maintaining a high level of purity in the gas mixture: most detectors constructed in this program have operated without interruption for several years.

OPERATING CHARACTERISTICS

A sensitive measure of linearity can be achieved by irradiating the detector with a beam of neutrons of uniform intensity per unit area, and measuring the position spectrum. Figure 8(a) and (b) show the X-axis and Y-axis responses, respectively, from the 20cm × 20cm detector. In the X-axis there is a small, residual modulation whose magnitude is small, about ±3%, and which is due to the inter-node charge division. The modulation has a periodicity equal to the node spacing and is extremely stable. In the Y-axis, there is a modulation due to the discrete anode wire locations. This again is very stable.

Position resolution has been determined by measuring detector response to a finely collimated beam of neutrons incident normally on the window. In Figure 2 are shown experimental measurements (solid circles) for three different propane pressures: 1.5, 2.5 and 6atmos. The first two are taken with the 20cm × 20cm detector with an additional 4atmos ^{3}He, and the third was taken with the 5cm × 5cm detector with an additional 8atmos ^{3}He. The solid curve in Figure 2 is the predicted resolution for a mixture of 6atmos ^{3}He with a range of propane pressures (since ^{3}He has a relatively small influence on proton range, its partial pressure has only a small effect on the predicted resolution). These measurements have been performed under normal operating conditions of the detectors, with almost no contribution to position resolution from electronic noise. There is good agreement between measured and predicted resolution. To our knowledge, the data point at 6atmos propane, corresponding to a FWHM <400μm, is the best resolution ever recorded in a gas-filled thermal neutron detector; the corresponding position distribution is shown in Figure 9.

A measure of absolute position accuracy is shown in Figure 10, which is the response of the large 50cm × 50cm detector to a raster scan of 36 primary beams of thermal neutrons. The excellent absolute accuracy is illustrated by the similarity of peaks in a particular row and a particular column.

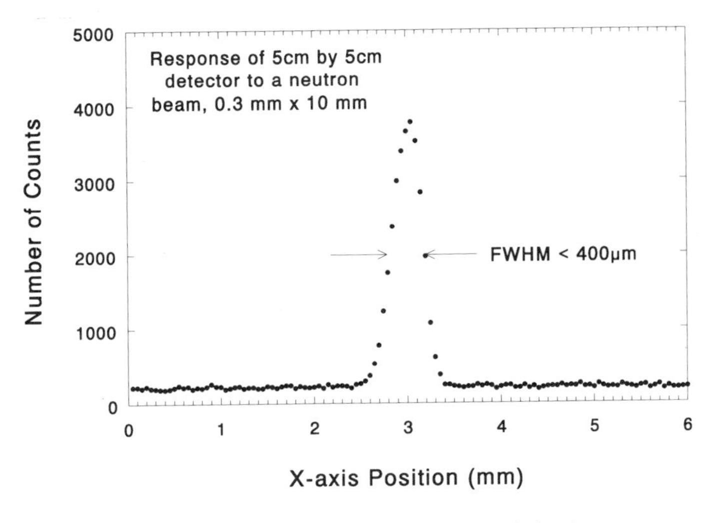

Figure 9. Position response from the very high resolution detector.

A useful summary of operating characteristics of gas filled proportional chambers for thermal neutron detectors is given in Table I. It should be noted that not all the properties listed can necessarily be achieved in one detector at the same time.

CONCLUDING REMARKS

A set of detectors has been designed and fabricated which provide high resolution, stable, and reliable performance to a range of structural biology experiments, in particular those at the Brookhaven's HFBR. In this ongoing program, new devices and techniques are under development; for example, methods to minimize parallax errors in planar detec-

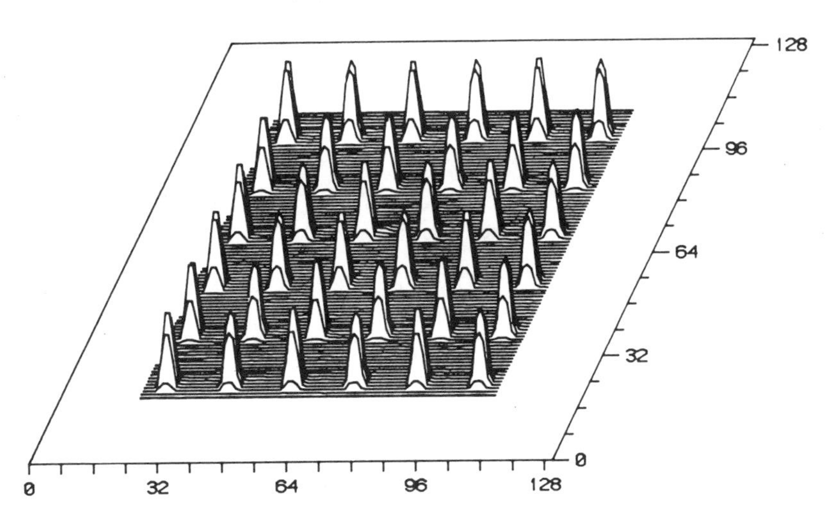

Figure 10. Response of the 50cm × 50cm detector to a raster scan of 36 primary neutron beams, illustrating the excellent absolution position accuracy of these detector systems (from Schoenborn *et al.*, 1986).

Table I. Performance figures of gas filled proportional chambers for thermal neutron detectors

Position Resolution (FWHM)	<0.4–2 mm
Number of Resolution Elements	from 128×128 to 1024×1024
Size	from 5×5 cm^2 to 50×50 cm^2
Wavelength Range	1–20 Å
Detection Efficiency	50–80 %
Counting Rate (total)	10^5–5×10^5 sec^{-1} (single readout)
Counting Rate (single peak)	5×10^4 sec^{-1}
Integral non-linearity	2×10^{-4} to 10^{-3}
Absolute Position Accuracy	30-100 μm
Stability of Origin	<50 μm
Stability of Response (efficiency)	<1%
Differential non-linearity	±3%
Dynamic Range	Single Neutron Detection
Timing Resolution	<250 ns

tors for small angle experiments are being investigated, and large, curved detectors for crystallography experiments are being studied. Finally, it should be emphasized that the proportional detectors described here have one important advantage over competing devices such as image plates and phosphor/CCDs: the proportional detector, with sub-microsecond timing resolution, is the only device from these three with adequate timing resolution for white beam experiments at spallation sources.

ACKNOWLEDGMENTS

We wish to acknowledge the invaluable contributions to this detector development program made by Joachim Fischer, who passed away earlier this year. We are indebted to Joe Mead and Frank Densing for fabrication and testing of the analysis electronics. We have had very helpful advice and feedback from colleagues working in Structural Biology, in particular Richard Korszun, Anand Saxena and Dieter Schneider at Brookhaven, and Benno Schoenborn at Los Alamos. This research was supported by the U.S. Department of Energy: Contract No DE-AC02-76CH00016.

REFERENCES

Anderson, H.H., & Ziegler, J.F., (1977). *Hydrogen Stopping Power and Ranges in All Elements*, Vol. 3, Pergamon Press, New York.

Boie, R.A., Fischer, J., Inagaki, Y., Merritt, F.C., Okuno, H., & Radeka, V., (1982). Two-dimensional high precision thermal neutron detectors. *Nucl. Instr. Methods,* 200:533–545.

Fischer, J., Radeka, V., & Boie, R.A., (1983). High Position Resolution and accuracy in ^{3}He two-dimensional thermal neutron detectors. Workshop on *The Position-Sensitive Detection of Thermal Neutrons,* ILL, Grenoble, France 11–12 October 1982. Proceedings edited by P. Convert and J.B. Forsyth (Academic Press, London), p129.

Garber, D.I., & Kinsey, R.R., (1976). Neutron cross sections Volume II, Curves. *Report No. BNL 325. Brookhaven National Laboratory*

Gatti, E., Longoni, A., Okuno, H., & Semenza, P., (1979). Optimum geometry for strip cathode or grids in MWPC for avalanche localization along the anode wires. *Nucl. Instr. Methods,* 163:83–92.

Goodwin, R.D., & Haynes, W.M., (1982). Thermophysical Properties of propane from 85 to 700°K at pressures to 70MPa. *U.S. Dept. of Commerce, National Bureau of Standards.*

Kopp, M.K., Valentine, K.H., Christophorou, L.G., & Carter, J.G., (1982). New gas mixture improves performance of ^{3}He neutron counters. *Nucl. Instr. Methods,* 201:395–401.

Mathieson, E., & Smith, G.C., (1989). Reduction in non-linearity in position-sensitive MWPC's. *IEEE Trans. Nucl. Sci.,* NS-36:305–310.

Radeka, V., & Boie, R.A., (1980). Centroid finding method for position-sensitive detectors. *Nucl. Instr. Methods,* 178:543–554.

Schoenborn, B.P., Schefer, J., & Schneider, D.K., (1986). The use of wire chambers in Structural Biology. *Nucl. Instr. Methods,* A252:180–187.

7

NEUTRON SCATTERING IN AUSTRALIA

Robert B. Knott

Australian Nuclear Science and Technology Organisation
Private Mail Bag
Menai NSW 2234, Australia

ABSTRACT

Neutron scattering techniques have been part of the Australian scientific research community for the past three decades. The High Flux Australian Reactor (HIFAR) is a multi-use facility of modest performance that provides the only neutron source in the country suitable for neutron scattering. The limitations of HIFAR have been recognized and recently a Government initiated inquiry sought to evaluate the future needs of a neutron source. In essence, the inquiry suggested that a delay of several years would enable a number of key issues to be resolved, and therefore a more appropriate decision made. In the meantime, use of the present source is being optimized, and where necessary research is being undertaken at major overseas neutron facilities either on a formal or informal basis. Australia has, at present, a formal agreement with the Rutherford Appleton Laboratory (UK) for access to the spallation source ISIS.

Various aspects of neutron scattering have been implemented on HIFAR, including investigations of the structure of biological relevant molecules. One aspect of these investigations will be presented. Preliminary results from a study of the interaction of the immunosuppressant drug, cyclosporin-A, with reconstituted membranes suggest that the hydrophobic drug interdigitated with lipid chains.

INTRODUCTION

Neutron scattering began in Australia in 1960 when the High Flux Australian Reactor, HIFAR, started operating routinely at full power. HIFAR is a medium flux ($10^{14}cm^{-2}sec^{-1}$) DIDO class reactor (Figure 1) that has been operating at 10MW(th) without major modification. HIFAR (Figure 2) is owned and operated by the Australian Nuclear Science and Technology Organisation (ANSTO), a statutory authority with the mission to 'ensure that its research, technology transfer, commercial and training activities in nuclear science and associated technologies will advance Australia's innovation, international

Neutrons in Biology, edited by Schoenborn and Knott
Plenum Press, New York, 1996

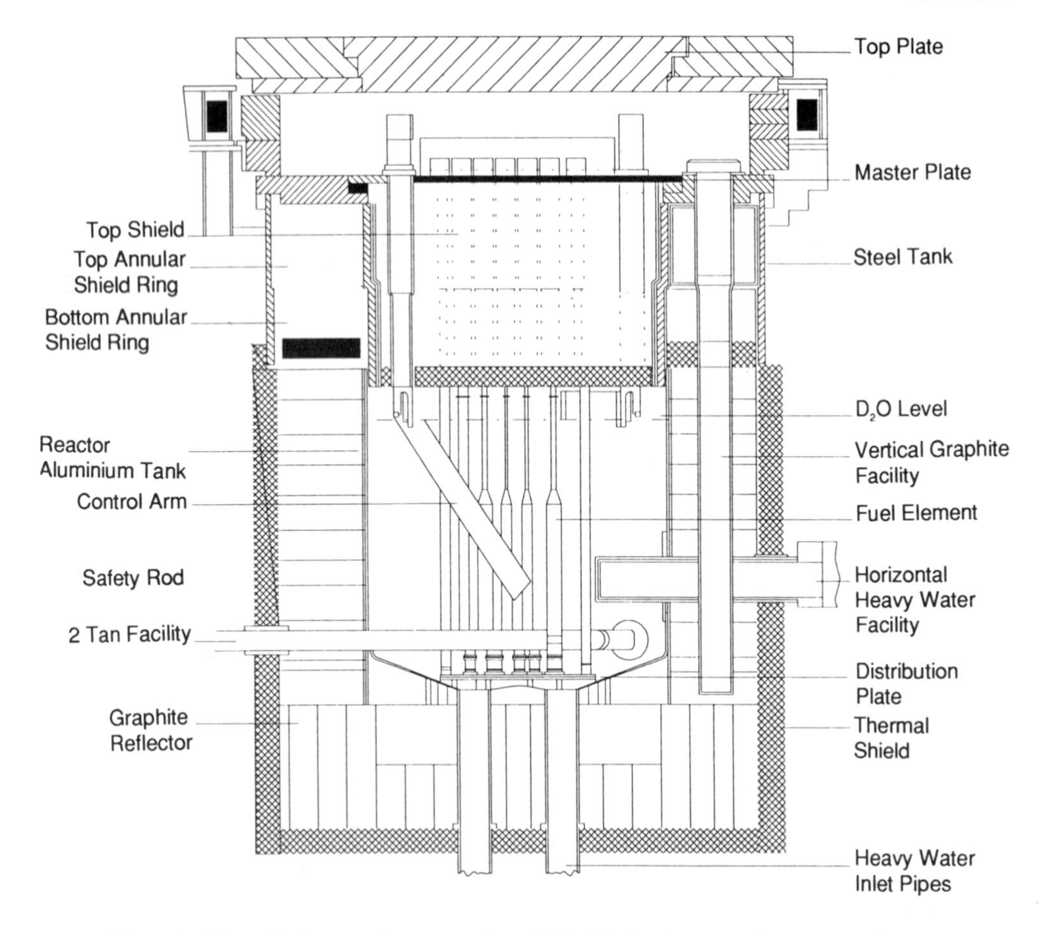

Figure 1. Schematic diagram of a cross-section of HIFAR indicating some of the essential features.

competitiveness and environmental and health management' (ANSTO Annual Report, 1995). ANSTO is locate approximately 30kms from Sydney, a major center of population.

The main aim of the HIFAR facility was initially materials testing in support of the government policy for a future nuclear power program in Australia. This remained of primary importance until the early 1970's when plans for a nuclear power program were deferred indefinitely. Since then the objectives of the research organisation have shifted in response to government policy to concentrate on more general matters related to nuclear science and technology, as reflected in ANSTO's present mission statement.

This history has had some impact on the neutron scattering effort especially in terms of competing priorities for resources. The influence of the broader neutron scattering community in other research laboratories and in Universities through the Australian Institute for Nuclear Science and Engineering (AINSE), has provided long term input and some stability to the effort.

Neutron Scattering Equipment

Figure 3 is a cartoon illustrating the historical development of neutron scattering instrumentation. To a large extent this demonstrates the major interests of the community.

Figure 2. View of the top of HIFAR with the top plate remove for routine maintenance.

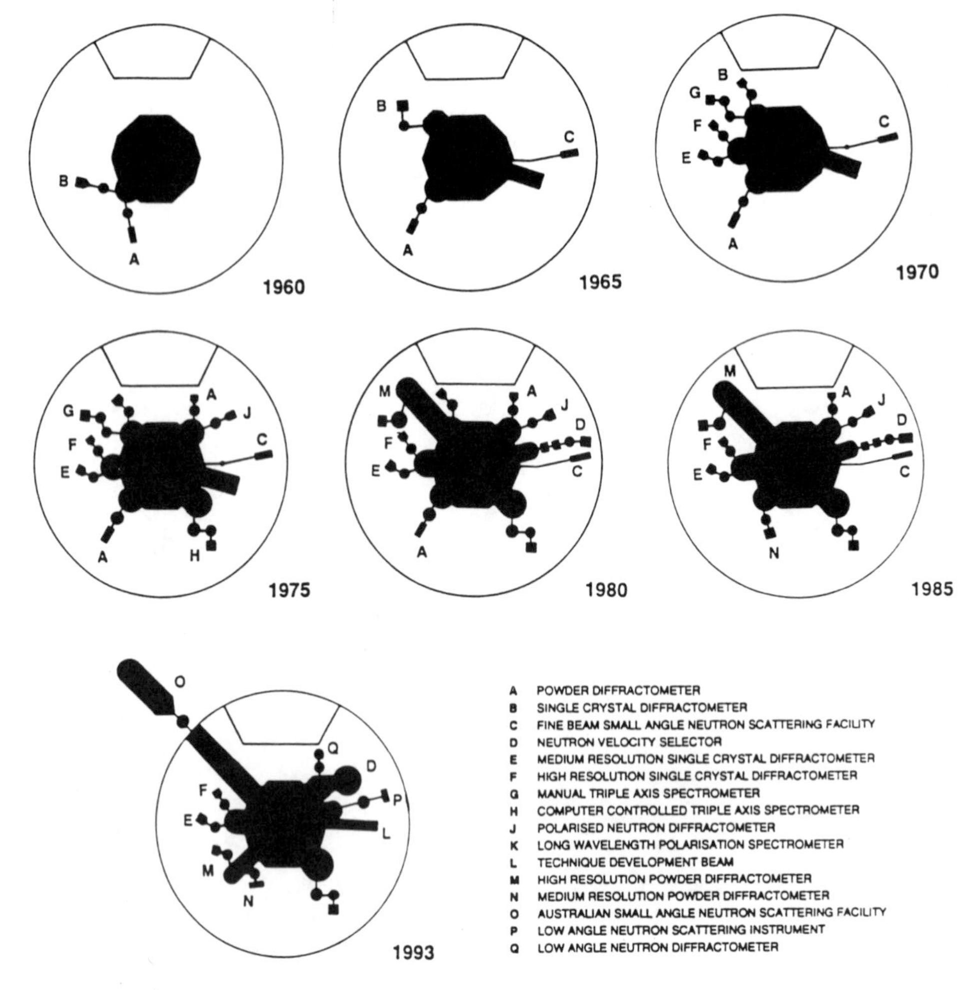

Figure 3. Shadow diagram of the neutron scattering instrument history of HIFAR.

Instruments for the investigation of problems in solid state physics and material science dominate the long term program. The first instruments were single crystal and powder diffractometers, followed by a triple axis spectrometer and then a range of instruments to address the perceived needs of the community. The present supported instruments include two single crystal diffractometers (2tanA, 2tanB), a polarized beam facility (LONGPOL), a triple axis spectrometer (TAS), two powder diffractometers (HRPD, MRPD) and a small-angle scattering (SANS) facility (Figure 4). A number of these instruments may be used for studies in structural biology.

The high resolution single crystal diffractometer is undergoing an upgrade. The goniometer has been replaced with Huber four circle system that is interfaced to locally developed software on an IBM-PC. A small 2D position sensitive detector (PSD) with single reflection capability will be installed in the near future. The primary shielding and optics will be upgraded at a future date. It is expected that, depending on the unit cell size, molecules of biological significance could be studied on this instrument. Since the sample is

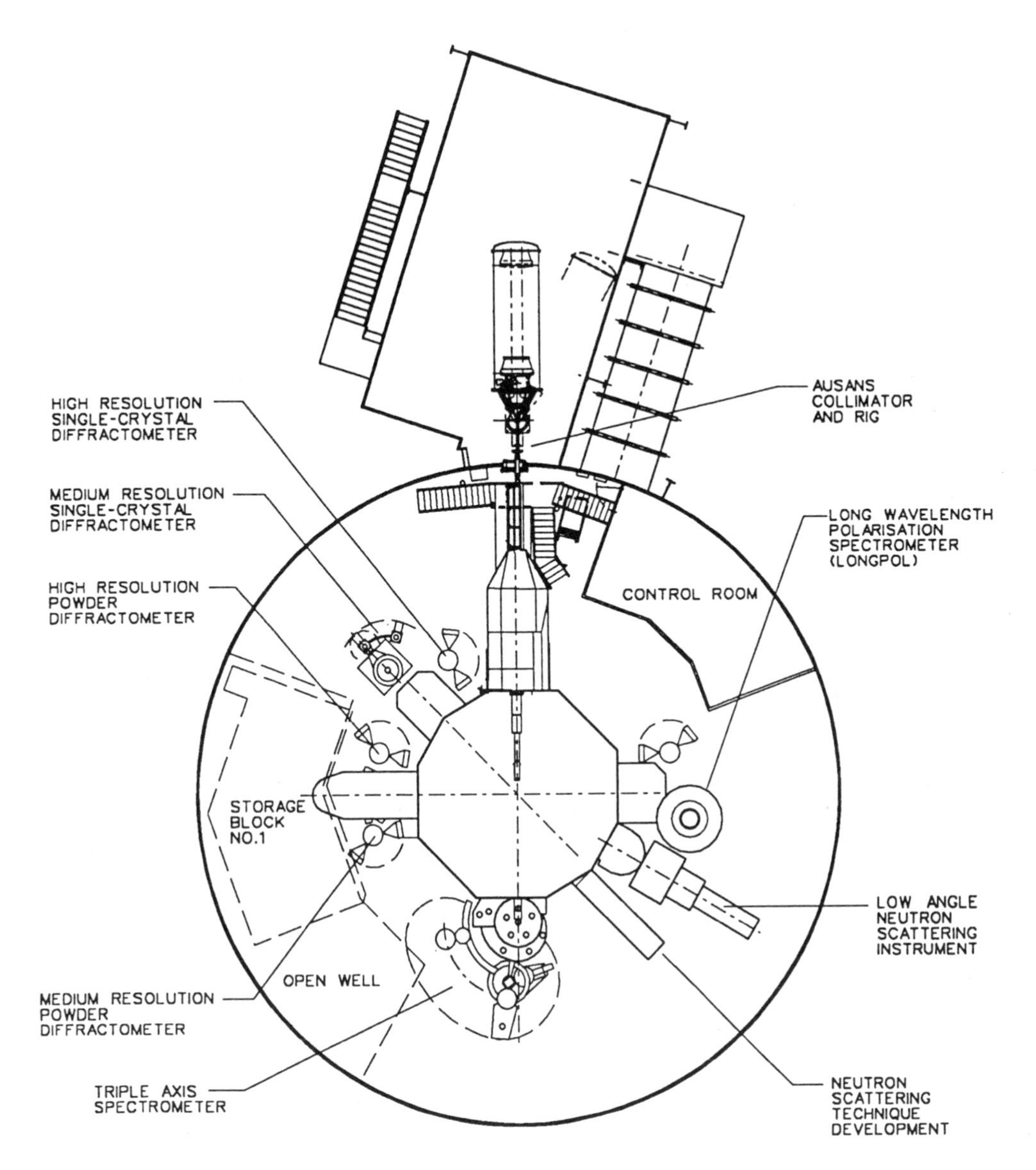

Figure 4. Schematic layout of the neutron scattering instruments presently located on HIFAR.

close to the neutron source, the flux at the sample position will be adequate ($\geq 5 \times 10^5 cm^{-2} sec^{-1}$), and being located on a tangential beamline, the background will be manageable.

A low angle diffractometer (beamline 4H5B) is particularly suited to studies on membrane structure. The original wavelength of 1.62Å has recently been increased to 2.36Å. The 1D PSD has an active volume 25mm in diameter and 300mm in length, and uses charge division for event encoding (Alberi *et al.,* 1975). The sample position will accommodate a number of sample environments including magnets, cryostats and controled humidity/temperature canisters. More details will be given later in this report.

When the polarized neutron diffractometer (LONGPOL) was relocated to another beamline, the vacated beamline (6HGR10) was upgraded to an intermediate q instrument.

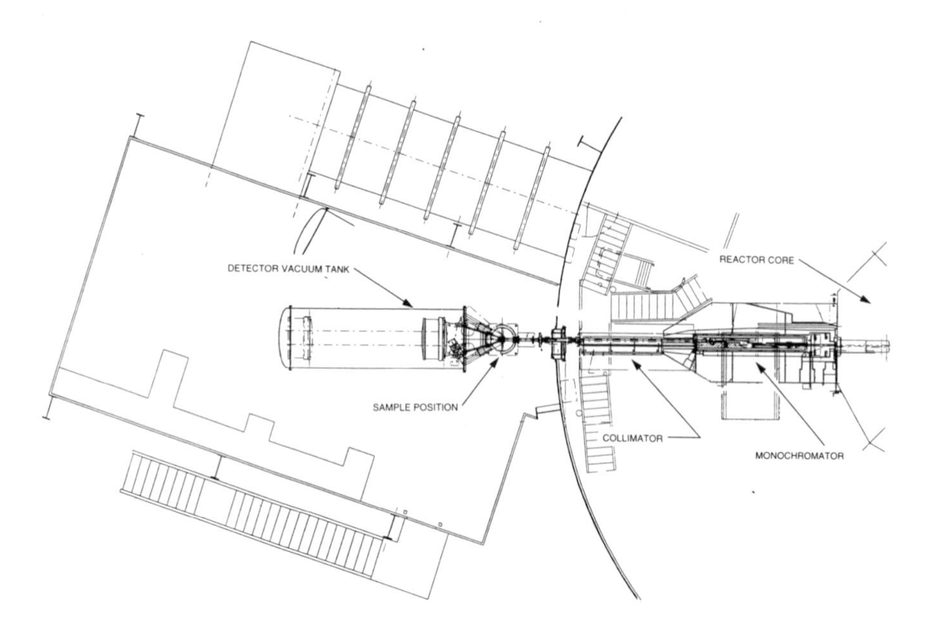

Figure 5. Plan of the small angle neutron scattering instrument.

Available space within the reactor containment building limited the instrument to a collimation length of 1.5 meters with a further 1.5 meter to the detector position. The dog-leg monochromator consists of double copper crystals. The first 2D PSD in Australia was installed on this beamline. It is a $18 \times 18 cm^2$ active area, high-resolution detector designed and built at BNL (Alberi, 1976; Boie *et al.,* 1982; Fischer *et al.,* 1983). It has an efficiency of 85% at 1.5Å and a spatial resolution of ~2mm. The original interface to a VAX/VMS computer system was redesigned to interface with an IBM-PC.

In the late 1980's the demand for a more competitive instrument lead to the design of a small angle neutron scattering (SANS) instrument. The SANS instrument (Figure 5) has a 5 meter collimation length and sample-to-detector distance range of 1.5 - 5 meters. The monochromator and first section of the collimator are located within the reactor containment building. The second section of the collimator, the sample position and the detector system are located in an external laboratory. Table Ia is a summary of the important design parameters. To maximize the neutron flux at the sample position, a tapered in-pile collimator and double multilayer monochromator system, based on the design developed at BNL (Lynn *et al.,* 1976; Saxena & Schoenborn, 1976; 1977a; 1988; Saxena & Majkrzak, 1984), were selected. At present the monochromators are single d-spacing, planar geometry used in reflection mode. Monochromators with d-spacing in the range 40 - 120Å will permit the selection of λ and $\Delta\lambda/\lambda$ over a considerable range. Geometric focussing and planned developments in multilayer technology will improve the neutron flux at the sample position. The measured and calculated neutron beam characteristics for the present configuration are shown in Figure 6 (ANSTO ANP Annual Report 1994; ANSTO/ANP - PR94). A wavelength-dependent attenuation term caused by air scatter is presently incorporated in this data. The multilayers used in this measurement have a d-spacing of 50Å, and, as yet, the $\Delta\lambda/\lambda$ for these measurement has not been determined experimentally. The measured flux profiles at 4Å are shown in Figures 6a and 6b, and the flux as a function of wavelength is shown in Figure 6c. The multilayer monochromator

Table I. (a) Design parameters for the small angle neutron scattering (SANS) facility located on a thermal beam on HIFAR. (b) Design parameters for the 2D position sensitive detector for the SANS facility.

(a) Instrument	
Wavelength range (λ)	2–7Å
Collimation length	5000mm
Sample-to-detector length range	1500–5000mm
q range (total)	0.005–0.1Å^{-1}
$\Delta\lambda/\lambda$	5–15%
Maximum flux (sample)	2×10^5cm^{-2}sec^{-1} (2Å)
(b) Detector	
Active area	640×640mm^2
Spatial resolution	5×5mm^2
Detection efficiency	>75% at 2Å
Maximum count rate (total)	10^5 sec^{-1}

system has an inherent variation in $\Delta\lambda/\lambda$ across the beam profile. Since the facility is located on a thermal neutron source, the flux is modest particularly at longer wavelengths. Nevertheless, considerable useful work can be undertaken on samples with high contrast and large physical dimensions (up to 40(H) × 50(V)mm^2).

The large 64 × 64cm^2 active area position sensitive detector (Figure 7) has been designed in collaboration with BNL and the ILL, and constructed at ANSTO. The design parameters are presented in Table Ib. The detector chamber has been developed along principles established at BNL, and the event readout system is based on the wire-by-wire principle with the PCOS (LeCroy) electronics used for event encoding. The system is interfaced to a DECstation 5000/200/PX under the ULTRIX operating system. Final testing of the detector is in progress.

The possible applications of this facility to problems in structural biology will be evaluated when the performance data in the extended q-flux space is available.

Future Developments

Given the present level of interest in neutron scattering techniques in Australia, it is to be expected that instrument development will continue at least at the present level. A quantum leap in activity could result from the decision to replace the present neutron source.

After considerable pressure from ANSTO and the broad spectrum of reactor users, the Australian Government established in 1992 a review process to investigate the need for a new research reactor. The Research Reactor Review (RRR) had wide powers and went to considerable lengths to seek out information on relevant issues both locally and internationally. The RRR was given the following general terms of reference:

i. whether, on review of the benefits and costs for scientific, commercial, industrial and national interest reasons, Australia has a need for a new research reactor.

ii. a review of the present reactor HIFAR, to include an assessment of the national and commercial benefits and costs of HIFAR operations, its likely remaining useful life and its eventual closure and decommissioning

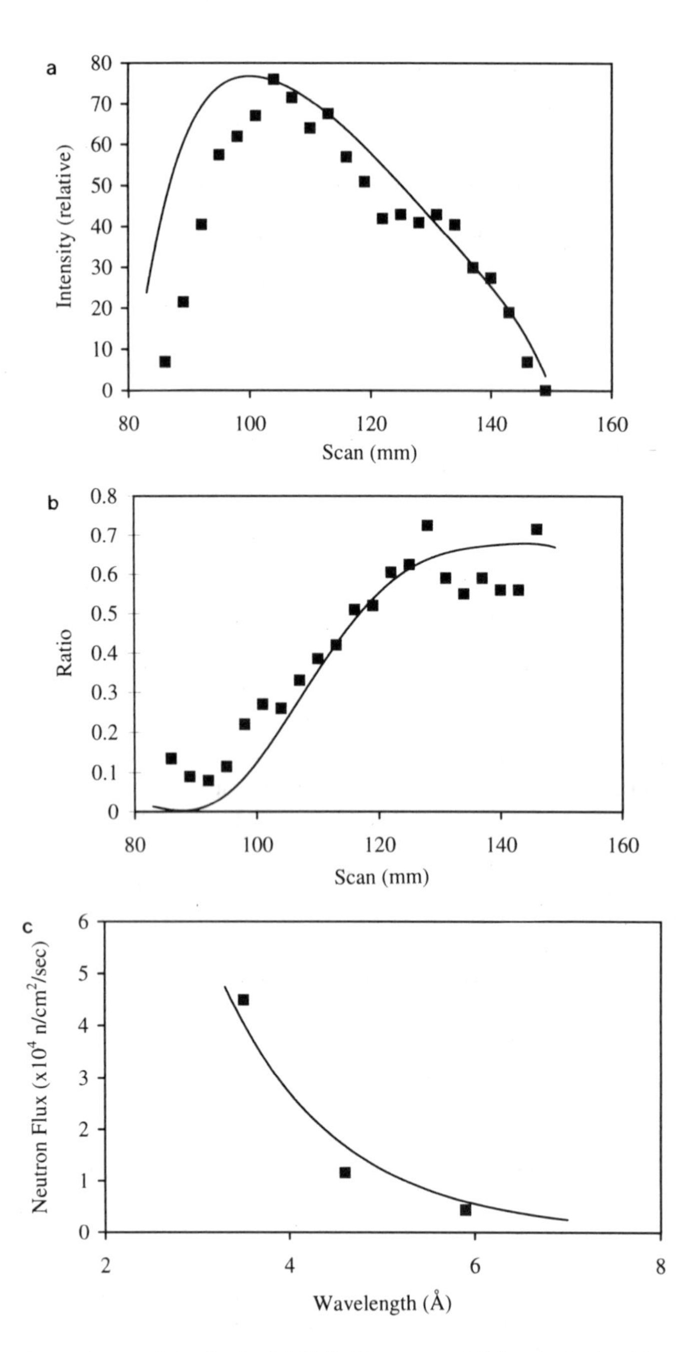

Figure 6. Neutron beam characteristics for the SANS facility measured (■) and calculated (——) at the monochromator exit. (a) Neutron flux profile in the horizontal plane. (b) Be ratio in the horizontal plane. (c) Neutron flux as a function of wavelength (from ANSTO ANP Annual Report, 1994; ANSTO/ANP - PR94).

iii. if the finding of (i) is that Australia has a need for a new nuclear research reactor, the Review will consider possible locations for a new reactor, its environmental impact at alternative locations, recommend a preferred location, and evaluate matters associated with regulation of the facility and organisational arrangements for reactor-based research.

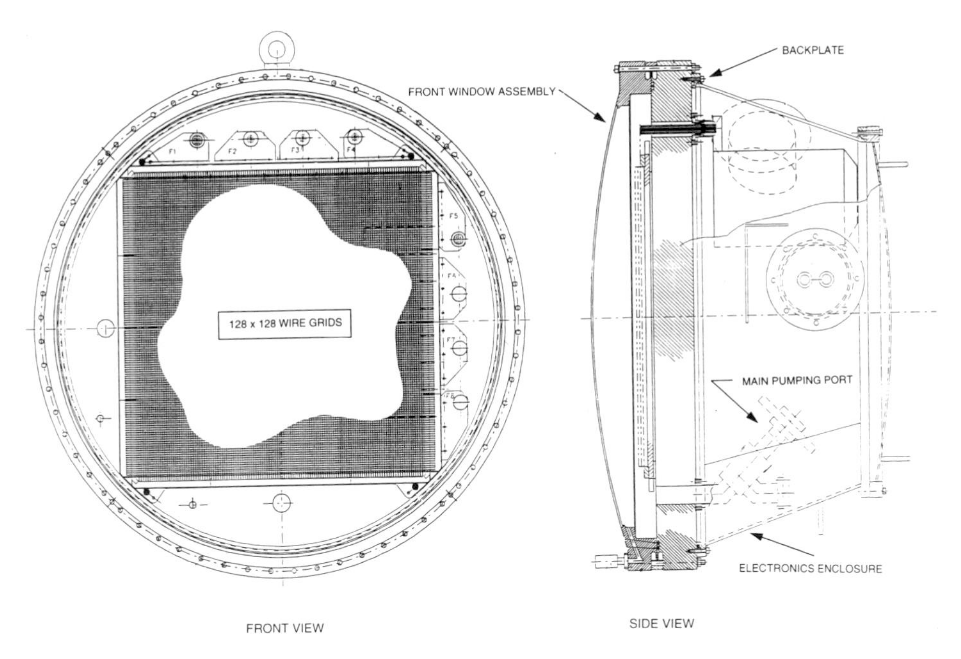

Figure 7. Engineering drawings of the 2D PSD for the small angle neutron scattering facility.

The three member Review panel had extensive experience in various fields of science, but no detailed previous knowledge of nuclear science and technology. It should be noted that the need for a neutron source in Australia is multi-facet with radioisotope production, materials irradiation and beam facilities having major importance. All interested parties (scientific, community, political, government etc) were invited to submit material to the RRR, and consultants were engaged as required. The response was substantial and considerable debate (both of the informed and uninformed varieties) took place during the period of the inquiry. After consideration of all options the scientific community put forward the unified proposal for a source based on the ORPHEE reactor (neutron flux $3 \times 10^{14} cm^{-2} sec^{-1}$)(Figure 8).

The recommendations of the RRR were presented to the national Government in August 1993. *In toto*, the recommendations are non-trivial but basically a number of issues need to become clearer before a decision on a new neutron source could be made. In the view of the panel, this should happen within an elapsed period of approximately five years. At the end of this time, a further investigation could well recommend the replace with another reactor, or with a spallation source. However, if the Government decided that the *national interest* dominated the broader technical/political arguments, then a decision could be taken at any time to replace HIFAR. It was clear in the RRR report that the social issues surrounding reactor technology were of concern to some - the political attractiveness of a non-reactor based source was noticeable. The present research reactor site is now on the outskirts of a major city with the original isolation all but disappeared with rapid residential development. Other issues considered important in the RRR report were that future development in technologies may not support a reactor as the preferred source; that a national nuclear waste repository must be established to address the present and future needs of our small inventory; and that the present scientific utilization of HIFAR must be

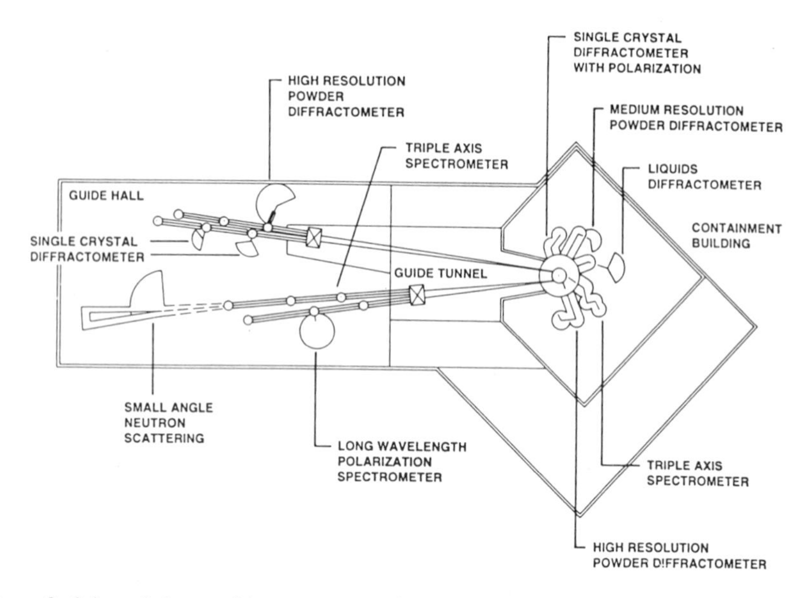

Figure 8. Schematic layout of the neutron scattering instruments located on the proposed new reactor.

enhanced and evidence of strong and diverse applications of neutron scattering to Australian science must be provided.

Subsequent to the RRR, ANSTO underwent another government initiated review to concentrate on the overall mission of the organisation. In many ways, this review focussed ANSTO's mission more directly on nuclear-related matters and felt that the future of ANSTO was conditional on the acquisition of a new reactor. The debate on a new neutron source now rests in the political arena. In the meantime, the technical assessment of the source, and its utilisation, continues in the development of the neutron scattering capability, as well as the other identified uses of a medium flux neutron source.

NEUTRON SCATTERING AND BIOLOGY

Interwoven in the HIFAR user program there were a number of isolated studies on biological material through the 1970's primarily single crystal diffraction studies of small molecules. Late in the decade a major effort on the structure of membranes was initiated and a superseded powder instrument (4H5B) was adapted for one-dimensional diffraction. The interaction of reconstituted membranes and local anaesthetics, gramacidin-A, ubiquinone, alcohols and cyclosporin-A were investigated. To meet the increasing demand, the instrument was upgraded in the early 1980's with the installation of the first 1D PSD in Australia.

Low Angle Diffraction from Reconstituted Membranes

Living organisms may be regarded as macrostructures of partitioned water. The partitioning is carried out effectively by the ubiquitous biological membrane. Membranes present a most complex challenge with their wide variety of morphology, composition and function. Despite the introduction of many sophisticated techniques and a substantial increase in knowledge of the composition, organisation, dynamic properties, and the role played in cellular physiology and cell-cell interactions, the understanding of the molecular mechanism that are implicit in the wide range of functions is far from complete. The current picture of biological membranes that has emerged is based on the fluid mozaic model (Singer & Nicholson, 1972). This model, whilst providing a general impression of a biological membrane, belies the complexities now recognized as features of many membranes. Rather than being a homogeneous 'sea' of lipid, membrane lipids exist in discrete domains which may differ substantially in their composition and properties.

Low angle diffraction from membrane structures has provided substantial quantities of useful information on a number of aspects of membrane structure and function (Blaurock, 1982; Franks & Levine, 1981; Gogol & Engelman, 1984; Schoenborn 1976; Torbet & Wilkins, 1976; Weir *et al.*, 1988; Worcester 1976; Zaccai *et al.*, 1975). Certain membranes occur naturally in stacks (eg myelin and the disk membranes in retinal rod outer segments) and can thus be studied intact (Yeager *et al.*, 1980; Chabre & Worcester, 1982); membrane fragments or large 'membrane' vesicles may be artificially stacked to give a diffraction pattern (Clark *et al.*, 1980; Montal *et al.*, 1982; Sadler & Worcester, 1982; Pachence *et al.*, 1983); and a majority of investigations have been carried out on highly ordered multibilayers of pure lipids or lipid/protein mixtures. Substantial evidence has been accumulated over the years that the lipid bilayer is a predominant component of natural membranes, and it is now generally accepted, although by no means proven in all cases, that lipids in membranes have a structure similar to that of lamellar bilayer phases observed in pure lipid-water preparations (Wilkins *et al.*, 1971; Tardieu *et al.*, 1973).

The Cyclosporin-A/Membrane Model. Information was sought on the behaviour of the immunosuppressant drug, cyclosporin-A (CsA), in pure lipid bilayers. CsA is a potent selective immunosuppressant drug with wide clinical application primarily for the prevention of graft rejection following heart, kidney and liver transplantation, and for treatment of acute graft-versus-host disease following allogenic bone marrow engraftment (White, 1982; Borel, 1986). Whilst having little advantage over other immunosuppressive therapy with respect to the frequency of post-operative viral infection, CsA has a distinct advantage in being non-toxic to bone marrow stem cells and dramatically decreasing the incidence of bacterial infection. The highly selective immunoregulation expressed by CsA coupled with low myelotoxicity, represents a basic innovation in immunology.

CsA ($C_{62}H_{111}N_{11}O_{12}$) is a cyclic peptide consisting of eleven amino acids, seven of which are N-methylated (Figure 9). Ten are known aliphatic amino acids. The novel amino acid in position 1 is (4R)-4-[(E)-2-butenyl]-4,N-dimethyl-L-threonine (MeBmt). The single crystal structure of CsA is shown in Figure 10 (Petcher *et al.*, 1976; Knott *et al.*, 1990). Residues 11 through 7 form a short antiparallel β-pleated sheet which has a type-II′ β-turn at residues 3 and 4. The substantial right-handed twist, and the hydrogen bonds 1–3 involved in the formation of the β-pleated sheet add to the structural stability of the molecule. Residues 7 through 11 form an open loop with a D-amino acid in position 8 (D-Ala-8) and a cis-amide bond between residues 9 and 10. The fourth intramolecular hydrogen bond N8-H8···O6 forms part of a seven member ring occasionally observed in tight γ-turns in protein structure (Chou *et al.*, 1982). The side chains of Abu-2, Val-5 and MeLeu-4-6-9-10 are in the staggered conformation with MeLeu chains extended. The side chain of MeBmt-1 is neatly folded into the ply of the β-pleated sheet which gives the molecule an overall compact globular shape in the single crystal environment.

Based on the observation that the binding affinities of CsA to T-lymphocytes and phospholipids are approximately the same, it was suggested that CsA may exert its immu-

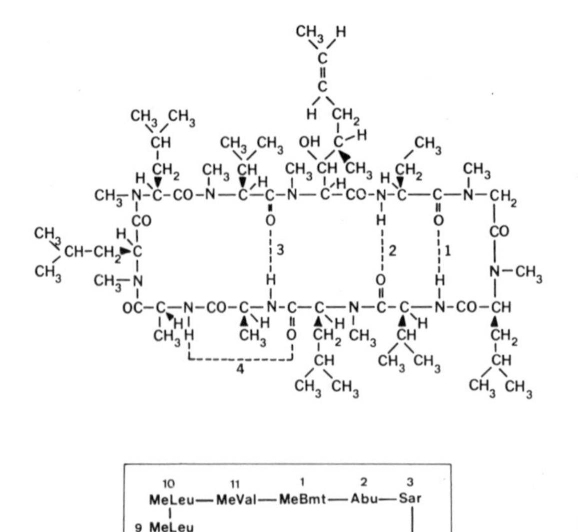

Figure 9. Chemical composition of the immunosuppressant drug cyclosporin-A. The four intramolecular hydrogen bonds are indicated.

nosuppressive effect through a nonspecific membrane-mediated mechanism (Legrue *et al.*, 1983). To investigate this proposal, the interaction between CsA and dimyristoylphosphatidylcholine (DMPC) multilamellar dispersions was determined using scanning calorimetry, and infrared and Raman spectroscopy (O'Leary *et al.*, 1986). It was found that the temperature and the maximum heat capacity of the lipid bilayer gel-to-liquid crystalline phase transition were reduced. Over realistic CsA mole fractions the transition temperature was reduced by approximately 0.5°C, and the maximum heat capacity by ~50%. CsA did not partition ideally between the two phases as indicated by the relationship between the shift in transition temperature with CsA concentration. A more pronounced effect was observed on the pre-transition where 1mol% CsA reduced the amplitude to <20% of the pure lipid value. Raman spectroscopy indicated that CsA induced effects on the phase transition were not accompanied by major structural rearrangements of the lipid

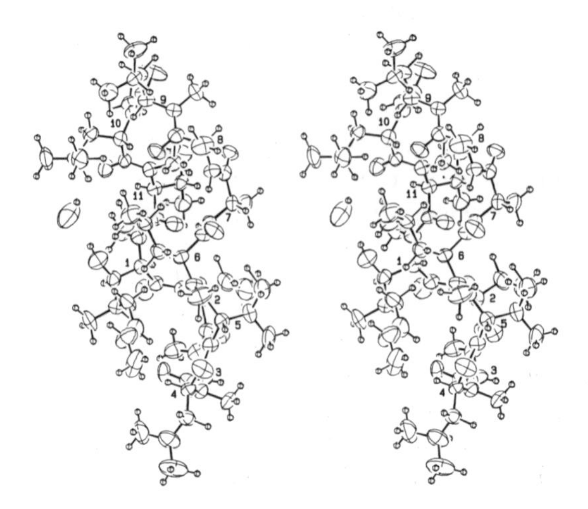

Figure 10. Single crystal structure of cyclosporin-A determined by neutron diffraction techniques (Knott *et al.*, 1990).

headgroup or hydrocarbon chains regions at temperatures remote from the transition temperature.

CsA is a compact structure if the MeBmt-1 side chain remains in the ply of the β-sheet segment. The low resolution structure of CsA resembles an ellipsoid of half axes of approximately 5Å by 5Å by 10Å. This compares with ~15Å for the length of a hydrocarbon chain. It was envisaged that the CsA molecule could pack into a lipid bilayer with its long axis parallel to the hydrocarbon chain.

Best diffraction from the bilayer structure is usually obtained below the lipid phase transition temperature with the hydrocarbon chains in the gel phase. Given the complexity of a natural membranes, investigating lipid systems below the phase transition may not be such a bad model in many instances. CsA appears to be a rigid molecule and its packing with other rather structurally-accommodating molecules, like the ubiquitous lipid, bears investigation.

The concentration of CsA used in these initial experiments was not meant to reflect a meaningful physiological value, but rather to cause the maximum effect to the bilayer structure commensurate with keeping the CsA/lipid interaction regions isolated. It was argued that having, on average, 20 lipid molecules associated with each CsA molecule would effectively screen individual interaction regions. The typical area for a typical phospolipid is 47–54$Å^2$ (Hauser *et al.*, 1981), and the projection of the CsA molecule down its long axis gives an area ~125$Å^2$. This implies that, on average, there was at least two rows of lipid molecules closely packed around each CsA molecule.

Theory

The theoretical basis for one dimensional neutron diffraction from a centrosymmetric bilayer is founded on:

$$\rho(x) = \sum_{n=1}^{\infty} (\pm) F(n) \cos(2\pi n x / d) \tag{1}$$

where ρ(x) is the scattering density as a function of x, the direction normal to the plane of the membrane, F(n) is the structure factor of order n, and d is the repeat spacing of the membrane structure. Accurate structural analysis depends foremost upon accurate values for the magnitudes of the F(n). Observed intensities, I(n), require correction for systematic effects which include sample absorption, extinction, disorder, multiple reflection, chromatic and geometric aberrations, background and the Lorentz correction. In general:

$$I_m(n) = L(n) A(n) |F(n)|^2 \tag{2}$$

where $I_m(n)$ is the measured integrated intensity of the nth order reflection, L(n) is the Lorentz correction and A(n) the sample absorption correction.

Absorption Correction. Absorption effects are compounded by the dependence of the beam path length through the sample, on the scattering angle. Classical absorption by nuclear capture is negligible. Incoherent scatter from hydrogen manifests itself in a manner similar to true absorption, causing attenuation of the diffracted intensity. For thin, planar specimens this absorption (designated type I) is readily calculated. The reduction of the diffracted intensity is given by:

$$\frac{1}{v}\int e^{-\mu t}dv \tag{3}$$

where μ is the linear absorption coefficient, t is the pathlength of neutrons through the sample and v is the sample volume. Integration gives:

$$A(n) = \frac{\sin\theta}{2\mu t_0}\left(1 - e^{-\frac{2\mu t_0}{\sin\theta}}\right) \tag{4}$$

where t_0 is the (average) sample thickness. The magnitude of the correction decreases with decreasing path length and so is largest for the low order Bragg reflections. Values for μ were calculated using a procedure outlined by Worcester and Franks (1976). Using a density of 1.014g/cm^3 for the bilayer (Johnson & Buttress, 1973), calculated values for μ at a neutron wavelength, λ, of 1.63Å are given by the expression:

$$\mu = (6.04 - 0.75\gamma_D)$$

where γ_D is the mole fraction of D_2O.

Extinction Correction. Extinction is the attenuation of the incident beam as the result of diffraction. It can result from multiple internal reflections in the sample (primary extinction) or from the reduction in the incident beam intensity due to diffraction from previous planes (secondary extinction). When diffraction is weak the effect is negligible. However, if the diffraction is intense (as for example in a well order sample containing D_2O) then parts of the sample will be substantially shielded from the incident beam. Corrections for extinction are not easily made. Plotting the structure factors against H_2O/D_2O ratio is the best estimate of extinction effects. If the correction factor for a given reflection (the first order in this case) was greater than 20%, the data was deleted.

Disorder. Disorder can be due to variation of the unit cell spacing (lattice disorder) or variation of the structure within the unit cell (substitution disorder). Methods for treating disorder in one dimensional crystals have been well documented. Lattice disorder was accounted for by fitting Gaussian peaks of variable width to each Bragg reflection.

Lorentz Correction. There is now general agreement that in the case of lamellar diffraction from membranes, the Lorentz correction, L, should be expressed as the product of two terms, L_1 and L_2.

The classical Lorentz correction, L_1, is due to the geometric parameters of the diffraction experiment. Briefly, the total integrated intensity diffracted depends on the angular velocity which the planes move through the diffraction position. Since the diffraction positions are smeared by the divergence and the λ spread of the incident beam, the effective velocity will vary with the position of the diffracted intensity in reciprocal space. For the restricted case of equatorial reflections from lamellar samples when the rotation axis is normal to the plane containing the incident and diffracted beams, the Lorentz correction is defined by (Arndt & Willis, 1966):

$$L_1(n) = 1/\sin(2\theta_n) \tag{5}$$

In the derivation of this expression, it was assumed that the detector was sufficiently large to accept all the diffracted intensity. If this is not the case, then an additional correction, L_2, is required to account for this loss in information. This effect, where the beam vertical divergence is convoluted with the sample mozaic spread (η_s), can be expressed in terms of measurable geometric parameters (Saxena & Schoenborn, 1977b). Briefly, the correction factor for a detector with a vertical acceptance, $2l$mm, is given by the relation:

$$L_2(n) = I_D(2l)/I_D(\infty) \tag{6}$$

where $I_D(\infty)$ is the detected intensity with $l = \infty$. It follows that for a detector sufficiently large to accept the entire diffracted beam, $L_2(n) = 1$ and no correction is required. However, if the contrary in true, then an expression for $L_2(n)$ can be derived in terms of η_s; the sample to detector distance, D; the scattering angle, $2\theta_n$; the detector acceptance height $2l$; beam divergence ξ; and beam width w.

$$I_D(2l) = \frac{C_1^0 C_2^0}{\left[(2\pi)^{1/2}\beta + w\right]} \int_{w/2-l}^{w/2+l} \operatorname{erf}\left(\frac{t}{\alpha\sqrt{2}}\right) + \frac{\beta}{(\alpha^2+\beta^2)^{1/2}} \left\{1 - \operatorname{erf}\frac{\beta t}{\alpha\left[2(\alpha^2+\beta^2)\right]^{1/2}}\right\} \times$$
$$\exp\left\{-\frac{t^2}{2(\alpha^2+\beta^2)}\right\} \tag{8}$$

where C_1^0 and C_2^0 are constants that disappear in the ratio; $\alpha = 2D\eta_s \sin\theta$; and $\beta = \xi D$. The integral expression was evaluated using the Gauss-Kronrod rules for numerical integration (IMSL Subroutine Library). Values of $L_2(n)$ were calculated as a function of the order of the reflection, for representative values of η_s. It is clear that the $L_2(n)$ correction is most important for the higher order reflections - the reflections that will provide the fine detail to the derived bilayer structure.

The Phase Problem Applying the above corrections, it is possible to accurately calculate the magnitude of F(n) from $I_m(n)$. Only one problem remains - the phase of each F(n) which was lost during data collection. In the centrosymmetric situation such as this, the problem is reduced to a choice between a value of 0 or π. Several methods of phase determination have been developed, two of which have proven particularly useful in this study.

i. If the scattering density of the water layer is systematically changed by isotopic substitution, then the structure factors F(n) will exhibit a linear dependence on the isotopic composition. It is usually reasonable to assume that the water is only located at the origin of the structure and therefore F(n) will increase or decrease with the addition of positive scattering density, depending on whether the F(n) is positive or negative respectively.

   This procedure also provides data that can be used to determine the water distribution profile for the bilayer structure. This is achieved by calculating the Fourier difference profile according to the relation:

$$\Delta\rho(x) = \sum_n \left[(\pm)F_{D_2O}(n) - (\pm)F_{H_2O}(n)\right] \cos\frac{2\pi nx}{d} \tag{7}$$

where $F_{D_2O}(n)$ and $F_{H_2O}(n)$ are the structure factors for the bilayer containing D_2O and H_2O, respectively.

ii. The second method is a conglomerate of empirical and intuitive arguments which can now be used with considerable confidence given the enormous volume of information available on bilayer structure. By making simple assumptions about the scattering density distribution, it is possible to select a phase combination from the 2^n possibilities which give Fourier profiles consistent with the assumptions. By comparing the ratios of scattering densities, it is possible to identify regions of known chemical composition.

Profile Scale Factor. In this study, it was necessary to compare, at least qualitatively, two physically different bilayer samples. The two samples, one with and one without, CsA were made under identical experimental conditions. The difficulty is that the experimental scattering density information for these samples is incomplete. The diffraction intensities are always collected in arbitrary units since $I_m(0)$ is coincident with the undiffracted beam and cannot be measured, and only a finite number of intensities are observed. To characterize any structural differences, the two scattering density profiles must be placed on either an absolute, or a normalized scale.

To maximize the information extracted from the experimental data, the scattering profiles may, in many cases, be placed on an absolute scale using the known scattering density data. A number of techniques are available (for example Worcester & Franks, 1976; Zaccai & Gilmore, 1979; McCaughan & Krimm, 1982; Gogol *et al.*, 1983; Blasie *et al.*, 1984; Knott & Schoenborn, 1986). A method of placing two almost identical bilayer profiles on the same normalized scale, is to compare directly the measured water profiles. A scale factor is obtained from the ratio of the water profile areas for samples with the same water content. In applying this method, it is assumed that no significant changes occur to the water distribution between the bilayers, and that the minimum in the water profile is due to zero $\rho(m)$.

Neither of these procedures will be necessary when a deuterated analogue of CsA is available. In that case a direct comparison of two chemically identical bilayer samples (one with hydrogenated CsA, the other with deuterated CsA) will be possible, and the difference in their neutron scattering density profiles will yield the location of the deuterium label directly.

Experimental

Equipment. The Low Angle Neutron Diffractometer (LAND) on the 4H5B beam line of HIFAR was used for data collection. The majority of data was collected using the 1-D PSD which integrated the pattern over a limited spatial range. Some data sets were collected using a 2D PSD (in an interim configuration) to acquire the entire diffraction pattern for comparison.

Sample Preparation. The sample preparation technique that proved most successful was an evaporative deposition protocol developed by Worcester (1976). The temperature and type of solvent used was important - if the solvent had too high a vapour pressure, then the sample would inevitably have a large η_s. The 4:1 v/v ethanol/water solvent proved most satisfactory. In order to optimize the signal-to-background ratio, it was desirable to prepare samples with η_s comparable in magnitude to the incident beam divergence. The diffracted intensity is proportional to the convolution of η_s and the beam divergence.

The background, on the other hand, is primarily due to the strong incoherent scattering from hydrogen atoms and is dependent only on the quantity of material in the sample and not its molecular order.

The lipids (DMPC, DPPC) were purchased from the Sigma Chemical Company (St Louis MO) and used without further purification. Pure powdered CsA was a gift from the Sandoz Pharmaceutical Company. Glass distilled H_2O was used throughout, and D_2O (>99.7% isotopic purity) was obtained from ANSTO. All chemicals used were at least AR grade. The pure lipid was dissolved in ethanol/H_2O solvent to a concentration of 20mg lipid/100μl solvent. The pure CsA was dissolved in ethanol/H_2O solvent to a concentration of 6.5mg CsA/100μl solvent. The 20:1 lipid/CsA molar ratio was obtained by mixing appropriate volumes. A diffraction sample was obtained by evaporative deposition of an appropriate quantity of lipid solution. Acid washed glass cover slip (50.(high) × 35.(wide) × 0.15(thick)mm) were used as substrates. The sample was mounted in a thin walled aluminium canister where temperature and RH were accurately controled. The canister had a water reservoir top and bottom through which water at a set temperature was circulated. All data was collected at 20°C. The design of the canister was such that the base was ~0.5°C cooler than the walls and the top. The RH was controled by a saturated salt solution (O'Brien, 1948) placed in a small container on the bottom of the canister. Important practical factors in the choice of the salt solution apart from the obvious RH, was the slope of the salt solubility and RH with temperature. In practice, three salt solutions proved most useful - NaCl, KCl and KNO_3 produced 76%, 86% and 96% RH respectively.

Data Collection. The bilayer sample was initially equilibrated with a saturated salt solution of D_2O. The PSD was fixed in a position to intercept the maximum number of reflections. The sample was stepped through ω with $\Delta\omega$ increments of 0.1°, starting several degrees in ω before the peak intensity of the first order reflection, and finishing several degrees after the eighth order reflection. Each ω step produced a 2θ-y pattern on the 2D PSD, and a 2θ profile on the 1D PSD. This raw data was stored for further analysis.

The first four reflections were strong and could be measured in 1 or 2 hours with a statistical accuracy of <1%. The higher order reflections were less intense and required measurement times longer by a factor of ~10. The salt solution was then changed for the next H_2O/D_2O ratio in a random sequence and allowed to equilibrate for at least 8 hours. It should be noted that there are subtle differences in the physicochemical properties of D_2O and H_2O. Biologically, some differences have a greater effect than others. However, it has been demonstrated (Coster *et al.*, 1982), that any differences induced in bilayer structure are below the detectable level when low angle diffraction techniques are employed. The exchange of H_2O with D_2O therefore represents a true isomorphous replacement. No account was taken for the small differences between the vapour pressures of H_2O and D_2O (Kirshenbaum, 1951), or the lower solubility of salts in D_2O (Miles & Menzies, 1937).

Data Reduction. Figure 11 is the result of summation of 100 ω steps (ie a complete scan) for the first four orders of reflection from a DMPC sample equilibrated with H_2O at 86%RH, collected on the 2D PSD. To include all this diffracted intensity, the data was integrated for each of the 100 ω steps, using an appropriate mask. Extrapolation of this mask to orders five to eight indicated that the detector acceptance angle was not sufficient to collect all the intensity for this sample. Arc integration of data within this circle segment produced a one-dimensional 2θ profile for each ω step. On the 1D PSD this integration was performed over the rectangular region created by the active volume. The Lorentz correction was different in each case. For the 2D PSD data, $L = L_1$ for the first four order re-

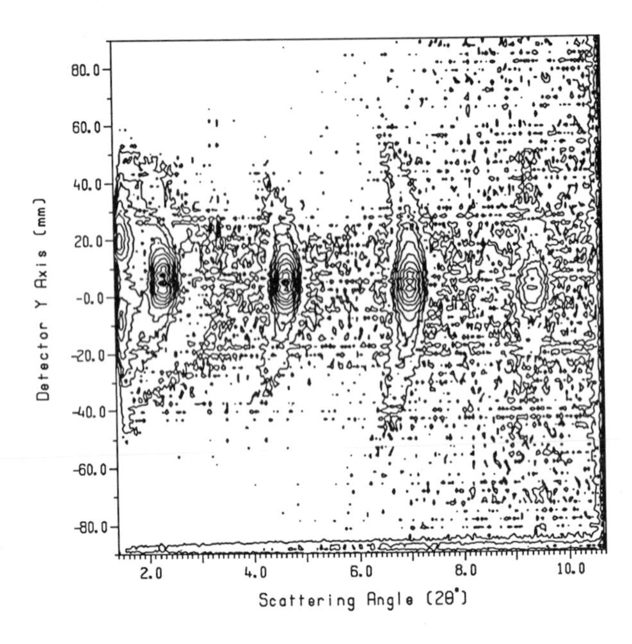

Figure 11. A pictorial representation of the sum of 100 ω scans for a DPPC sample collected on the two-dimensional position sensitive detector.

flections; and for the higher order 2D PSD data and the 1D PSD data, $L = L_1\ L_2$, with L_2 calculated for each sample.

An ω-2θ pattern could then be constructed for both sets of data (1D PSD and 2D PSD). Figure 12 is a typical example for four orders of reflection clearly indicating that η_s dominates the diffraction geometry. A plot of the intensity as a function of ω gives η_s of the sample (Figure 13). Two populations of sample were often observed. One population has a quite large η_s (>5° FWHM) and one with a quite small η_s (< 1° FWHM). The fact that the reflections are symmetric and invariant with 2θ is evidence that both populations

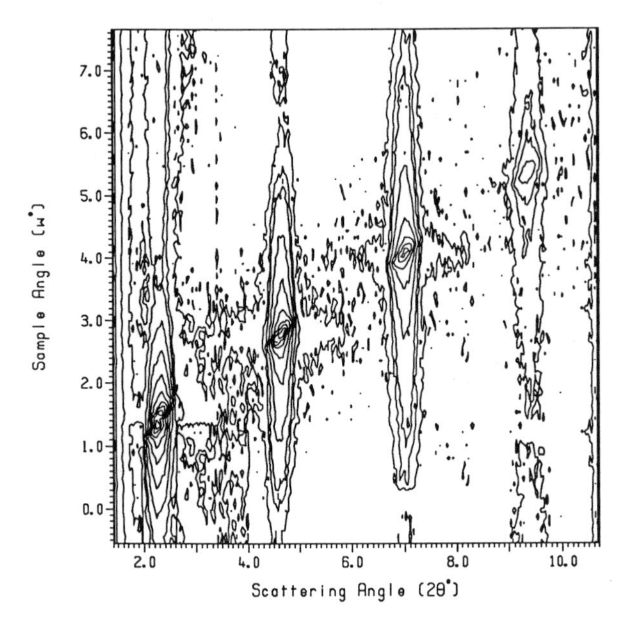

Figure 12. An example of an ω-2θ map for a DPPC sample generated by integration of the raw diffraction data.

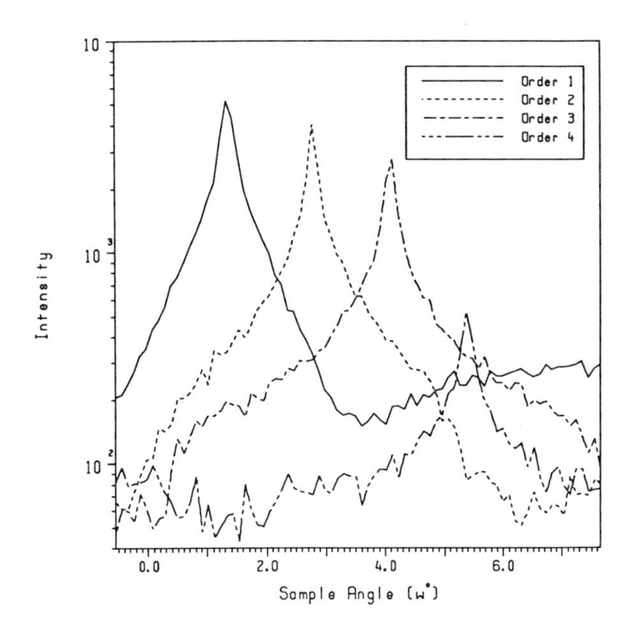

Figure 13. An example of the mozaic spread of a DPPC sample as a function of reflection order. The plot was generated from the ω-2θ map in Figure 12.

have the same characteristics at least at this resolution. If the edges of the sample were removed with a razor blade, the contribution of the large η_s was reduced in intensity. A difficulty introduced by this duality was the choice of the Lorentz correction, $L_2(n)$. As outlined previously, $L_2(n)$ was dependent on the sample η_s. No correction was necessary for the small η_s component, and the maximum correction to order 8 reflection for a (typical) 5° ~ η_s component was ~1.5. By order 4 reflection, smearing had made it virtually impossible to resolve the two components. If the apparent η_s was used in the reduction of the data sets, satisfactory agreement was obtained for data collected under identical experimental conditions.

The question then remained as to how best to obtain the most accurate values for the integrated intensity for each reflection from the ω-2θ maps (Figure 12). The first four order reflections were intense and presented little difficulty. The size and shape of the reflections were readily defined by the intensity contour level 2σ above background. Many of the higher orders (5–8) required summation to improve the signal-to-noise ratio before a clear indication of their shape and position could be established. A number of techniques of integration were investigated primarily to handle the less intense higher order reflections. Since these reflections provide detail to the structure reconstruction, it was important to be accurate.

Three pieces of information per reflection were required from this analysis - the position of the reflection $2\theta_n$, the total intensity $I_n(2\theta)$, and the width of the reflection $\Delta 2\theta_n$. Values of $2\theta_n$ were used to determine the d-spacing of the bilayer; the $I_n(2\theta)$ were used in the Fourier analysis; and the $\Delta 2\theta_n$ to check that disorder was not a major problem.

The first few steps of the analysis located the centre and the width of the reflection. Raw data was first smoothed using a convolution algorithm (Savitsky & Golay, 1964) and then differentiated. The position $2\theta_n$ enabled the calculation of the bilayer spacing using the Bragg relation in Equation 1. Fitting a Gaussian function to each reflection gave an unbiased value for $\Delta 2\theta_n$. Substantial variation in $\Delta 2\theta_n$ with n indicated lattice disorder and the data set was rejected. Values for $2\theta_n$ and $\Delta 2\theta_n$ were used to construct a mask in ω-2θ space to surround each reflection. The integrated intensities $I_n(2\theta)$, corrected for background, were used to generate structure factors by applying all the corrections outlined in above and then substituting in Equation 1. So far the major sources of error are changes in the bilayer structure and counting statistics particularly on the weaker reflections. The final task in Data Reduction was to establish sets of phases for all the structure factors.

Fourier Analysis. The accuracy of the F(n)'s obtained above was first verified by plotting the values as a function of the H_2O/D_2O ratio. For the large values of F(n) the expectation is that agreement should be excellent with the notable exception of F(1) at 100% D_2O. This underestimation was almost certainly due to extinction resulting from the extremely high contrast in scattering density between the lipid and D_2O layers.

F(n)'s with small values are clearly subject to larger errors. The plot immediately identified incorrect values for the measured F(n)'s; identified any systematic changes in bilayer structure; and verified the phase assignment. So far the experiments have been carried out in completely unknown and arbitrary intensity units. The scattering profiles were placed on a normalized scale using the water peak area. This gave an indication of qualitative differences between the scattering density distributions of bilayers with, and without the drug. It was not possible to compare the two samples mathematically.

A rigorous error analysis of the Fourier profiles may be carried out (for example Franks & Lieb, 1979). However, statistically accountable errors are not the major source of uncertainty in the scattering profiles. F(n)'s as a function of H_2O/D_2O ratio vary over quite a range in magnitude. The largest source of 'error' is due to series termination effects in the Fourier analysis which can seriously influence the physical interpretation of the re-constructed profiles.

Results and Discussion

The results for three bilayer systems are shown in Figures 14–17. The normalising scale factor made no attempt to account for the increased water content of the bilayer with increased hydration. The hydrations are labeled low (NaCl at 76%RH), intermediate (KCl at 86%RH) and high (KNO_3 at 96%RH). In each of the figures, the scattering profiles for the bilayers containing H_2O are presented, and the D_2O-H_2O difference (water) profiles are also included, but offset vertically for clarity. The origin of the profiles is defined as the centre of the hydrocarbon region. The error in the d-spacing was estimated from the variation in the positions of the larger F(n). A value of $\pm$ 0.3Å is typical for this data.

At this early stage of data analysis, the interaction of CsA with lipids can only be inferred in qualitative terms by observation of changes in regional scattering density, $\rho(m)$. There is evidence in the profiles calculated thus far for (i) additional scattering density in the hydrocarbon chain region, and (ii) changes in the water layer profile, that are consistent with the model for interdigitation of CsA with the lipids.

The increase in d-spacing with the addition of CsA to the DMPC bilayer at intermediate hydration (Figure 14) is small but significant. There are two possible effects which might account for this observation. Either the CsA was too long to fit neatly into one half of the DMPC bilayer, and has consequently elongated the lipid molecule; or, a re-structuring of the water layer has been induced to accommodate the hydrophobic CsA molecule. The maximum of the headgroup region has moved slightly toward the bilayer center with the addition of CsA. There are significant differences in the F(n)'s of the two bilayer samples. As a measure of these differences, consider the affect on the Fourier profiles. The water $\rho(m)$ in the CsA/DMPC bilayer is considerably lower than the hydrocarbon region

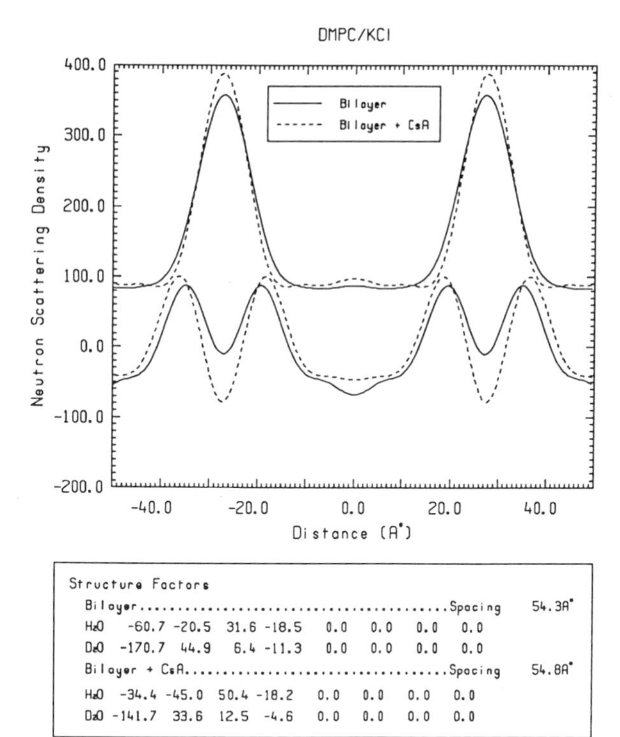

Figure 14. The Fourier profiles for a DMPC bilayer at intermediate hydration. The normalized structure factors are presented with the d-spacing in the insert.

$\rho(m)$, suggesting the water is close to the liquid value. The terminal methyl group $\rho(m)$ is no longer resolved in the CsA/DMPC bilayer, suggesting the hydrocarbon region is more disordered. The water profile is narrower with CsA, and there is some evidence for water in the hydrocarbon region. All this data is consistent with a model of the CsA molecule interdigitated between the lipid molecules, projecting into the headgroup region, and changing the structure of the water layer.

The d-spacing change in the DPPC bilayer at the low hydration (Figure 15) is small, and is subject to a similar interpretation as the DMPC bilayer. There are significant differences in the F(n)'s resulting in two quite different bilayer profiles. The water $\rho(m)$'s in the bilayer profiles are similar; the headgroup $\rho(m)$ is smaller with CsA; and the hydrocarbon $\rho(m)$ is larger. The two bilayer profiles are similar in shape suggesting little alteration to the lipid conformation with the addition of CsA. The water peak profile for the CsA/DPPC bilayer is broader suggesting water penetration into the headgroup region. This data is consistent with the model of a low density uniform distribution of CsA, interdigitated with the lipids, and symmetric about the bilayer centre.

Data for the same sample at intermediate hydration (Figure 16) indicate a significantly different model. The d-spacing change is now insignificant. The only statistically significant change in F(n)'s is for F(1). The water peak is now broader with CsA and the profile indicates substantial water $\rho(m)$ in the hydrocarbon region. At the high hydration (Figure 17), results are similar to the intermediate hydration results, however, the water peak for the CsA/DPPC bilayer has a significantly different shape.

In general, CsA induced d-spacing changes in a number of the bilayers investigated. It should be noted that the addition of cholesterol to EYL bilayers produced d-spacing changes readily accounted for by physically meaningful models (Worcester & Franks,

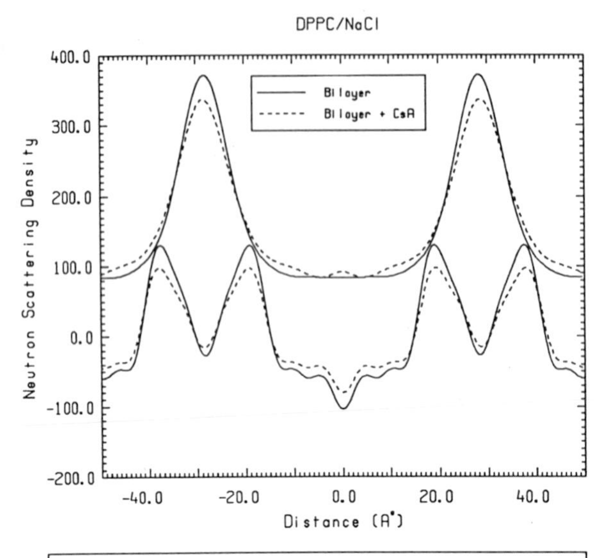

Structure Factors									
Bilayer								Spacing	56.6A°
H_2O	-76.8	-27.5	46.9	-32.3	-8.3	4.9	0.0	-10.8	
D_2O	-187.1	36.9	17.2	-20.5	-12.7	5.5	0.0	-10.5	
Bilayer + CsA								Spacing	57.1A°
H_2O	-60.2	-20.0	34.8	-23.9	-6.9	2.8	0.0	-7.8	
D_2O	-158.8	30.1	11.8	-13.8	-8.0	5.6	0.0	-7.1	

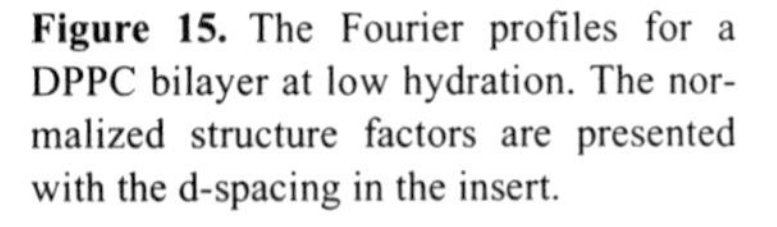
Figure 15. The Fourier profiles for a DPPC bilayer at low hydration. The normalized structure factors are presented with the d-spacing in the insert.

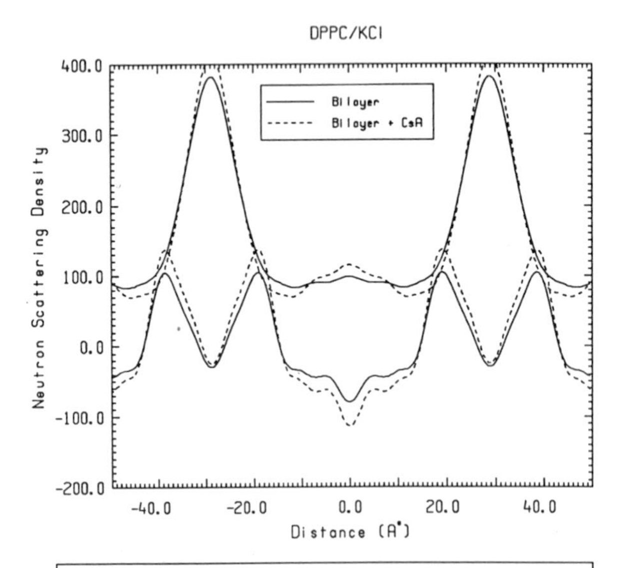

Structure Factors									
Bilayer								Spacing	57.7A°
H_2O	-54.8	-29.1	38.9	-22.5	-9.2	3.9	0.0	-7.6	
D_2O	-160.1	45.1	6.2	-9.1	-13.1	5.3	0.0	-5.8	
Bilayer + CsA								Spacing	57.7A°
H_2O	-67.5	-29.9	40.3	-23.4	-9.8	3.4	0.0	-8.8	
D_2O	-155.8	48.3	8.5	-10.3	-14.3	4.8	0.0	-6.4	

Figure 16. The Fourier profiles for a DPPC bilayer at intermediate hydration. The normalized structure factors are presented with the d-spacing in the insert.

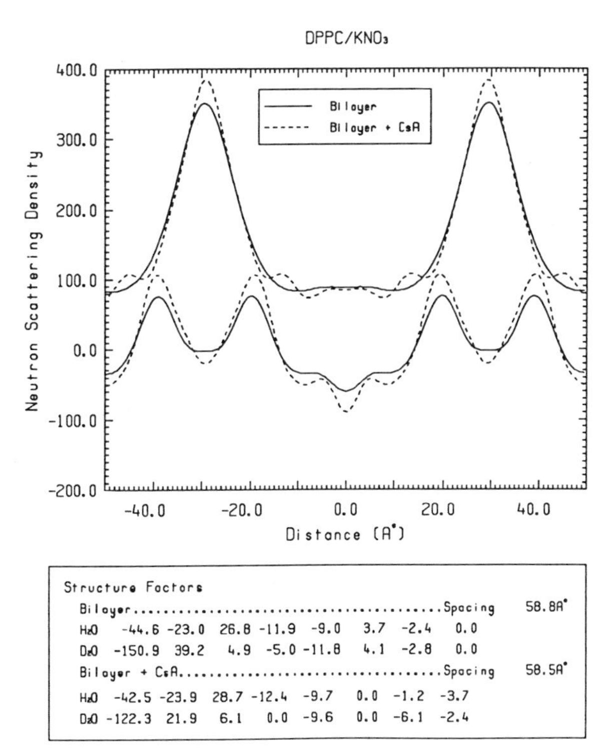

Figure 17. The Fourier profiles for a DPPC bilayer at high hydration. The normalized structure factors are presented with the d-spacing in the insert.

1976). However, the explanation for the interaction of alkanes with DOL bilayers, produced a most complex model (King *et al.*, 1985). In some cases, the bilayer appears able to accommodate additional molecules by adjustments in the direction parallel to the bilayer surface.

Conclusion

The results of these preliminary experiments provide qualitative evidence for the structural effects of the interaction of CsA with simple reconstituted membranes. The observed structural changes are most consistent with the model of CsA imbedded in a ordered manner in the hydrocarbon region of the lipid bilayer. This is a perfectly reasonable physical model for a hydrophobic molecule interacting with an amphiphilic environment. Clearly more detailed analysis of the complete suite of data is required, and further data sets collected on bilayers incorporating a deuterated labeled CsA analogue.

ACKNOWLEDGMENTS

The initial section of this paper was a summary of the efforts of many, too numerous to mention. I acknowledge the contribution of all neutron scattering scientists in Australia and our collaborators overseas. Without their commitment, neutron scattering in Australia would not have made the impact so clearly evident. The Report of the Research Reactor Review provides a comprehensive summary of a now not uncommon process. Limited numbers of this report are available from the author. The author wishes to acknowledge the enormous support given to neutron scattering in Australia by Prof Benno Schoenborn. The acquisition of much of the equipment for the application to biology was facilitated by Prof Schoenborn.

REFERENCES

Arndt, U.W., & Willis, B.T.M., (1966). *Single Crystal Diffractometry*. Cambridge University Press, UK.

Alberi, J., (1976). Development of large-area, position-sensitive neutron detectors. In *Neutron Scattering for the Analysis of Biological Structures*. (B.P. Schoenborn, editor) ppVIII24-VIII42. (Springfield, VA: Nat. Technical Inform. Serv.; US Dept. of Commerce).

Alberi, J., Fischer, J., Radeka, V., Rogers, L.C., & Schoenborn, B.P., (1975). A two-dimensional position-sensitive detector for thermal neutrons. *IEEE Trans. Nucl. Sci.*, NS-22:255–266.

Blasie, J.K., Pachence, J.M., & Herbette, L.G., (1984). Neutron diffraction and the decomposition of membrane scattering profiles into the scattering profiles of their molecular components. In *Neutrons in Biology* (B.P. Schoenborn, editor) pp201–210. Plenum Publishing Corporation, New York. ISBN 0–306–41508–9.

Blaurock, A.E., (1982). Evidence of bilayer structure and of membrane interactions from X-ray diffraction analysis. *Biochim. Biophys. Acta*, 650:167–207.

Boie, R.A., Fischer, J., Inagaki, Y., Merritt, F.C., Okuno, H., & Radeka, V., (1982). Two-dimensional high precision thermal neutron detector. *Nucl. Instr. Methods*, 200:533–545.

Borel, J.F., (1986). *Ciclosporin*. Progress in Allergy Series 38. Karger AG, Basel Switzerland. ISBN 3–8055–4221–6.

Chabre, M., & Worcester, D.L., (1982). X-ray and neutron diffraction of retinal rod outer segments. *Methods Enzymol.*, 81:593–604.

Chou, K.C., Pottle, M., Nemethy, G., Ueda, G., & Scheraga, H.A., (1982). Structure of β-sheets. Origin of the right-handed twist and of the increased stability of antiparallel over parallel sheets. *J. Mol. Biol.*, 162:89–112.

Clark, N.A., Rothschild, K.J., Luippold, D.A., & Simon, B.A., (1980). Surface-induced lamellar orientation of multilayer membrane arrays. Theoretical analysis and a new method with application to purple membrane fragments. *Biophys. J.*, 31:65–96.

Coster, H.G.L., Laver, D.R., & Schoenborn, B.P., (1982). Effect of $^2H_2O/H_2O$ replacement on the dielectric structure of lipid bilayer membranes. *Biochim. Biophys. Acta*, 686:141–143.

Franks, N.P., & Levine, Y.K., (1981). In *Membrane Spectroscopy*. (E. Grell, editor) pp437–487. Springer-Verlag, Berlin.

Franks, N.P., & Lieb, W.R., (1979). The structure of lipid bilayers and the effects of general anaesthetics. An X-ray and neutron diffraction study. *J. Mol. Biol.*, 133:469–500.

Fischer, J., Radeka, V., & Boie, R.A., (1983). High position resolution and accuracy in ^{3}He two-dimensional thermal neutron PSDs. In *Position-Sensitive Detection of Thermal Neutrons* (P. Convert and J.B. Forsyth, editors) pp129–140. Academic Press, London.

Gogol, E.P., Engelman, D.M., & Zaccai, G., (1983). Neutron diffraction analysis of cytochrome b_5 reconstituted in deuterated lipid multilayers. *Biophys. J.*, 43:285–292.

Gogol, E.P., & Engelman, D.M., (1984). Neutron scattering shows that cytochrome b_5 penetrates deeply into the lipid bilayer. *Biophys. J.*, 46:491–495.

Hauser, H., Pascher, I., Pearson, R.H., & Sundell, S., (1981). *Biochim. Biophys. Acta*, 650:21–51.

Johnson, S.M., & Buttress, N., (1973). The osmotic insensitivity of sonicated liposomes and the density of phospholipid-cholesterol mixtures. *Biochim. Biophys. Acta*, 307:20–26.

King, G.I., Jacobs, R.E., & White, S.H., (1985). Hexane dissolved in dioleoyllecithin bilayers has a partial molar volume of approximately zero. *Biochemistry*, 24:4637–4645.

Kirshenbaum, I., (1951). In *Physical Properties and Analysis of Heavy Water*. McGraw Hill, New York.

Knott, R.B., Schefer, J., & Schoenborn, B.P., (1990). Neutron structure of the immunosuppressant cyclosporin A. *Acta Cryst.*, C46:1528–1533.

Knott, R.B., & Schoenborn, B.P., (1986). Quantitation of water in membranes by neutron diffraction and X-ray techniques. *Methods Enzymol.*, 127:217–229.

LeGrue, S.J., Friedman, A.W., & Kahan, B.D., (1983). Binding of cyclosporine by human lymphocytes and phospholipid vesicles. *J. Immunol.*, 131:712–718.

Lynn, J.W., Kjems, J.K., Passell, L., Saxena, A.M., & Schoenborn, B.P., (1976). Iron-germanium multilayer neutron polarizing monochromators. *J. Appl. Cryst.*, 9:454- 459.

McCaughan, L., & Krimm, S., (1982). Biochemical profiles of membranes from X-ray and neutron diffraction. *Biophys. J.*, 37:417–426.

Miles, F.T., & Menzies, A.W.C., (1937). Solubilities of cupric sulfate and strontium chloride in deuterium water. *J. Am. Chem. Soc.*, 59:2392–2395.

Montal, M., Darszon, A., & Schindler, H., (1982). Functional reassembly of membrane proteins in planar lipid bilayers. *Q. Rev. Biophys.*, 14:1–79.

O'Brien, F.E.M., (1948). The control of humidity by saturated salt solutions. *J. Sci. Instrum.*, 25:73–76.

O'Leary, T.J., Ross, P.D., Lieber, M.R., & Levin, I.W., (1986). Effects of cyclosporin A on biomembranes. Vibrational spectroscopy, calorimetric and hemolysis studies. *Biophys. J.*, 49:795–801.

Pachence, J.M., Knott, R.B., Edelman, I.S., Schoenborn, B.P., & Wallace, B.A., (1983). Formation of oriented membrane multilayers of Na/K-ATPase. *Trans. NY Acad. Sci.*, 435:566–569.

Petcher, T.J., Weber, H.-P., & Rüegger, A., (1976). Crystal and molecular structure of an iodo-derivative of the cyclic undecapeptide cyclosporin A. *Helv. Chim. Acta,* 59:1480–1488.

Sadler, D.M., & Worcester, D.L., (1982). Neutron diffraction studies of oriented photosynthetic membranes. *J. Mol. Biol.*, 159:467–484.

Savitzky, A., & Golay, M.J.E., (1964). Smoothing and differentiation of data by simplified least squares procedures. *Anal. Chem.*, 36:1627–1639.

Saxena, A.M., & Majkrzak, C.F., (1984). Neutron optics with multilayer monochromators. In *Neutrons in Biology.* (B.P. Schoenborn, editor) pp143–157. Plenum Publishing Corporation, New York. ISBN 0–306–41508–9.

Saxena, A.M., & Schoenborn, B.P., (1976). Multilayer monochromators for neutron scattering. In *Neutron Scattering for the Analysis of Biological Structures* (B.P. Schoenborn, editor) ppVII30-VIII48. (Nat. Technical Infor. Serv.; US Dept. of Commerce, Springfield, VA).

Saxena, A.M., & Schoenborn, B.P., (1977a). Multilayer neutron monochromators. *Acta Cryst.*, 833:805–813.

Saxena, A.M., & Schoenborn, B.P., (1977b). Correction factors for neutron diffraction from lamellar structures. *Acta Cryst.*, A33:813–818.

Saxena, A.M., & Schoenborn, B.P., (1988). Multilayer monochromators for neutron spectrometers. *Material Science Forum,* 27/28:313–318.

Schoenborn, B.P., (1976). Neutron scattering for the analysis of membranes. *Biochim. Biophys. Acta,* 457:41–55.

Schoenborn, B.P., Saxena, A.M., Stamm, M., Dimmler, G., & Radeka, V., (1985). A neutron spectrometer with a two-dimensional detector for time resolved studies. *Aust. J. Phys.*, 38:337–351.

Singer, S.J., & Nicholson, G.L., (1972). The fluid mosaic model of the structure of cell membranes. *Science,* 175:720–731.

Tardieu, A., Luzzati, V., & Reman, F.C., (1973). Structure and polymorphism of the hydrocarbon chains of lipids: A study of lecithin-water phases. *J. Mol. Biol.*, 75:711–733.

Torbet, J., & Wilkins, M.H.F., (1976). X-ray diffraction studies of lecithin bilayers. *J. Theor. Biol.*, 62:447–458.

Weir, L.E., Cornell, B.A., & Knott, R.B., (1988). Neutron scattering of ubiquinone and gramicidin A in lipid bilayers. *Materials Science Forum,* 27&28:159–162.

White, D.J.H., (1982). *Cyclosporin A.* Proceedings of an International Conference on Cyclosporin A, Cambridge. Elsevier Biomedical Press, Amsterdam. ISBN 0–444–80410–2.

Wilkins, M.H.F., Blaurock, A.E., & Engelman, D.M., (1971). Bilayer structure in membranes. *Nature,* 230:72–76.

Worcester, D.L., (1976). In *Biological Membranes.* (D. Chapman and D.F.H. Wallach, editors) 3:1–46. Academic Press, New York.

Worcester, D.L., & Franks, N.P., (1976). Structural analysis of hydrated egg lecithin and cholesterol bilayers II. Neutron diffraction. *J. Mol. Biol.,* 100:359–378.

Yeager, M., Schoenborn, B.P., Engelman, D.M., Moore, P., & Stryer, L., (1980). Neutron diffraction analysis of the structure of rod photoreceptor membranes in intact retinas. *J. Mol. Biol.,* 137:315–348.

Zaccai, G., Blasie, J.K., & Schoenborn, B.P., (1975). Neutron diffraction studies on the location of water in lecithin bilayers. *Proc. Natl. Acad. Sci. USA,* 72:376–380.

Zaccai, G., & Gilmore, D., (1979). Areas of hydration in the purple membrane of *Halobacterium halobium*: A neutron diffraction study. *J. Mol. Biol.,* 132:181–191.

8

STRUCTURAL BIOLOGY FACILITIES AT BROOKHAVEN NATIONAL LABORATORY'S HIGH FLUX BEAM REACTOR

Z. R. Korszun, A. M. Saxena, and D. K. Schneider

Biology Department
Brookhaven National Laboratory
Upton, New York 11973

ABSTRACT

The techniques for determining the structure of biological molecules and larger biological assemblies depend on the extent of order in the particular system. At the High Flux Beam Reactor at the Brookhaven National Laboratory, the Biology Department operates three beam lines dedicated to biological structure studies. These beam lines span the resolution range from approximately 700Å to approximately 1.5Å and are designed to perform structural studies on a wide range of biological systems. Beam line H3A is dedicated to single crystal diffraction studies of macromolecules, while beam line H3B is designed to study diffraction from partially ordered systems such as biological membranes. Beam line H9B is located on the cold source and is designed for small angle scattering experiments on oligomeric biological systems.

INTRODUCTION

The High Flux Beam Reactor (HFBR) located at Brookhaven National Laboratory is a steady-state fission reactor currently operating at a power of 30 megawatts (MW) and is dedicated to basic research in biology, chemistry and physics. The Biology Department operates three beam lines and sponsors an external user program in structural biology. Beam line H3A is dedicated to macromolecular crystallography, while beam line H3B is dedicated to neutron scattering at intermediate angles from partially ordered systems and beam line H9B, which is located on the cold neutron source, is dedicated to small-angle scattering. The purpose of this article is to describe these biological facilities in terms of their capabilities and their technical specifications.

In Figure 1, we present the floor plan of the biological beam lines. The reactor core and moderator vessel are schematically represented on the left side of the figure. It is sur-

Neutrons in Biology, edited by Schoenborn and Knott
Plenum Press, New York, 1996

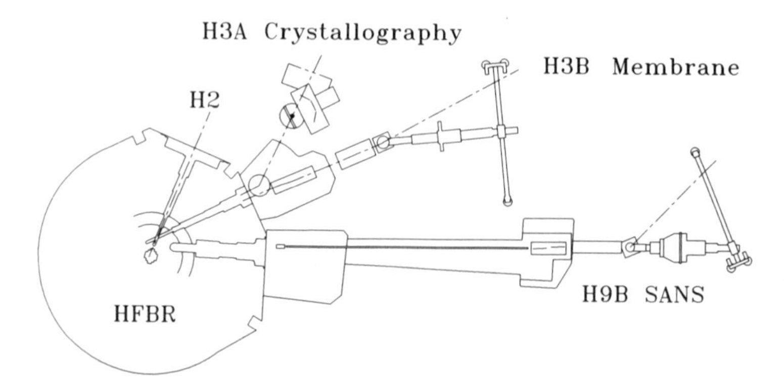

Figure 1. Floorplan showing the layout of the macromolecular crystallography, membrane, and small-angle neutron scattering beam lines for structural biology at the HFBR.

rounded by several meters of biological shielding composed of concrete, steel, and lead. The reactor vessel is filled with D_2O which serves as a moderator to thermalize the neutrons at a temperature of approximately 300K. Neutrons are extracted from the reactor vessel via aluminum thimbles that penetrate the vessel and the neutrons are transported via neutron guides to various optical elements and monochromatic neutrons are ultimately delivered to the spectrometers where diffraction and scattering experiments are performed on the experimental floor. From this figure, it can be seen that the H3 and H9 beams are extracted from thimbles mounted tangentially in the reactor vessel, while the beam on beam line H2 is extracted from a thimble which views the core radially. While the neutron intensity from a radially mounted thimble is higher than that obtained from a tangentially mounted thimble, the beam from a radial beam line is highly contaminated with gamma rays and epithermal (fast) neutrons which can damage the sample and significantly add to the background. In order to minimize this gamma and epithermal intensity the biology beam lines view the core tangentially.

The Cold Neutron Facility, which is located on the H9 beam lines, is also schematized in Figure 1. This facility utilizes liquid H_2 to produce cold neutrons in a small moderator vessel located at the tip of the thimble. The cold neutron spectrum is shifted to longer wavelength and peaks near 3.0Å. The calculated thermal spectrum and the cold spectrum from the HFBR are illustrated in Figure 2. The thermal flux spectrum has a Maxwellian velocity distribution that peaks at approximately 1.0Å, and as can be seen in the insert of Figure 2, the cold spectrum significantly enhances the long wavelength neutron intensity over the thermal tail. Given these source characteristics beam lines H3A and H3B are ideally suited for macromolecular crystallography and intermediate angle membrane scattering experiments, respectively, while beam line H9B is ideally suited for small-angle neutron scattering experiments. Each Structural Biology beam line will be described in more detail below.

In addition to the spectrometers, biochemical and mechanical support facilities are provided. The biochemical facilities include a new laboratory located at the HFBR, which contains an optical spectrophotometer, a centrifuge, and a complete set of benchtop appliances. Furthermore, complete biochemical laboratory facilities, including cold rooms, are located in the Biology Department. The machine shop, which is located on the experimen-

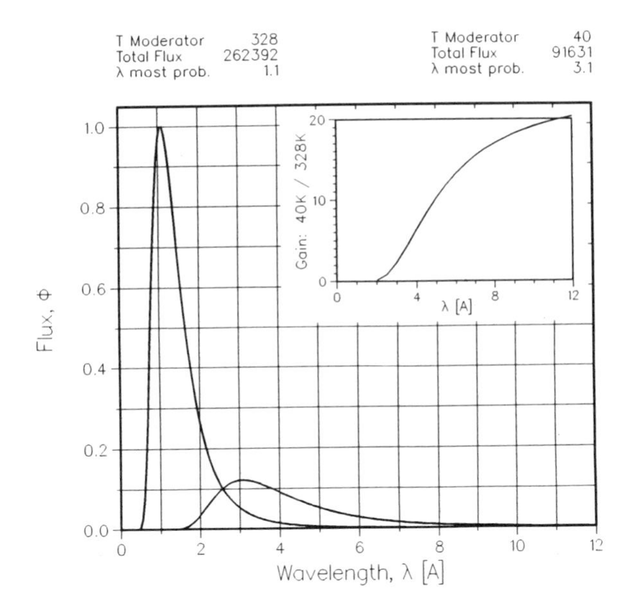

Figure 2. Calculated Maxwellian flux spectra versus wavelength at the HFBR. The cold neutron spectrum and its intensity enhancement over the corresponding thermal intensity in the region of 2 to 12Å are shown in the inset.

tal floor of the HFBR contains a mill, a lathe, drill press, and a band saw as well as a complete assortment of tools.

BEAM LINE H3A MACROMOLECULAR CRYSTALLOGRAPHY

The white neutron beam is monochromated by a single Zn crystal using the 002 reflection. The 1.61Å monochromatic beam, measuring approximately 6mm in diameter, is collimated through circular cadmium apertures that define the size of the beam. The intensity of the beam is approximately 1.0×10^6 neutrons/sec-cm^2 at a reactor power of 30MW. The specifications of the beam line are presented in Table I.

The H3A spectrometer is shown schematically in Figure 3. The footprint of the spectrometer is illustrated in the top half of the figure showing the monochromator tank, the collimation system, the 4-circle Eulerian cradle and the three detector assembly. The lower left panel shows a side view of the spectrometer and a back view of the detector en-

Table I. Characteristics of the H3A macromolecular crystallography station

Flux at sample	~1.0×10^6 n/cm^2-sec
Monochromator	(002) zinc
Wavelength λ	1.61Å
Beam Size at Sample (FWHM)	6mm diameter
	20′ divergence (variable)
Sample Scattering Angles	$-80° < 2\theta < 8°$
Detectors (3)	^{3}He filled 20cm × 18cm
	256 × 128 pixels
Detector-to-Sample Distance	600–1000mm
Counter Resolution	1.5mm
Efficiency	~80%

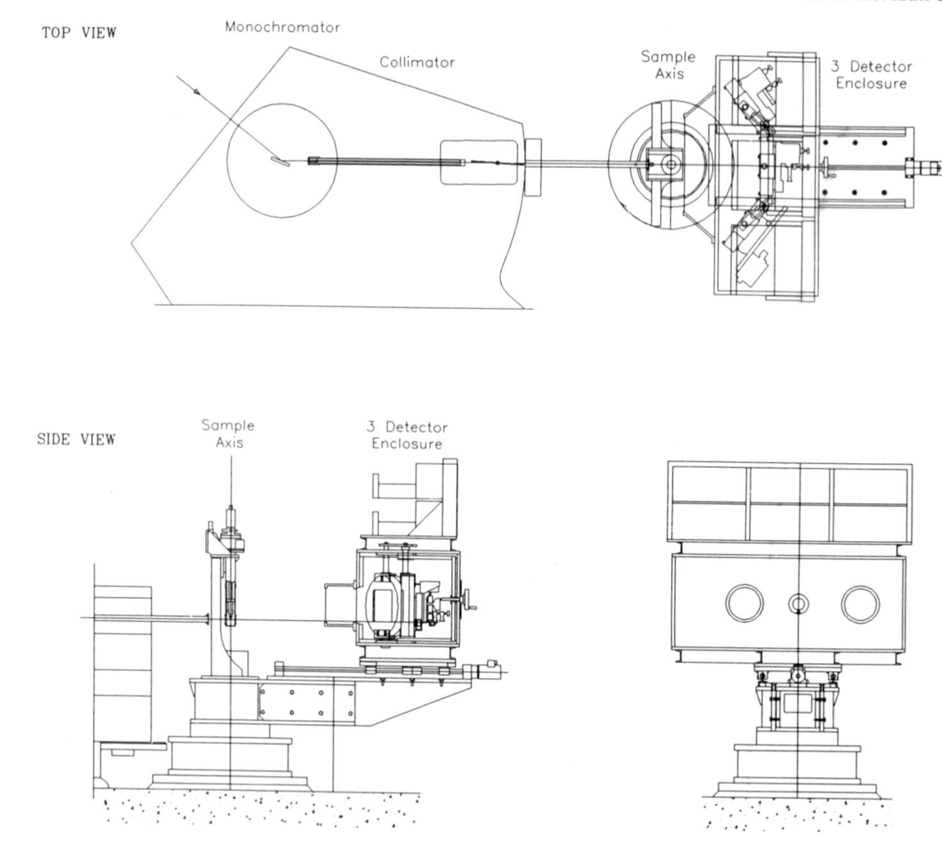

Figure 3. Schematic representation of beam line H3A.

closure is shown in the lower right panel. Crystals are mounted on a phi-axis stage for room temperature work, and a special phi-axis stage can be employed for cryogenic diffraction studies. The cold finger assembly consists of a 4 watt Displex refrigeration system, capable of prolonged running at controlled temperatures down to approximately 10K. The crystal is mounted on a goniometer head at the end of the cold finger. The bottom of the finger is made of a removable quartz tailpiece for easy mounting of the sample and the entire cold finger assembly is then evacuated to 10^{-6}torr.

The detector assembly consists of an aluminum housing with boron carbide shielding and houses three $20 \times 20\text{cm}^2$ area sensitive detectors. These detectors utilize ^{3}He as a neutron converter gas and operate at efficiencies approaching 80%. The three detectors are mounted in the enclosure on an adjustable trapezoidal mount which allows the detectors to maintain a confocal arrangement at all allowable crystal to detector distances. The detector readout electronics are mounted on the top of the enclosure. The entire enclosure is mounted on a slide and can be positioned anywhere from 600 to 1000mm from the sample. At the longest crystal to detector setting the blind spot between detectors is smaller than the active area of the detector so that by repositioning the detector enclosure at appropriate 2θ values all of reciprocal space can be collected using ω-scans at different χ settings.

The entire experiment is controlled by custom software running on a VAX 4200 computer using the VMS operating system. As data frames are accumulated, data are corrected and integrated offline using MADnes, the Munich Area Detector software. This configuration of data collection and data reduction has proven to work efficiently and allows data quality to be assessed very rapidly. Typically, hkl's, integrated intensities, and sigmas are obtained within a few hours of the completion of a scan.

BEAM LINE H3B MEMBRANE SCATTERING

The layout of the H3B beam line is shown in Figure 4 and its specifications are tabulated in Table II. Two apertures, positioned adjacent to the main shutter, define the cross-sectional area of the beam to be 1cm in the horizontal direction and 2.5cm in the vertical direction. The neutron beam is monochromated by reflections from two parallel thin-film multilayer monochromators with approximate d-spacings of 60Å. The first multilayer can be rotated around a vertical axis, while the second multilayer can also be translated along the primary beam. After reflections from the multilayers, the neutron beam emerges in a direction parallel to the original direction, but displaced by 4cm from it. A low-efficiency fission counter serves as a beam monitor; the data sets are often scaled with respect to the monitor counts.

The neutron beam is highly collimated for a typical experiment on lamellar samples. Characteristic beam dimensions are 0.1cm (H) × 1.5cm (V) and the beam divergence is less than 0.05°. This collimation is achieved by mounting slits on a high precision optical bench. The precision optical rail on which the slits are mounted is supported by two x-y translation stages that control the positions of the ends of the beam. The motions are automated so that a PC moves the motors and records the positions of the encoders.

A number of sample chambers are available that maintain the sample at a specified temperature and humidity. A recirculating bath that forces coolant through lines embedded in the walls of the chamber maintains the temperature of the sample, and the humidity is controlled by placing buckets containing appropriate saturated salt solutions in the chamber. Intensities of Bragg reflections are measured with a 20cm × 20cm area detector. The spectrometer and the data collection process are controlled by a VAX 4000-400 computer.

BEAM LINE H9B SMALL-ANGLE NEUTRON SCATTERING

The small-angle scattering spectrometer is shown in Figure 5 and the beam line specifications are presented in Table III. The spectrometer exploits the cold neutron spec-

Table II. Characteristics of the H3B membrane scattering station

Wavelength	2.0Å to 4.0Å
Wavelength Bandwidth ($\Delta\lambda/\lambda$)	0.03 to 0.08
Beam Dimensions	1.0cm(H) × 2.0cm(V)
Position Sensitive Detector:	
Detector area	18cm × 18cm
Data array	256(H) × 256(V) pixels
Resolution	1.3mm(H), 2.5mm(V)
Sample-to-detector Distance	50cm to 250cm
Angular Scanning Range	−5° to 40°
Q (Scattering Vector)	0.01Å^{-1} to 3.0Å^{-1}

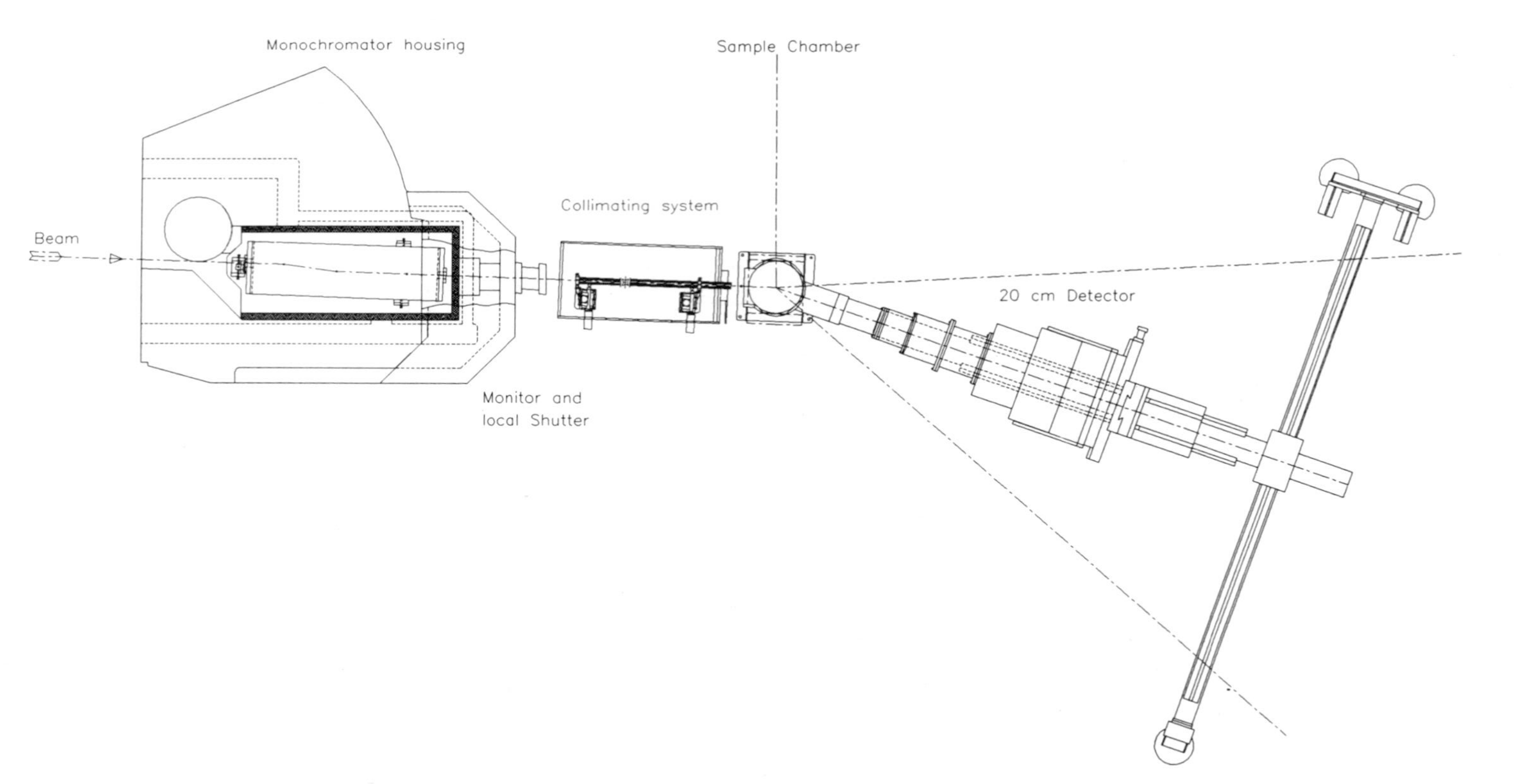

Figure 4. Schematic representation of beam line H3B.

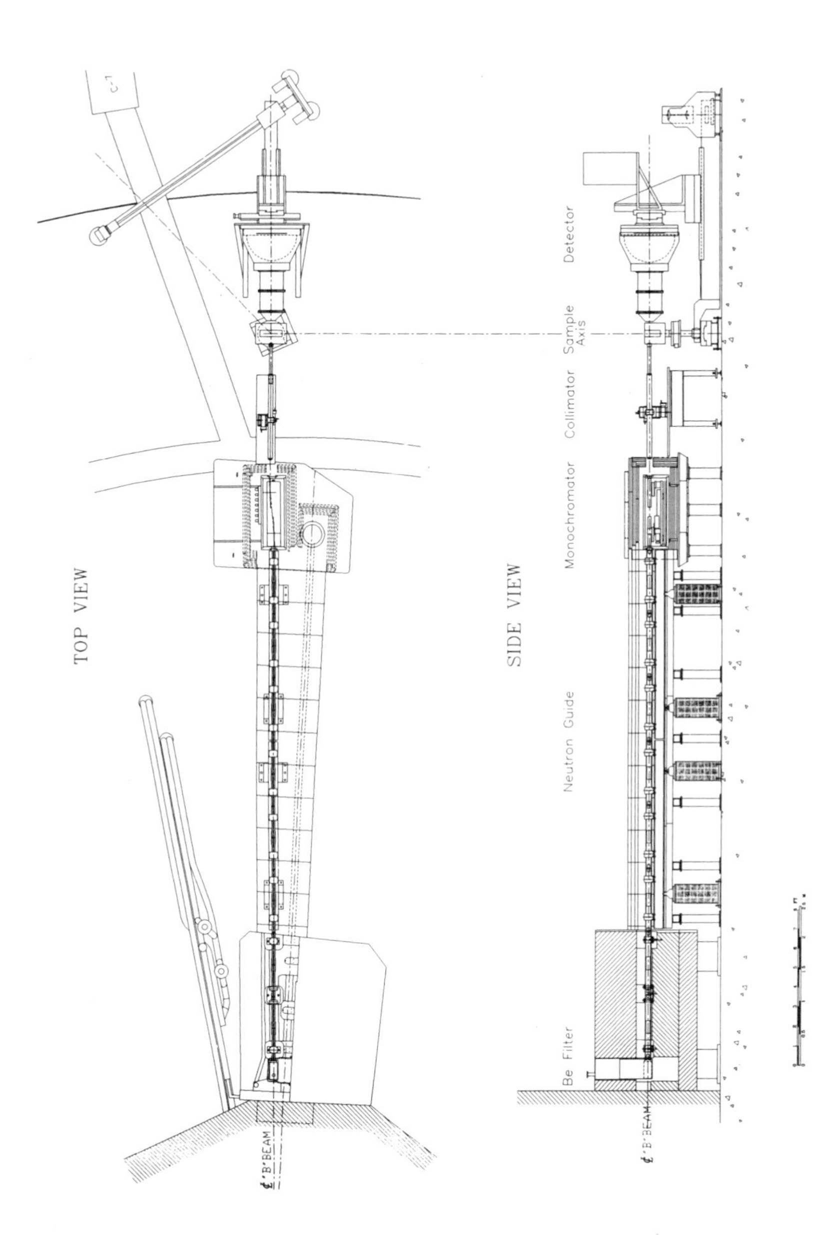

Figure 5. Schematic representation of beam line H9B.

Table III. Characteristics of the H9B small-angle neutron scattering facility

Wavelength	4.0Å to 8.0Å
Wavelength Spread (FWHM)	6-15%, 10% typical
Beam Divergence (FWHM horizontal)	14′
Flux at Sample (5Å, 6mm beam)	2×10^6n/cm^2-sec
Detector Area	50cm × 50cm
Detector Sampling	128 × 128 pixels
Sample-to-Detector Distance	50cm to 200cm
Angular Scanning Range	0–40°
Best Spectrometer Resolution	700Å
Minimum Q (5Å at 2m)	0.008Å^{-1}

trum from the H9 hydrogen-filled cold source through a series of straight neutron guides, that were recently extended to a length of 8 meters. Fast neutrons are filtered out by means of a beryllium block, which imposes a short wavelength cutoff at 4Å.

The monochromatization up to wavelengths of 8Å is achieved by means of a single Ni-Ti multilayer that deflects the beam by about 4°. Wavelengths are tuned either by rotation of the multilayer, or more simply at fixed angles, by selecting a multilayer of suitable spacing in the range of 60–120Å. The monochromator consists of two linked rotary stages and can also be operated as a double multilayer device.

The beam is collimated by a series of circular apertures of decreasing diameter down to the design beam diameter of 6mm at the sample. The wavelength spread is then about 10% at half height of an approximately triangular distribution. The flux at the sample typically exceeds 10^6 n/s-cm^2 at the current reactor power of 30MW.

A recently developed linear sample changer provides the repetitive handling capabilities required for solution scattering experiments. It shuffles up to 30 standard spectroscopy cuvettes at controlled temperatures between -10 and 40°C. With an insert, temperatures to 100°C can be reached. Alternatively, the sample axis may be equipped with the various auxiliary devices available at the HFBR, including cryostats and magnets.

Scattering is recorded with a locally built multiwire detector of 50 × 50 cm^2. When positioned closest to the sample, it can intercept scattering angles to 30°. At the other extreme, at maximal detector distance and wavelength, it can resolve first order diffraction peaks of 750Å. The detector can also be aimed at the sample from anywhere along a horizontal angular track of 45°.

A microVAX-3500 computer operating under VMS handles the spectrometer and controls a dedicated data acquisition system. A live image of the emerging scattering pattern is on display. Graphics terminals, X-terminals and a workstation are available for data reduction and analysis. BNLH9B is a fully networked node, and thus well connected for the import of programs or the export of data.

FUTURE DEVELOPMENTS

Future facility plans include replacement of the current Eulerian cradle on H3A by a diffractometer with the capability of rapid change from Eulerian to Kappa geometry. On H3B, the linear 2θ arm will be replaced by a radial arm which will make measurements at larger scattering angles possible. On beam line H9B, the sample to detector distance will be doubled, allowing scattering measurements to be made at a resolution approaching 1400Å.

ACKNOWLEDGMENTS

This work is supported by the Office of Health and Environmental Research of the United States Department of Energy, by an NSF neutron user grant MCB-9318839, and by an NSF multilayer development grant DIR-9115897 (A.M.S.).

9

NEUTRON INSTRUMENTATION FOR BIOLOGY

S. A. Mason

Institut Laue-Langevin
BP 156, 38042 Grenoble Cedex, France

ABSTRACT

In the October 1994 round of proposals at the ILL, the external biology review subcommittee was asked to allocate neutron beam time to a wide range of experiments, on almost half the total number of scheduled neutron instruments: on 3 diffractometers, on 3 small angle scattering instruments, and on some 6 inelastic scattering spectrometers. In the 3.5 years since the temporary reactor shutdown, the ILL's management structure has been optimised, budgets and staff have been trimmed, the ILL reactor has been re-built, and many of the instruments up-graded, many powerful (mainly Unix) workstations have been introduced, and the neighbouring European Synchrotron Radiation Facility has established itself as the leading synchrotron radiation source and has started its official user program. The ILL reactor remains the world's most intense dedicated neutron source. In this challenging context, it is of interest to review briefly the park of ILL instruments used to study the structure and energetics of small and large biological systems. A brief summary will be made of each class of experiments actually proposed in the latest ILL proposal round.

NEUTRON SCATTERING AND DIFFRACTION INSTRUMENT FOR STRUCTURAL STUDY ON BIOLOGY IN JAPAN

Nobuo Niimura

Advanced Science Research Center
Japan Atomic Energy Research Institute
Tokai-mura Ibaraki-ken 319–11, Japan

ABSTRACT

Neutron scattering and diffraction instruments in Japan which can be used for structural studies in biology are briefly introduced. Main specifications and general layouts of the intruments are shown.

NEUTRON SOURCE FACILITIES

In Japan we have three steady state reactor neutron source facilities, JRR-3M, JRR-2, and KUR and two pulsed neutron source facilities, Tohoku-LNS and KENS. The neutron scattering and diffraction instruments for structural study on biology are installed only in JRR-3M and KENS and are reported in this paper. The JRR-3M reactor was built in place of the old JRR-3 reactor. The JRR-3 was shutdown in 1983 to be replaced with an upgraded reactor JRR-3M. The construction of the new reactor was started in 1985. The whole reactor body of the old JRR-3 including the biological shielding weighing 2150 tons in total, was cut in one block out of the place where it had been and was carried out of the reactor building. It is now kept in a basement under the new guide hall.

Table I and Figure 1 show the main feature of neutron source and the whole layout of JRR-3M, respectively.

KENS has been operated since 1980 and upgraded in 1988. Table II and Figure 2 show the main feature of neutron source and the whole layout of KENS, respectively.

INSTRUMENTS AT JRR-3M

There are two instruments. Since the details of the instruments are presented at the symposium, reference is made to relevant papers.

Neutrons in Biology, edited by Schoenborn and Knott
Plenum Press, New York, 1996

Table I. The main features of neutron source of JRR-3M

Thermal power	20MW
Core size	60cm diameter × 75 cm height
Flux	2×10^{14} n/cm^2/sec
Beam tubes	7 horizontal tubes in the reactor hall
	2 horizontal tubes for neutron guides
Cold source	Liquid hydrogen
Neutron guides	Thermal guides ($\lambda^* = 2$Å) 2×20cm^2 ×2
Cold guides	($\lambda^* = 4$Å) 2×20cm^2 ×2
	($\lambda^* = 6$Å) 2×20cm^2 ×1

SANS-U

SANS-U is a small angle neutron scattering instrument, which was reported in the symposium by Dr. Imai (Ito *et al.*, 1995). The general features of the camera are similar to D11 at the ILL. Approximately one third of the beam time is allocated for studies in biology.

BIX

BIX is a neutron diffractometer for bio-crystallography, which was reported in the symposium (Niimura, 1995).

INSTRUMENTS AT KENS

SAN

SAN is a time-of-flight (TOF) type small angle neutron scattering instrument (Ishikawa *et al.*, 1986). It utilizes a wide band of incident neutron wavelength to cover a very wide range of Q by choosing appropriate scattering path length. The layout and main feature of SAN are shown in Figure 3 and Table III, respectively. The distinctive features

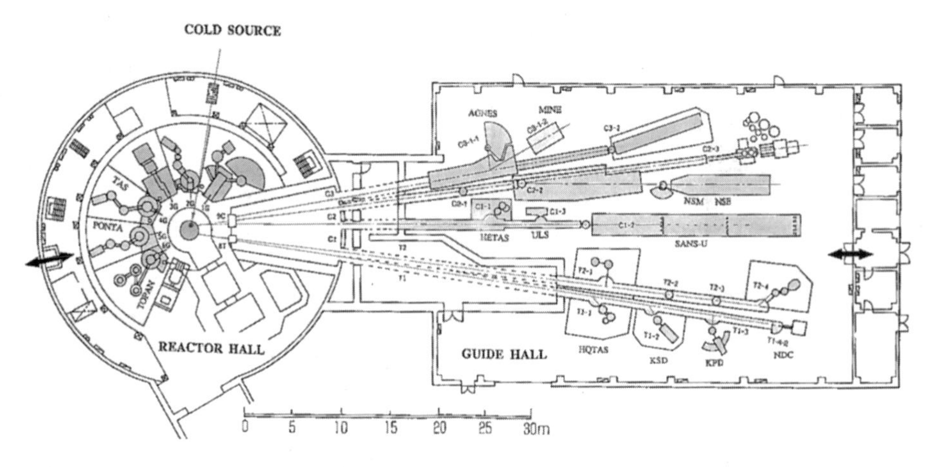

Figure 1. The whole layout of JRR-3M. (A color version of this figure appears in the color insert following p. 214.)

Table II. The main features of neutron source of KENS

Proton beam	
Energy (MeV)	500
Current (mA)	5
Pulse width (ns)	50
Repetition (pulses/sec)	20
No. of protons (protons/pulse)	1.5×10^{12}
Neutron beam	
Target material	Depleted uranium
Neutron yield (n/sec)	4.8×10^{14}
Cold moderator	Solid methane(20K)
Thermal moderator	Water

are as follows: (i) it is installed at the end of the curved guide tube even though being a time-of-flight type small angle neutron scattering instrument; (ii) a converging Soller-slit is used which enables the use of larger samples; and (iii) the area detector consists of a stack of 43 1D-PSDs.

WINK

WINK is a special instrument which covers very wide momentum transfer range ($0.01 < Q < 20Å^{-1}$) by combining a small angle instrument, a medium resolution powder

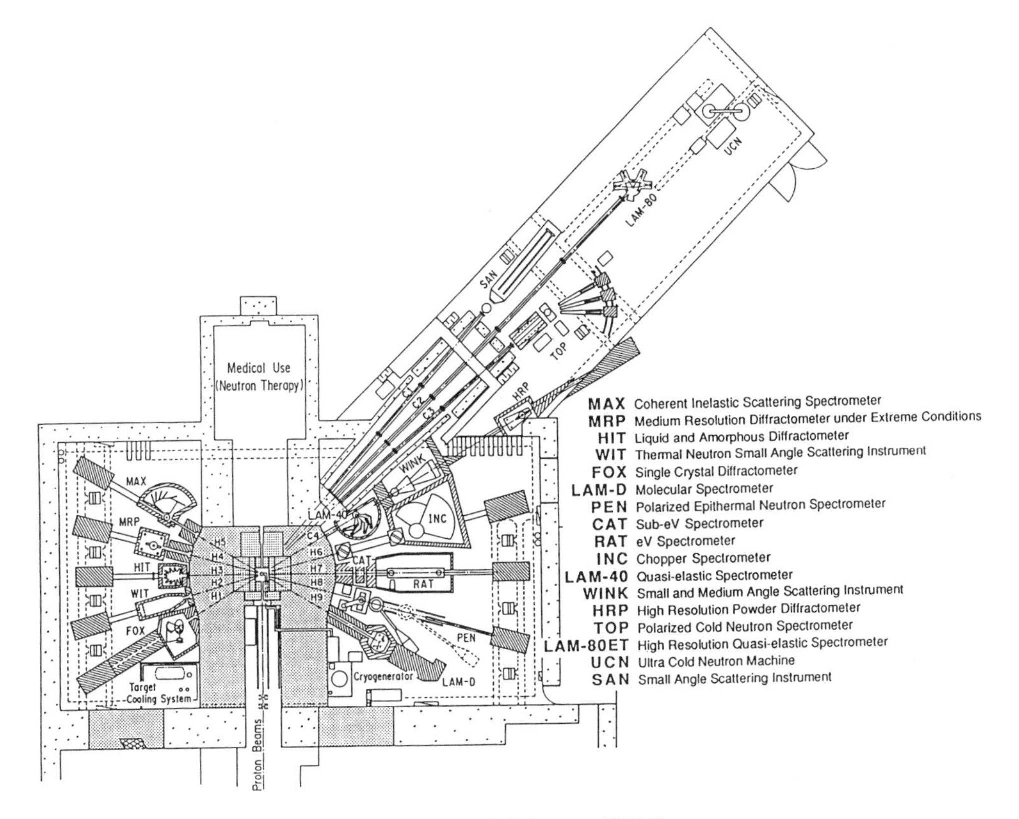

Figure 2. The whole layout of KENS.

Table III. The main features of instruments at KENS

SAN		
	Wavelength range	$3 < \lambda < 11$Å
	Q-range	$0.004 < Q < 0.6$Å^{-1}
	Flux at sample position	4×10^4 n/cm^2/sec
WIT		
	Wavelength range	$0.5<\lambda<4$Å
	Q-range	$0.01 < Q < 1.2$Å^{-1}
	Flux at sample position	1×10^4 n/cm^2/sec
WINK		
	Wavelength range	$0.5 < \lambda < 16$Å
	Q-range	$0.01<Q<12$Å^{-1}
	Flux at sample position	1×10^5 n/cm^2/sec
LAM-40		
	Q-range	$0.15<Q<2.6$Å^{-1}
	Energy resolution	200meV
LAM-80		
	Q-range	$0.1<Q<2.6$Å^{-1}
	Energy resolution	1.2-39meV
LAM-D		
	Maximum Energy	300meV
	Energy resolution	400meV

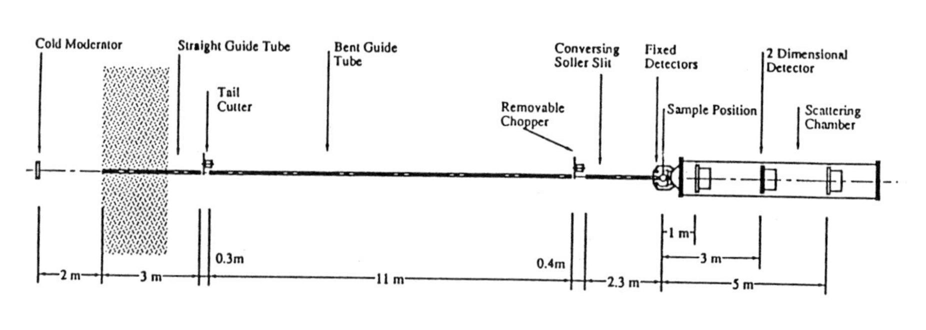

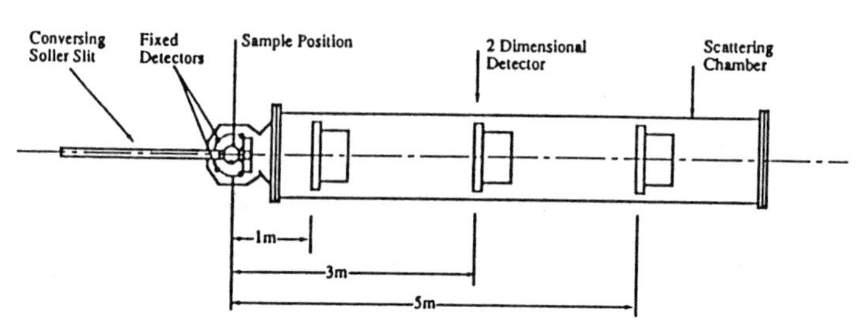

Figure 3. The layout of SAN.

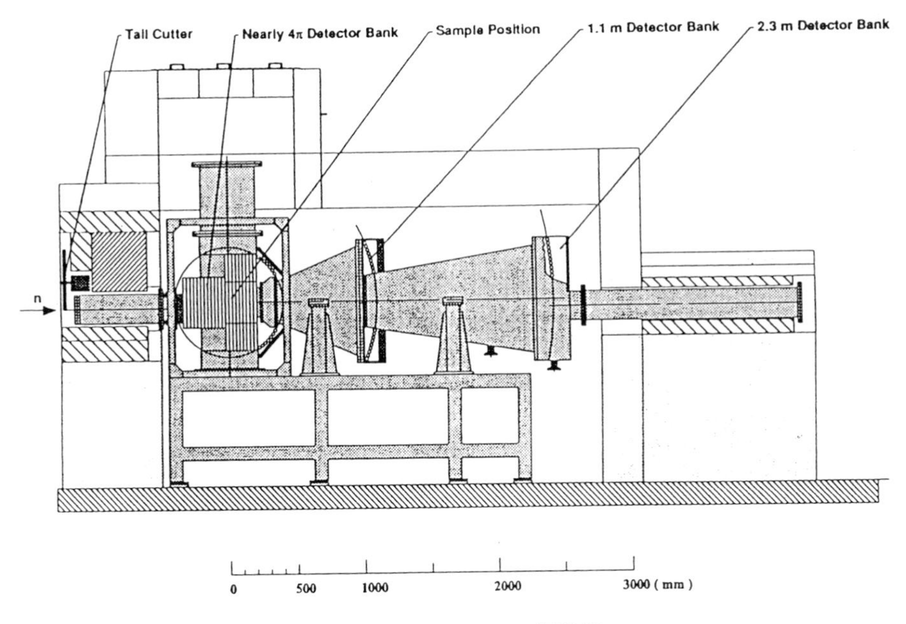

Figure 4. The side view of WINK.

diffractometer and a total scattering instrument (Furusaka *et al.*, 1993). The above Q range is covered by one measurement. Side view and main feature of WINK are shown in Figure 4 and Table III. Because of the detector system with wide coverage of solid angle and relatively relaxed resolution for the collimation system, counting efficiency of this instrument is very high.

WIT

WIT is a thermal neutron small angle scattering spectrometer (Niimura *et al.*, 1989). Annular detectors of glass scintillator are installed. A special beam stop, which provides both scattering and transmission measurements of a sample simultaneously, has been designed and installed. WIT is a complete user oriented machine which has no movable parts. The layout and main feature of WIT are shown in Figure 5 and Table III.

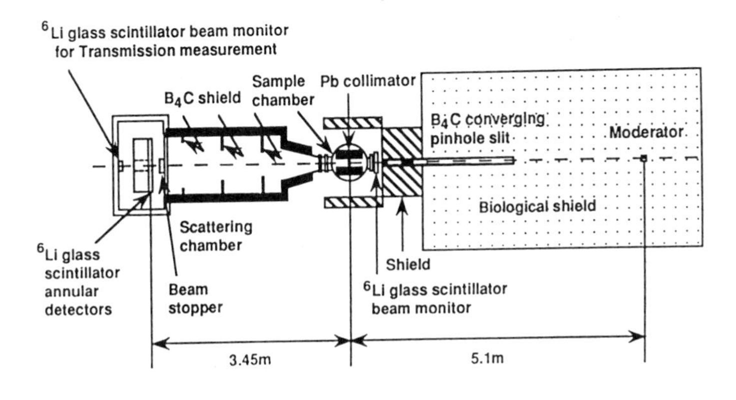

Figure 5. The layout of WIT.

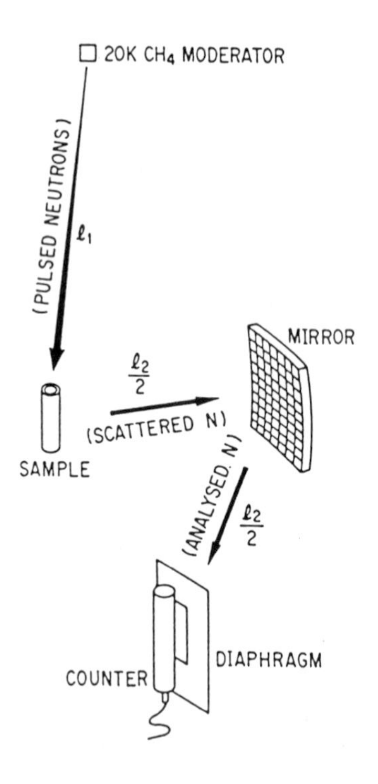

Figure 6. The adopted arrangement of the LAM spectrometer series.

LAM-80, LAM-40 and LAM-D

The LAM spectrometer series were constructed to measure the quasi- and low energy in-elastic scattering process in condensed materials (Inoue *et al.*, 1991). In order to carry out quasi elastic neutron spectroscopy using the pulsed cold source, it is most desirable that the neutron pulse emitted from the cold source should be utilized directly for energy analysis using the time-of-flight technique, without any auxiliary device for narrowing the pulse width. As to energy analyzing devices, a wide acceptance angle crystal analyzer mirror is the most suitable. The adopted arrangement of devices is shown in Figure 6.

In the case of time-of-flight measurements, the high resolution required is attained by selecting appropriately the length of a neutron flight path to the width of the neutron pulse, the Bragg angle and the material of the analyzer mirror.

LAM-80 is installed at the C-2 neutron guide hole, of which the distance from the cold source is about 26m. There are eight identical mirrors, each containing 400 mica crystal pieces of 12mm × 12mm × 3mm in size. Each analyzer mirror is mounted at four fixed angles: 15°, 51.7°, 78°, and 118°.

The shape of the sample is cylindrical (14mm in diameter and 80mm in height), to ensure identical geometrical conditions for every analyzer mirror.

The general principle of LAM-40 and LAM-D is the same as LAM-80.

With respect to LAM-40, the Bragg angle of the analyzer crystal is 40° instead of 80°. With respect to LAM-D, thermal neutrons are used instead of cold neutrons. The main features of the LAM spectrometer series are shown in Table III.

USER ACCESS

Users wishing to gain access to the instruments (SAN, WINK, WIT, LAM spectrometer series) at KENS, SANS-U and BIX at JRR-3M, should contact 'National Laboratory for High Energy Physics, Tsukuba, Ibaraki-ken 305, Japan', 'Neutron Scattering Facilities, Institute for Solid State Physics, Tokyo University, Tokai-mura, Naka-gun, Ibaraki-ken, 319–11, Japan' and 'Advanced Science Research Center, Japan Atomic Energy Research Institute, Tokai-mura, Naka-gun, Ibaraki-ken, 319–11, Japan'.

REFERENCES

Furusaka, M., Suzuya, K., Watanabe, N., Osawa, M., Fujikawa, I., & Satoh, S., (1993). WINK (small/medium angle diffractometer). KENS REPORT 9:25–27.

Inoue, K., Kanaya, T., Kiyanagi, Y., Ikeda, S., Shibata, K., Iwasa, H., Kamiyama, T., Watanabe, N., & Izumi, Y., (1991). A high-resolution neutron spectrometer using mica analyzers and the pulsed cold source. *Nucl. Instr. Methods*, A309:294–302.

Ishikawa, Y., Furusaka, M., Niimura, N., Arai, M., & Hasegawa, K., (1986). The time-of-flight small-angle scattering spectrometer SAN at the KENS pulsed cold neutron source. *J. Appl. Cryst.*, 19:229–238.

Ito, Y., Imai, M., & Takahashi, S., (1995). Small-angle neutron scattering instrument of Institute for Solid State Physics, the University of Tokyo (SANS-U) and its application to Biology. In *Neutron in Biology* (B.P. Schoenborn and R.B. Knott, editors), Plenum Press, New York.

Niimura, N., (1995). Neutron diffractometer for bio-crystallography (BIX) with an imaging plate neutron detector. In *Neutron in Biology* (B.P. Schoenborn and R.B. Knott, editors), Plenum Press, New York.

Niimura, N., Hirai, M., Ishida, A., Aizawa, K., Yamada, K., & Ueno, M., (1989). Thermal neutron small angle scattering spectrometer (WIT) using a 2D converging slit and annular glass scintillator detectors at KENS. *Physica B* 156 & 157:611–614.

SMALL-ANGLE NEUTRON SCATTERING INSTRUMENT OF INSTITUTE FOR SOLID STATE PHYSICS, THE UNIVERSITY OF TOKYO (SANS-U) AND ITS APPLICATION TO BIOLOGY

Yuji Ito,* Masayuki Imai,† and Shiro Takahashi

Neutron Scattering Laboratory
Institute for Solid State Physics
University of Tokyo
Tokai Naka Ibaraki 319–11, Japan.

ABSTRACT

A small-angle neutron spectrometer (SANS-U) suitable for the study of mesoscopic structure in the field of polymer chemistry and biology, has been constructed at the guide hall of JRR-3M reactor at the Japan Atomic Energy Research Institute. The instrument is 32m long and utilizes a mechanical velocity selector and pinhole collimation to provide a continuous beam with variable wavelength in the range from 5 to 10Å. The neutron detector is a $65 \times 65\text{cm}^2$ 2D position sensitive proportional counter. The practical Q range of SANS-U is 0.0008 to 0.45Å^{-1}. The design, characteristics and performance of SANS-U are described with some biological studies using SANS-U.

INTRODUCTION

In the field of biology, mesoscopic structures having the sizes of 10 to 1000Å play important roles for their functions and properties. The small-angle neutron scattering can provide unique information on the mesoscopic structures by varying the scattering contrast in biological substances. To facilitate such investigations, a small-angle neutron scattering spectrometer (SANS-U) has been constructed at the guide hall of 20MW JRR-3M reactor of Japan Atomic Energy Research Institute (Tokai). The instrument is composed of a mechanical velocity selector, collimator flight tube, sample chamber, two-dimensional

* Present address: Faculty of Liberal Arts and Education, Yamanashi University, Takeda Kofu 400, Japan.
† To whom correspondence should be addressed.

position sensitive detector and data acquisition system. In this article, the characteristics of the SANS-U and some examples of biological studies using the SANS-U are described.

SANS-U DESIGN

The SANS-U spectrometer is located at the cold neutron beam port C1-2 in the guide hall of 20MW JRR-3M reactor where a cold neutron flux is estimated to be 3×10^8 neutrons cm^{-2} s^{-1} at 20MW reactor power. The layout of the SANS-U is shown in both top and side views in Figure 1. The SANS-U is 34m long and is composed of velocity selector, collimator flight tube, sample stage, neutron detector and data acquisition system.

The neutron beam from a cold source is mechanically monochromated by the velocity selector having helical slits (Dornier) (Friedrich *et al.*, 1989). The mean wavelength of the monochromatic beam is determined by the rotation frequency of the selector in the

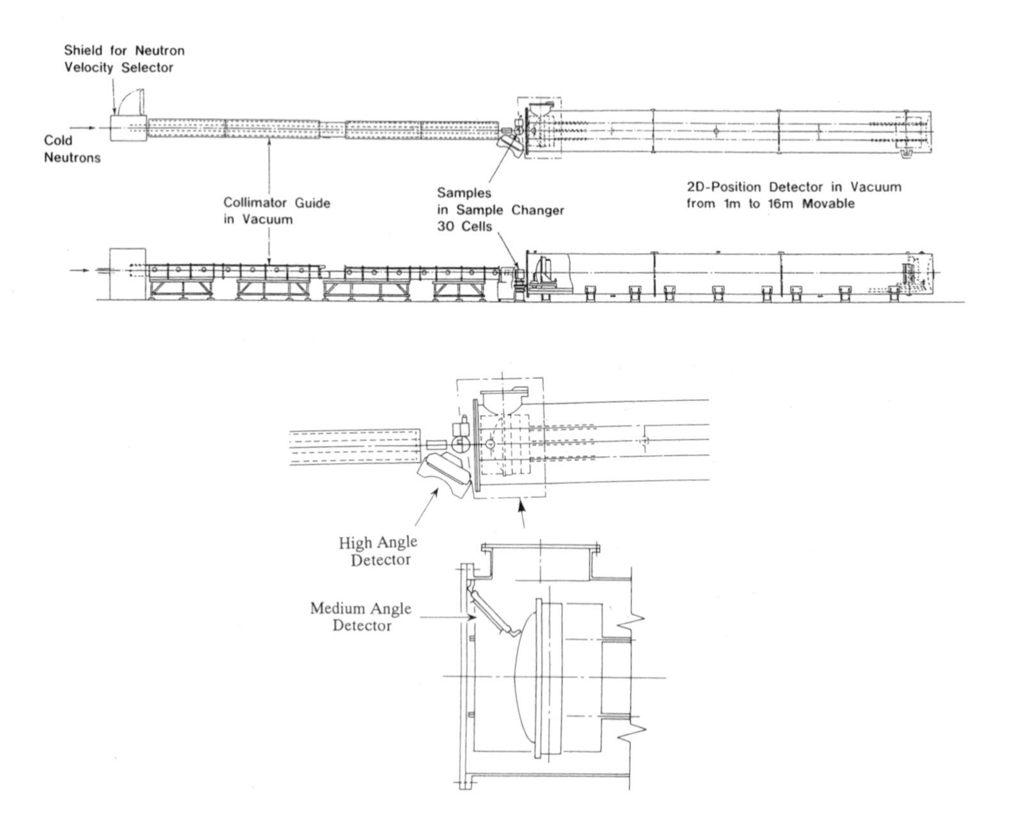

Figure 1. (a) Plan and elevation views of the SANS-U spectrometer. (b) Medium and high angle detectors are installed on the SANS-U.

range of 5 to 10Å (maximum rotating speed is 28,300 rpm) and the wavelength distribution of the beam, $\Delta\lambda/\lambda$, is 10%. Immediately following the velocity selector, a low-efficiency ^{3}He counter monitors the intensity of the monochromatic beam.

The monochromatic beam is collimated by the evacuated collimator flight tube with maximum length of 16m. In the collimator tube, six gates are equipped to define the beam collimation. For each gate there are a circular pinhole (20mm ϕ) and a rectangular opening (20 × 50mm). We can use suitable collimation length by selecting pinhole masks (conventional pinhole collimation).

At the sample position, a wide variety of sample supports are available. A standard sample chamber is available for the biological and polymer research in the temperature range from -20 to 80°C. This sample chamber is computer controlled to provide automatic change of samples in sample stands. One sample stand has five positions for quartz cells and a total of six stands (maximum 30 cells) can be mounted on the sample changer at once. For measurements at more extreme temperatures, the sample chamber can be replaced by either a cryostat or a furnace, both of which can be controlled by computer within an accuracy of 0.01°C (Noda *et al.*, 1983). Using these accessories we can measure SANS within a temperature range from -268 to 500°C.

Scattered neutrons are led to the evacuated scattering flight tube. In this flight tube a two-dimensional position-sensitive detector (2D-PSD: Ordela) is stationed (Borkowski & Kopp, 1978). The 2D-PSD has an active area of 65 × 65cm^2 with 5 × 5mm^2 resolution. The counting gas is ^{3}He-CF_4 at 270kPa. The detector can move along rails to vary the sample-to-detector distance continuously from 1 to 16m. The practical Q range of SANS-U, using the incident wavelength from 5 to 10Å, is 0.0008 to 0.45Å^{-1}. For transmission measurements, the ^{3}He counter is inserted in front of the 2D-PSD.

Data from the 2D-PSD are stored in histogramming memory modules under the control of a micro-VAX 3200 series computer. The real-time image of the data is displayed on an X-terminal in the form of a 2D contour, 3D bird view or 1D cross-section map. The SANS-U user can perform on-line data reduction and analysis to obtain 2D scattering functions or scattering profiles in absolute intensity scale using the micro-VAX or a Macintosh computer, which are linked to other computers via ethernet connection.

An additional feature of the SANS-U is a design to obtain wide Q range scattering profiles. The SANS-U has been constructed with a medium-angle detector array (1D-PSD) inside the flight tube, and a wide-angle detector array (1D-PSD) covering the Q range from 0.0008 to 2.5Å^{-1} as shown in Figure 1b. This detector system makes possible simultaneous measurement of wide, medium and small-angle neutron scattering.

The characteristics of the SANS-U are summarized in Table I.

Table I. Characteristics of SANS-U

Source	JRR3-M cold neutron guide (C1-2)
Monochromator	Mechanical Velocity Selector (Dornier)
Wavelength Range	5 to 10Å
Wavelength Resolution	10% $\Delta\lambda/\lambda$
Collimator Length	1 to 16m
Sample-to-detector Distance	1 to 16m continuously variable
Detector	65×65cm^2 ^{3}He 2D-PSD (0.5×0.5cm^2 spatial resolution)
Q Range	0.0008 to 0.45Å^{-1}
Sample Equipment	Automatic multi-specimen sample changer with temperature range −10 to 80°C
	Cryostats (10 to 300K)
	Furnace (300 to 800K)

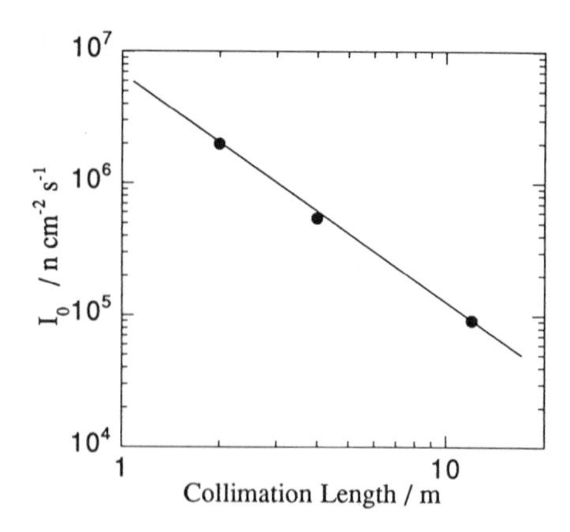

Figure 2. The neutron flux at the sample position as a function of collimator length.

INSTRUMENT CALIBRATION AND PERFORMANCE

Neutron Flux at Sample Position

The neutron fluxes at the sample position were conveniently estimated using incoherent scattering from calibrated polyethylene (Lupolen). For an incident beam with a mean wavelength of 6Å, the flux was measured as a function of the collimator length. The obtained flux was shown in Figure 2 is comparable to similar SANS instruments on reactors having 20MW power level (Table II).

Area Detector

The linearity of 2D-PSD was measured by illuminating a horizontal and vertical row of 3mm diameter pinholes, spaced 50mm apart, in a cadmium sheet placed directly over the pressure dome of the detector. The observed pinhole positions (in pixels, relative to the center of the detector) are plotted against the actual pinhole positions in Figure 3. The detector shows fairly good linearity in the observed area.

The uniformity of the efficiency of the 2D-PSD has been checked by measuring the scattering from the lupolen. The incoherent scattering per unit solid angle is constant at

Table II. Comparison of neutron current at the sample position for selected SANS instruments located on reactors with 20MW power level

Instrument	Qmin (Å^{-1})	$\Delta\lambda/\lambda$ (%)	λ (Å)	Neutron current (neutrons s^{-1})	Reference
SANS-U	0.01	10	6	4.7×10^5	
NIST 8m SANS	0.01	25	5.5	2.8×10^5	(a)
NIST 30m SANS	0.01	25	—	1.0×10^6	(b)
KWS I (Jülich)	0.01	25	5.5	2.5×10^6	(c)

References: (a) Glinka *et al.*, 1986. (b) 30-meter NIST/EXXON/U. of MINN. SANS Users Manual. (c) Schwahn *et al.*, 1990.

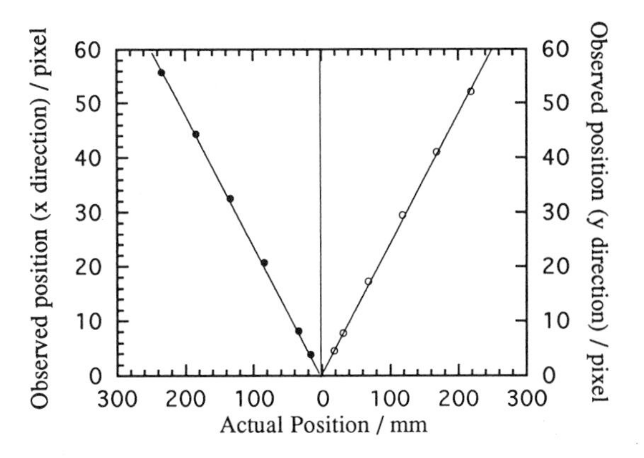

Figure 3. Measured spatial linearity of the SANS-U detector.

small angle and thus is a direct measure of the relative efficiency of the 2D-PSD. A typical measurement of the scattering from lupolen, corrected for the background scattering and planar geometry is shown in Figure 4 expressed in a circularly averaged fashion. From Figure 4 it can be seen that the detector efficiency is uniform over the most of the detector area.

SANS-U performance

In order to check the performance of SANS-U, we measured SANS from deuterium labelled poly(vinyl alcohol). The absolute scattered intensity I(Q) from a homogeneous mixture of deuterated and hydrogenated polymer chain having weight average degree of polymerization n is given by:

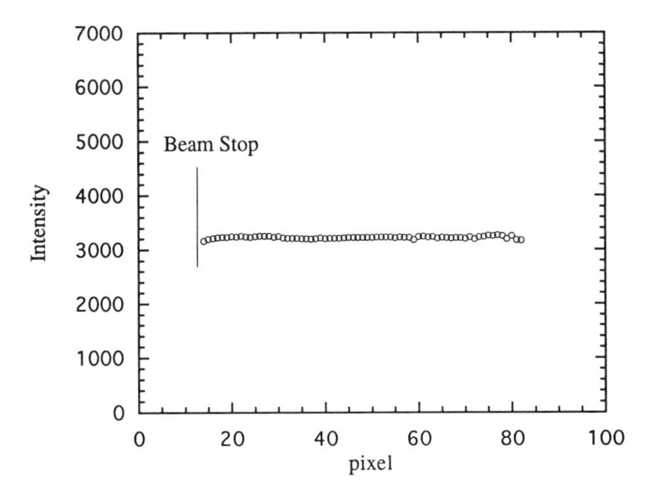

Figure 4. The uniformity of the SANS-U detector determined by the measurement of the scattering from standard polyethylene sample.

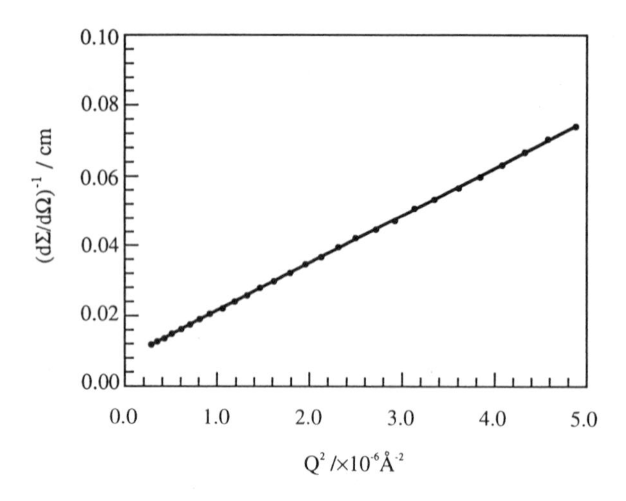

Figure 5. Plot of $I(Q)^{-1}$ versus Q^2 for a PVA mixture film.

$$I(Q) = K\phi_d\phi_h nP(Q) \tag{1}$$

where ϕ_i is the volume fraction of the i th component and P(Q) is the single chain structure factor, K is the contrast factor given by:

$$K = N[(a_d/v_d) - (a_h/v_h)]^2 v_d \tag{2}$$

where N is Avogadro's number and a_i and v_i are the scattering length and monomer volume of the i th component. With the random phase approximation (RPA) theory and assuming that the Flory's interaction parameter is zero, Equation 1 approximates to:

$$I(Q)^{-1} = \frac{K}{\phi_d\phi_h n}\left(1 + \frac{R_g^2}{3}Q^2\right) \tag{3}$$

where R_g is the radius of gyration of the chains. Using this relationship, one can obtain the weight average molecular weight M_w and R_g from the SANS profile. As the standard sample, we employed a 50/50 mixture of deuterated and hydrogenated poly(vinyl alcohol) (d-PVA and h-PVA) film supplied by Dr. Shibayama at Kyoto Institute for Technology (Shibayama *et al.*, 1990). The weight average molecular weight of d-PVA and h-PVA obtained by viscometry and GPC measurements are 3130 and 2875 respectively. The SANS measurement using SANS-U was performed with a sample to detector distance of 4m and a neutron wavelength of 7Å. The SANS data were corrected for transmission, incoherent scattering and sensitivity of detector and then normalized to the absolute scale using lupolen as the calibration standard. Figure 5 shows the plot of $I(Q)^{-1}$ versus Q^2 (Zimm plot) for the PVA mixture film. The obtained scattering profile shows fairly good linear relationship and a least-square fitting of the scattering profile gives M_w of 2990. This value is close to the value obtained by viscometry and GPC measurements.

SELECTED BIOLOGICAL STUDIES USING SANS-U

Recombinant yeast-Derived Human Hepatitis B Virus Surface Antigen Vaccine Particles

Recombinant human hepatitis B virus vaccine particle (yHBsAg) is an assembly of surface antigen proteins, lipids and hydrocarbons (Sato *et al.*, 1995a; 1995b). The information on the internal distribution of the surface antigen proteins in the particle is, therefore, of particular importance for a role of the vaccine.

Sato *et al.* (1995a; 1995b) investigated the internal structure using contrast variation method of yHBsAg solution with ^2H_2O contents of 0, 15, 30, 40, 60 and 100%.

Figure 6 shows the ^2H_2O contents dependence of SANS profiles of yHBsAg solutions. From these SANS measurements a contrast matching point was determined to be about 24% ^2H_2O content. This indicates that a large part of the vaccine particle is occupied by lipids and hydrocarbons from the host cells (yeast). The Stuhrmann plot for yHBsAg particles gives a linear relationship having positive slope as shown in Figure 7. This linear relationship indicates that the scattering density of the peripheral region is significantly higher than that in the central region, and its center of mass is consistent with that of the shape of the particle. Because the lipids and carbohydrates having lower scattering densities than the protein components form the spherical cluster within the particle, the cluster is considered to be positioned at the core of the particle and to be surrounded isotropically with yHBsAg antigen proteins to form a spherical vaccine particle. This is favorable to induce anti-virus antibodies at the surface of the vaccine. The radius of gyration and maximum dimension of the particle were estimated from the dataset of 100% ^2H_2O to be 105 and 340Å, respectively, which is in good agreement with the results from high-performance size-exclusion chromatography and molar mass measurements by low-angle laser light scattering.

Insect Lipophorins

Lipophorin is a circulating lipoprotein in insects and serves as a reusable shuttle to transport various lipids such as diacylglycerol and hydrocarbons between tissues (Katagiri

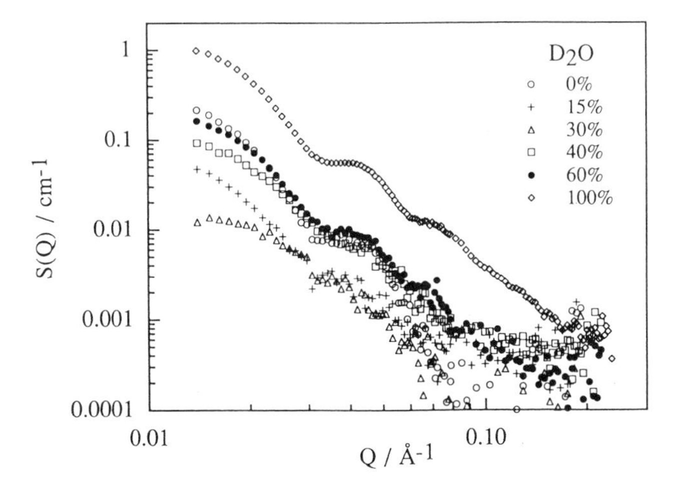

Figure 6. ^2H_2O contents dependence of SANS profiles of yHBsAg solutions.

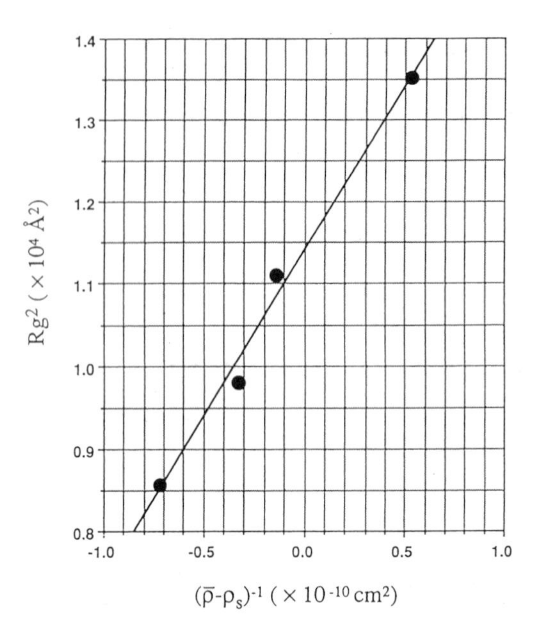

Figure 7. Stuhrmann plot of yHBsAg vaccine.

et al., 1993). Lipophorin forms a sphere-like particle with a particle weight of about 6.0×10^5 and contains two apoproteins and 40–50% (w/w) lipid. The structure of lipophorin was investigated by small-angle X-ray scattering (SAXS) and differential scanning calorimetry techniques. These studies suggested that the lipophorin is composed of three radially symmetrical layers; surface layer with phospholipid and apoLp-I, middle layer with diacylglycerol and apoLp-II and an inner core with hydrocarbons. In order to confirm this three layer model, Katagiri *et al.* (1993) carried out the SANS contrast variations measurements, because according to the theoretical calculation it is expected that the second maximum in the SANS profile is much smaller than that in SAXS profile. The obtained SANS profiles showed good agreement with the theoretical calculation for the symmetrical three-layer model as shown in Figure 8.

Cholesterol Effects on Ripple Phase of Dipalmitoylphosphatidylcholine

In dipalmitoylphosphatidylcholine (DPPC) dispersion system, ripple phase is observed between liquid-crystalline and gel phases (Adachi *et al.*, 1995). Incorporation of cholesterol into a phospholipid bilayer affects the nature of this ripple structure. For the distribution of cholesterol, three typical models have been proposed as shown in Figure 9. In Model (a) the cholesterol is localized in the valley of the ripple structure; in Model (b) two microdomains, a pure DPPC microdomain at the hills and a microdomain composed of DPPC 80mol% and cholesterol 20mol% in the valleys, are formed and two separate phases coexist; and Model (c) cholesterol and DPPC are homogenized in the membrane. In order to examine these structure models, Adati *et al.* (1995) performed the SANS measurements. The shape of the diffraction peaks was assumed to be represented by Gaussian function. The calculated neutron diffraction profile $I_{cal}(Q)$ was:

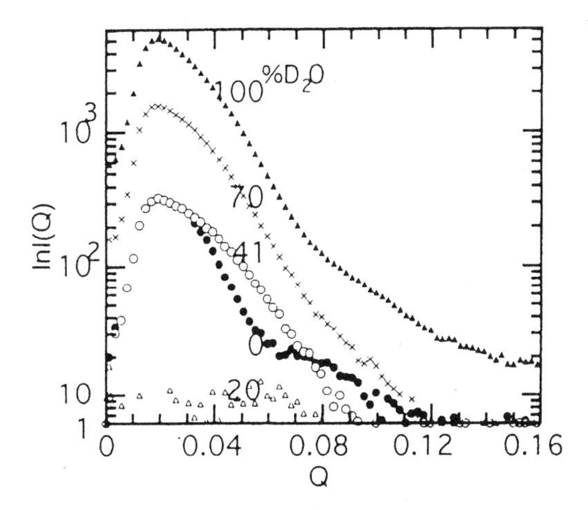

Figure 8. Small-angle neutron scattering of cockroach lipophorin in various ^2H_2O concentration.

$$I_{cal}(Q) = \sum_{h=-3}^{3} \sum_{k=-4}^{4} I(h,k) \exp\left(-\frac{(Q - q(h,k))^2}{2s^2} \right) \tag{4}$$

where I(h,k), q(h,k) and s are the intensity, position and the width of diffraction peak (h,k), respectively. I(h,k) and q(h,k) were calculated from the models. The width, s, was estimated from the observed profiles. This analysis was carried out for 100mol% deuterated and 85mol% deuterated cases, because the two cases have different distributions of neu-

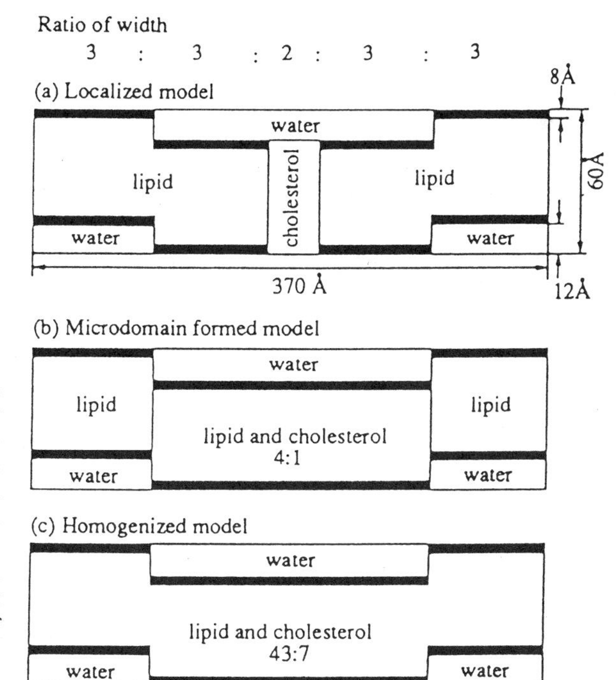

Figure 9. Three simple unit cell models of the neutron scattering length density profiles of the ripple structure for DPPC-cholesterol mixture. (a) cholesterol is localized (b) two microdomains are formed (c) cholesterol and DPPC are homogenized The lattice constants are a = 60Å, b = 370Å and γ = 90° which are obtained by X-ray diffraction. Thick lines represent headgroups of lipids. The thickness of water and headgroup are 12 and 8Å, repectively.

Table III. The value of L at the best fit as a measure of the quality of the model

Model	Localized	Microdomain formed	Homogenized
L	0.072	0.040	0.098

tron scattering length density, but have the same structure. Least-squares fitting factor, L, was defined by:

$$L = \frac{\sum_{Q}\left(I_{cal}^{100}(Q) - AI_{obs}^{100}(Q)\right)^2}{\sum AI_{obs}^{100}(Q)^2} + \frac{\sum_{Q}\left(I_{cal}^{85}(Q) - BI_{obs}^{85}(Q)\right)^2}{\sum_{Q} BI_{obs}^{85}(Q)^2} \tag{5}$$

where the parameters A and B mean the scale factor. The superscripts 100 and 85 correspond to 100 and 85mol% deuterated samples, respectively. The value of L is a measure of the quality of the fit between model and observed data. Those values at the best fit are listed in Table III. The model in which the cholesterol formed the microdomains gave a minimum value. These facts suggest that the microdomain composed of DPPC 80mol% and cholesterol 20mol% exist in the DPPC-cholesterol mixture. The calculated intensity profiles for Model (b) are shown in Figure 10 with the observed intensity profiles. There is a difference between the three models in spite of the simple models composed of rectangles. It is necessary construct a complicated model for the further progressive study.

Crystallization of Lysozyme

The success in the structure analysis of biological macromolecules using either X-ray or neutron depends on the availability of proper single crystals (Niimura *et al.*, 1994). Rational design of the crystal growth based on basic understanding of the growth mechanism is more and more called for. In order to clarify the nucleation process of the protein molecule, Niimura *et al.* (1994) performed the SANS experiments for the unsaturated solution of lysozyme. Figure 11 shows the SANS profiles obtained from unsaturated solution of lysozyme in ^2H_2O as a function of sodium chloride concentrations. As the sodium chloride was added to the solution, the scattering intensity in the low Q region was in-

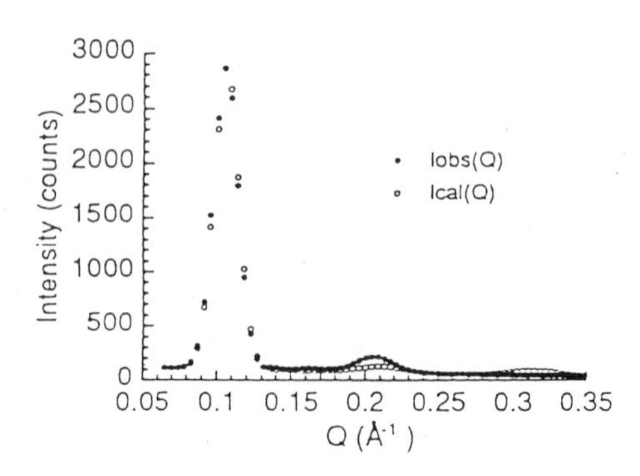

Figure 10. The comparison of observed and calculated neutron diffraction profiles for 100mol% deuterated DPPC-cholesterol mixture system.

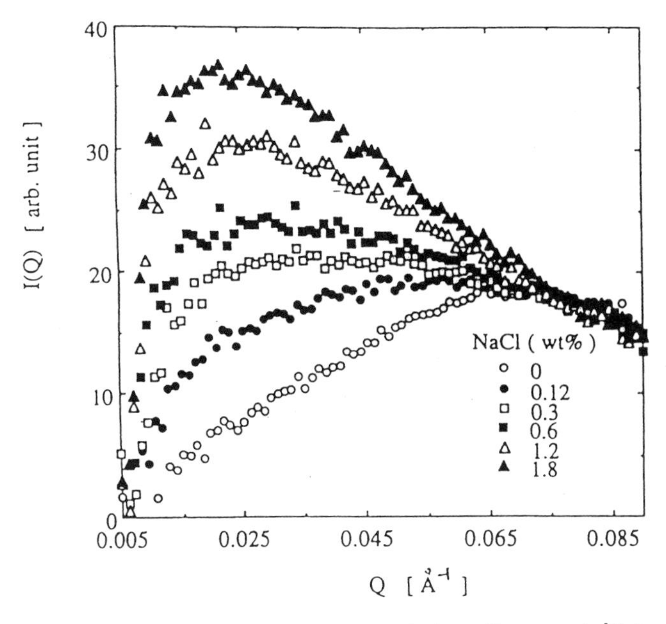

Figure 11. SANS profiles from unsaturated solutions of lysozyme in ^2H_2O.

creased and the peak position shifted to the low Q side. When the salt concentration is up to 1.8wt%, crystallization starts to grow. This result indicates that some aggregation process occurs before the crystallization.

CONCLUSIONS

The SANS-U at the 20MW JRR-3M reactor has been constructed for a general-purpose instrument covering a wide Q range (corresponding to structural inhomogeneities in the range of 10 to several 1000Å). The features of the SANS-U include relatively good Q resolution having $\Delta\lambda/\lambda = 10\%$, convenient sample changer system which can set maximum of 30 cells at once, and an interactive color graphic X-terminal. The SANS-U is suitable tool for biological research.

ACKNOWLEDGMENT

We thank Dr. D. Schwahn (IFF Julich) for helpful discussion on the SANS-U calibration.

REFERENCES

Adachi, T., Takahashi, H., & Hatta, I., (1995). Cholesterol effects on ripple structure studied by small angle neutron and X-ray diffraction. *Physica B* (in press).

Borkowski, C.J., & Kopp, M.J., (1978). Recent improvements to RC-line encoded position-sensitive proportional counters. *J. Appl. Cryst.,* 11:430–434.

Friedrich, H., Wagner, V., & Wille, P., (1989). A high performance neutron velocity selector. *Physica B,* 156&157:547–549.

Glinka, C.J., Rowe, J.M., & Larock, J.G., (1986). The small-angle neutron scattering spectrometer at the National Bureau of Standards. *J. Appl. Cryst.,* 19:427–439.

Katagiri, C., Ito, Y., & Sato, M., (1993). Small-angle neutron and X-ray scattering of insect lipophorins. *JAERI-M* 92–213:26–28.

Niimura, N., Minezaki, Y., Ataka, M., & Katsura., T., (1994). Small angle neutron scattering from lysozyme in unsaturated solutions to characterize the pre-crystallization process. *J. Cryst. Growth,* 137:671–675.

Noda, Y., Utiki, T., & Kajitani, M., (1983). Precise temperature control using micro-computers. (In Japanese). *Nihonkessho-gakkaishi,* 25:222–227.

Sato, M., Ito, Y., Kameyama, K., Imai, M., Ishikawa, N., & Takagi, T., (1995a). Small-angle neutron scattering study on recombinant yeast-derived human hepatitis B virus surface antigen. *Physica B* (in press).

Sato, M., Sato, Y., Kameyama, K., Ishikawa, N., Imai, M., Ito, Y., & Takagi, T., (1995b). Peripheral distribution of antigen proteins of recombinant yeast-derived human hepatitis B virus surface antigen vaccine particle: Structural characteristics shown by small-angle neutron scattering using contrast variation method. *J. Biol. Chem.,* (submitted).

Schwahn, D., Hahn, K., Streib, J., & Springer, T., (1990). Critical fluctuations and relaxation phenomena in the isotopic blend polystyrene/deuteropolystyrene investigated by small angle neutron scattering. *J. Chem. Phys.,* 93:8383–8391.

Shibayama, M., Kurokawa, H., Nomura, S., Roy, S., Stein, R.S., & Wu, W., (1990). Small-angle neutron scattering studies on chain asymmetry of coextruded poly(vinyl alcohol) film. *Macromolecules,* 25:1438–1443.

12

NEUTRON SCATTERING STUDIES ON CHROMATIN HIGHER-ORDER STRUCTURE

Vito Graziano, Sue Ellen Gerchman, Dieter K. Schneider, and
Venki Ramakrishnan

Biology Department
Brookhaven National Laboratory
Upton, New York 11973

ABSTRACT

We have been engaged in studies of the structure and condensation of chromatin into the 30nm filament using small-angle neutron scattering. We have also used deuterated histone H1 to determine its location in the chromatin 30nm filament. Our studies indicate that chromatin condenses with increasing ionic strength to a limiting structure that has a mass per unit length of 6–7 nucleosomes/11nm. They also show that the linker histone H1/H5 is located in the interior of the chromatin filament, in a position compatible with its binding to the inner face of the nucleosome. Analysis of the mass per unit length as a function of H5 stoichiometry suggests that 5–7 contiguous nucleosomes need to have H5 bound before a stable higher order structure can exist.

INTRODUCTION

The genome of higher organisms is organized as chromatin, a complex of DNA with proteins called histones. The fundamental unit of chromatin is the *nucleosome*, which consists of two copies each of the core histones H2a, H2b, H3 and H4, around which are wrapped approximately two turns of DNA (van Holde, 1989).

Ever since small-angle neutron scattering began to be used in biology - a state brought about by the advent of high-flux sources and area detectors - it has made important contributions to chromatin structure. In fact, contrast variation using neutron scattering on nucleosomes in solution (Hjelm *et al.*, 1977; Pardon *et al.*, 1975) provided the earliest physical evidence that DNA was on the outside of the histone octamer in the nucleosome.

The linker histone H1, or its counterpart H5 in avian erythrocyte chromatin, binds to the outside of the nucleosome, and is required for the organization of nucleosomes into a

Neutrons in Biology, edited by Schoenborn and Knott
Plenum Press, New York, 1996

higher-order structure called the 30nm filament. The filament becomes highly compact at ionic strengths that approach the physiological range. A definite model for how nucleosomes are organized into the 30nm filament was first proposed by Finch & Klug (1976), based on electron microscopy. In this model, called the 'solenoidal model', nucleosomes are arranged in a helical array, with 6–8 nucleosomes per turn of the helix. Each nucleosome is connected by linker DNA to its neighbor in the helix. In this model, linker DNA and histone H1 are proposed to be inside the filament. Since then, other models for the structure of the 30nm filament have been proposed, based on various experimental data. The continuously bent linker model of McGhee *et al.* (1983) differs from the solenoidal model in that the linker DNA, and histone H1/H5 can alternate between the inside and outside. In the helical ribbon model of Woodcock *et al.* (1984), neighboring nucleosomes make a zigzag pattern to form a ribbon. The entire zigzag ribbon then winds into a helix to form the 30nm filament. The crossed-linker model of Staynov (1983) proposes that a single helical array of nucleosomes has straight linker DNA that threads back and forth through the interior of the fiber. In the crossed-linker model of Williams *et al.* (1986), two arrays of nucleosomes are connected by a zigzag pattern of linker DNA; the double array twists about its axis to form the fiber, so that the linker DNA goes back and forth through the interior.

The models differ in some important measurable properties. For example, the solenoidal class of models propose a mass per unit length that is almost half that of the helical ribbon and crossed-linker models. Another important difference is the predicted location of the linker histone H1 in the filament.

Neutron scattering on long filaments can reveal directly the mass per unit length and cross-sectional radius of gyration of the filaments. Moreover, if it is possible to remove the linker histones H1/H5 and replace them by a deuterated counterpart, then by contrast variation one could determine the location of the deuterated component relative to the rest of the structure. In this article, we shall review some of our results in this area. All our neutron scattering measurements were done using the small-angle spectrometer on beamline H9B of the High Flux Beam Reactor at Brookhaven National Laboratory (Schneider & Schoenborn, 1984).

NEUTRON SCATTERING TO STUDY CHROMATIN FIBERS

By lightly digesting nuclei with micrococcal nuclease, it is possible to obtain chromatin fragments that have in excess of 100 nucleosomes. These filaments are highly extended. The scattering from dilute solutions of such filaments may be analyzed using the rod approximation (Porod, 1982). In this approximation, one considers chromatin filaments to be highly elongated rods, and at low angles, the scattered intensity obeys the relation:

$$I(q) \propto \frac{\mu}{q} \exp\left(-\frac{q^2 R_x^2}{2}\right) \tag{1}$$

The scattering vector $q = 4\pi\sin\theta/\lambda$, where 2θ is the scattering angle and λ is the wavelength. The scattering is related to the mass per unit length μ and the cross-sectional radius of gyration R_x. Thus cross-sectional Guinier plots, which are plots of $ln[qI(q)]$ vs. q^2, should be linear at low angles, and the cross-sectional radius of gyration and mass per unit length of the filaments can be estimated respectively from the slope and intercept of a linear fit to the data. Such an analysis was applied to chromatin filaments using both X-rays (Sperling & Tardieu, 1976) and neutrons (Baudy & Bram, 1978; Suau *et al.*, 1979). In our work, we have relied almost exclusively on this type of analysis. The advantage of neutrons in this context is the ability to deuterate specific components and vary the contrast, or the difference between the scattering length densities of macromolecule and solvent. To analyze the data at various contrasts, we have used the formalism developed by Stuhrmann and his coworkers (Ibel & Stuhrmann, 1975) for the three-dimensional case and applied it to the two-dimensional case of the rod approximation.

Mass Per Unit Length and Cross-Sectional Radius of Gyration as a Function of Ionic Strength

In our first study, we asked how the mass per unit length and cross-sectional radius of gyration of chromatin filaments changed with the ionic strength of the solvent. To some extent, this question had been addressed using neutron scattering before by Suau *et al.* (1979), who analyzed chromatin at both high and low ionic strength. However, we wanted to resolve the controversy regarding the mass per unit length of chromatin filaments by doing scattering and scanning transmission electron microscopy (STEM) experiments on the same sample. Further, electron microscopy (Thoma *et al.*, 1979) and later scattering experiments (Bordas *et al.*, 1986) suggested that at low ionic strength, chromatin already had an organization into a higher-order structure. This meant that it was important to study chromatin at several ionic strengths to determine its course of condensation.

We compared the results from scattering and STEM studies on native chicken erythrocyte chromatin as a function of ionic strength (Gerchman & Ramakrishnan, 1987). In a second study (Graziano *et al.*, 1988), we asked whether it was possible to reconstitute depleted chromatin with pure histone H5 and obtain scattering curves that behaved like native chromatin. The results from these two studies are summarized in Figures 1 and 2. Figure 1a shows scattering curves from native chicken erythrocyte chromatin in NaCl concentrations ranging from 10 to 60mM. As can be seen, both the slope and the intercept of the linear region at low angles increase, suggesting that the fibers are becoming thicker and more compact. The steep inner slopes yield radii of gyration that range from 80–125 Å, which is characteristic of a higher-order structure. Figure 1b shows the scattering curves for chromatin in 80, 100 and 120mM NaCl. There is little difference between these curves, showing that chromatin has reached a limit of compaction. We then examined depleted chromatin, which was made by selectively removing linker histones H1 and H5 from chromatin under conditions in which nucleosome sliding does not take place. In Figure 1e, which shows the scattering curves for depleted chromatin, we see that in the absence of H1/H5, there is no steep inner slope, and the outer shallow slope is characteristic of a '10nm' filament, which represents nucleosomes without any higher level of organization.

In Figures 1c and 1d, we address the question of whether it is possible to add back linker histone under appropriate conditions and get back the structure of native chromatin. As can be seen in Figure 1c, the scattering curves for depleted chromatin reconstituted with H5 are very similar to the corresponding curves for native chromatin shown in Figure 1a. Similarly, reconstituted chromatin reaches a limiting structure (Figure 1d) just as native chromatin does.

The mass per unit length can be calculated from the intercepts made by the linear fits in Figure 1, using a calibration for the absolute intensity (Jacrot & Zaccai, 1981). The

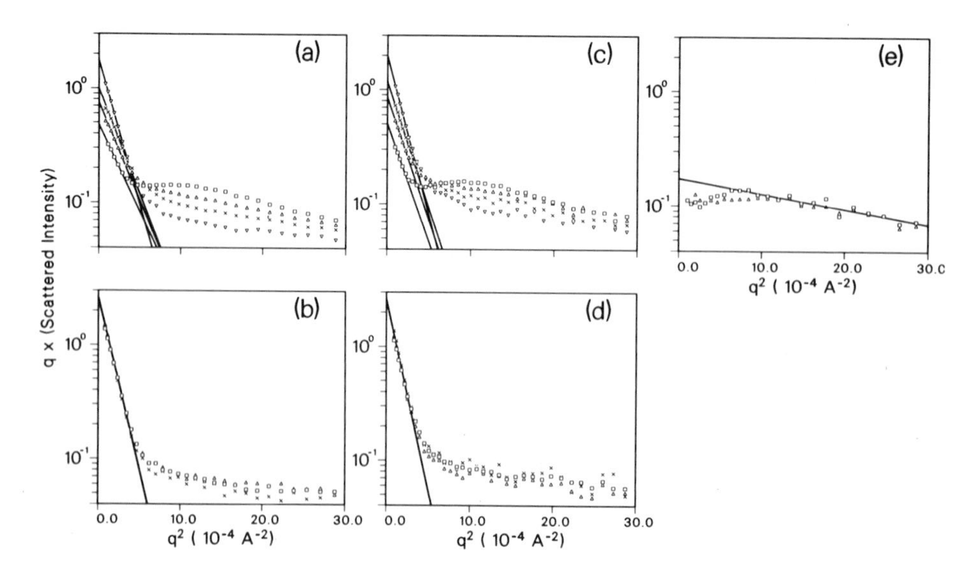

Figure 1. Cross-sectional Guinier plots of the low-angle region of scattering curves from native, reconstituted and depleted chromatin. Cross-sectional radii of gyration and the mass per unit length were calculated respectively from the slope and intercept at $q = 0$ of fits to the linear portion of the curve at low angles, represented by the straight lines shown. (a) Native chromatin in (□) 10mM NaCl; (Δ) 20mM NaCl; (×) 30mM NaCl; (∇) 60mM NaCl. (b) Native chromatin in (□) 80mM NaCl; (Δ) 100mM NaCl; (×) 120mM NaCl. (c) and (d) Reconstituted chromatin in the same conditions as in (a) and (b). (e) Depleted chromatin in (□) 0mM NaCl; (Δ) 60mM NaCl. The buffers all contained 5mM Hepes, 0.2mM EDTA, pH 7.5. (From Graziano *et al.*, 1988).

values are expressed in units of 'nucleosomes per 11nm'. This is because the nucleosome is approximately 11nm on edge, so the repeating unit along a filament in which nucleosomes are in contact is of this order. Thus the 11nm corresponds to a 'turn' of the helix in several models for chromatin.

In Figure 2, we show the mass per unit length of native and reconstituted chromatin as a function of the NaCl concentration in the solvent. Clearly, the mass per unit length levels off at a value of 6–7 nucleosomes/11nm, and the values for reconstituted chromatin

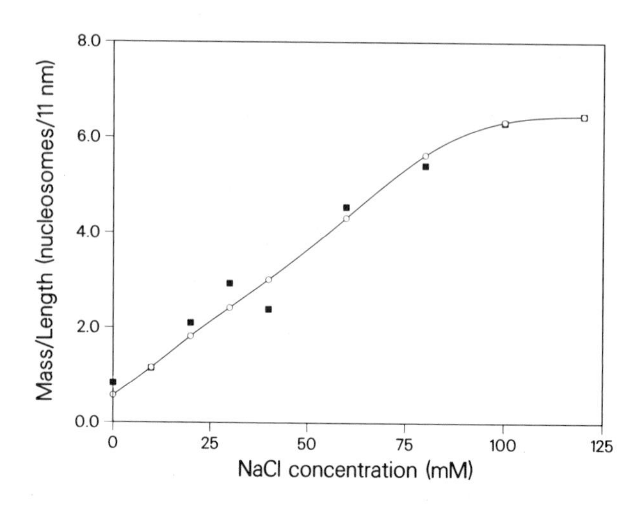

Figure 2. The mass per unit length of chromatin as a function of the NaCl concentration in the buffer. The buffer also contained 5mM Hepes, 0.2mM EDTA, pH 7.5. Line connecting open circles represents data for native chromâtin, while the black squares are the values for chromatin reconstituted with pure histone H5. (From Graziano *et al.*, 1988).

are in good agreement with those of native chromatin. This value for the plateau is inconsistent with models for the 30nm filament that require much higher mass per unit length, and is in agreement with the prediction of the solenoidal class of models. Similar results were obtained when chromatin was reconstituted with pure chicken H1 rather than H5 (Graziano & Ramakrishnan, 1990). These experiments establish that we are able to obtain reasonable scattering curves for chromatin and that we can reconstitute depleted chromatin with linker histone to get scattering behavior indistinguishable from that of native chromatin.

DEUTERATION OF LINKER HISTONE

Deuteration of eukaryotic proteins has been greatly facilitated by the development of systems for overexpression of recombinant genes in *E.coli*. One of the most successful systems for expressing proteins in *E.coli* is the T7 expression system (Studier *et al.*, 1990). Unfortunately, neither this system nor various other systems are capable of expressing full-length chicken H5 in *E.coli* (Gerchman *et al.,* 1994). However, the gene for histone H1a, which is one of the six chicken histone genes, can be expressed well in *E.coli* using the T7 system. We were able to fully deuterate H1 by growth of *E.coli* in D_2O in a fully deuterated minimal medium with deuterated succinate as the carbon source. The resulting H1 was judged to be greater than 99% deuterated from electro-spray ionization mass spectrometry. We were also able to produce 65% deuterated H1 by growth of *E.coli* in 80% D_2O using protonated glucose as the carbon source.

CONTRAST VARIATION EXPERIMENTS ON CHROMATIN CONTAINING FULLY OR PARTIALLY DEUTERATED H1

We now describe the results of our experiments to locate H1 in the 30nm filament by using deuterated H1 and contrast variation (Graziano *et al.*, 1994). We studied native chromatin (N), chromatin reconstituted with protonated recombinant H1 (H), chromatin recombinated with partially deuterated H1 (DH) and chromatin reconstituted with fully deuterated H1 (D). These samples were all studied at 85mM NaCl, because our previous studies had shown that chromatin is almost fully compact at this ionic strength, but is still reasonably soluble. Each sample was dialyzed into H_2O and D_2O buffers and the appropriate H_2O/D_2O mixtures made by mixing.

Figure 3 shows cross-sectional Guinier plots on N- and H-chromatin (Figure 3a) and D-chromatin (Figure 3b). The curves for N- and H- chromatin are nearly identical in both 30% and 80% D_2O. This was true for the other contrasts measured as well, showing that the scattering curves for native chromatin and chromatin reconstituted with recombinant H1 are indistinguishable even as a function of contrast. The slopes for 30% and 80% D_2O are very similar. On the other hand, the slopes for D-chromatin in 30% and 80% D_2O are quite different.

Figure 4 shows the variation of the normalized intercept $[qI(q)]_0^{1/2}$ as a function of the per cent D_2O in the solvent. Again, the curves for N- and H-chromatin are indistinguishable, and the match-point of 49.8% D_2O is close to that calculated for chicken erythrocyte chromatin. The match-point for D-chromatin is shifted to a value of 63.1% D_2O, again in excellent agreement if one assumes that the H1 is fully deuterated. The match point for

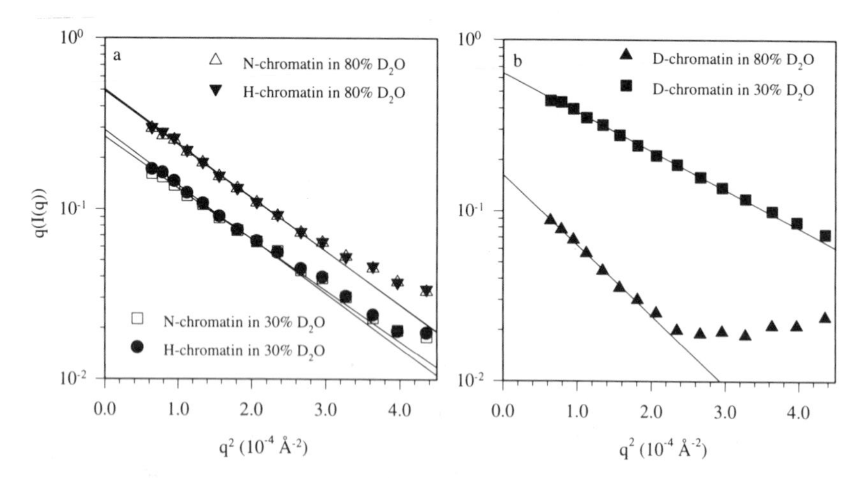

Figure 3. Cross-sectional Guinier plots on (a) native chromatin and chromatin reconstituted with unlabeled recombinant H1 and (b) chromatin reconstituted with deuterated H1. (From Graziano *et al.*, 1994).

DH-chromatin has an intermediate value of 58.6% D_2O which is exactly what one would expect for H1 that is 65% deuterated.

In Figure 5, the variation of R_x^2 as a function of inverse contrast (Stuhrmann plot) is shown. The steep negative slope for D-chromatin and the insignificant slope for N- and H-chromatin shows clearly that the deuterated material is on the inside of the filament. Again, the values for DH-chromatin lie intermediate between those for H- and D-chromatin. The lack of a significant curvature suggests that there is little difference (in projection down the fiber axis) between the centers of mass of the H1 and chromatin components.

An estimate of the separate cross-sectional radii of gyration of H1 and the rest of chromatin can be obtained by plotting R_x^2 as a function of f, the fractional excess scatter of H1, which will depend on the contrast (Serdyuk, 1975). The contrast in turn, will depend on both the per cent D_2O in the solvent, and the level of deuteration. In Figure 6, we see that R_x^2 decreases linearly with f. As expected, the points for both partially deuterated chromatin and fully deuterated chromatin fall on the same line. From the extrapolated val-

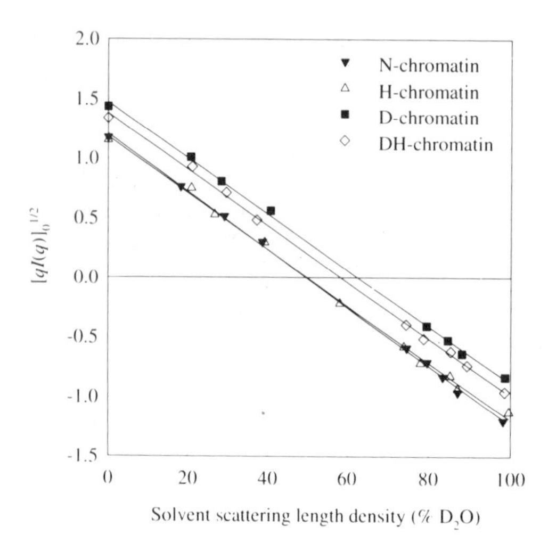

Figure 4. A plot of the forward scattering amplitude $[qI(q)]_0^{½}$ as a function of the percent D_2O in the buffer for native (N) chromatin, and chromatin reconstituted with unlabeled (H), 65% deuterated (DH) and fully deuterated (D) H1.

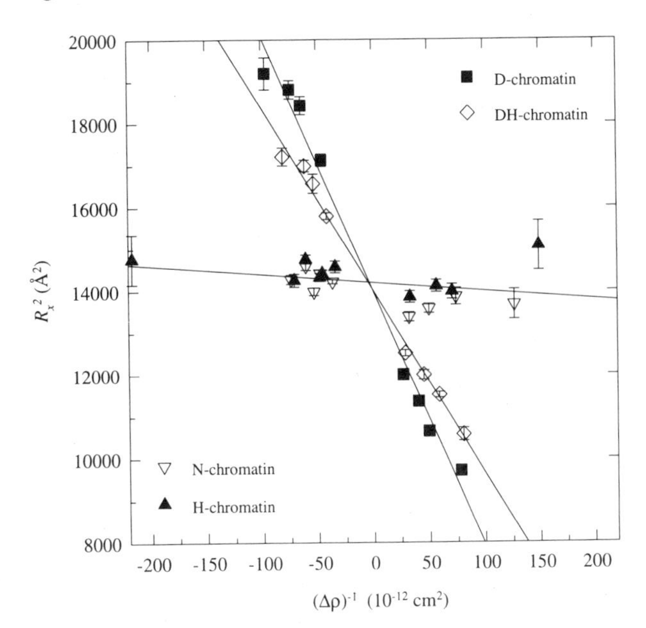

Figure 5. Stuhrmann plot for the cross-sectional radii of gyration of the various types of chromatin. See text for details.

ues at f=1 and f=0, we can estimate the values for the cross-sectional radius of gyration of H1 and the rest of chromatin respectively. If one uses the parallel axis theorem and the measured values of 30Å and 45Å for the radii of gyration of H1 and the nucleosome respectively, then one estimates that the chromatin component (largely nucleosomal) lies about 115Å from the fiber axis, whereas the H1 component lies about 60–65Å from the fiber axis. If one draws a box of roughly the dimensions of a nucleosome, then it can be seen from Figure 7 that the center of mass of H1 roughly coincides with the inner face of the nucleosome.

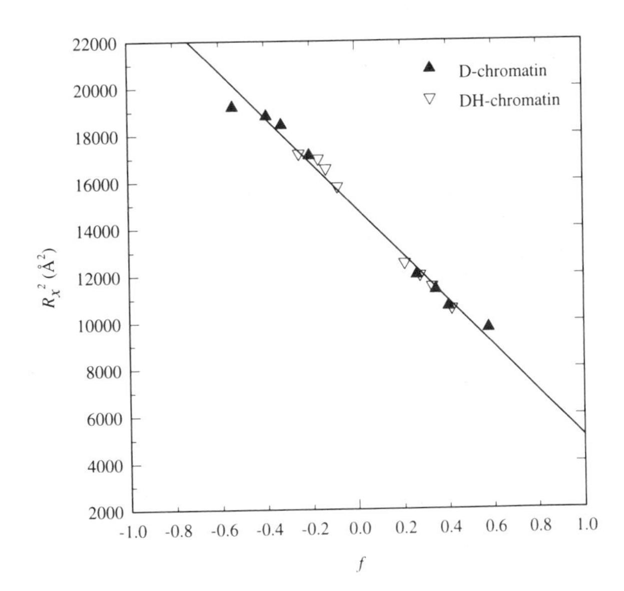

Figure 6. Plot of the square of the cross-sectional radius of gyration of chromatin as a function of the fractional excess scatter of H1. The plot allows one to estimate by extrapolation the contribution of the H1 component alone and that of the rest of chromatin.

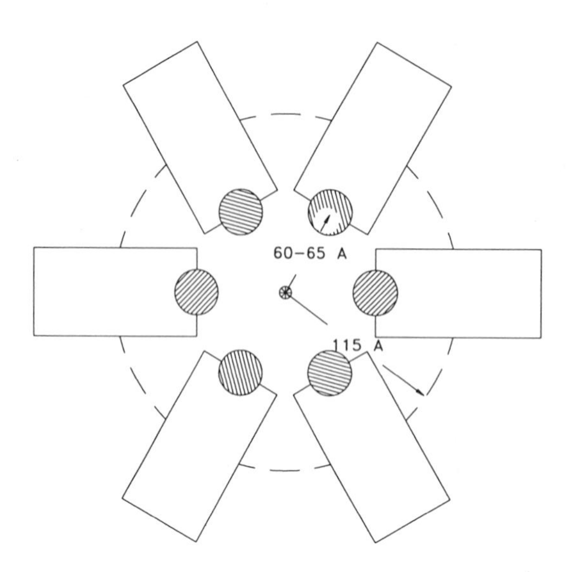

Figure 7. Model for the location of H1 and the 30nm filament, based on neutron scattering data. The model suggests that there are approximately 6 nucleosomes per turn of the helix, and that the center of mass of H1, shown by a shaded circle, is at about the same radial location as the inner face of the nucleosome, represented by rectangles.

DISCUSSION

A number of caveats must be applied to the kind of analysis used in this work. The first is that chromatin filaments are not infinitely long, rigid and uniform rods. Thus the Porod approximation used in the analysis is not rigorously true. Hjelm (1985) has studied the effect of finite length and non-uniform cross-section on the rod approximation. In general, the results of this study should be close to the 'real' values, but it is not clear how much the deviation is at low ionic strengths when the filaments are highly extended. Another point is that chromatin filaments are intrinsically heterogeneous. It is not merely that the cross-section is not uniform, but it changes from one section of the filament to the next. There may also be whole classes of filaments whose structure deviates from the mean significantly. This and other issues have led to some question of whether in fact there is a specific structure for the 30nm filament (Giannasca & Horowitz, 1993; Zlatanova *et al.*, 1994). The values that we measure are root-mean-squared values, and there may be considerable variation between individual filaments. However, there is no question that regardless of individual variation in the filaments, on average, the location of H1 is interior relative to the nucleosome.

Although we have determined that the center of mass of the linker histone H1 is inside relative to the nucleosome, this does not by itself place the individual parts of H1 with great accuracy. The reason is that all linker histones have a tri-partite structure. They consist of a central globular domain that binds to the nucleosome, and are flanked by extended, highly basic arms. In particular, the C-terminal arm makes up about half the protein, and almost half the residues in the arm consist of lysine or arginine. Thus our study cannot say anything about the relative disposition of the C-terminal and globular domains. However, if the arms could be preferentially deuterated relative to the globular domain, then it might be possible to determine their location using the methods described here.

While small-angle scattering, especially with neutrons, is useful to answer global questions, it cannot, unfortunately, provide definitive answers to the problem of chromatin higher-order structure. The reason is that the models differ mainly in the connectivity of DNA: What is the path of the linker DNA as it goes from one nucleosome to the next? Is this path topologically the same in all filaments or does it vary? What is the chemical nature of the interaction between linker histones and chromatin? To answer these questions, it is likely that one will have to construct uniform filaments that consist of tandem repeats of a defined sequence of DNA that positions histone octamers precisely, and obtain fiber diffraction patterns or even single crystals from these filaments. Even if it is possible to construct such filaments and obtain their structure, it will not necessarily settle the question of the structure of chromatin in the cell nucleus, as there is some evidence that the structure formed by refolding *in vitro* is not the same as that in the cell nucleus (Giannasca & Horowitz, 1993). Despite these reservations, it is our opinion that neutron scattering experiments have provided some definite answers in a field that is riddled with controversy.

ACKNOWLEDGMENTS

This work was supported by the Office of Health and Environmental Research of the U.S. Department of Energy, and grant GM 42796 from the NIH.

REFERENCES

Baudy, P., & Bram, S., (1978). Chromatin fiber dimensions and nucleosome orientation: a neutron scattering investigation. *Nucl. Acids Res.,* 5:3697–3714.

Bordas, J., Perez-Grau, L., Koch, M.H.J., Vega, M.C., & Nave, C., (1986). The superstructure of chromatin and its condensation mechanism. *Eur. Biophys. J.,* 13:157–173.

Finch, J.T., & Klug, A., (1976). Solenoidal model for superstructure in chromatin. *Proc. Natl. Acad. Sci. USA,* 73:1897–1901.

Gerchman, S.E., Graziano, V., & Ramakrishnan, V., (1994). Expression of chicken linker histones in *E.coli:* Sources of problems and methods for overcoming some of the difficulties. *Protein Expr. Purif.,* 5:242–251.

Gerchman, S.E., & Ramakrishnan, V., (1987). Chromatin higher-order structure studied by neutron scattering and scanning transmission electron microscopy. *Proc. Natl. Acad. Sci. USA,* 84:7802–7806.

Giannasca, P.J., & Horowitz, R.A., (1993). Transitions between *in situ* and isolated chromatin. *J. Cell Sci.,* 105:551–561.

Graziano, V., Gerchman, S.E., & Ramakrishnan, V., (1988). Reconstitution of chromatin higher-order structure from histone H5 and depleted chromatin. *J. Mol. Biol.,* 203:997–1007.

Graziano, V., & Ramakrishnan, V., (1990). Interaction of HMG14 with chromatin. *J. Mol. Biol.,* 214:897–910.

Graziano, V., Gerchman, S.E., Schneider, D.K., & Ramakrishnan, V., (1994). Histone H1 is located in the interior of the chromatin 30-nm filament. *Nature,* 368:351–354.

Hjelm, R.P., (1985). The small-angle approximation of X-ray and neutron scatter from rigid rods of non-uniform cross-section and finite length. *J. Appl. Cryst.,* 18:452–460.

Hjelm, R.P., Kneale, G.G., Suau, P., Baldwin, J.P., & Bradbury, E.M., (1977). Small angle neutron scattering studies of chromatin subunits in solution. *Cell,* 10:139–151.

Ibel, K., & Stuhrmann, H., (1975). Neutron scattering from myoglobin in solution. *J. Mol. Biol.,* 93:255–265.

Jacrot, B., & Zaccai, G., (1981). Determination of molecular weight by neutron scattering. *Biopolymers,* 20:2413–2426.

McGhee, J.D., Nickol, J.D., Felsenfeld, G., & Rau, D.C., (1983). Higher order structure of chromatin: orientation of nucleosomes within the 30 nm chromatin solenoid is independent of species and linker length. *Cell,* 33:831–841.

Pardon, J. F., Worcester, D.L., Wooley, J.C., Tatchell, K., van Holde, K.E., & Richards, B.M., (1975). Low-angle neutron scattering from chromatin subunit particles. *Nucl. Acids Res.,* 2:2163–2176.

Porod, G., (1982). In *Small Angle X-ray Scattering* (O. Glatter & O. Kratky, editors) pp17–33, Academic Press, London

Schneider, D.K., & Schoenborn, B.P., (1984). In *Neutrons in Biology* (B.P. Schoenborn, editor) vol. 27, pp119–142, Plenum Press, New York.

Serdyuk, I.N., (1975). Electromagnetic and neutron scattering from the 50S subparticle of *E.coli* ribosomes. *Brookhaven Symp. Biol.,* 27:IV:49–60.

Sperling, L., & Tardieu, A., (1976). The mass per unit length of chromatin by low-angle X-ray scattering. *FEBS Lett.*, 64:89–91.

Staynov, D., (1983). Possible nucleosome arrangements in the higher order structure of chromatin. *Int. J. Biol. Macromol.*, 5:3–9.

Studier, F.W., Rosenberg, A.H., Dunn, J.J., & Dubendorff, J.W., (1990). Use of T7 RNA polymerase to direct expression of cloned genes. *Meth. Enzymol.*, 185:61–89.

Suau, P., Bradbury, E.M., & Baldwin, J.P., (1979). Higher-order structure of chromatin in solution. *Eur. J. Biochem.*, 97:593–602.

Thoma, F., Koller, T., & Klug, A., (1979). Involvement of histone H1 in the organization of the nucleosome and of the salt dependent superstructures of chromatin. *J. Cell Biol.*, 83:403–427.

van Holde, K.E., (1989). *Chromatin*, Springer-Verlag, New York.

Williams, S.P., Athey, B.D., Muglia, L.J., Schappe, R.S., Gough, A.H., & Langmore, J.P., (1986). Chromatin fibers are left-handed helices with diameter and mass per unit length that depend on linker length. *Biophys. J.*, 49:233–248.

Woodcock, C.L.F., Frado, L.-L., & Rattner, J.B., (1984). The higher-order structure of chromatin: evidence for a helical ribbon arrangement. *J. Cell. Biol.*, 99:42–52.

Zlatanova, J., Leuba, S.H., Yang, G., Bustamante, C., & van Holde, K., (1994). Linker DNA accessibility in chromatin fibers of different conformations: A reevaluation. *Proc. Natl. Acad. Sci. USA*, 91:5277–5288.

13

THE STRUCTURE OF THE MUSCLE PROTEIN COMPLEX $4Ca^{2+}$·TROPONIN C· TROPONIN I

Monte Carlo Modeling Analysis of Small-Angle X-Ray Data

Glenn Allen Olah and Jill Trewhella

Chemical Science and Technology Division
Loa Alamos National Laboratory
Mail Stop J586 Los Alamos, New Mexico 87545

ABSTRACT

Analysis of scattering data based on a Monte Carlo integration method was used to obtain a low resolution model of the $4Ca^{2+}$·troponin C·troponin I complex. This modeling method allows rapid testing of plausible structures where the best fit model can be ascertained by a comparison between model structure scattering profiles and measured scattering data. In the best fit model, troponin I appears as a spiral structure that wraps around $4Ca^{2+}$·troponin C which adopts an extended dumbbell conformation similar to that observed in the crystal structures of troponin C. The Monte Carlo modeling method can be applied to other biological systems in which detailed structural information is lacking.

INTRODUCTION

Muscle action results when thick and thin protein filaments move past each other (for a review, see Zot & Potter, 1987). Thick filaments are composed of myosin, while thin filaments are made from a helical assembly of actin monomers. The contractile force is generated by the cyclic attachment and detachment of the myosin heads, S1, to specific sites on actin monomers and the power stroke is driven by actin-S1-myosin ATPase activity. Tropomyosin is another protein associated with the thin filament and is polymerized head to tail in the grooves of the actin helix. Each tropomyosin has one troponin complex bound to it. The troponin complex regulates the muscle contraction/relaxation event and imparts the calcium sensitivity to the switching mechanism. Troponin has three subunits: troponin C (TnC) which binds calcium, troponin I (TnI) which inhibits the actin-S1-myosin ATPase activity, and troponin T (TnT) which binds troponin to tropomyosin. When TnC binds calcium, a signal is transmitted via TnI which releases its inhibition of the actin/myosin interaction via TnT and

Neutrons in Biology, edited by Schoenborn and Knott
Plenum Press, New York, 1996

tropomyosin. A high resolution crystal structure of the TnC component exists showing it to have an unusual dumbbell shape, with C- and N- terminal globular domains connected by an extended 8–9 turn solvent-exposed α-helix (Herzberg *et al.*, 1985; Sundaralingam *et al.*, 1985). The C-terminal domain contains two high-affinity Ca^{2+}/Mg^{2+} binding sites, while the N-terminal domain contains two low-affinity Ca^{2+} binding sites (Potter & Gergely, 1975) which are not occupied in the crystal structure. It is Ca^{2+} binding to the two low affinity sites in the N-terminal domain which regulates the contractile event. Low resolution X-ray crystallographic and electron micrograph data of the troponin complex bound to the tropomyosin show it to have an elongated shape with TnC and TnI forming a globular domain while TnT forms a long tail which interacts with tropomyosin (Flicker *et al.*, 1982; White *et al.*, 1987). Further structural and functional details on the troponin complex or on the TnC·TnI domain are lacking.

We are interested in obtaining structural information on the TnC·TnI domain which is central in the functioning of the calcium-regulatory switch. Therefore, we completed small-angle X-ray and neutron scattering experiments on $4Ca^{2+}$·TnC (partially deuterated) complexed with TnI (non-deuterated) in order to determine structural information such as the radius of gyration (R_g), the pairwise length distribution function [P(r)], the maximum linear dimension (d_{max}) for the overall complex and its components, the separation of the centers-of-mass of the components, the symmetry of the components, and the relative dispositions of symmetry axes. This information provides constraints that limit the possible models for the complex, especially in cases where the structures are highly asymmetric.

In general, one-dimensional data obtained from small-angle scattering is not sufficiently constraining to uniquely define a three-dimensional structure. Further compounding this problem, modeling of scattering data has usually depended on a trial-and-error approach that did not always, because of required computer time, completely test all of parameter space defining a structure. To avoid this problem when modeling $4Ca^{2+}$·TnC·TnI complex, we developed and automated a Monte Carlo integration modeling method that allowed us to systematically and rapidly test plausible models for the complex against the scattering data. In the case of the $4Ca^{2+}$·TnC·TnI complex the combination of constraints from the scattering data, the previously determined crystal structure of TnC, and known molecular volumes allowed a low-resolution structure to be derived.

MATERIALS AND METHODS

A brief description of the scattering experiment, analyses and modeling method are outlined in this section. Details of this work have been published elsewhere (Olah & Trewhella, 1994; Olah *et al.*, 1994).

Experimental

Neutron scattering experiments were performed at the low-Q-diffractometer (LQD) at the Manuel Lujan Jr. Scattering Center (LANSCE) (Seeger *et al.*, 1990) and at the 30m SANS instrument (NG3) at the National Institute of Standards and Technology (NIST). X-ray scattering data were collected using the scattering instrument at Los Alamos which is described elsewhere (Heidorn & Trewhella, 1988). Neutron scattering measurements involved measuring a 'contrast series' in which a complex of deuterated $4Ca^{2+}$·TnC and nondeuterated TnI is solubilized in solvents with different D_2O/H_2O ratios. 78% of the nonexchangeable hydrogens of the TnC component have been deuterated placing the total complex match point at 71% D_2O. Partial deuteration of the TnC component allowed measurement of two contrast points above the match point, *viz* 90% and 100% D_2O, and three contrast points below the match point, *viz* 0%, 20% and 40%. Non-deuterated TnI was prepared as described elsewhere (Wilkinson & Grand, 1975; 1978). Assuming only 45% of exchangeable hydrogens for each protein readily exchange, the match points were calculated from the chemical compositions of the components as 42% D_2O for TnI and 101% D_2O for $4Ca^{2+}$·TnC.

D_2O-induced aggregation was present in samples with $D_2O\% > 40\%$ and was eliminated by adding 2 to 3M urea to the solvent and keeping sample concentrations low (1.5 - 3mg/ml). It is known that TnC complexed with TnI forms a stable complex under saturating Ca^{2+} conditions even in the presence of 6M urea (Grabarek *et al.*, 1981).

Data Analysis

Three different analyses were used to extract structural information from the scattering data. All three analyses required measurement of a 'contrast series' with at least three contrast points defining the series. In this study, nine measurements at six contrast points (five from neutron scattering and one from X-ray scattering) were completed.

Analysis (1) (Olah et al., 1994). If we ignore internal scattering density fluctuations, then the scattering intensity as a function of the scattering vector, $|Q|$ ($= 4\pi \sin\theta/\lambda$; θ is the scattering angle and λ is the wavelength of the incident and scattering X-ray or neutron), can be written at each contrast point as:

$$I(Q, \Delta\rho_C, \Delta\rho_I) = \Delta\rho_C^2 I_C(Q) + \Delta\rho_C \Delta\rho_I I_{CI}(Q) + \Delta\rho_I^2 I_I(Q) \quad (1)$$

where the subscripts C and I refer to $4Ca^{2+}$·TnC and TnI. $\Delta\rho_C$ and $\Delta\rho_I$ are the scattering density above the average solvent scattering density for $4Ca^{2+}$·TnC and TnI, respectively. $I_C(Q)$ and $I_I(Q)$ represent the 'basic scattering functions' for $4Ca^{2+}$·TnC and TnI in the complex, respectively, and $I_{CI}(Q)$ is a cross-term. $\Delta\rho_C$ and $\Delta\rho_I$ are readily calculated from chemical and isotope compositions of the two components of the complex and the solvent. Equation 1 defines a set of linear equations from which $I_C(Q)$ and $I_I(Q)$ are extracted. Guinier analysis of $I_C(Q)$ and I_I (Q) give R_g for $4Ca^{2+}$·TnC and TnI, respectively. The inverse Fourier transforms of $I_C(Q)$ and $I_I(Q)$ give the pairwise distribution functions, $P_C(r)$ and $P_I(r)$, respectively, which are the frequency of all interatomic vectors within each component and hence goes to zero at a d_{max} value for each component. The case where $\Delta\rho_C$ and $\Delta\rho_I$ are equal gives R_g and d_{max} for the entire complex.

Analysis (2) (Moore, 1981). For a two-component system, the square of R_g for each component and for the entire complex can be obtained using the 'parallel axis theorem':

$$R_g^2 = f_C' R_C^2 + f_I' R_I^2 + f_C' f_I' D^2 \quad (2)$$

where $f_C' = \Delta\rho_C V_C / (\Delta\rho_C V_C + \Delta\rho_I V_I)$ and $f_I' = 1 - f_C'$. R_C and R_I are the radius of gyration of the $4Ca^{2+}$·TnC and TnI components, respectively. D is the separation of the centers-of-mass of the two components in the complex.

Analysis (3) (Ibel & Stuhrmann, 1975). R_g^2 dependence on the scattering contrast can be written as:

$$R_g^2 = R_m^2 + \alpha/\Delta\rho - \beta/\Delta\rho^2 \tag{3}$$

where R_m is the R_g at infinite contrast and $\Delta\rho$ is the average scattering density for the entire complex above the average solvent scattering density. For a two-component system, we have the expressions:

$$R_m^2 = f_C R_C^2 + f_I R_I^2 + f_C f_I D^2 \tag{4a}$$

$$\alpha = (\Delta\rho_C - \Delta\rho_I) f_C f_I \left[R_C^2 - R_I^2 + (f_I^2 - f_C^2) D^2\right] \tag{4b}$$

$$\beta = (\Delta\rho_C - \Delta\rho_I)^2 f_C^2 f_I^2 D^2 \tag{4c}$$

where $f_C = V_C/(V_C + V_I)$ and $f_I = 1 - f_C$.

The coefficient α is the second moment of the scattering density fluctuations about the mean value for the total scattering particle, while β is related to the square of the first moment of the density fluctuations about the mean. If the sign of α is positive, then the lower scattering density component is located more toward the inside of the complex than the higher scattering density component. A negative α indicates the reverse case. β is proportional to the square of the separation of the centers-of-mass of the two components in the complex.

Modeling

The Monte Carlo approach has previously been used to evaluate models against scattering data (Heidorn & Trewhella, 1988; Olah *et al.*, 1993), but this application required automation of the software to rapidly and systematically test models while independently varying all the structural parameters defining a model. The method involves defining the molecular boundaries of a model structure and placing that structure within a box. Random points are generated within the box, and those points which fall within the molecular boundary are saved. The number of saved points is proportional to the scattering density multiplied by the volume of the object in question. The sign of the scattering density at a point is also stored. For a two component system such as $4Ca^{2+}$·TnC·TnI complex, the number of saved random points representing TnI was set at 700, sufficient to guarantee statistical variations are negligible, and the appropriate number of saved random points in the TnC component was then determined depending on the relative contrast of the two components. Distances between every pair of saved points representing a model are determined and weighted by the product of the sign of each point. The histogram of all pair distances is the calculated P(r) function for the model. Model scattering profiles are readily calculated from the P(r) functions by a Fourier inversion (Olah & Trewhella, 1994; Olah *et al.*, 1993; Moore, 1980). The calculated data is scaled to I(0) of the measured data and a comparison made by calculating a reduced-χ^2 value. Minimization of the reduced-χ^2 value when varying parameters defining a given model was used as a 'best fit' criteria. Initially, we used the TnC crystal structure to define the molecular volume and boundary of the $4Ca^{2+}$·TnC component when constructing test models (see Olah

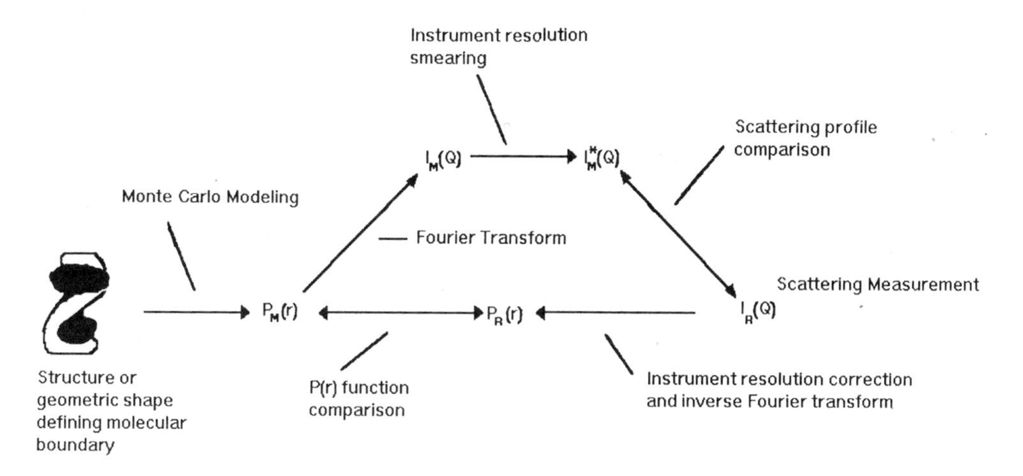

Figure 1. Schematic showing steps involved when comparing scattering profiles and P(r) functions calculated from a model with measured data. The subscript M stands for model and R for raw data.

& Trewhella, 1994 for details, and the RESULTS section below). Only comparisons at contrast points corresponding to TnI ($I_I(Q)$ derived from Equation 1) and $4Ca^{2+}$·TnC·TnI (X-ray scattering data) were directly made in order to decrease computation time. Equation 1 was used to obtain the model scattering profiles at other contrast points. Comparison between calculated and measured P(r) functions can also be made. Figure 1 shows a schematic of the steps required for comparing calculated model scattering profiles against the measured data.

RESULTS

Table I summarizes the structural parameters derived from the X-ray and neutron scattering experiments using the three different types of analyses described in Data Analysis Section.

Analyses (2) and (3) indicate the separation of $4Ca^{2+}$·TnC and TnI centers-of-mass, D, in the complex are approximately coincident (<10Å). Analysis (3) gave a negative value for α indicating $4Ca^{2+}$·TnC is more toward the interior of the complex relative to TnI. From Analysis (1), d_{max} values indicate the TnI component is significantly more extended than $4Ca^{2+}$·TnC, with d_{max} equal to that of the entire complex. In addition, Guinier analysis for rod-like particles (Glatter, 1982) gives average radius of gyration of cross-section, $R_{c'}$, values for each component and the complex that indicate their long axes are all approximately coincident (Olah *et al.*, 1994). In Figure 2c, a comparison between $P_C(r)$ functions for TnC calculated from the crystal structure and from the measured data obtained from Analysis (1) show them to be very similar. We therefore concluded that the structure of $4Ca^{2+}$·TnC in the complex like the crystal structure has the helix connecting the two globular domains fully extended. The scattering data, therefore, imposes considerable constraints on possible models for the complex to structures in which the TnC component is in an extended configuration, similar to the crystal structure, with TnI encompassing and extending beyond TnC, approximately symmetrically, at both ends of the long axis of the complex.

Table I. Comparison of structural parameters derived from the scattering data and the final 'best fit' model

		R_g (Å)	$R_{c'}$ (Å)	d_{max} (Å)
$4Ca^{2+}$•TnC				
	Experiment	23.9 ± 0.5	10.7 ± 1.0	72 ± 3
	Model	24.1	10.3	73
TnI				
	Experimental	41.2 ± 2.0	20.5 ± 2.0	118 ± 4
	Model	40.1	20.5	114
$4Ca^{2+}$•TnC•TnI				
	Experimental	33.0 ± 0.5	16.2 ± 1.5	115 ± 4
	model	33.4	15.7	117

Separation of the centers of mass of $4Ca^{2+}$•TnC and TnI

Experimental	< 10Å[b]
Model	4.2Å

Angle between long axes of $4Ca^{2+}$•TnC and TnI

Experimental	0°[c]
Model	6.4°

[a]Experimental parameters from Olah *et al.* (1994].
[b]From analysis of contrast dependence of R_g (Olah *et al.*, 1994).
[c]From analysis of $R_{c'}$ values (Olah *et al.*, 1994).

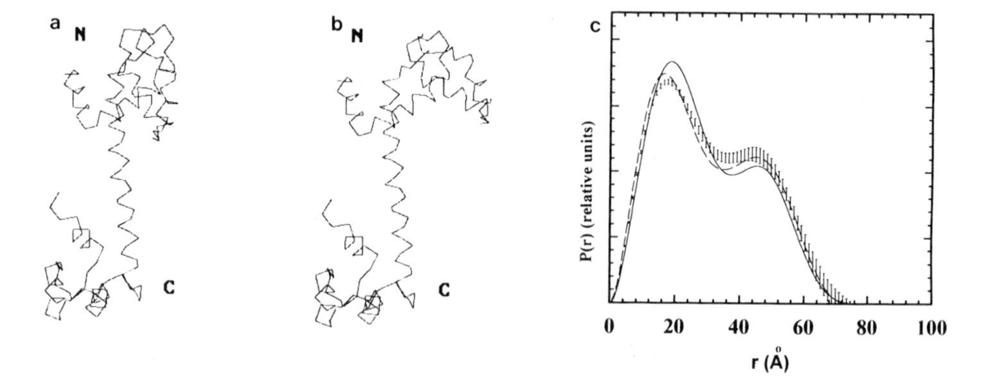

Figure 2. Structure (a) represents the TnC crystal structure which did not have Ca^{2+} bound to the N-terminal domain. TnC in the scattering experiments had all four Ca^{2+} binding sites occupied; therefore, as represented by structure (b), a hinge was defined in the N-domain and the hydrophobic 'cup' was opened up to simulate the effect of Ca^{2+} binding. Only C_α traces are shown for clarity. In panel (c), the $P_C(r)$ function was calculated from the measured scattering data and is represented by one standard deviation error bars. The solid line is the P(r) function calculated from the crystal structure. The similarity between the P(r) functions calculated from the crystal structure and the measured data justify the use of the crystal structure in the initial model search. Final model searches were optimized by opening up the N-domain in the crystal structure. The dashed line represents the 'best fit' to $P_C(r)$ after opening up the N-domain 'cup'.

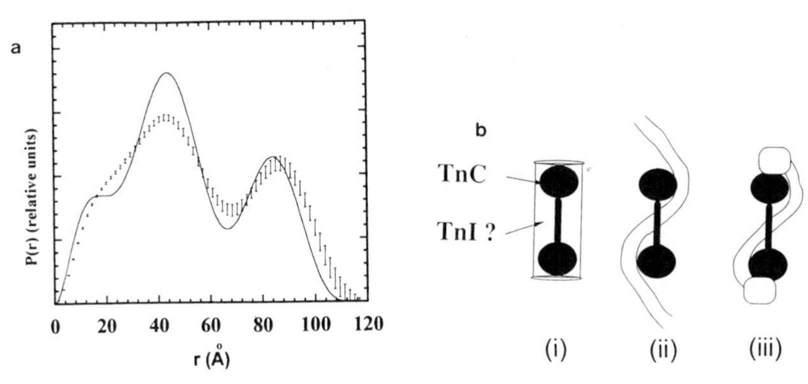

Figure 3. (a) $P_I(r)$ function for the TnI component in the complex was calculated from Equation 1 and is represented by one standard deviation error bars. The solid line was obtained from the 'best fit' model. (b) TnC is represented by a dumbbell shape structure similar to the crystal structure. (i) A cylindrical TnI encompassing TnC would have too large a volume. (ii) From the periodic features in the $P_I(r)$ function, a spiral structure could encompass the TnC and have the correct volume. (iii) d_{max} and the amplitude of the second peak at 86Å in the $P_I(r)$ function suggest that the central spiral is capped by two end regions.

Furthermore, inspection of the $P_I(r)$ function obtained from Analysis (1) shows a 43Å periodicity (multiple peaks at 43Å and 86Å), a shoulder at ~15Å and a d_{max} approximately 40Å longer than TnC indicating that TnI forms some sort of regular spiral-like structure that winds from one end of TnC to the other, and extends approximately 20Å beyond $4Ca^{2+}$·TnC (see Figure 3a and Olah *et al.*, 1994 for details). The volume constraint on TnI and the dimensions of $4Ca^{2+}$·TnC restrict turns of the spiral structure to approximately two or less, while d_{max} value for TnI restricts the product of its pitch and number of turns. In addition to a central spiral region, the amplitude of the 86Å peak in the $P_I(r)$ function and the d_{max} value for TnI suggest that there are two regions at each end of the TnI central spiral (see Figure 3b). A number of classes of models were systematically tested against the scattering data in which each class differed in the geometric shapes defining these two end regions. At the resolution of the scattering data, the possible shapes for the end regions were approximated by semi-ellipsoids, cylinders, or toroids. It was found early on that only the toroid end caps could simultaneously fit all the measured scattering data. Therefore, eight parameters were used to define the TnI component: the pitch (P), number of turns (N), minor radius (R_{min}), major radius (R_{max}), and the length (L) of the TnI central spiral, and the minor radius ($R^{cap}{}_{min}$), major radius ($R^{cap}{}_{max}$) and fraction of circle (N_{cap}) of the toroid caps.

The TnC crystal structure was taken to represent the $4Ca^{2+}$·TnC in the complex. In order to test models of the complex, the TnC and TnI components had to be positioned with respect to each other. Since the scattering data constrains the long axes of the two components and their centers-of-mass to be approximately coincident, the centers-of-mass of the N- and C- domains of TnC were placed on the same axis as the TnI central spiral. After the two components were positioned, the parameters defining the model were systematically varied and each model compared against the scattering data. A summary of the model refinement is given in Table II (Olah & Trewhella, 1994).

A further refinement was done involving modifications of the TnC component. TnC in the scattering experiments have all four calcium binding sites occupied while this was not the case in the crystal structure which only had the C-terminal domain binding sites

Table II. Summary of model refinement

(a) Variables determined in the initial fine and coarse searches, and the respective search range and increments. The parameters are defined in the text and are given in units of Å except for N and N_{cap} which are fractions of 1 complete turn. The values in parentheses are for the initial coarse search.

Variable	Best fit value	Search range	Increment
N	1.0	0.7–1.2	0.1
		(0.5–2.0)	(0.5)
R_{min}	6.0	5.5–7.5	0.5
		(5.0–15.0)	(2.0)
R_{max}	13.0	12.0 16.0	0.5
		(12.0–22.0)	(2.0)
L	69	69	0
		(66.0–72.0)	(3.0)
R^{cap}_{min}	6.0	5.5–7.5	0.5
		(5.0–15.0)	(2.0)
R^{cap}_{max}	24.5	23.0–26.0	0.5
		(12.0–28.0)	(4.0)
N_{cap}	0.9	0.8–1.0	0.1
		(0.0–1.0)	(0.2)

(b) Reduced-χ^2 values for the best fit model against the scattering data (I(Q) vs Q): (i) after the fine search, (ii) after the final refinement steps involving opening the N-terminal domain of TnC and optimizing the position of TnI with respect to TnC.

	(i)	(ii)	Q-range ($Å^{-1}$)
$4Ca^{2+}$•TnC	1.4	0.6	0.02–0.20
TnI	6.3	6.3	0.02–0.17
	0.9	0.9	0.02–0.15
$4Ca^{2+}$•TnC•TnI	1.1	1.0	0.02–0.20

occupied. A comparison of the structurally similar N- and C-terminal domains indicates the effect of Ca^{2+} binding probably involves the opening of a ‘cup-shaped’ domain, exposing hydrophobic residues on the inner surface of the cup (Herzberg *et al.*, 1986). We therefore defined a hinge in the N-terminal domain and modified it so it is more open like the Ca^{2+} bound C-terminal domain as shown in Figure 2b. A final refinement step was to optimize the position of TnC with respect to TnI by allowing the axis connecting the N- and C- domains of TnC to translate within 10Å of and within 8° parallel to the central TnI spiral axis. The improvement in the fit to the TnC scattering data is evident by the decrease in the reduced-χ^2 value (Table IIb) and by a slightly better agreement between the model $P_C(r)$ function and the $P_C(r)$ function calculated from the measured data shown in Figure 2c.

The scattering profiles from the final refined structure are shown along with the measured scattering data in Figure 4 and a stereoview of the structure is illustrated in Figure 5 (Olah & Trewhella, 1994).

CONCLUSIONS

We derived a low-resolution solution model of $4Ca^{2+}$·TnC·TnI based on X-ray and neutron scattering data in combination with other known physical constraints using a

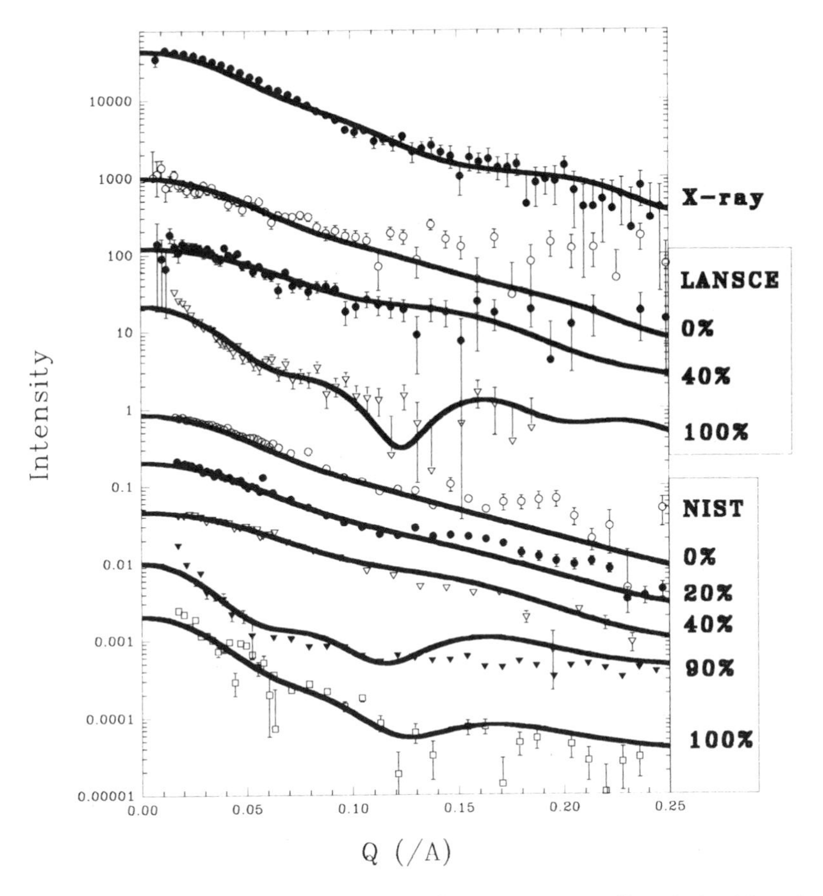

Figure 4. The nine measured scattering profiles compared with the scattering profiles calculated from the final refined structure. The data have been displaced for clarity and each profile (contrast point) f is labeled on the right side of the figure by the percent D_2O in the solvent.

Monte Carlo integration method. At the resolution of our scattering data and because of the fortuitous constraints derived from the unusual symmetries and asymmetries in this structure, we believe the basic features of the structure are correct. In particular, the presence of the central spiral region of TnI and its interaction with the two hydrophobic 'cup' regions of $4Ca^{2+}$·TnC is firmly established. This interaction further suggests a molecular basis for the Ca^{2+}-sensitivity of the troponin switch responsible for regulating the muscle

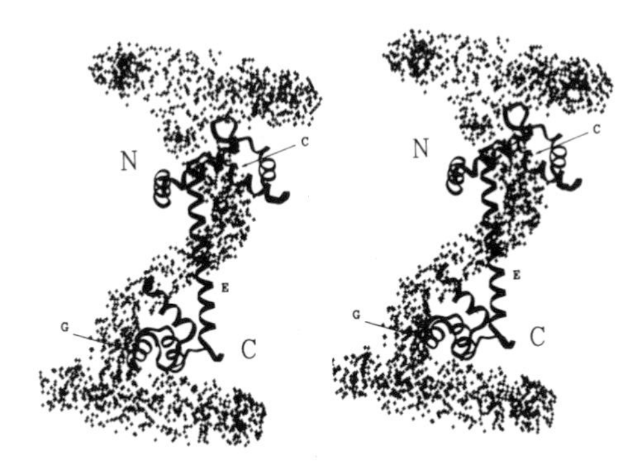

Figure 5. Stereoview of the final refined structure showing the central spiral of TnI (black crosses) winding around $4Ca^{2+}$·TnC (ribbon). The N- and C- domains, as well as helices C, E, and G are labeled. These three helices are known to interact with TnI.

contraction/relaxation cycle. In the presence of Ca^{2+}, each end of the TnI central spiral is anchored to one of the hydrophobic 'cups' in each domain of TnC preventing the inhibitory action of TnI. Removal of Ca^{2+} from the low affinity sites in the N-domain of TnC disrupts the interaction with TnI and provides enough flexibility for TnI to interact with its binding sites on actin.

From the perspective of $4C^{2+}$·TnC·TnI as a case study, we propose that the Monte Carlo integration modeling approach in combination with sufficient physical constraints, will be applicable to a wide range of biological systems that can be studied by small-angle scattering. We have recently applied this approach to the scattering data from fd gene 5 protein/single-stranded DNA complex (Olah *et al.*, 1995). Modeling of the scattering data showed the complex forms a left-handed superhelical structure with the protein wrapping around the DNA which is located on the inside of the complex. Similar to the $4Ca^{2+}$·TnC·TnI complex, other structural information, predominantly from electron microscopy and X-ray crystallography, sufficiently constrained the conformation space and allowed a low-resolution model of the superhelical structure to be obtained.

ACKNOWLEDGMENTS

This work was performed under the auspices of the Department of Energy (Contract W-7405-ENG-36), and was supported by DOE/OHER Project KP-04–01–00–0 (J.T.), NIH Project GM40528 (J.T.) and in part by an appointment of G.A.O. to the Alexander Holleaender Distinguished Postdoctoral Fellowship Program sponsored by DOE/OHER, and administered by the Oak Ridge Institute for Science and Education.

REFERENCES

Flicker, P.F., Phillips, G.N. Jr., & Cohen, C., (1982). Troponin and its interactions with Tropomyosin. An electron microscope study. *J. Mol. Biol.*, 162:485–501.

Glatter, O., (1982). In *Small Angle X-ray Scattering* (O. Glatter and O. Kratky, editors) Chapter 4, Academic Press, New York.

Grabarek, Z., Drabikowski, W., Leavis, P.C., Rosefeld, R.S., & Gergely, J., (1981). Proteolytic fragments of Troponin C. *J. Biol. Chem.*, 256:13121–13127.

Heidorn, D.B., & Trewhella, J., (1988). Comparison of the crystal and solution structures of Calmodulin and Troponin C. *Biochemistry,* 27:909–915.

Herzberg, O., Moult, J., & James, M.N.G., (1985). Structure of the calcium regulatory muscle protein Troponin C at 2.8Å resolution. *Nature,* 313:653–659.

Herzberg, O., Moult, J., & James, M.N.G., (1986). A model for the Ca^{2+}-induced conformational transition of Troponin C. *J. Biol. Chem.*, 261:2638–2644.

Ibel, K., & Stuhrmann, H.B., (1975). Comparison of neutron and X-ray scattering of dilute myoglobin solutions. *J. Mol. Biol.*, 93:255–265.

Moore, P. B., (1980). Small-angle scattering. information content and error analysis. *J. Appl. Cryst.*, 13:168–175.

Moore, P.B., (1981). On the estimation of the radius of gyration of the subunits of macromolecular aggregates of biological origin *in situ*. *J. Appl. Cryst.*, 14:237–240.

Olah, G.A., & Trewhella, J., (1994). A model structure of the muscle protein complex $4Ca^{2+}$ Troponin C Troponin I derived from small-angle scattering data: implications for regulation. *Biochemistry*, 33(43):12800–12806.

Olah, G.A., Mitchell, R.D., Sosnick, T.R., Walsh, D.A., & Trewhella, J., (1993). Solution structure of the cAMP-dependent protein kinase catalytic subunit and its contraction upon binding the protein kinase inhibitor peptide. *Biochemistry*, 32(14):3649–3657.

Olah, G.A., Rokop, S.E., Wang, A.C.-L., Blechner, S.L., & Trewhella, J., (1994). Troponin I encompasses an extended Troponin C in the Ca(2+)-bound complex: a small-angle X-ray and neutron scattering study. *Biochemistry*, 33(27):8233–8239.

Olah, G.A., Gray, D.M., Gray, C.W., Kergil, D.L., Sosnick, T.R., Mark, B.L., Vaughan, M.R., & Trewhella, J., (1995). Structures of fd gene 5 protein·nucleic acid complexes: A combined solution scattering and electron microscopy study. *J. Mol. Biol.,* 249:576–594.

Potter, J.D., & Gergely, J., (1975). The calcium and magnesium binding sites on Troponin and their role in the regulation of Myofibrillar Adenosine Triphosphatase. *J. Biol. Chem.,* 250:4628–4633.

Seeger, P.A., Hjelm, R.P., & Nutter, M.J., (1990). The low-Q diffractometer at the Los Alamos Neutron Scattering Center. *Mol. Cryst. Liq. Cryst.* 108A:101–117.

Sundaralingam, M., Bergstrom, R., Strasburg, G., Rao, S.T., Roychowdhury, P., Greaser, M., & Wang, B.C., (1985). Molecular structure of Troponin C from chicken skeletal muscle at 3-Angstrom resolution. *Science,* 227:945–948.

White, S.P., Cohen, C., & Phillips, G.N. Jr., (1987). Structure of co-crystals of Tropomyosin and Troponin. *Nature,* 325:826–828.

Wilkinson, J.M., & Grand, R.J.A., (1975). The amino acid sequence of Troponin I from rabbit skeletal muscle. *Biochem. J.,* 149:493–496.

Wilkinson, J.M., & Grand, R.J.A., (1978). Comparison of amino acid sequence of Troponin I from different striated muscles. *Nature,* 271:31–35.

Zot, A.S., & Potter, J.O., (1987). Structural aspects of Troponin·Tropomyosin regulation of skeletal muscle contraction. *Ann. Rev. Biophys. Biophys. Chem.,* 16:535–559.

14

STRUCTURAL MODEL OF THE 50S SUBUNIT OF *E. COLI* RIBOSOMES FROM SOLUTION SCATTERING

D. I. Svergun,[1*] M. H. J. Koch,[1] J. Skov Pedersen,[2] and I. N. Serdyuk[3]

[1] EMBL
Hamburg Outstation, Notkestraβe 85, D-22603 Hamburg, Germany
[2] Department of Solid State Physics, Risø National Laboratory
DK-4000 Roskilde, Denmark
[3] Institute of Protein Research, Russian Academy of Sciences
142292 Poustchino, Moscow Region, Russia

ABSTRACT

The application of new methods of small-angle scattering data interpretation to a contrast variation study of the 50S ribosomal subunit of *Escherichia coli* in solution is described. The X-ray data from contrast variation with sucrose are analyzed in terms of the basic scattering curves from the volume inaccessible to sucrose and from the regions inside this volume occupied mainly by RNA and by proteins. From these curves models of the shape of the 50S and its RNA-rich core are evaluated and positioned so that their difference produces a scattering curve which is in good agreement with the scattering from the protein moiety. Based on this preliminary model, the X-ray and neutron contrast variation data of the 50S subunit in aqueous solutions are interpreted in the frame of the advanced two-phase model described by the shapes of the 50S subunit and its RNA-rich core taking into account density fluctuations inside the RNA and the protein moiety. The shape of the envelope of the 50S subunit and of the RNA-rich core are evaluated with a resolution of about 40Å. The shape of the envelope is in good agreement with the models of the 50S subunit obtained from electron microscopy on isolated particles. The shape of the RNA-rich core correlates well with the model of the entire particle determined by the image reconstruction from ordered sheets indicating that the latter model which is based on the subjective contouring of density maps is heavily biased towards the RNA.

* On leave from the Institute of Crystallography Russian Academy of Sciences Leninsky pr. 59 117333 Moscow, Russia.

Neutrons in Biology, edited by Schoenborn and Knott
Plenum Press, New York, 1996

INTRODUCTION

Establishing the three-dimensional structure of ribosomes, the supramolecular protein-RNA complexes which carry out protein synthesis in all organisms, is one of the central problems in structural molecular biology. Although the structure of the eucaryotic and procaryotic ribosomes and their subunits has been investigated by various methods over almost three decades, many questions concerning the three-dimensional structure and mutual arrangement of the ribosomal proteins and RNA remain open.

The large (50S) subunit of procaryotic ribosomes contains two ribosomal RNA molecules (23S and 5S rRNA) with total molecular mass 980kDa and 34 proteins (L1-L34) with total molecular mass 460kDa (Nierhaus, 1982). The most important features of visually derived electron microscopic models of isolated 50S particles (eg Lake, 1976; Kiselev *et al.*, 1982; Vasiliev *et al.*, 1983; Stöffler & Stöffler-Meilicke, 1986) are well described by a 'consensus model' (Lake, 1985) which is roughly hemispheric with dimensions around $250 \times 230 \times 150Å^3$ and displays a characteristic asymmetric crown view. Similar gross features are also conserved in more recent models based on three-dimensional image reconstruction (Hoppe *et al.*, 1986; Radermacher *et al.*, 1987) which also display finer details. In contrast, the model obtained by processing electron microscopic images from two-dimensional sheets of 50S (Yonath *et al.*, 1987) has smaller dimensions (around $180 \times 170 \times 150Å^3$) and a significantly different shape.

Various methods were used to specifically investigate the structure of the protein and the rRNA moieties. In particular, May *et al.* (1992) used neutron small-angle scattering (SAS) on samples of 50S subunit with selective deuteration of specific proteins to measure the inter-protein distances. Immunoelectron microscopy was extensively used in conjunction with chemical cross-linking to establish the positions of individual proteins on the surface of the 50S subunit (Lake, 1985; Stöffler-Meilicke & Stöffler, 1990). The structure of the ribosomal RNA was specifically visualized by electron spectroscopic imaging (Korn *et al.*, 1983b). The density map of the 70S ribosome presented in the electron microscopic study of Frank *et al.* (1991), also contains information about the shape of the 50S subunit and its RNA.

Comparisons of different structural models as well as comprehensive references can be found in the papers above and in several reviews (eg Nierhaus, 1982; Wittmann, 1983; Yonath & Berkovitch-Yellin, 1993).

Among the methods already mentioned, SAS is the only one which provides low-resolution structural information about the shape and inner structure of *native* particles in solution. In SAS studies of biopolymers, dilute monodisperse solutions are used (Feigin & Svergun, 1987) and the SAS intensity is an isotropic function proportional to the scattering from a single particle averaged over all orientations. The main advantage of using solutions is that the particles are kept in their native state. The price to pay is the drastic loss of structural information caused by the random orientation of particles.

Extensive SAS studies of the 50S subunit were made in the past using both X-rays and neutrons. In particular, in the X-ray studies of Tardieu & Vachette (1982) and Meisenberger *et al.* (1984) model calculations were done using several electron microscopic models. Neutron contrast variation studies of Stuhrmann and co-workers (reviewed in Koch & Stuhrmann, 1979) led to a low-resolution model of the structure of 50S. Serdyuk (1979) investigated the inner structure of the 50S subunit by using a combination of X-ray, neutron and light scattering. In all these studies data interpretation was done in terms of integral parameters and the shape modeling was, with one exception (Stuhrmann *et al.*, 1977), based on the electron microscopic models.

More detailed interpretation of solution scattering data is always ambiguous as it is in general not possible to obtain a unique three-dimensional model from a one-dimensional isotropic scattering pattern. The information content of the SAS data can be significantly enriched by varying the contrast (Stuhrmann & Kirste, 1965) which is usually done by modifying the scattering density of the solvent. In X-ray studies, this is achieved by addition of salts, sucrose or glycerin; with neutrons by isotopic replacement of H_2O by D_2O (Feigin & Svergun, 1987; Chapter 4). Each of the contrasting techniques has its specific advantages and shortcomings. In X-ray synchrotron studies, for instance, one has a high flux and the instrumental smearing is negligible but only a narrow range of contrasts is available. With neutrons the range of contrasts is much broader but the instrumental conditions are usually poorer.

Recently, novel data analysis methods using the multipole expansion (Svergun & Stuhrmann, 1991; Svergun, 1991; 1994) were developed which perform a general model search and allow to combine the advantages of the different contrast variation techniques by simultaneously fitting the results of X-ray and neutron contrast variation experiments. The present paper describes the application of these methods to a joint X-ray and neutron contrast variation study of the 50S ribosomal subunit. In this study we derive a model of the structure of the 50S ribosomal subunit at a resolution around 40Å which fits a large body of X-ray and neutron data and suggests a way to resolve the controversy between electron microscopic models.

MATERIALS AND METHODS

Sample Preparation

50S ribosomal subunits of *Escherichia coli* MRE600 bacteria were obtained and analyzed using the method of Gavrilova *et al.* (1976). Before measurements the preparations were dialyzed against a buffer containing 20mM Tris-HCl (pH 7.4), 5mM $MgCl_2$, 100mM NH_4Cl and 1mM dithiothreitol. For X-ray contrast variation, samples in buffers containing 0, 7, 14, 24, 32 and 38 weight percent sucrose were used, and in the neutron measurements the buffers contained 0, 14, 40, 67 and 97% D_2O. The samples with the highest sucrose and D_2O concentrations were prepared by dialysis, those with intermediate concentrations by mixing. The sucrose and D_2O contents were controlled by density and transmission measurements. Biological activity of the samples was tested using poly(U) dependent poly(Phe) synthesis as described by Serdyuk, Grenader & Koteliansky (1977). Additional control of the homogeneity of the solutions of 50S subunits was done by analytical ultracentrifugation.

Scattering Experiments and Data Reduction

The synchrotron radiation data were collected following standard procedures using the X33 camera (Koch & Bordas 1983; Boulin *et al.*, 1986; 1988) of the EMBL on the storage ring DORIS III of the Deutsches Elektronen Synchrotron (DESY) and multiwire proportional chambers with delay line readout (Gabriel & Dauvergne, 1982). The wavelength was λ = 1.5Å and the range of momentum transfer $[s_{min}, s_{max}]$ = [0.008, 0.153]$Å^{-1}$ (here $s = (4\pi/\lambda)\sin\theta$, where 2θ is the scattering angle). The data were normalized to the intensity of the incident beam, corrected for the detector response, the scattering of the buffer was subtracted and the difference curves were scaled for concentration and transmission. The statistical error propagation was calculated using the program SAPOKO (Svergun & Koch, unpublished).

Neutron scattering experiments were performed at the Risø SANS facility using the cold source of the DR3 reactor (Lebech, 1990). Two experimental settings were used: sample-detector distance 3m, average wavelength 8Å (setting 1 covering range of momentum transfer $0.009 < s < 0.070 Å^{-1}$) and sample-detector distance 1m, average wavelength 6Å (setting 2, $0.04 < s < 0.20 Å^{-1}$). The neutrons were monochromatized by a mechanical velocity selector giving a wavelength distribution with full-width-half-maximum $\Delta\lambda/\lambda = 0.18$. The data were normalized to the intensity of the incident beam and corrected for the response of the area detector by dividing by the scattering of pure water. The spectra were radially averaged and corrected for the background scattering from buffer, cuvette and other sources using conventional procedures. The difference curves were scaled for concentration and sample transmission.

The concentration of the 50S subunit was measured spectrophotometrically assuming 1 A_{260} unit to correspond to 66μg of 50S. Solvents with the 50S concentrations 4–12mg/ml were used; concentration series measurements displayed no noticeable effects, neither for X-ray nor for neutron scattering measurements.

Contrast Variation

The excess scattering length density of a particle in solution can be written as (Stuhrmann & Kirste, 1965):

$$\rho(\vec{r}) = \bar{\rho}\rho_c(\vec{r}) + \rho_s(\vec{r}) \tag{1}$$

where $\rho_c(\vec{r})$ describes the particle shape ($\rho_c(\vec{r}) = 1$ inside the particle and zero elsewhere), whereas $\rho_s(\vec{r})$ corresponds to the internal density fluctuations with respect to the average value $\bar{\rho} = \langle\rho(\vec{r})\rangle - \rho_0$ (ρ_0 is the scattering length density of the homogeneous solvent). The scattering from an ensemble of dissolved particles is then:

$$I(s,\bar{\rho}) = \bar{\rho}^2 I_c(s) + \bar{\rho} I_{cs}(s) + I_s(s) \tag{2}$$

Here $I_c(s)$ is the shape scattering curve, $I_s(s)$ the scattering from internal structure and $I_{cs}(s)$ the cross term. For the forward scattering, $I_c(0) = \bar{\rho}^2 V_c$, $I_{cs}(0) = I_s(0) = 0$, where V_c is the invariant particle volume inaccessible to the solvent. The radius of gyration of the particle as a function of contrast is written as:

$$R_g^2(\bar{\rho}) = R_c^2 + \alpha/\bar{\rho} - \beta/\bar{\rho}^2 \tag{3}$$

where R_c is the radius of gyration at infinite contrast, and the constants α and β are defined by Ibel & Stuhrmann (1975), and the Porod volume as:

$$V^{-1}(\bar{\rho}) = Q(\bar{\rho})/2\pi I(0,\bar{\rho}) = V_c^{-1} + (\bar{\rho}V_c)^{-2}\int \rho_s^2(\vec{r})d^3\vec{r} \tag{4}$$

where

$$Q(\bar{\rho}) = \int_0^\infty I(s,\bar{\rho})s^2 ds$$

is the Porod invariant.

If the particle consists of two distinct components with different scattering densities its scattering can be expressed as:

$$I(s,\bar{\rho}) = I(s,\bar{\rho}_a,\bar{\rho}_b) = \bar{\rho}_a^2 V_a^2 I_a(s) + 2\bar{\rho}_a\bar{\rho}_b I_{ab}(s) + \bar{\rho}_b^2 V_b^2 I_b(s) \quad (5)$$

where $\bar{\rho}_a$, V_a, $\bar{\rho}_b$, V_b are the contrasts and volumes of the components, respectively, $I_a(s)$ and $I_b(s)$ their normalized intensities and $I_{ab}(s)$ the cross-term [here, $I_a(0) = I_b(0) = I_{ab}(0) = 1$].

Data Processing

The individual scattering curves were analyzed separately using the indirect transform program GNOM (Svergun *et al.*, 1988; Svergun, 1992) which uses the regularization method to evaluate the distance distribution function p(r) of the particle related to the scattering intensity by the Fourier transform:

$$I(s) = \int_0^{D_{max}} p(r)\frac{\sin(sr)}{sr}dr \quad (6)$$

where D_{max} is the maximum particle size. The data sets obtained at different contrasts were processed simultaneously using the program CGNOM, a generalized version of GNOM. The program fits the Fourier transform of the three real space functions:

$$p(r,\bar{\rho}) = \bar{\rho}^2 p_c(r) + \bar{\rho}p_{cs}(r) + p_s(r) \quad (7)$$

to the entire set of contrast variation data (Svergun, 1991; 1994). Each of the three basic scattering curves is then evaluated using Equation 6. The same program was used to analyze the data in terms of the scattering from two components according to Equation 5.

Multipole Expansion

This section contains the main equations of the formalism of the multipole expansion used for data interpretation. A more detailed description can be found in the original papers of Stuhrmann (1970a; 1970b).

A particle density distribution ρ(r) can be represented as a series:

$$\rho(\vec{r}) \approx \rho_L(\vec{r}) = \sum_{l=0}^{L}\sum_{m=-l}^{l}\rho_{lm}(r)Y_{lm}(\omega) \quad (8)$$

where (r,ω) = (r,θ,φ) are spherical coordinates, and

$$\rho_{lm}(r) = \int_{\omega} \rho(\vec{r}) Y^*_{lm}(\omega) d\omega \tag{9}$$

are the radial functions and $Y_{lm}(\omega)$ spherical harmonics. The truncation value L describes the resolution of the representation of the particle structure [$\rho_L(r) \rightarrow \rho(r)$ when $L \rightarrow \infty$]. With this expansion the particle SAS intensity is expressed as (Harrison, 1969; Stuhrmann, 1970a):

$$I(s) = 2\pi^2 \sum_{l=0}^{L} \sum_{m=-l}^{l} |A_{lm}(s)|^2 \tag{10}$$

Here the partial amplitudes $A_{lm}(s)$ are given by the Hankel transforms of the radial functions:

$$A_{lm}(s) = i^l (2/\pi)^{1/2} \int \rho_{lm}(r) j_l(sr) r^2 dr \tag{11}$$

where $j_l(sr)$ are the spherical Bessel functions.

Shape Scattering

The formalism of the multipole expansion is effectively applied to the contrast variation data analysis as follows. The shape of a wide variety of globular particles is conveniently described by the angular boundary (envelope) function $F(\omega)$:

$$\rho_c(\vec{r}) = \begin{cases} 1, 0 \leq r < F(\omega) \\ 0, r \geq F(\omega) \end{cases} \tag{12}$$

This function can also be developed into the series:

$$F(\omega) \approx F_L(\omega) = \sum_{l=0}^{L} \sum_{m=-l}^{l} f_{lm} Y_{lm}(\omega) \tag{13}$$

where the multipole coefficients are complex numbers:

$$f_{lm} = \int_{\omega} F(\omega) Y^*_{lm}(\omega) d\omega$$

The set of f_{lm} coefficients describes the shape of the particle at the given resolution. Representing the spherical Bessel function as a power series:

$$j_l(sr) = \sum_{p=0}^{p_{max}} d_{lp} (sr)^{l+2p} \tag{14}$$

where

$$d_{lp} = \frac{(-1)^p}{2^p p![2(l+p)+1]!!}$$

and substituting this series into Equation 11, one obtains (Stuhrmann, 1970b):

$$A_{lm}(s) = i^l (2/\pi)^{1/2} \sum_{p=0}^{p_{max}} d_{lp} s^{l+2p} \int_{\omega} Y_{lm}^*(\omega) d\omega \int_{r=0}^{F(\omega)} r^{l+2p+2} dr$$
$$= (is)^l (2/\pi)^{1/2} \sum \frac{d_{lp} f_{lm}^{(l+2p+3)}}{l+2p+3} s^{2p} \qquad (15)$$

where

$$f_{lm}^{(q)} = \int [F(\omega)]^q Y_{lm}^*(\omega) d\omega \qquad (16)$$

The coefficients of the power series of the partial amplitudes and hence the SAS intensity are thus expressed as a non-linear combination of the multipole coefficients $f_{lm}^{(q)}$ of the shape function. Svergun & Stuhrmann (1991) developed the procedure to perform the shape search by minimizing the functional:

$$R = \frac{\int_0^{s_1} |I_c(s) - I_t(s)| s^2 ds}{\int_0^{s_1} I_c(s) s^2 ds} + \chi |r_0|^2 \qquad (17)$$

Here $I_c(s)$ is the shape scattering curve determined from the experimental data in the interval $[0, s_1]$, $I_t(s)$ the theoretical curve from the given shape, $\mathbf{r}_0$ the position of the center of mass of the particle (readily evaluated from the f_{lm} coefficients) and $\chi \approx 10^{-2}$ is the weighting factor. The second term keeps the center of the particle close to the origin.

Inhomogeneous Particles

At sufficiently high contrasts, especially for single-component particles (eg proteins), the shape scattering normally dominates the other components in Equation 2 at low momentum transfers. At higher resolution, however, the scattering from inhomogeneities plays a significant role even for 'almost homogeneous' particles and has to be taken into account. Scattering from inhomogeneities inside the given shape can be described by the density fluctuations produced by the spherical harmonics with higher orders ($l \geq 7$) and represented as the linear superposition:

$$I_f(s) = (\Delta\rho_f)^2 \sum_l \Lambda_l^2 I_l(s) \qquad (18)$$

where $\Delta\rho_f$ is the maximum amplitude of fluctuations, $I_l(s)$ and $0 \leq \Lambda_l \leq 1$ are the normalized scattering intensity and magnitude of the fluctuations with characteristic size $\delta r = \pi R_c/l$, respectively. The intensities $I_l(s)$ are expressed in terms of the shape coefficients (Svergun, 1994). Scattering from the fluctuations is additive to the shape scattering and depends on the magnitudes Λ_l which can be included as additional unknown parameters in the minimization procedure for shape determination.

Particles consisting of two or more components are of particular interest for contrast variation studies. Let the particle have two distinct components A and B; Equation 5 then takes the form:

$$I(s,\bar{\rho}) = 2\pi^2 \sum_{l=0}^{L} \sum_{m=-l}^{l} \{\bar{\rho}_a^2 |A_{lm}(s)|^2 + \bar{\rho}_a \bar{\rho}_b |A_{lm}(s)B_{lm}^*(s) + A_{lm}^*(s)B_{lm}(s)| + \bar{\rho}_b^2 |B_{lm}(s)|^2\} \tag{19}$$

where $A_{lm}(s)$ and $B_{lm}(s)$ are the partial amplitudes of the two components. If these amplitudes are known, the scattering intensity will depend only on the relative position of the components. Fixing the first component, the relative position is described by the rotation of the second component by the Euler angles α,β,γ followed by a displacement along the vector $\mathbf{u} = (u,\omega_u)$. The new partial amplitudes of the second component can be analytically expressed as (Svergun, 1991):

$$B_{lm}^{(s)}(s) = 4\pi(-1)^m \sum_{p=0}^{L-l} i^p j_p(su) \sum_{q=-p}^{p} Y_{pq}^*(\omega_u) \sum_{k=|l-p|}^{l+p} \begin{pmatrix} l & p & k \\ 0 & 0 & 0 \end{pmatrix} \sum_{t=-k}^{k} \left[\frac{(2l+1)(2p+1)(2k+1)}{4\pi}\right]^{1/2} \begin{pmatrix} l & p & k \\ -m & q & m-q \end{pmatrix} \mathscr{D}_{m-q,t}^{(k)}(\alpha\beta\gamma) B_{kt}(s) \tag{20}$$

where

$$\begin{pmatrix} l & p & k \\ m & q & t \end{pmatrix}$$

are 3j Wigner coefficients and $\mathscr{D}_{m,m}^{(l)}(\alpha\beta\gamma)$ is the operator of finite rotations (Edmonds, 1957). The scattering intensity at each contrast (Equation 19) is thus expressed in terms of six parameters which can be evaluated by minimizing the overall discrepancy (eg sum of the functionals similar to Equation 17) and this allows to determine the mutual positions of the components.

If the partial amplitudes of the components are not known, they should also be involved in the minimization procedure. Two practically important cases can be distinguished. If A and B are separate domains as in Figure 1a, each of them can be described independently by a shape function $F(\omega)$ and $G(\omega)$, respectively, which is developed into the series (Equation 13) and the amplitudes are expressed *via* the corresponding shape coefficients. The core-shell case where the particle consists of a compact core surrounded by a shell (Figure 1b) is of particular practical importance. This situation is also described with the help of two shape functions: $F(\omega)$ corresponding to the whole particle and $G(\omega)$ corresponding to the core. Then the two components entering Equation 19 are defined as follows: component A is the particle of shape $F(\omega)$ and density $\rho_a = \rho_{shell}$, whereas the component B is the particle of shape $G(\omega)$ and density $\rho_b = \rho_{shell}\text{-}\rho_{core}$.

For the two cases, the total intensity at each contrast (Equation 19) depends on the shape coefficients f_{lm} and g_{lm} which are obtained by minimizing the overall discrepancy. The rotational and translational parameters are relevant only for the two-domain case. In the core-shell case they are taken into account by the function $G(\omega)$ [its center of mass

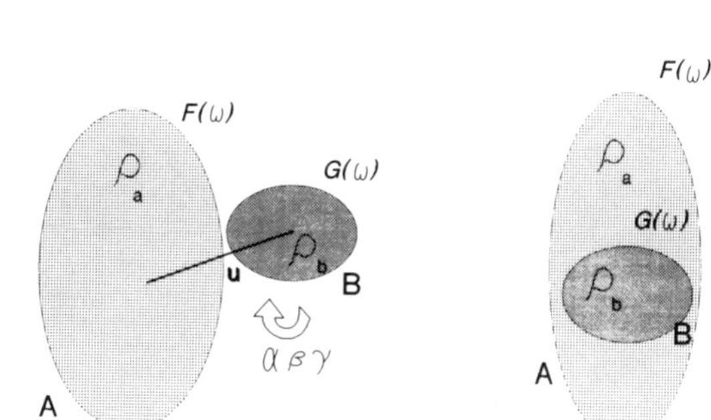

Figure 1. Types of two-component particle: (a) two separate domains; (b) core-shell model. For explanations see text.

does not necessarily coincide with that of F(ω)]. Note that by definition of this model $F(\omega) \geq G(\omega)$ for all ω, which means that the sets f_{lm} and g_{lm} cannot be independent from each other.

RESULTS

General Parameters

The experimental X-ray and neutron scattering curves after data reduction and buffer subtraction are shown in Figures 2a and 2b. The dependence of the integral parameters on contrast provides the first consistency check. For this, each data set was treated separately and the values of the zero-angle intensity, the radius of gyration and the Porod volume were evaluated. The matching points of the 50S were estimated from the linear fit of the normalized forward scattering to be at $\langle\rho\rangle \cong 0.455 e/Å^3$ for X-rays and 59% D_2O for neutrons (Figures 3a and 3c). Using these matching points, the contrasts for each data set are evaluated. The plots of the squared radius of gyration *versus* reciprocal contrast are

Table I. Geometrical parameters of the 50S subunit and its models

Model	Radius of Gyration R_g(Å)	Maximum Size D_{max}(Å)	Volume $V(10^6 Å^3)$
50S[1)]	80	250	3.1
RNA[1)]	65	200	1.8
Protein[1)]	97	250	1.1
50S[2)]	78	260	2.6
RNA[2)]	64	200	1.5
Protein[2)]	94	260	1.1
50S[3)]	79	260	2.9
50S[4)]	73	270	2.1
50S[5)]	77	250	2.5
50S[6)]	64	220	1.5

[1)]and [2)] contrast variation in sucrose and water, this work;
[3)]Vasiliev's model (Vasiliev *et al.*, 1983);
[4)]Frank's model (Radermacher *et al.*, 1987);
[5)]model of Lake (1976) as evaluated by Tardieu & Vachette (1982);
[6)]Yonath's model (Yonath *et al.*, 1987)

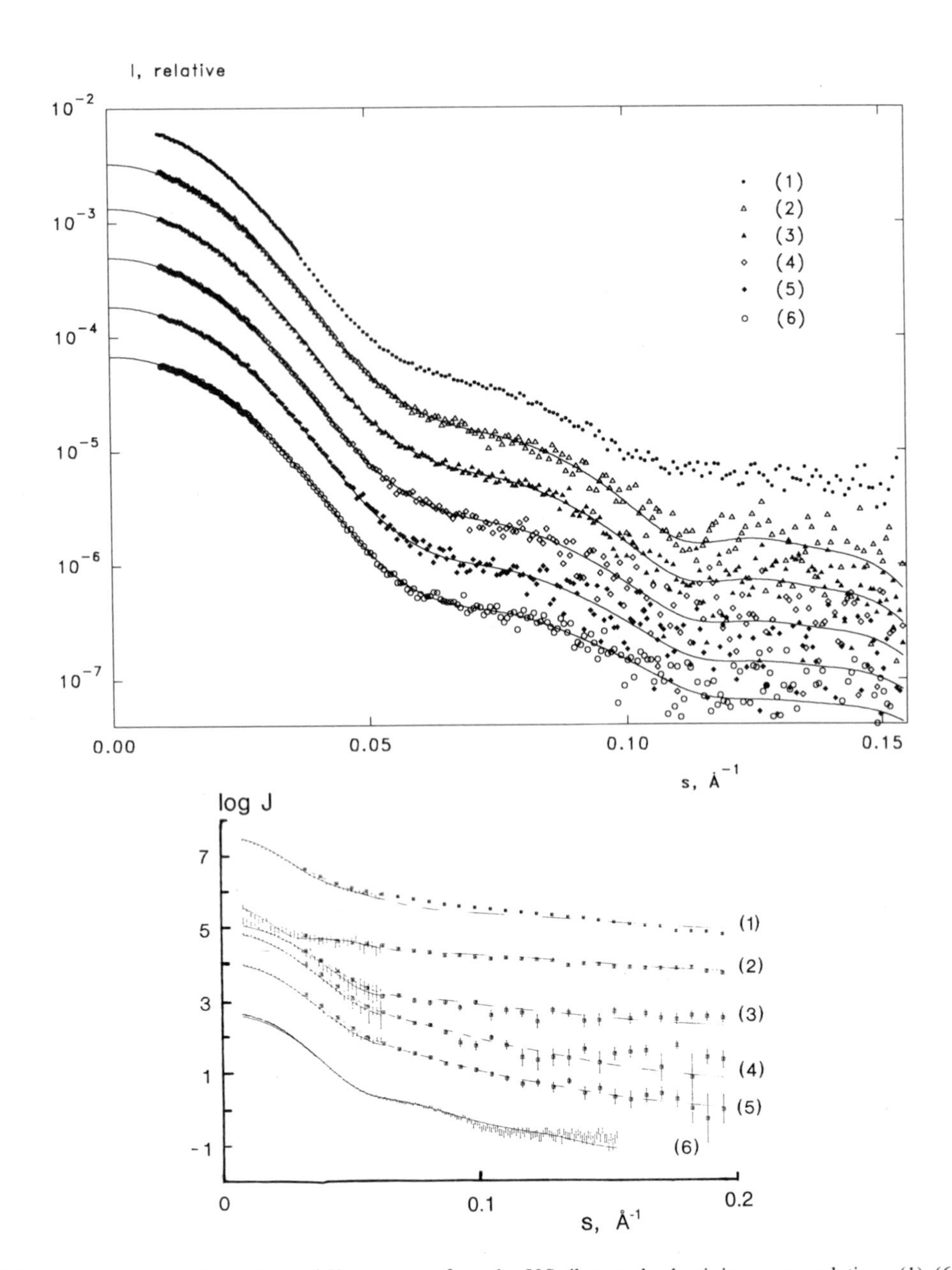

Figure 2. (a): processed experimental X-ray curves from the 50S ribosomal subunit in sucrose solutions. (1)–(6) correspond to 0, 7, 14, 24, 32, 38 weight percent sucrose. Successive curves are divided by 2 for better visualization. For the data sets (2)–(6) the fit by the indirect transform program CGNOM is shown in full lines. (b): processed contrast variation data from aqueous solutions. (1)–(5) neutron data in 0, 14, 40, 67 and 96% D_2O, respectively. The data from the first setting are shown as error bars, from the second setting - rectangles with error bars. The fit to the first setting is drawn in full lines, to the second setting in dashed lines. (6) X-ray scattering in water: experimental data (rectangles with the error bars) and fit (full line).

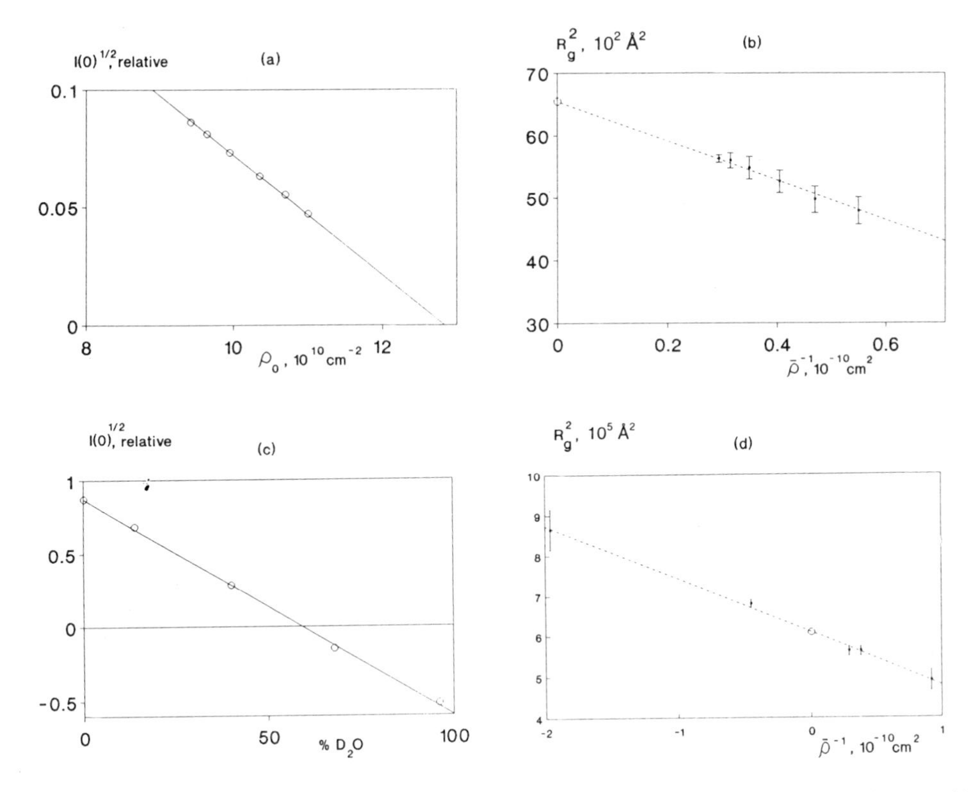

Figure 3. Contrast dependence of the forward scattering (a,c) and of the radius of gyration (b,d). Figures (a-b) correspond to the X-ray data, (c-d) to the neutron data. Associated errors in I(0) are smaller than the diameter of the open circles in (a,c). The linear least squares fit is shown as dashed line in (b,d), R_c is marked by the open circles.

presented in Figures 3b and 3d and give R_c = 80Å as a preliminary estimate of the radius of gyration at infinite contrast for the X-rays and 78Å for neutrons. The maximum size of the 50S particle was found to be 250Å. The radii of gyration of the 50S and its components (Table I) are in a good agreement with the earlier studies of Serdyuk & Grenader (1977) and Tardieu & Vachette (1982). They also support the generally accepted hypothesis that the RNA and proteins in 50S are well segregated (ie a compact RNA-rich core is surrounded by the protein moiety).

The point corresponding to the X-ray data in the buffer without sucrose (leftmost point in Figures 3a and 3b) in the contrast dependence of I(0) and R_g somewhat deviates from the others. This is even more pronounced in the plot of the Porod volume (Figure 4). Here, the points corresponding to the sucrose buffers are well fitted by a straight line whereas the volume corresponding to the X-ray scattering in water is noticeably smaller. Moreover, from numerous SAS studies (Serdyuk 1979; Tardieu & Vachette, 1982; Meisenberger *et al.,* 1984) it is known that the volume of the hydrated 50S is $V_c = 2.5$ -2.6×10^6Å^3, and the volume fraction of protein is about $v_{pr} = 0.37$. The scattering length densities of hydrated proteins and RNA are approximately $\rho_{pr} = 11.6 \times 10^{10}$ and $\rho_{rna} = 13.5 \times 10^{10}$cm^{-2}, respectively (as we simultaneously use X-ray and neutron data, the scattering length densities and contrasts will be expressed in 10^{10}cm^{-2}; 1.0e/Å^3 = 28.2×10^{10}cm^{-2}). Using these values, the mean square fluctuation:

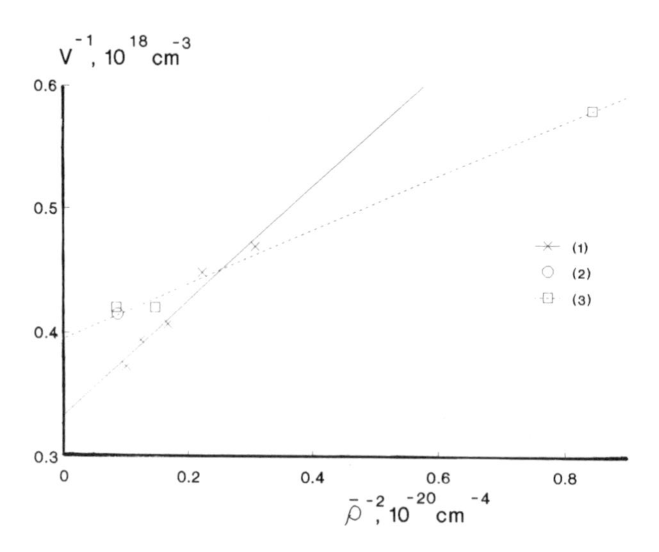

Figure 4. Porod volume as a function of contrast. (1) X-ray data in sucrose solutions; (2) X-ray data in buffer without sucrose; (3) neutron data: the points (from left to right) correspond to buffers with 0%, 14% and 40% D_2O.

$$\langle \rho_s^2 \rangle = \int \rho_s^2(r) d^3r \cong V_c\left[(1-v_{pr})(\rho_{rna} - \langle \rho \rangle)^2 + v_{pr}(\langle \rho \rangle - \rho_{pr})^2\right]$$

is estimated to be 210–220cm^{-1}. The linear dependence for X-rays yields significantly larger values: V_c = 3.0 × 10^6Å^3 and $\langle \rho_s^2 \rangle = 430$ cm^{-1}. These differences are readily explained by noting that the point corresponding to the X-ray scattering in water fits well to the linear regression of the Porod volume dependence for the neutron data (Figure 4). Quantitatively, V_c = 2.6 × 10^6Å^3 and $\langle \rho_s^2 \rangle = 150$ cm^{-1} are obtained for neutron and X-ray scattering in water whereas the theoretical value for neutrons is $\langle \rho_s^2 \rangle = 145$ cm^{-1}. This means that the neutron contrast variation data together with the X-ray scattering in water correspond well to the two-phase model with an invariant volume of 2.6 × 10^6Å^3. At the same time, the scattering in sucrose reveals a larger invariant volume which is the volume inaccessible to the sucrose molecules. The mean square fluctuation for this case is increased due to the presence of water as the third phase inside the invariant volume. The most important conclusion is that the X-ray scattering curve in water should not be used together with the X-ray data from the solutions containing sucrose; instead, it should be analyzed together with the neutron H_2O/D_2O exchange data.

Structure Modeling, X-Ray Data

The X-ray experimental data were treated simultaneously in terms of the three basic characteristic functions (*cf* Equations 2, 5 and 7). The development into the 'shape-inhomogeneities' functions yields the distribution functions and their propagated errors shown in Figure 5a and the backtransformed basic scattering functions in Figure 5b. The corresponding fit to the experimental data is presented in Figure 2a. The formal decomposition into the scattering from RNA and protein assuming the contrasts and volume fractions of the components as written above results in the distribution functions in Figure 6a and the scattering curves in Figure 6b. The shape scattering and the scattering from the RNA part can be further interpreted in terms of their shapes, since at low resolution both 50S and its RNA-rich core can be considered as compact globular particles.

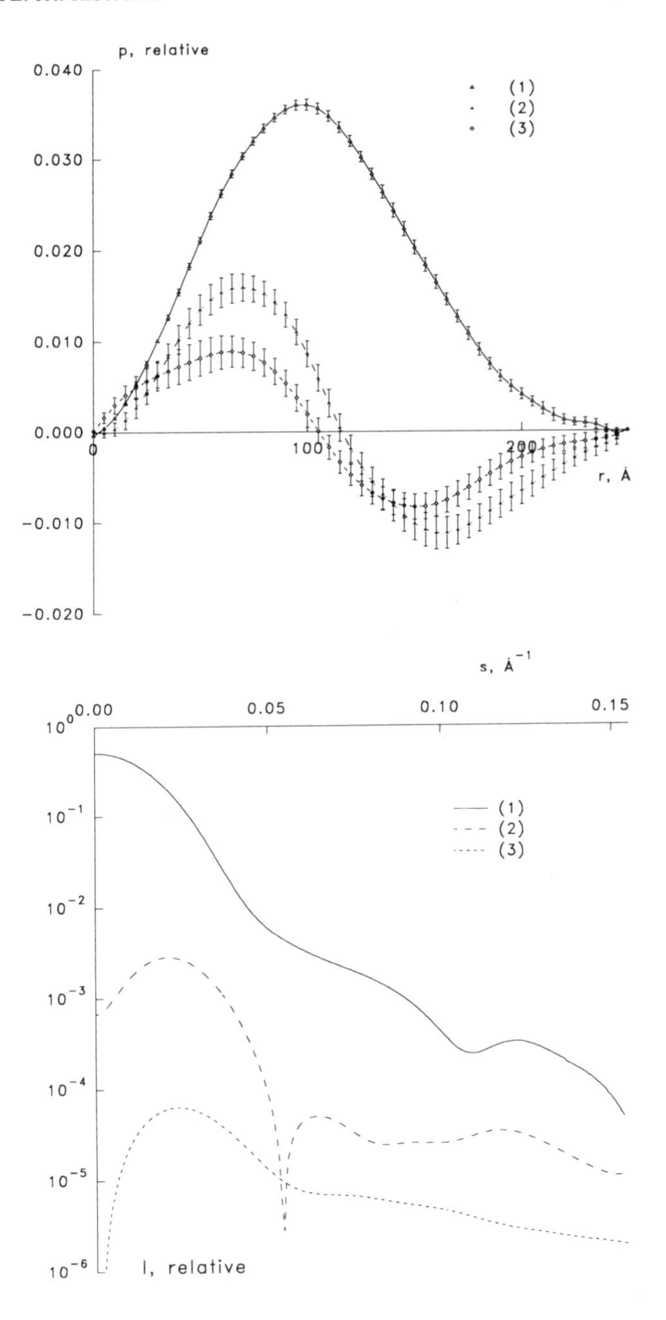

Figure 5. Basic scattering functions for X-rays in real (a) and reciprocal (b) space: (1) shape, (2) cross-term, (3) inhomogeneities. In reciprocal space, logarithms of the absolute values of intensities are presented; the cross term (2) is negative for $s > 0.055 Å^{-1}$.

The shape modeling followed the procedures described by Svergun & Stuhrmann (1991). The scattering curves were scaled to the evaluated Porod volumes and the functional (Equation 17) was minimized. In both cases, the best fit triaxial ellipsoids were used as initial approximation (half axes $113 \times 56 \times 125 Å^3$ for the 50S subunit, $91 \times 45 \times 92 Å^3$ for the RNA-rich core). Starting with $L = 3$ and $s_1 = 0.05 Å^{-1}$, the resolution was gradually increased to $L = 7$ and $s_1 = 0.11 Å^{-1}$. The tails of the scattering curves (from $s_1 = 0.11 Å^{-1}$ to s_{max} were not included in the functional to be minimized and were used to select the best

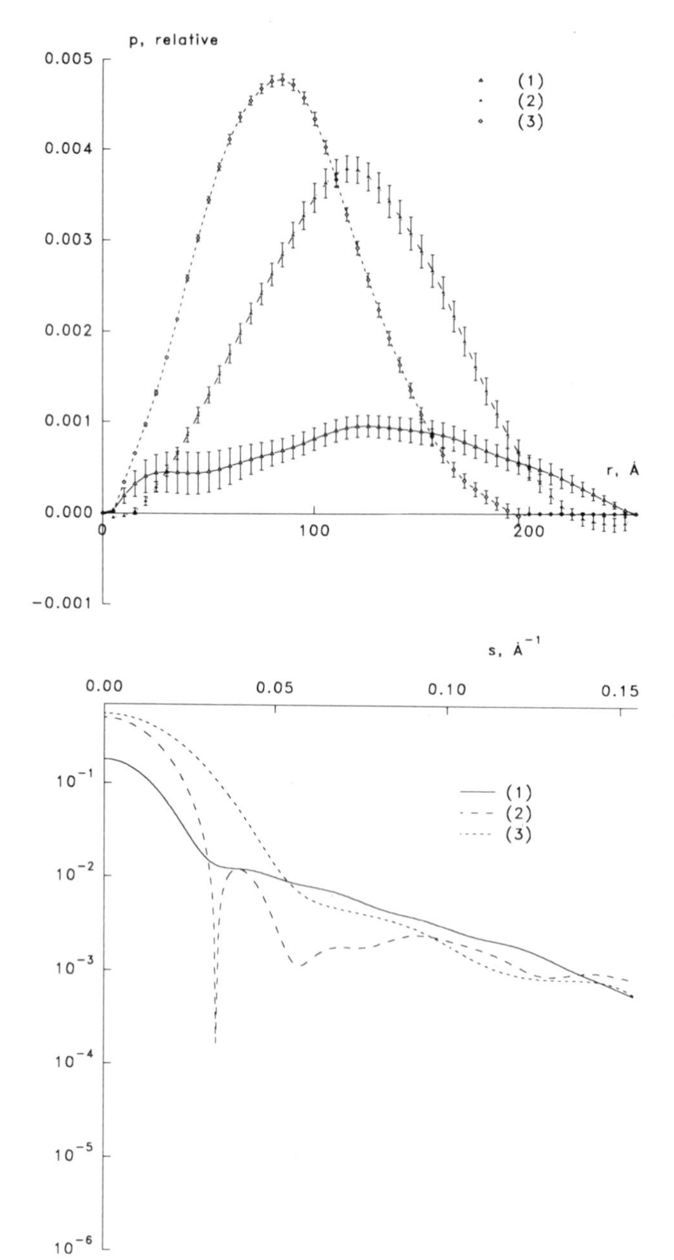

Figure 6. Decomposition into the basic functions of protein and RNA for X-rays in real (a) and reciprocal (b) space: (1) $I_{pro}(s)$, (2) cross-term (negative for $s > 0.03Å^{-1}$), (3) $I_{rna}(s)$.

solution at higher resolution (L = 7), where the interpretation is not unique. Figures 7a and 7b presents the basic functions together with the fits by the initial three-axial ellipsoid and by the restored shapes at L = 5 (Figures 8a and 8d) and L = 7 (Figures 8b and 8c). Final R-factors at L = 7 are 7×10^{-3} for the 50S and 2×10^{-3} for the RNA part with $\chi = 0.01$.

If one now puts the two shapes together and assumes the first body (50S) to have a contrast $\bar{\rho}_a = 1$, and the second body (the RNA-rich core) a contrast $\bar{\rho}_b = -1$, the hollow particle thus obtained should correspond to the region occupied mainly by proteins. The six rotational and translational parameters which describe the relative position of the RNA-rich core were found so as to fit the basic scattering curve $I_{pro}(s)$ scaled to the difference of the Porod volumes of 50S and RNA part. The functional to be minimized was similar to Equation 17 with $\chi = 0$ and $s_1 = 0.09Å^{-1}$. Figure 7c presents the fits obtained using the best superposition of two triaxial ellipsoids (established in the same way) and the shapes of 50S and RNA-rich core at L = 5 and L = 7 positioned as shown in Figure 8. The distance between the centers of mass of the 50S and its RNA-rich core in this model is only 10Å. For L = 7, the R-factor in the range $[0, s_1]$ is 0.04, in the entire range $[0, s_{max}]$ it is 0.17.

Structure Modeling, Neutron Data

The model derived from the X-ray data has several limitations. Thus, the shape scattering curve corresponds to the volume inaccessible to sucrose rather than to the actual shape of the particle. The excess volume (about 20%) also biases the scattering curves of the RNA and protein component. Moreover, the scattering from the protein moiety cannot be determined reliably because the range of contrasts available in the X-ray studies is distant from the matching point of RNA. Note also that the two-phase model assumes a constant scattering length density inside each phase and thus does not take into account the inhomogeneities inside the RNA and protein components.

This model was refined using the neutron contrast variation data. As reliable desmearing of the neutron data is difficult, fitting of the unprocessed experimental data was preferred as a data analysis strategy. The data sets obtained at different contrasts were processed simultaneously in the frame of the advanced two-phase model of Svergun (1994). The shape of the entire particle was described using the angular function $F(\omega)$ and the shape of the RNA-rich core by the function $G(\omega)$ (see Figure 1b). The SAS intensity at a given contrast $\bar{\rho}$ is written using Equation 19, where $A_{lm}(s)$ and $B_{lm}(s)$ are the partial amplitudes of the particle shape and the RNA shape expressed via the coefficients f_{lm} and g_{lm}, respectively, $\bar{\rho}_a = \bar{\rho}_{rna}$, $\bar{\rho}_b = \bar{\rho}_{rna} - \bar{\rho}_{pro}$ is the difference between the contrasts of RNA and protein, $\bar{\rho} = v_{pro}\bar{\rho}_{pro} + (1 - v_{pro})\bar{\rho}_{rna}$, v_{pro} is the volume fraction of protein. The scattering from inhomogeneities Equation 18 of the RNA-rich core was evaluated for the shape $G(\omega)$ and that of the protein component for the difference shape function $H(\omega) = F(\omega) - G(\omega)$. The magnitudes Λ_l describing these contributions were included as additional unknown parameters in the minimization procedure.

The experimental data sets were normalized in such a way that:

$$I(0, \bar{\rho}_{50S}) = \left[v_{pro}\bar{\rho}_{pro} + (1 - v_{pro})\bar{\rho}_{rna}\right]^2 V_c^2 \qquad (21)$$

where the values of the forward scattering are obtained by GNOM and V_c is the invariant particle volume. The contrasts of the 50S and its components were taken as (Moore *et al.*, 1974; Serdyuk, 1979):

$$\left.\begin{aligned} \bar{\rho}_{pro} &= 2.29 - 5.62y \\ \bar{\rho}_{rna} &= 4.13 - 6.07y \\ \bar{\rho}_{50S} &= 3.45 - 5.90y \end{aligned}\right\} \qquad (22)$$

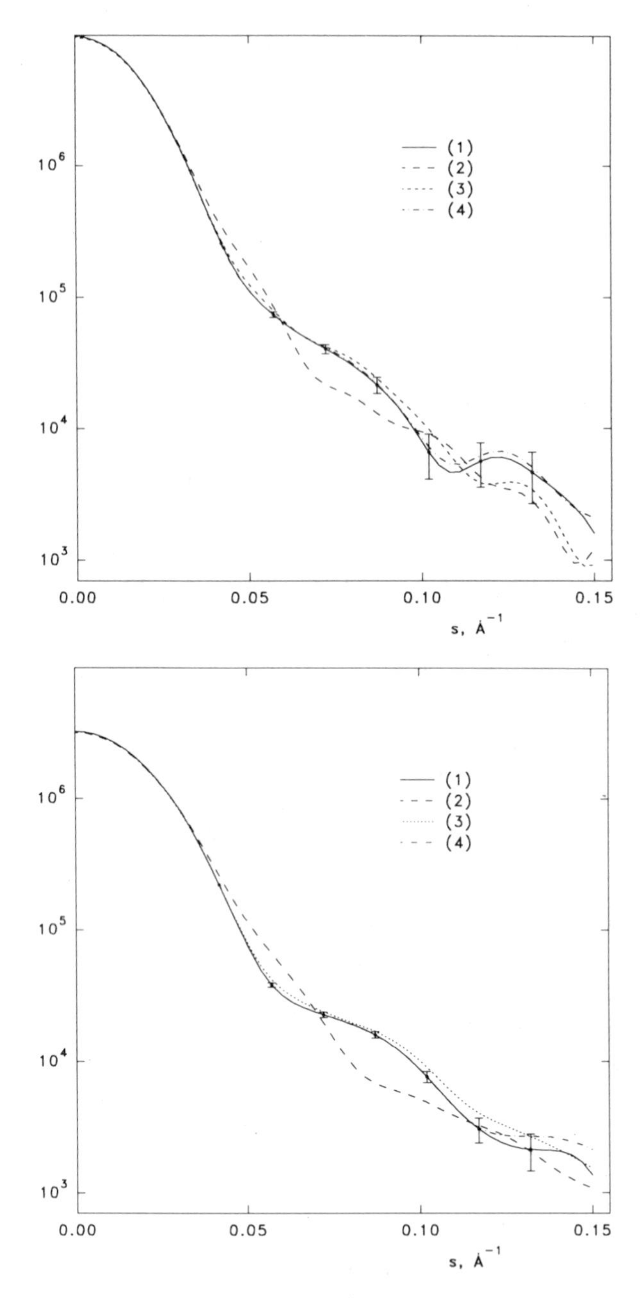

Figure 7. Fit of the X-ray scattering from the 50S, (a) its RNA-rich core (b) and the protein moiety (c): (1) basic scattering curves, (2) best fit with triaxial ellipsoids, (3) and (4) scattering from the restored shapes up to resolution of L = 5 and L = 7, respectively. For the protein moiety, the scattering of the best superposition of the corresponding models of the 50S and its RNA-rich core is shown. Error bars indicate the standard deviations evaluated from the real space functions.

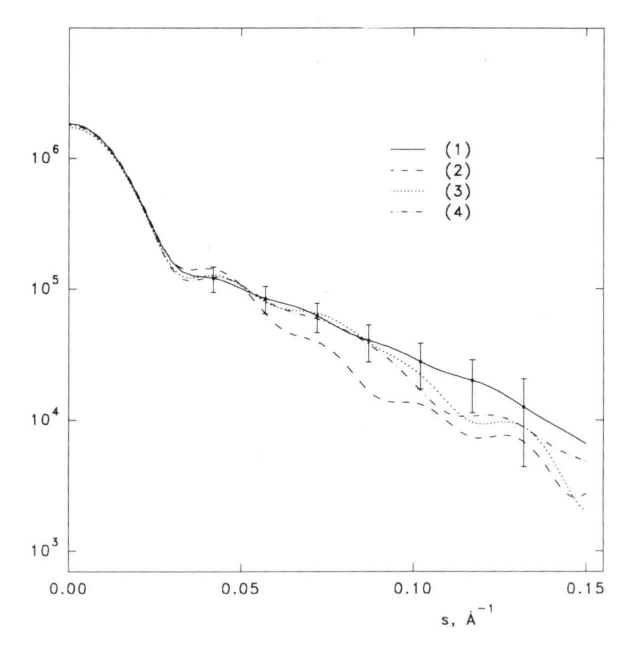

Figure 7. (*Continued*)

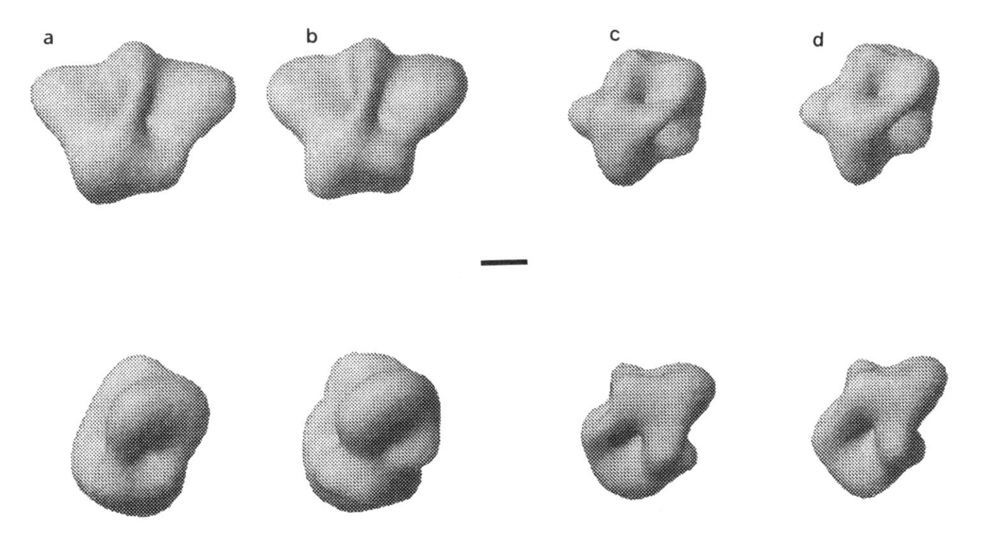

Figure 8. Shaded models of the shapes of the 50S subunit (a,b) and its RNA-rich core (c,d) inaccessible to sucrose. Columns (a) and (d) correspond to L = 5, (b) and (c) to L = 7. For the 50S subunit, crown and kidney views are shown; the position of the RNA model corresponds to the best fit to the protein scattering (see text).

where $0 \le y \le 1$ is the D_2O fraction of the solvent. The maximum possible amplitudes of the density fluctuations inside each phase were estimated as follows. In the protein or RNA in the partially deuterated solvent with a D_2O content y, the maximum amplitude of the density fluctuations will be $\Delta\rho_f = \max\left[\bar{\rho}_c(y), y\rho_{D_2O} - \rho_{H_2O}\right]$, where $\bar{\rho}_c(y)$ denotes the contrast of the component (RNA or protein). Using Equation 21 and the values of the scattering length densities of water and heavy water yields for the two components:

$$\left.\begin{aligned}\Delta\rho_f^{pro}(y) &= \max\left[2.29 - 5.62y, 6.97y + 0.56\right]\\ \Delta\rho_f^{rna}(y) &= \max\left[4.13 - 6.07y, 6.97y + 0.56\right]\end{aligned}\right\} \quad (23)$$

The shape coefficients were determined by a non-linear minimization procedure simultaneously fitting the experimental data sets at all contrasts (Svergun, 1994). For each instrumental setting, the R-factor between the experimental data and the theoretical curve was taken as:

$$R_K^2 = \frac{\sum_{j=1}^{M(K)}\left\{\int_{s_{min}^{(K)}}^{s_{max}^{(K)}}\left[\frac{J_e(s,\bar{\rho}_j) - J_t(s,\bar{\rho}_j)}{\sigma(s,\bar{\rho}_j)}\right]^2 ds\right\}}{\sum_{j=1}^{M(K)}\left\{\int_{s_{min}^{(K)}}^{s_{max}^{(K)}}\left[\frac{J_t(s,\bar{\rho}_j)}{\sigma(s,\bar{\rho}_j)}\right]^2 ds\right\}} \quad (24)$$

where K is the setting number, $s_{min}^{(K)}$ and $s_{max}^{(K)}$ are the limiting values of the momentum transfer, M(K) is the number of contrasts $\bar{\rho}_j$ [j = 1,...M(K)] recorded for this setting, $\sigma(s,\bar{\rho}_j)$ represents the experimental errors in the j-th data set $J_e(s,\bar{\rho}_j)$. The theoretical fit $J_t(s,\bar{\rho}_j)$ was evaluated according to Equation 19 and smeared using the resolution function R(s,s') as:

$$J(s) = \int I(s')R(s,s')ds' \quad (25)$$

where J(s) denotes the smeared curve. Fast smearing of the theoretical curves was achieved using the approach of Pedersen *et al.* (1990) with R(s,s') approximated by a Gaussian function of variable width depending on the actual value of momentum transfers.

In total eleven experimental curves were fitted simultaneously. Two neutron settings (1 and 2) contained each five scattering curves at different contrasts (M(1) = M(2) = 5). The X-ray scattering in water was analyzed together with the neutron data and was considered to be the third experimental setting consisting of one curve (M(3) = 1). A formal D_2O content of 17% giving the same ratio of contrasts of protein and RNA as for the X-rays (Serdyuk, 1979) was ascribed to this curve.

Additional information was included in the minimization procedure in the form of penalty functions. For the 50S and the RNA-rich core, restrictions on their radii of gyration and volumes were imposed. The deviation from the desired values of the radius of gyration R_0 and the particle volume V_0 was described by:

$$P(R_g,V) = (1 - R_g/R_0)^2 + \left[1 - (V/V_0)^{1/3}\right]^2 \quad (26)$$

where R_g and V are the current values. The penalty due to the negativity of the shape function $F(\omega)$ was expressed as:

$$N(F) = \int_{\omega^-} F(\omega)^2 d\omega \,/ \int F(\omega)^2 d\omega \tag{27}$$

where the first integral is taken over the range of negative $F(\omega)$ ($N(F) = 0$ if $F(\omega)$ is always positive). As $F(\omega) \geq G(\omega) \geq 0$ by definition, the non-negativity was imposed for all three functions $F(\omega)$, $G(\omega)$ and $H(\omega)$. The current values of R_g and V for the 50S and RNA-rich core are readily evaluated from the corresponding shape coefficients (Svergun & Stuhrmann, 1991); fast estimation of the integrals in Equation 27 was done using an angular grid based on Fibonacci numbers (Svergun, 1994).

As the R-factors for the different settings are relative quantities, each setting was weighted with the empirical fidelity factor W_K so that the functional to be minimized was:

$$\Phi(f_{lm}, g_{lm}, \Lambda_l) = \sum_{K=1}^{3} W_K R_K^2 \,/ \sum_{K=1}^{3} W_K + \chi[P(R_c, V_c) + P(R_{rna}, V_{rna}) + N(F) + N(G) + N(H)] \tag{28}$$

A relative weight of two was ascribed to each curve in the first neutron setting, of one to the second neutron setting and five to the X-ray curve, leading to fidelities of W_1 = 0.5, $W_2 = W_3 = 0.25$. The higher weight of the X-ray curve reflects the wider angular range and the absence of the instrumental smearing in the synchrotron measurements. The penalty term in the square brackets in Equation 28 ensures that the geometrical parameters of the two shapes do not significantly deviate from the desired values and that the shape functions remain non-negative (the weighting factor χ is equal to 0.5).

The values of the contrasts evaluated from Equation 22 are presented in Table II. The model-independent geometrical parameters the 50S subunit and its RNA-rich core, estimated from the experimental data in the previous section were taken as desired values in Equation 26. The minimization was started from the shapes of the 50S subunit and the RNA-rich core up to the resolution L = 5 shown in Figures 8a and 8d and rescaled to the volumes V_c and V_{rna}, respectively.

The model was refined in two steps. First, only the low-angle neutron setting 1 was included into the functional Equation 10 and the preliminary shapes $F(\omega)$ and $G(\omega)$ were estimated assuming a pure two-phase model. Later, settings 2 and 3 were added and the scattering from the density fluctuations within each phase was taken into account.

Table II. Contrasts and R-factors at different D_2O contents

y	$\bar{\rho}_{50S}$ $10^{10}cm^{-2}$	$\bar{\rho}_{pro}$ $1010cm^{-2}$	$\bar{\rho}_{rna}$ $1010cm^{-2}$	R1	R2	R3
0	3.45	2.29	4.13	0.035	0.030	—
14	2.62	1.50	3.28	0.039	0.037	—
40	1.09	0.04	1.70	0.062	0.080	—
67	−0.51	−1.48	0.06	0.166	0.092	—
97	−2.22	−3.10	−1.70	0.069	0.185	—
X-rays	2.57	1.45	3.22	—	—	0.069

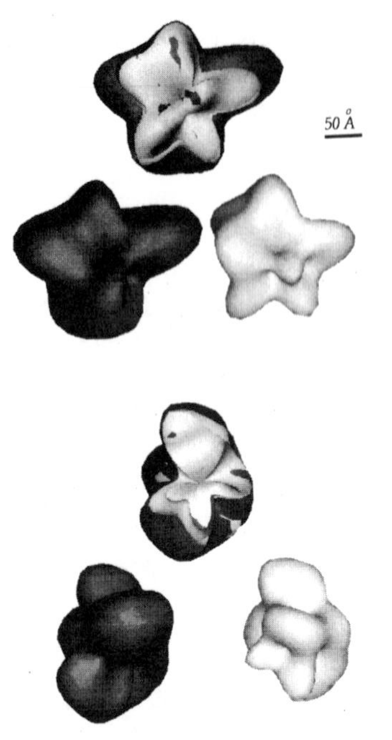

Figure 9. Refined model of the 50S subunit (the particle envelope is shown in dark grey, the RNA-rich core in light grey). Top: crown view, bottom: kidney view. Middle column presents the interior view (cross-section by the plane through the paper).

Figure 2b shows the best fit to the experimental data with the R-factors summarized in Table II. Minor systematic deviations in the fit to the X-ray data (curve 6) most probably originate from differences in the invariant volume for X-rays and neutrons due to H/D exchange. The refined shapes of the 50S subunit and the RNA-rich core are presented in Figure 9. Their spatial resolution estimated as $\pi b_0/(L+1)$, where b_0 is the radius of the equivalent sphere (Svergun & Stuhrmann, 1991) is 45Å for the 50S shape and 38Å for the shape of the RNA-rich core.

The geometrical characteristics of the model are given in Table I; the volume fraction of proteins is 0.4 and the separation between the centers of mass of protein component and RNA-rich core is 12Å. Figures 10a, 10b and 10c present the ideal (not smeared) scattering intensities of the 50S subunit, and its components calculated from the model along with the scattering from the density fluctuations inside the RNA-rich core and the protein moiety evaluated as linear superpositions of functions (Equation 18) with the amplitudes of fluctuations shown in Figure 10d.

DISCUSSION

Both the preliminary (Figure 8) and the refined (Figure 9) models display the 'crown view' and the 'kidney view' of the overall shape of 50S. It is worth noting here that these models were obtained directly using data from the native particles in solution, without any *a priori* information from other methods. The density fluctuations inside the RNA-rich core are relatively small and correspond to an average size of about 23Å (ie the distance between the RNA strands). In contrast, in the protein region strong scattering from inhomogeneities with an average size of 35Å which can be considered as an estimate of the average size of an individual protein molecule is detected.

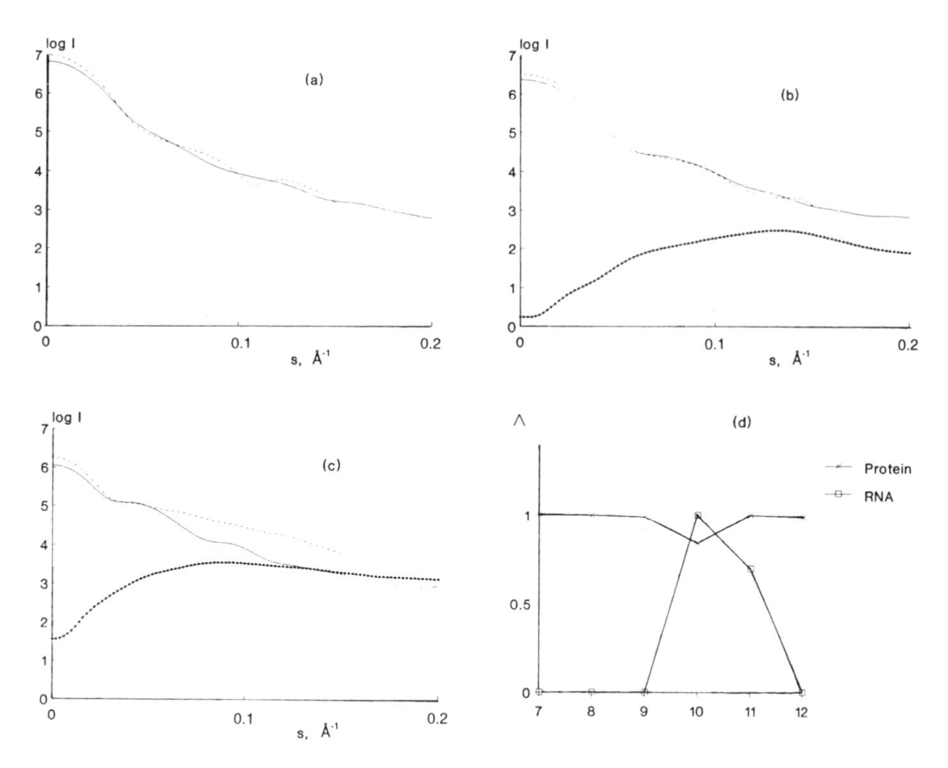

Figure 10. Scattering from the models of the 50S subunit (a), its RNA-rich core (b) and the protein moiety (c); full lines: scattering from the model in Figure 9, dashed lines: scattering curves obtained from X-ray contrast variation with sucrose; symbols: scattering from inner inhomogeneities inside the RNA-rich core and the protein moiety with the amplitudes of the fluctuations presented in (d).

We now compare our results with the electron microscopic models of Vasiliev *et al.* (1983), Radermacher *et al.* (1987) and Yonath *et al.* (1987) referring to them for short as Vasiliev's, Frank's and Yonath's model, respectively. Vasiliev's model, typical for the visually derived models, was obtained on freeze-dried and shadowed samples, Frank's model by three-dimensional reconstruction of a conical tilt series from particles with a well defined orientation relative to the grid. Negatively stained 50S subunits from *E.Coli* were studied in both cases. The general outlines of both models agree well with the 'consensus' model although Frank's model displays finer details. Yonath's model was obtained by image reconstruction from two-dimensional crystals of the 50S subunit from *B.stearothermophilus* and displays major discrepancies with the previous models. All the models were obtained from the authors in the form of data files describing the particle envelopes and their geometrical parameters are given in Table I. For comparison, boundary functions $F(\omega)$ for the models were evaluated and expanded in terms of spherical harmonics up to $L = 7$ corresponding to a resolution of about 30–40Å. This smooths finer details on the surface but preserves the gross features which are relevant in the present comparison. The deviation between two shapes $F_1(\omega)$ and $F_2(\omega)$ was characterized by the root mean square:

$$R_\omega = \left\{ \int_\omega \left[F_1(\omega) - F_2(\omega) \right]^2 d\omega \right\}^{1/2} \tag{29}$$

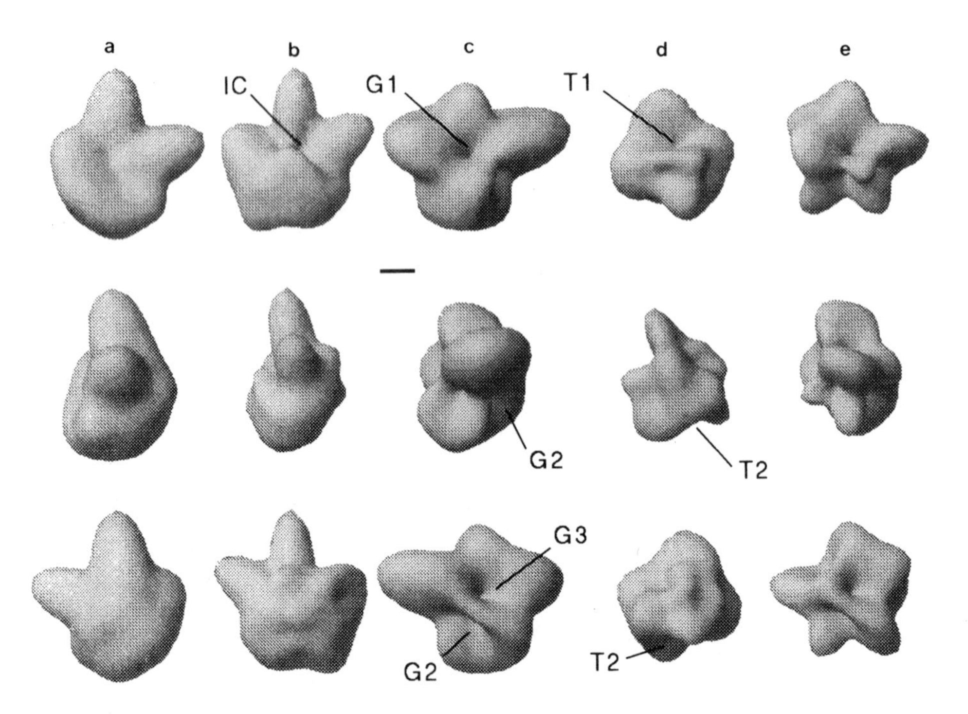

Figure 11. Shaded models of the 50S subunit : (a) Vasiliev's model; (b) Frank's model; (c) and (e) envelopes of 50S and its RNA-rich core, this work; (d) Yonath's model. Top row: crown view, middle row: kidney view, bottom row: view from the back (crown view rotated 180°). Bar length, 50Å. Models (a), (b) and (d) were expanded in spherical harmonics up to L = 7.

Vasiliev's and Frank's model (Figure 11a and 11b) were rotated so as to approximately fit our model (Figure 11c) in the crown orientation. They display a good qualitative agreement to our shape of 50S (quantitatively, R_{ω} is equal to 22Å for both models). Their integral parameters presented in Table I also correlate well with those of our model and of the model of Lake (1976). The flatter appearance of Frank's model results from specimen flattening during sample preparation. Recently, this group reported the results of a study of frozen hydrated 50S specimens (Radermacher *et al.*, 1992). On photographs their new model (which has not yet been distributed) displays similar major features but appears to be more rounded and thus in better agreement with our results. Our model has a pronounced groove in the interface side (marked G1) which correlates well with the 'interface canyon' of Frank's model (marked IC) and another groove (G2) in the lower part on the back side which was also observed by Korn *et al.* (1983a). These two grooves could make the path of the nascent chain from the interface side to the secretory domain (Lake, 1985). There is also a deep groove G3 on the back side which opposes to G1 and is almost connected to it thus corresponding well to the most pronounced hole in Frank's model when represented with a higher density threshold (Frank *et al.*, 1990).

Yonath's model lacks the crown view and its orientation for comparison can thus not easily be selected. The main feature of this model is a tunnel, which was rotated to make its ends (T1 and T2) coincide with the grooves G1 and G2 (Figure 11d). This model gives, both qualitatively and quantitatively, a significantly better agreement with our model of the RNA-rich core (Figure 11e, R_ω = 22Å) than with that of the entire 50S (R_ω = 29Å). The integral parameters evaluated from Yonath's model (Table I) are far too small to be attributed to the whole 50S subunit but also well coincide with those of RNA-rich core. This all indicates that Yonath's model must be strongly biased towards the RNA. The most likely explanation (also suggested by Frank *et al.*, 1991) is that a too high threshold was selected by Yonath *et al.* (1987) for the tracing of the contour lines in the image reconstruction.

The shape of the RNA-rich core (Figure 11e) displays several finger-like protuberances and protrudes at several sites of the particle surface. These features are in agreement with the direct electron microscopic observations of Korn *et al.* (1983b) and Frank *et al.* (1991) as well as with the results obtained by Kühlbrandt & Unwin (1982) for the crystalline eucaryotic ribosomes. Space for the proteins is left mostly in the regions of the two projecting side arms and at the bottom of the particle. In general this correlates well with the immunoelectron microscopy model of Walleszek *et al.* (1988), the main difference being that in the latter model no proteins are located near the bottom. Such a total absence of proteins at the bottom of the particle would, however, lead to a significant separation of the centers of mass of the protein moiety and the RNA-rich core which is incompatible with the well established fact that this distance does not exceed 20–30Å (Koch & Stuhrmann, 1979; Serdyuk, 1979) and with our X-ray and neutron results. Note also that, according to Lake (1985), at least two proteins are located at the bottom of the 50S subunit.

Comparing the models in Figure 11 one should bear the limitations inherent to the methods used in their construction in mind. Thus, integral parameters and appearance of Frank's and Yonath's models depend significantly on the density threshold selected for contouring the particle boundary. In fact, in the original paper of Radermacher *et al.* (1987), several models constructed with different density thresholds were presented, and the data set available to us had an enclosed volume of 2.1×10^6Å^3 whereas the model displayed in the illustrations had a volume of 2.34×10^6Å^3. The same is true for Yonath's model which had a volume of 1.5×10^6Å^3 compared to the 1.8×10^6Å^3 stated in the original article of Yonath *et al.* (1987).

In our two-phase scheme we assume sharp phase boundaries and full segregation between the regions occupied by RNA and proteins. Although well justified as this level of resolution (see eg Serdyuk *et al.*, 1979; Kühlbrandt & Unwin, 1982; Frank *et al.*, 1991), this is a simplified model, and systematic deviations in the fit of the 97% D_2O data (Figure 2, curve 1) which is most sensitive to the cross term between RNA and proteins may result from this simplification. Truncation of the number of harmonics leads to a termination effect whereby the lower order multipoles try to compensate for the lack of higher ones. Consequently, some features and especially those corresponding to the limiting harmonic L = 5 (eg the projecting arms), may appear artificially enhanced. Obviously, the enantiomorphic structure would produce the same fit to the scattering data.

Let us also consider the validity of our model in terms of information content. The number of multipole coefficients used to describe both shapes was $2 \times (L + 1)^2 = 72$. Additionally, $2 \times 6 = 12$ parameters described the scattering from the density fluctuations. This gives a total number of 84 parameters for the final model. The number of independent parameters is, however, smaller. Thus, six parameters describe the orientation and the position of the center of mass of the 50S which can be taken arbitrarily. Restrictions on the values of R_g and V both for RNA-rich core and 50S decrease the number of unknowns by four. The amplitudes of fluctuations L_l inside RNA-rich core are equal to zero for l = 7,8,9,12 and can thus be omitted. This leaves 70 parameters which are still correlated because of the restrictions $F(\omega) \geq 0$, $G(\omega) \geq 0$ and especially $H(\omega) \geq 0$ The exact reduction resulting from this correlation is difficult to establish formally. According to Shannon's theorem which applies to the function sI(s), the experimental range contains about $K = s_{max}D_{max}/\pi \cong 16$ sampling points (Moore, 1980, Taupin & Luzzati, 1982) which gives an estimate of the number of independent parameters. The contrast variation data set then contains $3K + K_l = 54$ parameters (here $K_l = 6$ is the number of contrasts used, Svergun, 1994). These estimates suggest that a representation in terms of 70 not entirely independent parameters is not unreasonably different from what is allowed by a strict application of Shannon's theorem.

The results of this study clearly illustrate the potential of the new approaches for data interpretation of the data from combined X-ray and neutron solution scattering. The model of the 50S ribosomal subunit reconciles the electron microscopic models and should also provide the necessary information for phasing the low-angle reflections in the crystallographic studies of ribosomes (Yonath, 1992). The developed SAS data analysis methods which have also successfully been used to study protein complexes (eg König *et al.*, 1993) and are now being applied to the contrast variation study of the 70S ribosome.

ACKNOWLEDGMENTS

The authors are indebted to Dr V. Vasiliev, Dr J. Frank and Prof. A. Yonath for providing their models. They also thank Dr J. Frank for the helpful discussions, Miss P. Brouillon and Dr Z. Sayers for their help in preparing the samples for the experiments. The neutron scattering experiments at Risø National Laboratory were supported by the Commission of the European Community through the Large Installation Plan. The work was also supported by the NATO Linkage Grant LG 921231.

REFERENCES

Boulin, C., Kempf, R., Koch, M.H.J., & McLaughlin, S.M., (1986). Data appraisal, evaluation and display for synchrotron radiation experiments: hardware and software. *Nucl. Instrum. Methods*, A249:399–407.

Boulin, C.J., Kempf, R., Gabriel, A., & Koch, M.H.J., (1988). Data acquisition systems for linear and area X-ray detectors using delay line readout. *Nucl. Instrum. Methods*, A269:312–320.

Edmonds, A.R., (1957). *Angular Momentum in Quantum Mechanics*. Princeton: Princeton Univ. Press.

Feigin, L.A., & Svergun, D.I., (1987). *Structure Analysis by Small-Angle X-ray and Neutron Scattering*. Plenum Press, New York.

Frank, J., Verschoor, A., Radermacher, M., & Wagenknecht, T., (1990). Morphologies of eubacterial and eucaryotic ribosomes as determined by three-dimensional electron microscopy. In *The Ribosome - Structure, Function and Evolution* (W.Hill editor) ASM, Washington, DC, pp. 107–113.

Frank, J., Penczek, P., Grassucci, R., & Srivastava, S., (1991). Three-dimensional reconstruction of the 70S *Escherichia coli* ribosome in ice: the distribution of ribosomal RNA. *J. Cell Biol.*, 115:597–605.

Gabriel, A., & Dauvergne, F., (1982). The localization method used at EMBL. *Nucl. Instrum. Methods*, 201:223–224.

Gavrilova, L.P., Kostiashkina, O.E., Koteliansky, V.E., Rutkevich, N.M., & Spirin, A.S., (1976). Factor-free ('Non-enzymic') and factor-dependent systems of translation of polyuridylic acid by *Escherichia coli* ribosomes. *J. Mol. Biol.*, 101:537–552.

Harrison, S.C., (1969). Structure of tomato bushy stunt virus. I. The spherically averaged electron density. *J. Mol. Biol.*, 42:457–483.

Hoppe, W., Oettl, H., & Tietz, H.R., (1986). Negatively stained 50S ribosomal subunits of *Escherichia coli*. *J. Mol. Biol.*, 192:291–322.

Ibel, K., & Stuhrmann, H.B., (1975). Comparison of neutron and X-ray scattering of dilute myoglobin solutions. *J. Mol. Biol.*, 93:255–265.

Kiselev, N.A., Stelmashchuk, V.Ya., Orlova, E.V., Vasiliev, V.D., & Selivanova, O.M., (1982). Strand-like structures and their three-dimensional organization in the large subunit of the *Escherichia coli* ribosome. *Molec. Biol. Rep.*, 8:191–197.

König, S., Svergun, D.I., Koch, M.H.J., Hübner, G., & Schellenberger, A., (1993). The influence of the effectors of yeast pyruvate decarboxylase (PDC) on the conformist of the dimers and tetramers and their pH-dependent equilibrium. *Eur. Biophys. J.*, 22:185–194.

Koch, M.H.J., & Bordas, J., (1983). X-ray diffraction and scattering on disordered systems using synchrotron radiation. *Nucl. Instrum. Methods*, 208:461–469.

Koch, M.H.J., & Stuhrmann, H.B., (1979). Neutron scattering studies of ribosomes. In *Methods Enzymol.* (K. Moldave and L. Grossman editors) Academic Press, New York, vol. LIX, pp670–706.

Korn, A.P., Elson, D., & Spitnik-Elson, P., (1983a). A survey of 50S ribosomal subunits by dark field electron microscopy. *Eur. J. Cell Biol.*, 31:325–333.

Korn, A.P., Spitnik-Elson, P., Elson, D., & Ottensmeyer, F.P., (1983b). Specific visualization of ribosomal RNA in the intact ribosome by electron spectroscopic imaging. *Eur. J. Cell Biol.*, 31:334–340.

Kühlbrandt, W., & Unwin, P.N.T., (1982). Distribution of RNA and protein in crystalline eucaryotic ribosomes. *J. Mol. Biol.*, 156:431–448.

Lake, J.A., (1976). Ribosome structure determined by electron microscopy of *Escherichia coli* small subunits, large subunits and monomeric ribosomes. *J. Mol. Biol.*, 105:131–159.

Lake, J.A., (1985). Evolving ribosome structure: domains in archaebacteria, eubacteria, eocytes and eucaryotes. *Ann. Rev. Biochem.*, 54:507–539.

Lebech, B., (1990). Neutron scattering facilities at Risø. *Neutron News*, 1:7–13.

May, R.P., Nowotny, V., Nowotny, P., Voss, H., & Nierhaus, K.H., (1992). Inter-protein distances within the large subunit from *Escherichia coli* ribosomes. *EMBO Journal*, 11:373–378.

Meisenberger, O., Pilz, I., Stöffler-Meilicke, M., & Stöffler, G., (1984). Small-angle X-ray study of the 50S ribosomal subunit of *Escherichia coli*. A comparison of different models. *Biochem. Biophys. Acta*, 781:225–233.

Moore, P.B., Engelman, D.M., & Schoenborn, B.P., (1974). Asymmetry of the 50S ribosomal subunit of *Escherichia coli*. *Proc. Natl. Acad. Sci. USA*, 71:172–176.

Moore, P.B., (1980). Small-angle scattering. Information content and error analysis. *J. Appl. Cryst.*, 13:168–175.

Nierhaus, K.H., (1982). Structure, assembly and function of ribosomes. *Current Topics in Microbiology and Immunology*, 97:81–155.

Pedersen, J.Skov, Posselt, D., & Mortensen, K., (1990). Analytical treatment of the resolution function for small-angle scattering. *J. Appl. Cryst.*, 23:321–333.

Radermacher, M., Srivastava, S., & Frank, J., (1992). The structure of the 50S ribosomal subunit from *E.coli* in frozen hydrated preparation reconstructed with secret. Proceedings of EUREM 92, Granada, pp.19–20.

Radermacher, M., Wagenknecht, T., Verschoor, A., & Frank, J., (1987). Three-dimensional structure of the large ribosomal subunit from *Escherichia coli*. *EMBO Journal*, 6:1107–1114.

Serdyuk, I.N., & Grenader, A.K., (1977). On the distribution and packing of RNA and protein in ribosomes. *Eur. J. Biochem.*, 79:495–504.

Serdyuk, I.N., Grenader, A.K., & Koteliansky, V.E., (1977). Study of 30S ribosomal subparticle protein-deficient ribonucleoprotein derivative by X-ray diffusion scattering. *Eur. J. Biochem.*, 79:505–510.

Serdyuk, I.N., Grenader, A.K., & Zaccai, G., (1979). Study of the internal structure of *Escherichia coli* ribosomes by neutron and X-ray scattering. *J. Mol. Biol.*, 135:691–707.

Serdyuk, I.N., (1979). A method of joint use of electromagnetic and neutron scattering: a study of internal ribosomal structure. In *Methods Enzymol.* (K. Moldave and L. Grossman editors), Academic Press, New York, vol. LIX, pp. 750–775.

Stöffler-Meilicke, M., & Stöffler, G., (1990). Topography of the ribosomal proteins from *Escherichia coli* within the intact subunits as determined by immunoelectron microscopy and protein-protein cross-linking. In *The Ribosome, Structure, Function and Evolution* (W.Hill editor) ASM, Washington, pp123–133.

Stöffler, G., & Stöffler-Meilicke, M., (1986). Immuno electron microscopy of *Escherichia coli* ribosomes. In *Structure, Function and Genetics of Ribosomes* (B.Hardesty and G.Kramer, editors), Springer, New York, pp28–46.

Stuhrmann, H.B., (1970a). Interpretation of small-angle scattering of dilute solutions and gases. A representation of the structures related to a one-particle scattering functions. *Acta Cryst.*, A26:297–306.

Stuhrmann, H.B., (1970b). Ein neues Verfahren zur Bestimmung der Oberflächenform und der inneren Struktur von gelösten globulären Proteinen aus Röntgenkleinwinkelmessungen. *Zeitschrift für Physikaliche Chemie Neue Folge*, 72:177–184; 185–198.

Stuhrmann, H.B., & Kirste, R.G., (1965). Elimination der intrapartikulären Untergrundstreuung bei der Röntgenkleinwinkelstreuung am kompakten Teilchen (Proteinen). *Zeitschrift für Physikaliche Chemie Neue Folge*, 46:247–250.

Stuhrmann, H.B., Koch, M.H.J., Parfait, R., Haas, J., Ibel, K., & Crichton, R.R., (1977). Shape of the 50S subunit of *Escherichia coli* ribosomes. *Proc. Natl. Acad. Sci. USA*, 74:2316–2320.

Svergun, D.I., (1991). Mathematical methods in small-angle scattering data analysis. *J. Appl. Cryst.*, 24:485–492.

Svergun, D.I., (1992). Determination of the regularization parameter in indirect transform methods using perceptual criteria. *J. Appl. Cryst.*, 25:495–503.

Svergun, D.I., (1994). Solution scattering from biopolymers: advanced contrast variation data analysis. *Acta Cryst.*, A50:391–402.

Svergun, D.I., & Stuhrmann, H.B., (1991). New developments in direct shape determination from small-angle scattering. 1. Theory and model calculations. *Acta Cryst.*, A47:736–744.

Svergun, D.I., Semenyuk, A.V., & Feigin, L.A., (1988). Small-angle scattering-data treatment by the regularization method. *Acta Cryst.*, A44:244–250.

Tardieu, A., & Vachette, P., (1982). Analysis of models of irregular shape by solution X-ray scattering: the case of the 50S ribosomal subunit from *E.coli*. *EMBO Journal*, 1:35–40.

Taupin, D., & Luzzati, V., (1982). Informational content and retrieval in solution scattering studies. I. Degrees of freedom and data reduction. *J. Appl. Cryst.*, 15:289–300.

Vasiliev, V.D., Selivanova, O.M., & Ryazantcev, S.N., (1983). Structure of the *Escherichia coli* 50S ribosomal subunit. *J. Mol. Biol.*, 171:561–569.

Walleczek, J., Schüler, D., Stöffler-Meilicke, M., Brimacombe, R., & Stöffler, G., (1988). A model for the spatial arrangement of the proteins in the large subunit of the *Escherichia coli* ribosome. *EMBO Journal*, 7:3571–3576.

Wittmann H.G., (1983). Architecture of procaryotic ribosomes. *Ann. Rev. Biochem.*, 52:35–65.

Yonath, A., Leonard, K.R., & Wittmann, H.G., (1987). A tunnel in the large ribosomal subunit revealed by three-dimensional image reconstruction. *Science*, 236:813–816.

Yonath, A., & Berkowitch-Yellin, Z., (1993). Hollows, voids, gaps and tunnels in the ribosome. *Curr. Opin. Struct. Biol.*, 3:175–181.

Yonath, A., (1992). Approaching atomic resolution in crystallography of ribosomes. *Annu. Rev. Biophys. Biomol. Struct.*, 21:77–93.

15

PROBING SELF ASSEMBLY IN BIOLOGICAL MIXED COLLOIDS BY SANS, DEUTERATION, AND MOLECULAR MANIPULATION

R. P. Hjelm,[1*] P. Thiyagarajan,[2] A. Hoffman,[3] C. Schteingart,[3] and Hayat Alkan-Onyuksel[4]

[1] Los Alamos Neutron Scattering Center
Los Alamos National Laboratory
Los Alamos New Mexico 87545
[2] Intense Pulsed Neutron Source
Argonne National Laboratory
Argonne Illinois 60436
[3] Department of Medicine
University of California, San Diego
La Jolla California 92093-0813
[4] Department of Pharmaceutics
University of Illinois, Chicago
Chicago, Illinois 60612

ABSTRACT

Small-angle neutron scattering was used to obtain information on the form and molecular arrangement of particles in mixed colloids of bile salts with phosphatidylcholine, and bile salts with monoolein. Both types of systems showed the same general characteristics. The particle form was highly dependent on total lipid concentration. At the highest concentrations the particles were globular mixed micelles with an overall size of 50Å. As the concentration was reduced the mixed micelles elongated, becoming rodlike with diameter about 50Å. The rods had a radial core-shell structure in which the phosphatidylcholine or monoolein fatty tails were arranged radially to form the core with the headgroups pointing outward to form the shell. The bile salts were at the interface between the shell and core with the hydrophilic parts facing outward as part of the shell. The lengths of the rods increased and became more polydispersed with dilution. At sufficiently

* Send Correspondence to: Rex P. Hjelm, H805, Los Alamos National Laboratory, Los Alamos, New Mexico 87545-1663. Tel: 505-665-2372; FAX: 505-665-2676; E-mail: Hjelm@LANL.GOV

Neutrons in Biology, edited by Schoenborn and Knott
Plenum Press, New York, 1996

low concentrations the mixed micelles transformed into single bilayer vesicles. These results give insight on the physiological function of bile and on the rules governing the self assembly of bile particles in the hepatic duct and the small intestine.

INTRODUCTION

The transport of lipophilic materials in biological systems requires carrier-detergent systems to emulsify relatively insoluble molecules that would otherwise precipitate in the aqueous environment. Bile is one such system. It is responsible for the transport of insoluble material, such as cholesterol, from the liver to the intestine. Once in the intestine, bile takes on a second function in emulsifying dietary fats. Bile aids in the action of lipases in hydrolyzing dietary triglycerides and facilitates the adsorption of insoluble digestion products into the intestinal lumen (Borgström *et al.,* 1986). Small-angle neutron scattering (SANS), is providing information essential to unraveling the particle morphology and the nature of self assembly in these important biological mixed colloids.

Bile is a complex mixture, the major components of which are phosphatidylcholine, the common component of cell membranes, and bile acid (Cabral & Small, 1989). The bile acids consist of a cholesteric core, containing hydroxy groups at two or three positions. In humans the carboxylic acid group is commonly conjugated with either glycine or taurine. The resulting detergents are cholylglycine and cholyltaurine, for the trihydroxy forms and chenodeoxycholylglycine and chenodeoxycholyltaurine, for the dihydroxy-forms. Other constituents of bile are bilirubin and cholesterol. Bile components are excreted by the liver. The self assembled mixture then moves down the hepatic bile duct to the gall bladder, where it is concentrated. On hormonal and mechanical stimulus from the stomach the bile is diluted into the complex milieu in the duodenum. Thus this mixed colloid is subjected to a cycle of concentration and dilution in the normal course of physiological functioning.

Once in the intestine the material is in a complex environment and is in continuous equilibrium with dietary fatty materials and digestion products (Borgström *et al.,* 1986). Among these are the hydrolysis products of dietary triglycerides, monoolein and oleic acid (Dreher *et al.,* 1967; Hofmann, 1963). Thus the make up of colloid particles formed in the intestine changes continuously. Further, the volume of dietary fats is too large to be emulsified at once by the available bile; thus the system is characterized by a complex exchange of material between emulsified and phase separated components. As the dietary mixture moves down the small intestine the fatty components are removed by adsorption into the endothelial cells of the intestine. Distally in the intestine the bile salt and cholesterol are reabsorbed. The bile salt is cycled to the liver for reuse.

In modeling such a series of environments and physiological processes for physical chemical measurement we have to simplify the system considerably, taking the point of view that the features seen in mixtures of the major components determine the important characteristics of particles formed *in situ*, with the other components and environments being perturbations on the major theme. Thus, studies necessarily center on mixtures of one of the common human bile salts with phosphatidylcholine. Further, the cycle of assembly, concentration, dilution and disassembly all are likely to involve equilibria between differing proportions and concentrations of components. Thus studies on the concentration dependence and component ratios in solution are important not just for understanding self assembly in these colloids, but also for understanding the physiological functional forms of bile colloid particles.

Studies of bile salts with monoolein are important to understanding the physiology of bile, as monoolein is part of the intestinal milieu. In addition comparative studies of particle morphology, will give information on the structural determinants of self assembly. Monoolein, for example, has a nonionic head group and a single fatty acid tail. This differs from phosphatidylcholine, which is a two-tailed, zwitterionic surfactant. Thus the comparison will be important in understanding the importance of the different moieties in determining the structure.

The first evidence towards understanding of the self assembly of these systems comes from the studies of Don Small and collaborators (Small, 1967; 1971; Small *et al.*, 1966), who devised the first phase map of the ternary water, cholate, phosphatidylcholine system. This map is quite complex with a number of liquid crystal (hexagonal, lamellar and bicontinuous) phases. The map shows, however, that the mixtures are isotropic at physiologically-relevant compositions. Since its discovery, the isotropic region has been extensively studied by dynamic light scattering (DLS), which showed a strong dependence on particle size and morphology on composition and lipid concentration (Mazer *et al.*, 1980; 1984; Schurtenberger *et al.,* 1985). The conclusions from these studies were: (i) mixed micelles of bile salt and phosphatidylcholine are present in the isotropic region at sufficiently high total lipid concentrations. (ii) when concentrated isotropic solutions are diluted, the mixed micelle size increases. (iii) when the systems are diluted sufficiently, there is a transition to vesicles. The DLS studies left open the question of the particle form and arrangement of components in the mixed micelle and vesicle phases.

It turns out that SANS provides the ideal measurements to determine the concentration dependent particle form and internal structure in these mixed colloid systems (Hjelm *et al.,* 1988; 1990a; Hjelm *et al.,* 1990b; Hjelm *et al.,* 1992; Hjelm *et al.,* 1995). Two factors combine to make this so. First, the length scales probed by SANS measurements correspond well to the overall mixed micelle and vesicle sizes and at the same time to the molecular sizes of the components. Second, the strong difference in scattering between hydrogen and deuterium provides an important source of contrast for neutron scattering. Thus by studying the systems in D_2O, sufficient contrast results so that a wide range of concentrations can be easily accessed, even with modest neutron sources. Also, labeling specific parts of components with deuterium allows a determination of the distribution of the label.

Here we review our work on the particle morphology of bile salt-phosphatidylcholine and bile salt-monoolein mixed aqueous colloids. We describe how SANS was used to establish micelle shape and how the arrangement of components in the micelles was determined using specific deuteration. Finally, we present some new work on mixtures of bile salts with monoolein.

METHODS

Preparation of Mixed Colloids

Mixed colloids of bile salt and phosphatidylcholine were prepared by coprecipitation in which the separate components dissolved in ethanol were mixed in the required molar ratio, then dried. The mixture was taken up in a D_2O buffer, pD 7.4, containing 0.15M NaCl to a final concentration of 50g/l. This stock was diluted to the final concentrations by the addition of D_2O buffer. Similar methods were used to make the bile salt-monoolein mixtures.

Small-Angle Neutron Scattering

SANS measurements are conducted at either the Small-angle Neutron Diffractometer (SAD) at Argonne National Laboratory, or the 30m small angle neutron camera, NG7SANS, at the National Center for Cold Neutron Research of the National Institutes of Standards and Technology, Gaithersburg, Maryland. Scattered intensity, I, as a function of scattering angle, 2θ, and incident wavelength, λ, was converted to differential scattering cross section per unit mass of lipid (mg^{-1}) as a function of momentum transfer, $Q = (4\pi/\lambda) \sin\theta$. The method for doing this for data taken on time-of-flight instruments, such as SAD, is described elsewhere (Hjelm, 1988). Estimation of the errors on I were computed by propagating the variances on the raw data using standard methods. These are reported as root mean square (rms) deviations about the means.

Analysis

The analysis of particle shape from SANS data starts with the well known Guinier approximation for small angle scattering intensity as a gaussian when $QR_g < 1$, where R_g is the radius of gyration about the particle scattering length density distribution centroid. This approximation is written as:

$$I(Q) = \Delta M_0 \exp\left(\frac{1}{3} Q^2 R_g^2\right) \tag{1}$$

where ΔM_0 is the contrast-weighted scattering mass of the particle.

Modifications of the Guinier approximation are used to analyze scattering from particles with one or two dimensions sufficiently large that the particle boundaries do not contribute to the vignetted view of the scattering measurement. In these cases the gaussian approximation is changed by the inclusion of a power law in Q with the power corresponding to the inverse of the dimension of the particle. The Guinier approximation for rodlike forms is given by (Hjelm, 1985; Luzzati, 1960):

$$I(Q) = Q^{-1} \Delta m_0 \exp\left(\frac{1}{2} Q^2 R_c^2\right) \tag{2}$$

Here R_c is the radius of gyration of the rod cross section with respect to the rod axis, and Δm_0 is the contrast weighted scattering mass per unit length along the rod axis (see Equation 4). The Guinier approximation for a sheet is given by (Knoll *et al.,* 1981; Porod, 1982):

$$I(Q) = Q^{-2} \Delta\mu_0 \exp(Q^2 R_d^2) \tag{3}$$

Here the prefactor, $\Delta\mu_0$, is the contrast-weighted scattering mass per unit area, and R_d is the radius of gyration of the sheet scattering length density taken relative to the centroid plane of the sheet.

The prefactor, contrast dependent term in Equations 1 through 3 carries important information on the composition density of the particles. In particular we can write Δm_0 of Equation 2 for a mixed colloid particle as (Hjelm *et al.,* 1992):

$$\Delta m_0 = n_L \frac{N_0}{\left[\Gamma^{-1} W_B + W_L\right]} \left[\Delta\rho_B V_B \gamma^{-1} + \Delta\rho_L V_L\right]^2 \tag{4}$$

where the units are m/g for Δm_0. In Equation 4 N_0 is Avogadro's number, $\Delta\rho$ is the contrast and V is the molecular volume. The W are gram molecular weight equivalents, Γ is the mole ratio of phosphatidylcholine or monoolein, L, to bile salt, B, in the solution, with γ the molar ratio of the two species in the particles. By using Equation 4 we can calculate the linear density of insoluble amphiphile, n_L, for a rodlike micelle from the values obtained using the approximations of Equation 2.

A full analysis of the particle shape is carried out using suitable (unitless) spherically-averaged form factors $\langle|F(Q)|^2\rangle$. In this work on mixed colloids the useful form factors are that for a cylinder of radius R and height, H, given as:

$$\left\langle |F(Q)|^2 \right\rangle = 16 \int_0^{2\pi} \frac{\sin^2(QH/2\cos\theta)}{(QH\cos\theta)^2} \frac{J_1^2(QR\sin\theta)}{(QR\sin\theta)^2} \cos\theta d\theta \tag{5}$$

and that for a vesicle or radius R and thickness t = R-r,

$$\left\langle |F(Q)|^2 \right\rangle = 9 \left[\frac{R^2 j_1(QR) - r^2 j_1(Qr)}{Q(R^3 - r^3)} \right]^2 \tag{6}$$

In Equations 5 and 6 J_1 is a first order Bessel function of the first kind and j_1 is the corresponding spherical Bessel function.

In real self assembling colloid solutions the form factors of a particle with a simple size and shape can only rarely be fit to a real scattering curve. Rather a population of particles with different sizes and shapes must be taken into consideration. In this case we define an average scattering function:

$$\overline{\left\langle |F(Q)|^2 \right\rangle} = \frac{\int f(p_1,p_2)[\Delta\rho(p_1,p_2)]^2[V(p_1,p_2)]^2 \left\langle |F(Q(p_1,p_2))|^2 \right\rangle dp_1 dp_2}{\int f(p_1,p_2)[\Delta\rho(p_1,p_2)]^2[V(p_1,p_2)]^2 dp_1 dp_2} \tag{7}$$

In Equation 7 p_1 and p_2 are model parameters (such as R and H in Equation 5), V is the particle volume and $\Delta\rho$ is the particle contrast. The probability density function (PDF) of a particle having a value of p_1 between p_1 and p_1+dp_1 simultaneously with it having a value of p_2 between p_2 and p_2+dp_2 is $f(p_1,p_2)dp_1dp_2$.

Finally, the measured intensity also includes the contribution of instrument resolution ΔQ. For a typical SANS measurement $\Delta Q/Q \approx 0.1$. If we define the resolution function, R(Q), then the measured intensity is:

$$I(Q) = R(Q) \otimes \overline{\left\langle |F(Q)|^2 \right\rangle} (N_0 M^{-1}) \tag{8}$$

where the convolution operation is indicated by $\otimes$ and M is the total mass of lipid in the beam. In our analysis we attempt to invert both Equations 8 and 7 using the maximum entropy algorithm (Hjelm *et al.*, 1990b) to obtain the joint PDF, $f(p_1,p_2)$, given that the solution consistent with cylindrical mixed micelles.

A complete analysis of the scattering would also include the effects of interparticle interactions (Kotlarchyk & Chen, 1993; Long *et al.*, 1994) so that

$$I(Q) = S'(Q)\overline{\langle |F(Q)|^2 \rangle},$$

where $S'(Q)$ is the structure factor taking into account these interactions, and rescaled to account for the effects of particle anisotropy and polydispersity. A proper form for the structure factor is difficult to find, however, and because we believe that $S'(Q)$ does not affect our analysis to any great extent over the Q-domain and concentrations studied we do not include in our analysis.

RESULTS

We illustrate the behavior of these systems with the bile salts cholylglycine (Figure 1) and cholyltaurine (Figure 2) mixed with egg yolk phosphatidylcholine (EYPC). These two bile salts are the common ones found in humans. The behavior with dilution from 16.7g/l to 1.0g/l total lipid of both of these mixtures is essentially the same. We draw the following general conclusions from the form of the scattering curves (Hjelm *et al.*, 1988; 1990a; Hjelm *et al.*, 1990b; Hjelm *et al.*, 1992; Hjelm *et al.*, 1995). (i) At the highest concentrations the scattering is consistent with progressive growth of the particles from a globular form at the highest concentrations to elongated forms. (ii) With sufficient dilution the scattering is indicative of long rods. With sufficient dilution there is a concentration dependent transition to vesicles. (iii) The vesicle sizes decreases on further dilution. (iv) There appears to be a coexistence region of rodlike micelles and vesicles. We outline the analysis that bring us to these general conclusions.

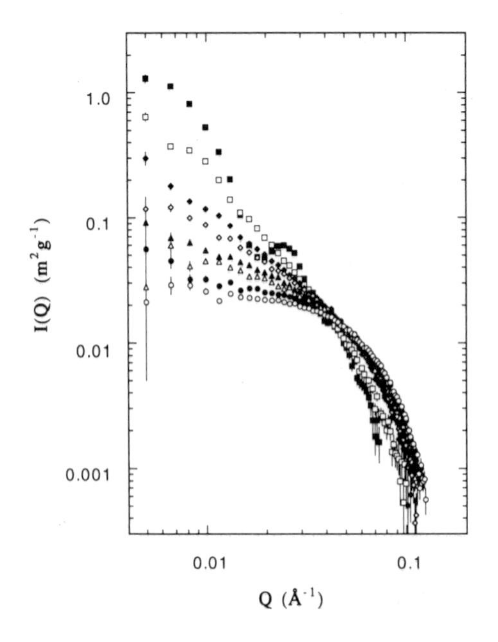

Figure 1. The concentration-dependent scattering from solutions of cholylglycine with egg yolk phosphatidylcholine: Total lipid concentrations in each solution: ●, 10g/l; ¤, 8.3g/l; Δ, 7.1g/l; ▲, 6.3g/l; ◊, 5.0g/l; ◆, 4.2g/l; □, 3.3g/l; ■, 2.0g/l. Γ = 0.5.

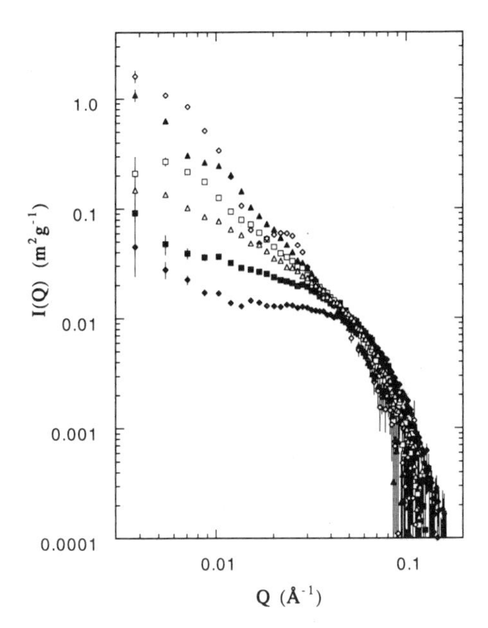

Figure 2. The concentration-dependent scattering from solutions of cholyltaurine with egg yolk phosphatidylcholine: Total lipid concentrations in each solution: ◆, 16.7g/l; ■, 7.2g/l; Δ, 4.2g/l; □, 3.3g/l; ▲, 2.5g/l; ◊, 1.7g/l. $\Gamma = 0.5$.

Figure 3 shows a Guinier analysis according to Equation 1 for one of the cholylglycine-EYPC mixtures in Figure 1, and the results of the analysis is shown in Table I. These values must be considered as an approximation as we have not evaluated $S'(Q)$. Even so these values for R_g and ΔM_0 and those published previously on the cholylglycine mixtures (Hjelm *et al.*, 1988; 1990a; Hjelm *et al.*, 1990b; Hjelm *et al.*, 1992; Hjelm *et al.*, 1995) are very close to those calculated when attempts are made to model $S'(Q)$ by hard core interactions (Long *et al.*, 1994a; Long *et al.*, 1994b).

The result of dilution of the high concentration solutions is elongation of the particles along a single axis. This is documented by fitting the scattering curves to the cylindrical form factor of Equation 5. The problem, however, is that no single model parameter

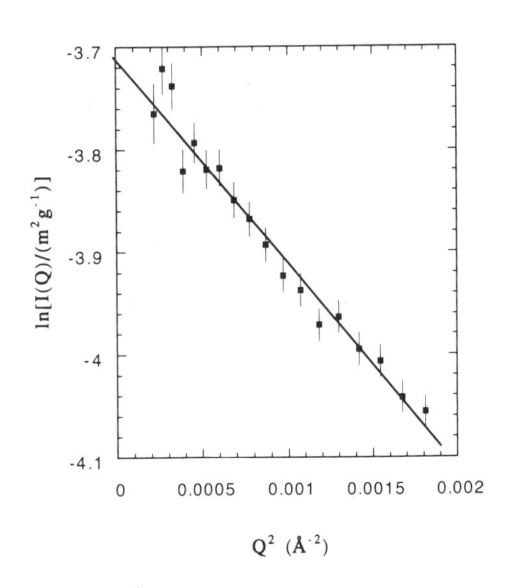

Figure 3. Guinier analysis of SANS data from solutions of cholylglycine with egg yolk phosphatidylcholine: ■, data; ———, fit to Equation 1. Total lipid concentration is 16.7g/l.

Table I. Structural parameters from scattering analysis of globular mixed micelles

Mixture	C_{lipid} (g/l)	ΔM_0 ($10^{-4}m^2/g$)[a]	R_g (Å)
EYPC-cholylglycine[b]	16.7	240 (10)	24.1 (0.3)
Monoolein-cholylglycine	10.0	188 (1)	22.9 (0.4)

[a]Numbers in parentheses are root mean square uncertainty in the regression.
[b]See text for abbreviations.

can be made to fit the scattering fully satisfactorily - provision must be made for particle polydispersity (Hjelm *et al.*, 1990b). This can be done using the method of maximum entropy (Hjelm *et al.*, 1990b) to invert Equations 7 and 8 to determine a population PDF, f(R,H). When this is done we find that the scattering can be explained by a population of prolate cylinders with radius of about 27Å and varying heights. The mean heights of the cylinders increase with decreasing concentration (Table II), as does the polydispersity characterized by the rms (Table II). Maximum entropy tends to find solutions that maximize the variance in the PDF, and this tends to PDF's that have multiple modes with values of H = 50, 100, and 150Å (Hjelm *et al.*, 1990b). Due to the non-uniqueness of the inversion, other PDF's are certainly possible. Thus, the result, though not proving that the particles formed have discrete lengths that are multiples of the size of the globular particles found at the highest concentrations, show that the data is consistent with this possibility.

The continuation of particle elongation is to form a long rodlike micelle (Hjelm *et al.*, 1990a; Hjelm *et al.*, 1990b; Hjelm *et al.*, 1992; Hjelm *et al.*, 1995). This conclusion is documented using the Guinier analysis modified for a rodlike morphology (Figure 4), which shows the approach to the form expected by Equation 2 as the concentration is reduced. Only the low-Q part of the scattering curve changes with dilution. This is the behavior expected for a system of elongated or rodlike particles which grow along the rod axis, but with the radius remaining constant. The value of the rod radius from this data is 27Å, in good agreement with our earlier determinations (Hjelm *et al.*, 1988; 1990a; Hjelm *et al.*, 1990b; Hjelm *et al.*, 1992; Hjelm *et al.*, 1995). The rodlike nature of the mixed micelles has been confirmed by DLS, static light scattering (Egelhaaf & Schurtenberger, 1994) and with SANS (Long *et al.*, 1994b) by other workers. The light scattering studies show, furthermore, that the rods are wormlike (Egelhaaf & Schurtenberger, 1994).

Having determined the form of the particles present in these mixtures, we turn our attention to the arrangement of components within the rodlike mixed micelles. In general there are two ways that the rodlike phosphatidylcholines can associate within the rodlike

Table II. Statistical parameters for PDF is calculated using maximum entropy

Lipid	R_g	Radius			Height		
(g/l)	(Å)	mean (Å)	mode (Å)	rms (Å)[a]	mean (Å)	mode(s) (Å)	rms (Å)[a]
16.7	24.5	28	31	5	38	40	12
10.0	28.3	27	27	3	57	50 100	24
8.3	36.1	27	28	3	60	50 100	26

[a]Root mean square deviation of the PDF's about the mean.

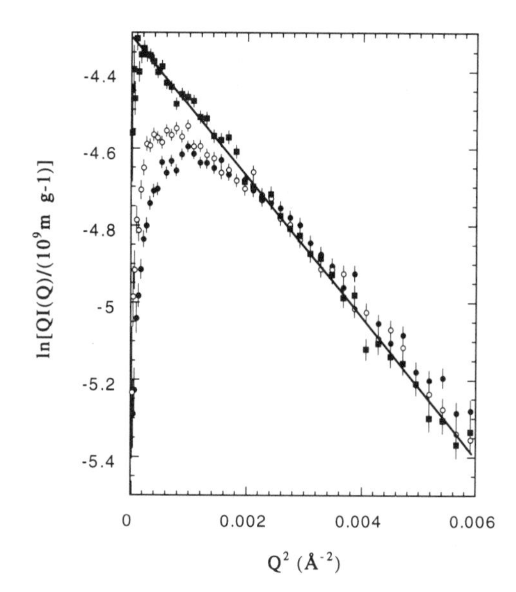

Figure 4. Guinier analysis modified for a rodlike form factor of SANS data from solutions of cholylglycine with egg yolk phosphatidylcholine: Data: •, 7.1 g/l; ¤, 6.3 g/l; ■, 5.0g/l, total lipid concentration. Fit; ———, to Equation 2 for the 5.0g/l sample.

micelle structure - perpendicular or parallel to the rod axis. Indeed both types of arrangement have been proposed.

One, which we call the stacked disk model, envisions a basic disk-like micelle, present at higher concentrations consisting of a small disk-like bilayer of phosphatidylcholine. The detergent, the bile salt, acts to stabilize this structure by forming a ribbon around the edge of the disk. The hydrophobic side of the bile salt being associated with the fatty tails of the phospholipid at the edge of the disk, leaving the hydrophilic side of the hydroxylated cholesterol core and the head groups in contact with the aqueous medium (Mazer *et al.*, 1980; 1984; Small, 1967; 1971; Small *et al.*, 1966). This structure would be approximately 50Å high. Particle elongation, according to this model, occurs by stacking of the disks. The phospholipid would be oriented parallel to the long axis of the rods in this model.

The other model, which we refer to as the radial shell model, has the phospholipid arranged radially to the long axis of the rod (Nichols & Ozarowski, 1990). Thus there would be a hydrophobic core of the fatty acid chains and an outer shell of the phosphatidylcholine head groups in contact with the surrounding water. The bile salt in this scenario, is located at the interface between the core and shell of the structure. The long axis of the cholesterol core is roughly parallel to the micelle axis with the hydrophilic parts of the molecule being part of the shell region of the micelle. The finite length of the micelle is stabilized by the presence of end capping bile salts. In this model, growth is the result of repartitioning of the relatively soluble bile salt into the aqueous phase on dilution. This leaves less material available for endcapping. Consequently the micelles are no longer stable as separate entities and stick together to form a longer structure.

To discriminate between these two models we use the phosphatidylcholine, dipalmitoylphosphatidylcholine, DPPC and compare the scattering with the choline deuterated or not (Hjelm *et al.*, 1992). Because we measure the scattering in D_2O, it is anticipated that if the radial-shell model applied the value of R_c in Equation 2 would be smaller for the material with deuterated DPPC than with protonated material. Indeed, as shown in Figures 5 and 6, this is the case. More detailed calculations support this view (Hjelm *et al.*, 1992;

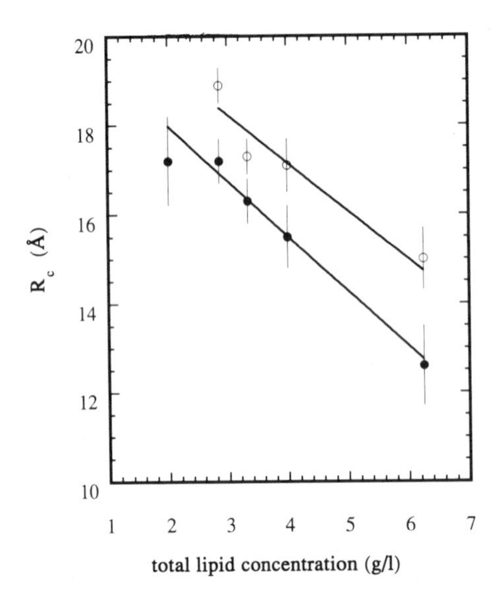

Figure 5. Guinier analysis modified for a rodlike form factor of SANS data from solutions of cholyltaurine with dipalmitoylphosphatidylcholine and with d_9-deutero-dipalmitoylphosphatidylcholine. ¤, with DPPC; •, with d_9-DPPC.

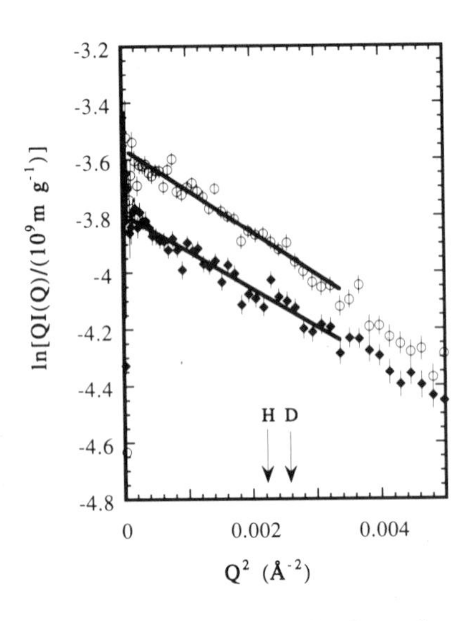

Figure 6. Cross-sectional radii of gyration calculated from SANS data from solutions of cholyltaurine with dipalmitoylphosphatidylcholine and with d_9-deutero-dipalmitoylphosphatidylcholine. ¤, with DPPC; •, with d_9-DPPC.

Hjelm *et al.,* 1995). Without deuteration it is not possible to use neutron scattering to discriminate between the two models.

We have devised a model (Hjelm *et al.,* 1992), shown in Figure 7, which explains one way that the rods are assembled and which accounts for the SANS observations and for the measurements of the molar ratio of PC and bile salts in the mixed micelles (Duane, 1977; Spink *et al.,* 1982). In this the packing density, n_L, of the PC is one molecule per ångstrom. The question arises as to the generality of this model. This is tested by using Equation 4 to calculate n_L. These values are shown for the DPPC-cholyltaurine mixtures and different EYPC-cholylglycine and EYPC-chenodeoxycholylglycine solutions in Table III. The values have been revised slightly upward from those published earlier (Hjelm *et al.,* 1992), as somewhat smaller values of the molecular volume of the bile salt in the range 520 to 600Å^3 are used in this calculation. We find that for all cases considered, n_L is about 1 phosphatidylcholine molecules per ångstrom, consistent with the model in Figure 7. This value is higher than that determined from light scattering studies of 0.77Å^{-1} (Egelhaaf & Schurtenberger, 1994).

There is a biphase region of long rods and vesicles at intermediate concentrations (Hjelm *et al.,* 1995; Long *et al.,* 1994b; Walter *et al.,* 1991). The scattering here is characterized by the dominance of the characteristic scattering for a rod given by Equation 2, coexisting with the first maximum of vesicle-like scattering (Figure 8). The position of this maximum is the same as that seen for the largest vesicles.

The transition to vesicles is marked by a sudden increase in the observable absolute scattering intensity at $Q = 0$Å^{-1} and a transition to the unmistakable form factor for a large spheroid object (Figures 1, 2 and 8). That the form is a vesicle is seen by comparison with the scattering expected for a spherical shell from Equation 6, taking into account the in-

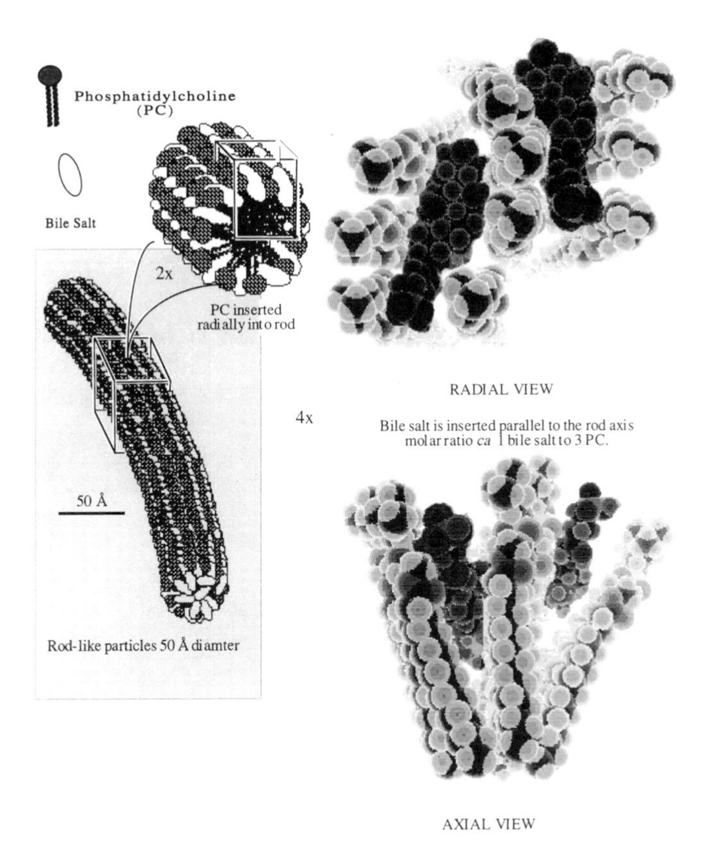

Figure 7. Space-filling model illustrating one way that phosphatidylcholine and bile salts can be arranged in rod-like mixed micelles. Schematic of a semi-rigid rodlike mixed micelle inside greyed area. Box indicates a segment of a radial-shell mixed micelle rod expanded 2-fold. Possible molecular arrangement of a few PC (light shaded) and bile salt (cholyltaurine; dark-shaded) molecules. The fatty acid chains shown in extended conformations are oriented perpendicularly to the rod axis. The head groups face outward in the solvent. The choline methyls are outermost. These are the groups that were deuterated. The arrangement of the bile salt with its axis parallel to the rod axis is inferred by measured values of γ (Duane, 1977; Spink *et al.*, 1982) that suggest that approximately three DPPC's are associated with each bile salt. The bile salt molecules act as wedges in only one direction. In this model the side of the bile salt with the three hydroxyl groups faces outward.

strument resolution factor, R(Q). This is shown in Figure 9, and the results of the analysis are summarized in Table IV. The sizes of the vesicles are greatest at the highest concentration where vesicles are present, and decrease with greater dilution beyond the transition (Figure 9, Table IV).

Unfortunately, direct measurements to show the arrangement of material in the vesicles have not be devised. However, it is possible to conclude from the scattering that the particles are single bilayer vesicles. First of all, Equation 3 can be used to analyze the scattering shown for the 5.0g/l sample of Figure 8, the results of which are shown in Table V (Hjelm *et al.,* 1988; 1990a; Hjelm *et al.,* 1990b; Hjelm *et al.,* 1995). The value of the thickness is consistent with a bilayer. Second, the fitted values for vesicle form factors in Table IV show similar values for the thickness (Hjelm *et al.,* 1988; 1990a; Hjelm *et al.,* 1990b; Hjelm *et al.,* 1992; Hjelm *et al.,* 1995). Finally, if one scales the values for the vesicle radius, R, with the values for I(0) in Table IV, one finds that $I(0) \propto R^2$ - exactly the

Table III. Phosphatidylcholine and total lipid density in rodlike micelles found in mixtures of phosphatidylcholines and bile salts

Mixture	Γ	Lipid (g/l)	γ^{-1}	R_c (Å)	n_L[a] (Å^{-1})	n_0[a] (Å^{-1})
Dppc-cholyltaurine[b]	0.6	4.0	0.35	17.1 (0.7)	0.9	1.1
	0.6	3.33	0.33	17.3 (0.6)	1	1.5
	0.6	2.86	0.31	18.6 (0.4)	1.2	1.5
EYPC-cholylglycine[b]	0.5	5.0	0.45	19.1 (0.2)	1	1.45
	0.56	5.0	0.40	19.2 (0.2)	1	1.4
	0.8	10	0.30	19.0 (0.2)	1	1.3
	0.8	8.3	0.34	18.4 (0.2)	1	1.3
	0.8	7.1	0.28	19.2 (0.2)	1	1.3
	0.9	10	0.29	19.0 (0.2)	0.9	1.2
EYPC-chenodeoxycholylglycine	0.8	1.7	0.22	19.0 (0.3)	1.1	1.3
Monoolein-cholylglycine	1.2	4.0	1.32	15.9 (0.2)	1	1.7

[a]Numbers in parentheses are root mean square uncertainty in the regression
[b]See text for abbreviation

result expected if the vesicles consist of the thin shell (Hjelm *et al.*, 1988; 1990a; Hjelm *et al.*, 1990b; Hjelm *et al.*, 1992; Hjelm *et al.*, 1995). How the bile salt is arranged in these structures is not yet evident, but it is clear that they constitute a considerable amount of the mass of the vesicles. The vesicles become smaller with dilution beyond the mixed micelle to vesicle phase transition (Hjelm *et al.*, 1988; 1990a; Hjelm *et al.*, 1990b; Hjelm *et al.*, 1992; Hjelm *et al.*, 1995). This mass can only be removed by repartitioning of the bile salt. The phospholipid is much too insoluble to be in the aqueous phase to any significant extent.

An approach towards understanding the determinants of self assembly in the bile salt-swelling lipids is to consider the particle structure mixed colloids of bile salt and

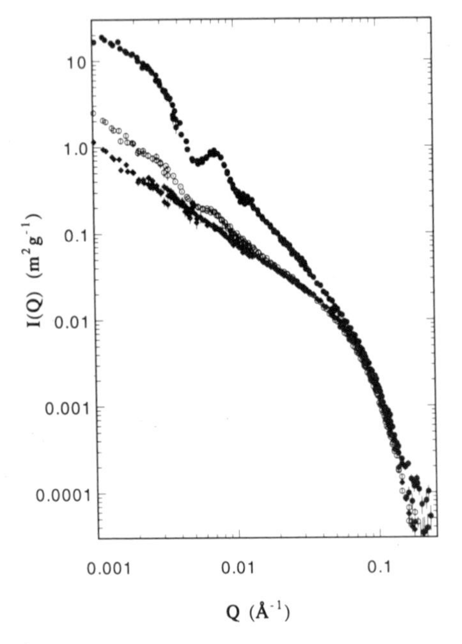

Figure 8. The concentration-dependent scattering from solutions of cholylglycine with egg yolk phosphatidylcholine in the transition between rodlike mixed micelles and vesicles. Total lipid concentrations in each solution: •, 5.0g/l; ¤, 5.6g/l; ◆, 6.5g/l. $\Gamma = 0.8$.

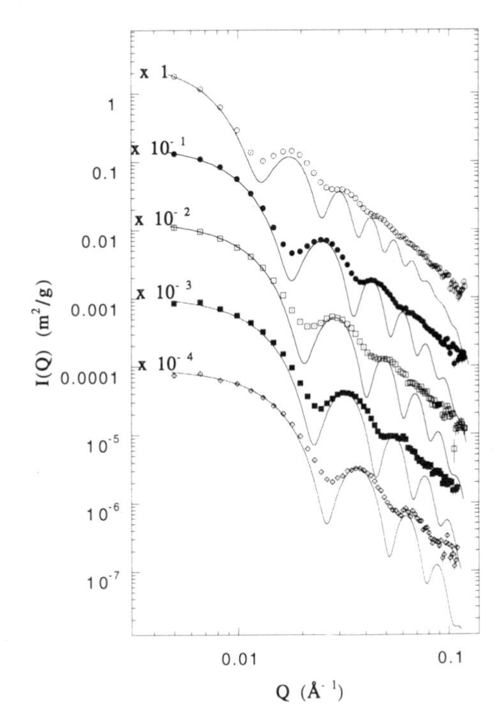

Figure 9. Vesicle form factors fitted to scattering data. EYPC-CG solutions at ◘, 2.5g/l; ●, 2.0; □, 1.76g/l; ■, 1.43g/l; ◊, 1.0g/l. ———, data fitted to vesicle form factors using the parameter given in Table IV.

monoolein. Monoolein differs from the long chain PC's in being a neutral lipid with a single oleate chain. The physiological importance of these mixtures stems from the observation that monoolein is formed in the small intestine from the action of pancreatic lipase on dietary triglyceride; thus these studies are also aimed towards an understanding the physiological importance of colloid particle form and structure in bile. We consider samples at compositions in the isotropic domain of the phase map, in which previous NMR and DLS studies of the cholyltaurine-monoolein-water system (Svärd *et al.*, 1988) have revealed globular micelles and vesicles.

In Figure 10 are shown some examples of scattering observed in cholylglycine-monoolein mixtures at $\Gamma = 1.2$. This ratio is chosen as it gives the same stoichiometry of fatty acid moieties to bile salt molecules as found in the DPPC mixtures used in the deuteration studies cited above. There are striking similarities between the SANS observed in the cholylglycine solutions and the cholylglycine-EYPC counterparts.

At the highest concentrations the shape and magnitude of the scattering from the two types of mixtures are similar, suggesting the same size and shape of the micelles. This im-

Table IV. Fitted parameters for vesicles

Lipid (g/l)	$\Delta I\ M_0$ (m^2/g)	Radius (Å)	Shell thickness (Å)
2.5	320	277	40
2.0	180	202	44
1.67	140	178	44
1.43	94	144	46
1.0	94	144	46

Table V. Vesicle wall parameters from Guinier analysis modified for sheet-like forms

Mixture	$\Delta\mu_0$ $(10^{14}g^{-1})^a$	R_d (Å)	Apparent shell thickness (Å)
EYPC-cholylglycine[b]	5.92 (0.02	11 (1)	38
EYPC-cholylglycine[b]	7.96 (0.01)	11 (1)	38
monoolein-cholylglycine	6.6 (0.1)	9 (1)	30

[a]Numbers in parentheses are root mean square uncertainty in the regression
[b]See text for abbreviation

pression is strengthened by comparison of the R_g and ΔM_0 values obtained from the Guinier analysis of Equation 1, given in Table I.

At lower concentrations the mixtures pass into what appears to be a coexistence regime of rodlike forms and vesicles. This is inferred from the existence of two distinct domains in the scattering. At low Q the scattering is suggestive of a vesicle. At higher Q the scattering is more like that expected for a rod. The R_c measured for the rod is slightly less than that measured for the EYPC systems, being 16Å (Table III). This corresponds to a radii of about 23Å, smaller than 25 to 27Å (Hjelm *et al.*, 1988; 1990a; Hjelm *et al.*, 1990b; Hjelm *et al.*, 1992) from the R_c of the PC mixtures (Table III). This observation is understandable in terms of the radial organization of PC or monoolein discussed above, as monoolein has a shorter overall length than PC. We derive a packing density of monoolein of about $1.0Å^{-1}$ using Equation 4 and the value for γ of 1.38 from the solubility studies of Hofmann (Hofmann, 1963). This is the same as observed in the EYPC systems (Table III).

The characteristic scattering signature of vesicle is evident when the sample is diluted to 2.5g/l. The Guinier analysis modified for sheet-like forms according to Equation 3 suggests that the thickness is 30Å, slightly smaller than that observed for the corresponding phosphatidylcholine mixtures (Table V). The $\Delta\mu_0$ derived from this analysis is consistent with those obtained in EYPC vesicles (Hjelm *et al.*, 1990a; Hjelm *et al.*, 1990b;

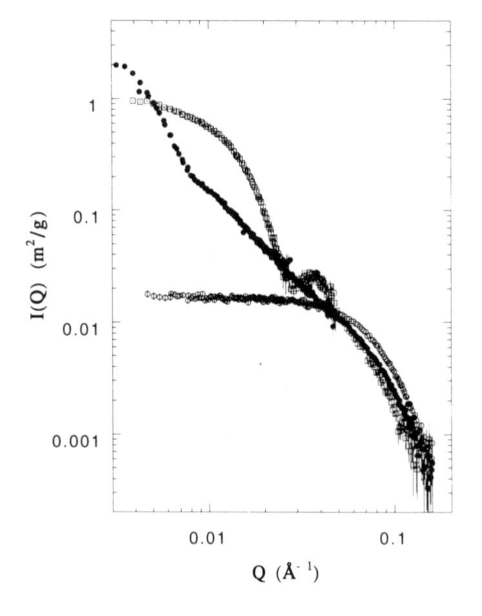

Figure 10. The concentration-dependent scattering from solution of cholylglycine with monoolein. Γ = 1.2. ¤, 10g/l; •, 4.0g/l; □, 2.0g/l.

Hjelm *et al.*, 1992) (Table V), suggesting again that the packing of lipids is very similar when either the EYPC or monoolein is present.

We draw two conclusions on the self assembly of swelling lipid-cholylglycine particles from these results. The first is that neither the two-tailed nature nor the zwitterionic character of the PC is a determinant of the form or structure of the particles in these systems. The second is that the packing densities of lipid in the rodlike forms observed in these systems appear to be the same. Finally, that the forms and structures of cholylglycine-monoolein mixed particles are the same as those of the cholylglycine-EYPC particles suggests that the particle form and structure may be important in the physiological function of bile.

ACKNOWLEDGMENTS

This work was conducted under the auspices of the United States Department of Energy. This work benefited from the use of the Low-Q Diffractometer at the Los Alamos Neutron Scattering Center of the Los Alamos National Laboratory which is supported by the Office of Basic Energy Sciences of the United States Department of Energy under contract W-7405-ENG-36 to the University of California. This also work benefited from the use of the Small Angle Neutron Diffractometer at the Intense Pulsed Neutron Source at Argonne National Laboratory. The facilities at IPNS are funded by the U.S. Department of Energy, BES-Materials Science, under Contract W-31–109-Eng-38. Work at UCSD was supported in part by NIH Grant DK 21506 (to AFH). Measurements on monoolein-cholylglycine were done on NG7SANS at the National Institutes of Standards and Technology. We thank Dr. C. Glinka (NIST) for expert help in carrying out these measurements

REFERENCES

Borgström, B., Barrowman, J.A., & Lindström, M., (1986). *Bile Acids in Intestinal Lipid Digestion and Adsorption*. In *Sterols and Bile Acids*, H. Danielsson and J. Sjovall, editors. Elvsevier. pp405–425.

Cabral, D.J., & Small, D.M., (1989). *Physical Chemistry of Bile*. In *Handbook of Physiology, The Gastrointestinal System III*, Section 4, S.G. Sultz, J.G. Forte and B.B. Rauner, editors. Waverly Press. pp621–662.

Dreher, K.D., Schulman, J.H., & Hofmann, A.F., (1967). Surface chemistry of the monoglyceride-bile salt system: Its relationship to the function of bile salts in fat absorption. *J. Coll. Interface Sci.*, 25:71–83.

Duane, W.C., (1977). Taurocholate and taurodeoxycholate-lecithin micelles: the equilibrium of bile salt between aqueous phase and micelle. *Biochem. Biophys. Res. Comm.*, 74:223–229.

Egelhaaf, S.U., & Schurtenberger, P., (1994). Shape transformations in the lecithin-bile salt system: from cylinders to vesicles. *J. Phys. Chem.*, 98:8560–8573.

Hjelm, R.P., (1985). The small-angle approximation of X-ray and neutron scatter from rigid rods of non-uniform cross section and finite length. *J. Appl. Cryst.*, 18:452–460.

Hjelm, R.P., (1988). The resolution of TOF Low-Q Diffractometers: instrumental, data acquisition and reduction factors. *J. Appl. Cryst.*, 21:618–628.

Hjelm, R.P., Thiyagarajan, P., & Alkan, H., (1988). A small-angle neutron study of the effects of dilution on particle morphology in mixtures of glycocholate and lecithin. *J. Appl. Cryst.*, 21:858–863.

Hjelm, R.P., Thiyagarajan, P., & Alkan, H., (1990a). Small-angle neutron scattering studies of mixed bile salt - lecithin colloids. *Mol. Cryst. Liq. Cryst.*, 180A:155–164.

Hjelm, R.P., Thiyagarajan, P., Sivia, D., Lindner, P., Alkan, H., & Schwahn, D., (1990b). Small-angle neutron scattering from acqueous mixed colloids of lecithin and bile salt. *Prog. Coll. Polymer Sci.*, 81:225–231.

Hjelm, R.P., Thiyagarajan, P., & Alkan-Onyuksel, H., (1992). Organization of phosphatidylcholine and bile salt in rodlike mixed micelles. *J. Phys. Chem.*, 96:8653–8661.

Hjelm, R.P., Thiyagarajan, P., Hofmann, A.F., Schteingart, C., Alkan-Onyusel, M.H., & Ton-Nu, H.-T., (1995). Structure of mixed micelles present in bile and intestinal content based on studies of model systems. Falk Symposium, 80 (in press).

Hofmann, A.F., (1963). The function of bile salts in fat absorption. *Biochem. J.*, 89:57–68.

Kotlarchyk, M., & Chen, S.-H., (1983). Analysis of small angle neutron scattering spectra for polydispersed interacting colloids. *J. Chem. Phys.*, 79:2461–2469.

Long, M.A., Kaler, E.W., & Lee, S.P., (1994a). Structural characterization of the micelle-vesicle transition in lecithin - bile salt solutions. *Biophys. J.*, 67:1733–1742.

Long, M.A., Kaler, E.W., Lee, S.P., & Wignall, G.D., (1994b). Characterization of lecithin-taurodeoxycholate mixed micelles using small-angle neutron scattering and static and dynamic light scattering. *J. Phys. Chem.*, 98:4402–4410.

Luzzati, V., (1960). Interpretation des mesures absolues de diffusion centrale des rayons X en collimation ponctuelle ou lineaire: Solutions de particules globulaires et de bâtonnets. *Acta Cryst.*, 13:939–945.

Knoll, W., Haas, J., Stuhrmann, H.B., Fuldner, H.-H., Vogel, H., & Sackmann, E., (1981). Small-angle neutron scattering of aqueous dispersions of lipids and lipid mixtures. A contrast variation study. *J. Appl. Cryst.*, 14:191–202.

Mazer, N.A., Benedek, G.B., & Cary, M.C., (1980). Quasielastic light scattering studies of aqueous biliary lipid systems. *Biochemistry,* 19:601–615.

Mazer, N.A., Schurtenberger, P., Cary, M.C., Preisig, R., Weigand, K., & Kanzig, W., (1984). Quasielastic light scattering studies of native hepatic bile from the dog: comparison with aggregative behaviour of model biliary lipid systems. *Biochemistry,* 23:1994–2005.

Nichols, J.W., & Ozarowski, J., (1990). Sizing of lecithin-bile salt mixed micelles by size-exclusion high-performance liquid chromatography. *Biochemistry,* 29:4600–4606.

Porod, G., (1982). In *Small-Angle Scattering of X-rays*, O. Glatter and O. Kratky, editors. Academic Press, New York. pp17–52.

Schurtenberger, P., Mazer, N.A., & Kanzig, W., (1985). Micelle to vesicle transition in aqueous solutions of bile salt and lecithin. *J. Phys. Chem.*, 89:1042–1049.

Small, D.M., (1967). Physicochemical studies of cholesterol gallstone formation. *Gastroenterology,* 52:607–610.

Small, D.M., (1971). *The Physical Chemistry of the Cholanic Acids*. In *The Bile Lipids: Chemistry, Physics and Metabolism*, P.P. Nair and D. Kritchevsky, editors. Plenum, New York. pp247–354.

Small, D.M., Bourgès, M.C., & Dervichian, D.G., (1966). The biophysics of lipidic association I. The ternary systems lecithin-bile salt-water. *Biochim. Biophys. Acta,* 125:563–580.

Spink, C.H., Müller, K., & Sturtevant, J.M., (1982). Precision scanning calorimetry of bile salt - phosphatidylcholine micelles. *Biochemistry,* 21:6598–6605.

Svärd, M., Schurtenberger, P., Fontell, K., Jönsson, B., & Lindman, B., (1988). Micelles, vesicles, and liquid crystals in the monoolein-sodium taurocholate-water system. Phase behaviour, NMR, self-diffusion, and quasi-elastic light scattering studies. *J. Phys. Chem.*, 92:2261–2270.

Walter, A., Vinson, P.K., & Talmon, Y., (1991). Intermediate structures in the cholate-phosphatidylcholine vesicle-micelle transition. *Biophys. J.*, 60:1315–1325.

16

NEUTRON DIFFRACTION STUDIES OF AMPHIPATHIC HELICES IN PHOSPHOLIPID BILAYERS

J. P. Bradshaw,[1,2] K. C. Duff,[2] P. J. Gilchrist,[1] and A. M. Saxena[3]

[1] Department of Preclinical Veterinary Sciences, University of Edinburgh
Summerhall, Edinburgh EH9 1QH, United Kingdom
[2] Department of Biochemistry, University of Edinburgh Medical School
Hugh Robson Building, George Square, Edinburgh EH8 9XD, United Kingdom
[3] Biology Department, Brookhaven National Laboratory
Upton, New York 11973

ABSTRACT

The structural feature which is thought to facilitate the interaction of many peptides with phospholipid bilayers is the ability to fold into an amphipathic helix. In most cases the exact location and orientation of this helix with respect to the membrane is not known, and may vary with factors such as pH and phospholipid content of the bilayer. The growing interest in this area is stimulated by indications that similar interactions can contribute to the binding of certain hormones to their cell-surface receptors. We have been using the techniques of neutron diffraction from stacked phospholipid bilayers in an attempt to investigate this phenomenon with a number of membrane-active peptides. Here we report some of our findings with three of these: the bee venom melittin; the hormone calcitonin; and a synthetic peptide representing the ion channel fragment of influenza A M2 protein.

AN INTRODUCTION TO AMPHIPATHIC HELICES

An amphipathic helix may be defined as an α-helix with opposing polar and non-polar faces oriented along the long axis of the helix (Segrest *et al.,* 1974). Such a molecule is capable of interacting with phospholipid bilayers in either of two principal orientations (see Figure 1). In one, the helix lies partially embedded in the bilayer surface, with its axis parallel to the bilayer. In this position the polar surface remains exposed to the aqueous medium while the hydrophobic residues lie on that portion of the peptide which penetrates into the hydrocarbon region of the bilayer. In the alternative orientation, the axis of the he-

Neutrons in Biology, edited by Schoenborn and Knott
Plenum Press, New York, 1996

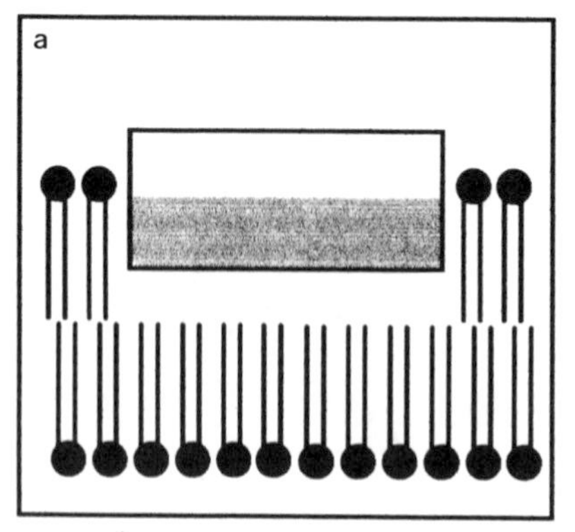

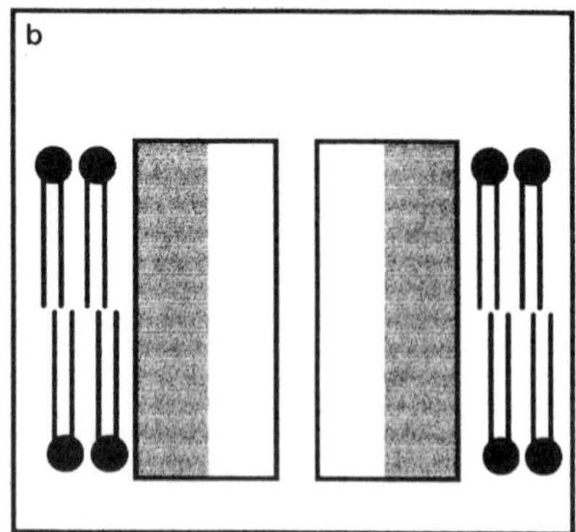

Figure 1. The two principal modes of interaction of amphipathic helices with phospholipid bilayers. In (a) the peptide is oriented with the axis of the helix parallel to the bilayer surface enabling the hydrophobic amino acids to penetrate into the hydrocarbon layer. In scenario (b) the peptide self-associates as tetramers to form a cylinder with hydrophobic outside and hydrophilic inside.

lix lies along the normal to the bilayer, spanning its entire width. In order to adopt this orientation without exposing polar residues to hydrophobic fatty-acyl chains, it is necessary for the peptide to aggregate into trimers, tetramers or more. The resulting structure is effectively a tube with a polar core which may, or may not, have channel properties according to its physical properties and dimensions. There are many possible variations on this orientation, based on the number of monomers in the aggregate structure, its angle relative to the bilayer and details of the peptide structure such as uneven distribution of large and small amino acid sidechains, or breaks in the helix caused by proline, for example.

Neutron diffraction, complimented by judicious use of deuterium labeling, is a powerful technique for distinguishing between the two possible orientations of amphipathic helices. To date, we have used it to study three different peptides. These are: melittin, one of the principal constituents of bee (*Apis melifera*) venom, 26 amino acid residues long; calcitonin, a peptide hormone of 32 residues; and M2 peptide, a synthetic 25 residue molecule representing the putative trans-membrane segment of influenza A M2 protein. All three are illustrated schematically in Figure 2. In this paper I will address the general issue of the study of amphipathic helices by neutron diffraction then focus in on one of the peptides to illustrate, in more detail, the sort of information which such studies may reveal.

THE USE OF AMPHIPATHIC HELICES IN NEUTRON EXPERIMENTS

Many biological phospholipids will orient spontaneously into stacked bilayer structures when rehydrated from a humid atmosphere. This fact has formed the basis of their use in diffraction experiments for many years, initially by X-rays and more recently by neutrons. These systems have been used as models of biological membranes of increasing complexity as different phospholipids, small molecules, peptides and even proteins have been inserted into them. We have been using this approach to study the interactions of amphipathic helices with phospholipid bilayers.

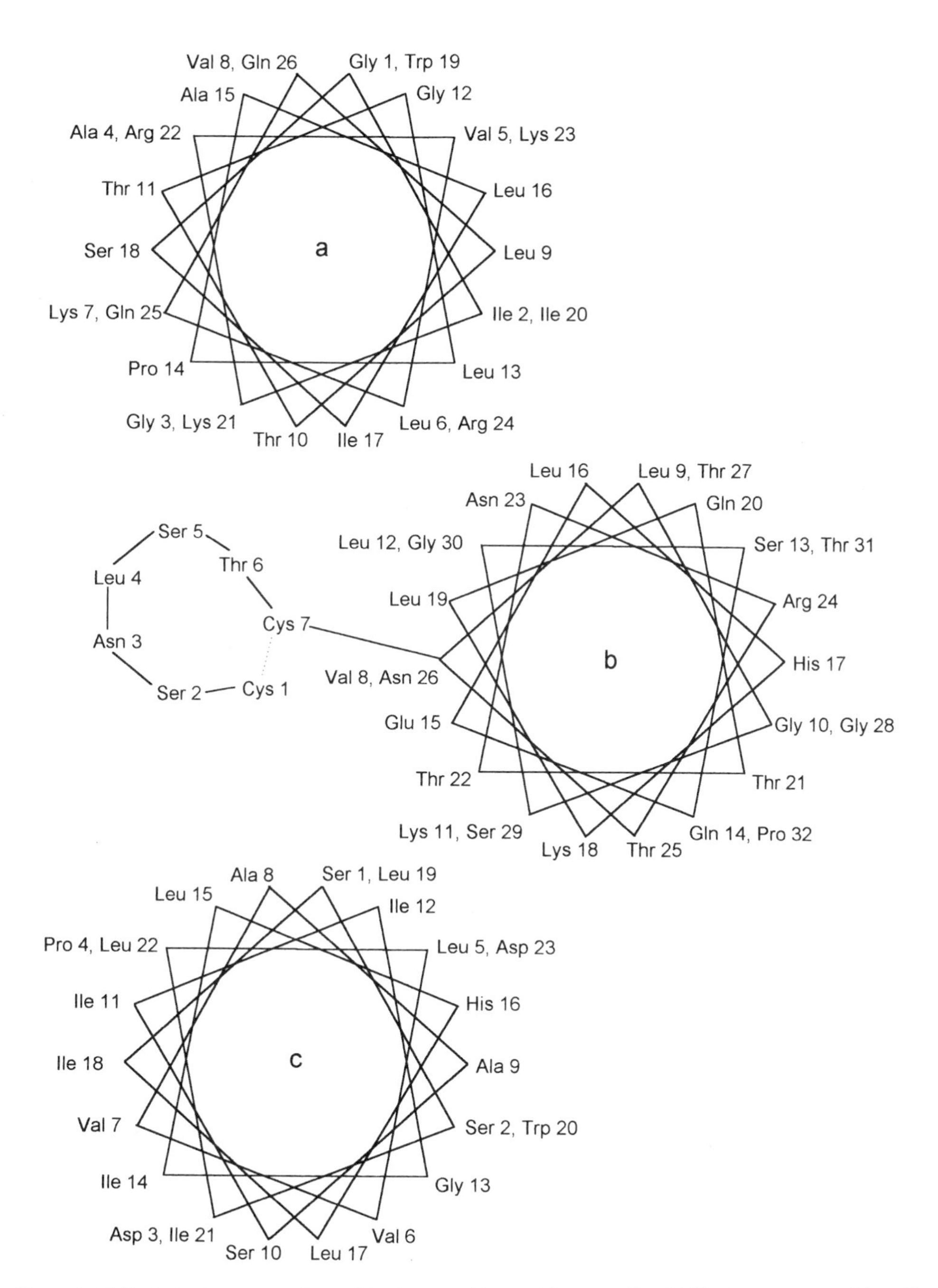

Figure 2. Helical wheel representations of the three peptides used in our studies. (a) The bee venom peptide melittin; owing to the doubtful detectability of single deuterons in membrane-bound peptide by neutron diffraction, an Ala-12 analogue of melittin was synthesised to allow greater incorporation of label. (b) Salmon calcitonin, a peptide hormone, showing the disulphide bridge between residues 1 and 7. (c) A synthetic peptide representing the transmembrane domain (residues 22 to 46) of the M2 protein of influenza A virus.

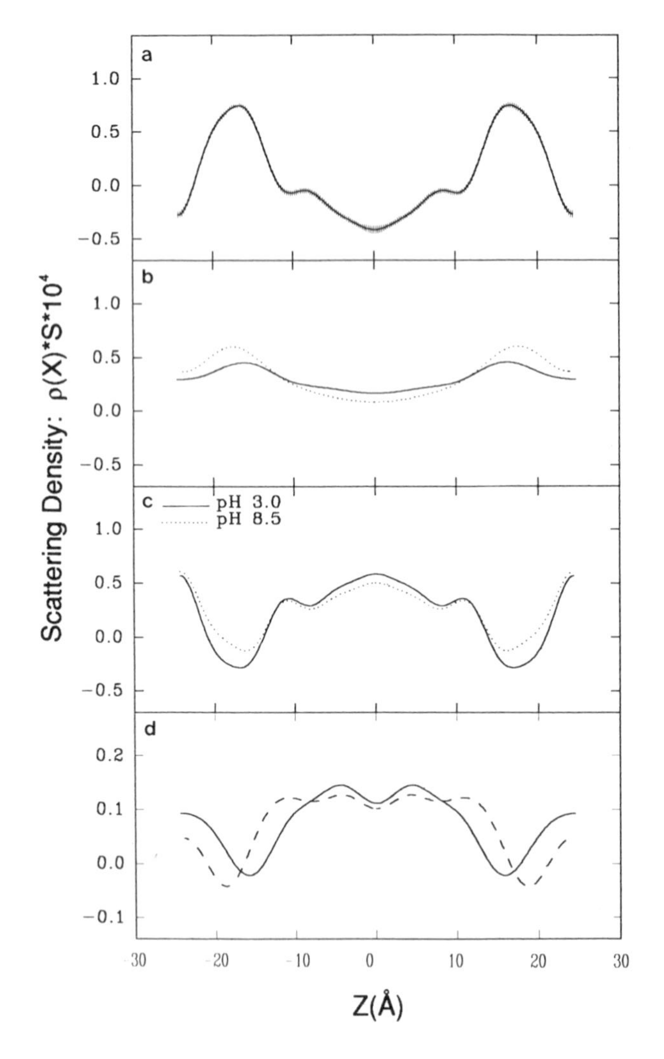

Figure 3. Neutron scattering density profiles of bilayers of (a) DOPC and (b) DOPC with 5 % (mol) melittin prepared at either pH3.0 (solid line) or pH8.5 (dotted line). All at 37°C and 100% relative humidity (H_2O). The profiles are plotted on the 'relative absolute' scale of Wiener *et al.* (1991). Plot (c) is a difference profile calculated by subtracting (a) from each of the profiles shown in (b) and is related to the distribution of melittin in the bilayers of DOPC. (d) Difference neutron scattering density profiles showing the distribution of deuterium labels on 5 % (mol) ($^2H_3$-Ala 15)-melittin in bilayers of DOPC at either pH3 (solid line) or pH8.5 (dotted line). Note the pH dependence of the label distribution, resulting from shift of the peptide between the surface and transbilayer locations brought about by protonation of N-terminus. In each plot, the horizontal scale represents distance from the centre of the bilayer along the bilayer normal and the vertical scale is neutron scattering density. The error bars in plot (a) delineate the maximum variation in a large number (1,000) of Fourier profiles generated by Monte Carlo sampling of the structure factors and their associated errors.

Oriented phospholipid bilayer stacks are prepared as follows. Samples comprising 20mg DOPC (1,2-dioleoyl-sn-glycero-3-phosphocholine), with or without peptide, are dissolved in methanol and applied to one side of a quartz glass microscope slide by pasteur pipette or airbrush. The solvent is allowed to evaporate before the slide is dried in vacuo for several hours. The samples are then rehydrated (with ^1H_2O, ^2H_2O or some mixture of the two) by placing in a humid atmosphere, again for several hours, before being transferred to the sample cell of the neutron instrument. Each sample is judged to have achieved equilibrium when there is no further shift in the angle of diffraction of the first order lamellar repeat, and the calculated lamellar repeat distance corresponds to that predicted from X-ray diffraction experiments. All samples are run at 37°C and 100% relative humidity (experience has shown that lower relative humidities, produced by replacing the water bath by one containing a saturated salt solution, are slow to stabilise). A typical experiment produces eight orders of diffraction from a lamellar repeat distance of approximately 50Å.

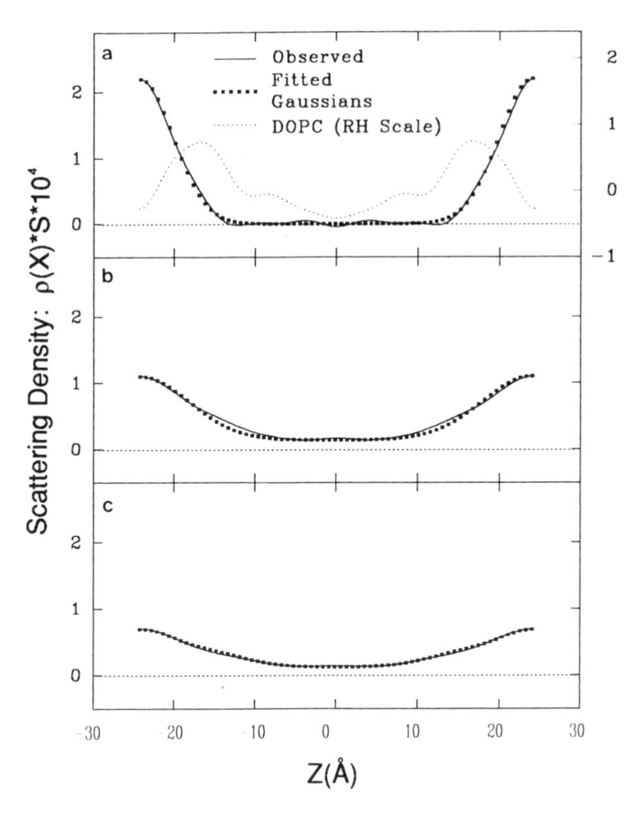

Figure 4. Difference neutron scattering density profiles showing the distribution of water in bilayers of (a) DOPC; (b) DOPC containing 5 mol% melittin prepared at pH3; and (c) DOPC containing 1 mol% M2 peptide. All samples were held at 37°C and 100% relative humidity (H_2O). Each of the best fit lines indicated represents the sum of two or four Gaussians, determined using the method of Bradshaw *et al.* (1994) over four (M2 peptide) or eight (DOPC, Melittin) orders of diffraction. The Gaussians have the following parameters: (a) $Z_W = \pm 2.288 \pm 0.53$Å, A_W (1/e-halfwidth) = 5.41 ±0.44Å; (b) $Z_W = \pm 23.77 \pm 0.53$Å, $A_W = 8.54 \pm 0.67$Å; (c). $Z_W = \pm 23.74 \pm 0.2$Å, $A_W = 5.56$ ±0.2Å, 53 % of total and $Z_W = \pm 16.46 \pm 0.1$Å, $A_W = 6.57 \pm 0.1$Å, 47 % of total. The Gaussians are fitted by minimising A_{err} in the expression:

$$A_{err} = \frac{\Sigma|F_{obs}(h) - F_{calc}(h)|}{\Sigma|F_{obs}(h)|}$$

In the simplest form of neutron experiment the amphipathic helix is unlabeled. The results of a typical experiment are shown in Figure 3b, in which melittin has been incorporated into stacked bilayers composed of DOPC. When compared with profiles of pure phospholipid bilayers, it is clear that the peptide penetrates deep into the lipids. This could be interpreted as indicating a transbilayer orientation, but the low contrast makes it difficult to be conclusive. A more detailed picture starts to form when the difference method is used (Figure 3c). The scattering density distribution of the whole peptide can be visualised by transforming difference structure factors in which structure factors from pure phospholipid bilayers are subtracted from those collected from bilayers containing the peptide. However, the resulting profile is still not totally unambiguous since it is likely to contain changes in the lipid and water components of the system as well as the peptide itself. A more fruitful approach has proved to be to use the difference method to reconstruct the water, or more precisely, the heavy water, distribution. When the resulting profile can be put on an absolute scale (or a 'relative absolute' scale as described by Wiener *et al.* (1991) by determining the H_2O:lipid ratio, the orientation of the melittin is clarified by the shape of the water distribution. The peptide's transbilayer orientation reveals itself in a continuous spread of ^2H_2O scattering density right across the bilayer (Figure 4). This could represent water within the channel structure or protons (deuterons) exchanged onto some amino acid sidechains. This latter alternative is particularly attractive for proton channel peptides, as is the case with two of the amphipathic helices we have studied.

Considerably more information becomes accessible when deuterium labels are inserted into the peptide itself. We have achieved this by both chemical modification and by incorporating deuterated amino acids during the synthesis. Of the two approaches, the latter is to be preferred for the true isomorphous nature of its substitution. Figure 3d shows an example of the results of such an experiment, in this case melittin labeled with $^2H_3$-alanine as residue 15 (Bradshaw *et al.,* 1994). In order to study the consequences of protonation of the N-terminus upon the interaction of this peptide with phospholipid bilayers, analogues of melittin, some of which were specifically deuterated at either Ala 12 or Ala 15, were synthesised. These peptides were incorporated into bilayers of DOPC at either low pH (N-terminus protonated) or high pH (N-terminus unprotonated). X-ray and neutron diffraction data were collected from ordered stacks of these bilayers and from peptide-free controls. Changes in the water, melittin and deuterium label distributions were consistent with a model in which the melittin has two possible bilayer locations. One is at the surface, the other towards the centre of the bilayer. These two positions may be interpreted in terms of the two possible interactions of amphipathic helices with phospholipid bilayers, as described earlier. The distribution of the peptide between these two equilibrium sites appears to be pH-dependent, with a larger population of surface melittin when the N-terminus is unprotonated.

However, one shortcoming of this type of study is that the label peaks in a typical difference map tend to be relatively wide and overlapping. The picture is complicated in the melittin data by the simultaneous presence of peptide in both transbilayer and parallel

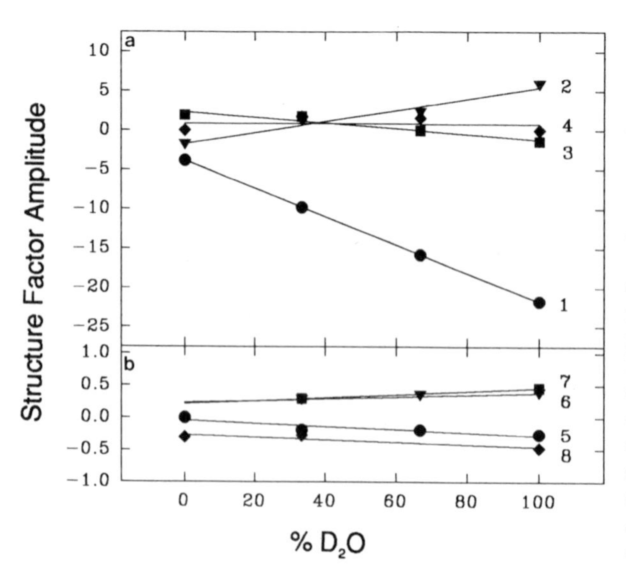

Figure 5. An example of structure factor data, in this instance DOPC with 1 mol% M2 peptide at 37°C. Four sets of structure factors are shown, corresponding to the four concentrations of ^2H_2O used in subsequent experiments on the same multi-bilayer sample. Each order is fitted (least-squares) to a straight line with negative slope for the odd orders and positive slope for the even orders. This relationship breaks down at higher order number, indicating that as the resolution increases, the water (^2H_2O) distribution can no longer be accurately described by a single Gaussian. For clarity, the eight orders are split between two plots; (a) shows orders 1 to 4 and (b) shows orders 5 to 8, on an expanded vertical scale.

orientations. Consequently, we sought to address this ambiguity by reducing the peptide content in subsequent studies, whilst increasing the size of the label. The M2 study detailed below, for example, used $^2H_8$-valine to give a signal which is nearly three times stronger, allowing a corresponding reduction in peptide concentration. At the same time we adopted the practice of collecting four sets of data from each sample, at 0, 33, 67 and 100% ^2H_2O; an example is shown in Figure 5. This not only simplifies the phase problem enormously, but also increases the accuracy of the experiment. Each structure factor at 0% ^2H_2O is fixed by four independent measurements, instead of just one if only two ^2H_2O concentrations are used. The first subject of this approach was a peptide representing part of one of the proteins of influenza virus.

INFLUENZA VIRUS

Immediately after the end of the Great War, in 1918, there was an outbreak of influenza. World-wide, twenty one million people died. In Britain the virus was responsible for more deaths than the war had claimed. In Alaska, whole townships were destroyed. Such devastation is not new, an epidemic in 412BC was chronicled by both Hippocrates and Livy. Subsequent to this, there are records of numerous pandemics, including one in 1889–90 in which one percent of the population of Europe died. Since I am currently based in Edinburgh, I could not talk about the epidemiology of the virus without saying that Mary Queen of Scots is said to have 'fallen acquainted' with 'flu on her arrival in Edinburgh from France in 1562.

Charles Cockburn, head of the virology section of the World Health Organisation wrote in 1973: 'The influenza virus behaves just as it seems to have done for five hundred or a thousand years, and we are no more capable of stopping epidemics or pandemics than our ancestors were'. This could be a new concept, the idea that viruses have behaviour. However, it is becoming apparent that they do have physiology, including ion channel activity.

To date vaccine and drug therapy directed against influenza have only been spectacular in their failure, although recently there have been reports of promising work on a sialidase inhibitor in Australia (Vonitzstein *et al.,* 1993). Another drug, amantadine (1-amino-adamantane hydrochloride), has been licensed in the UK since 1976. It is effective in the prophylaxis and treatment of influenza, but its prescription has been limited due to its unpleasant side-effects. We have used X-ray and neutron diffraction to study the bilayer distribution of amantadine in both charged and uncharged forms (Duff *et al.,* 1993)

M2 PROTEIN

The influenza virus has a segmented RNA genome. Segment 7 of the genome of influenza A codes for two proteins, called M1 and M2. M1 is found in large amounts in the virion where it forms a layer directly under the viral envelope. M2 is a 97 amino acid protein which was originally thought to only be expressed in infected cells, but has since been found in virions at a level of 14–68 molecules per virus (Zebedee & Lamb, 1988).

As a result of work carried out mostly by Lamb (USA), Hay (UK) and, more recently, ourselves, M2 appears to be a proton channel which is functional during two stages of the replicative cycle of the virus. Immediately after viral endocytosis, and concomitant with pH-induced fusion between the endosomal and viral membranes, M2 is thought to conduct protons into the interior of the virion. The resulting reduction in pH induces the release of ribonucleoprotein from the matrix protein (M1) and eventual nuclear infection (Martin & Helenius, 1991). When haemagglutinin is transported to the cell surface in post-Golgi vesicles, M2 may also facilitate influenza synthesis and assembly by countering any vesicular acidification (Sugrue *et al.,* 1990). The relative importance of these events, which occur at different stages in the replicative cycle, depends upon the particular viral strain (Belshe & Hay, 1989). Amantadine is thought to operate by impeding proton flow during these events (Hay, 1989). Some mutants of influenza A appear to become resistant to amantadine by single amino acid substitutions in the membrane-spanning domain of M2 (Hay *et al.,* 1985).

This paper will address itself only to our neutron work on M2. The details of our other work, which showed that the peptide was indeed α-helical (Duff *et al.,* 1992), capable of translocating protons and reversibly inhibited by amantadine (Duff & Ashley, 1992), may be found in the references cited.

In order to study the membrane interaction of M2 and the molecular mechanism of amantadine action on M2, a 25 amino acid peptide was synthesised, its sequence corresponding to the predicted transmembrane segment of M2 protein which is conserved in several native strains of influenza A. This segment stretches from residue 22 to residue 46 of the viral protein; in the synthetic peptide the next three residues were included at each end to help define this membrane spanning region. An amantadine-resistant mutant sequence was also synthesised in which the valine at position 27 (residue 6 in the peptide) was replaced by alanine. Some of the native peptide incorporated $^2H_8$-valine at the same

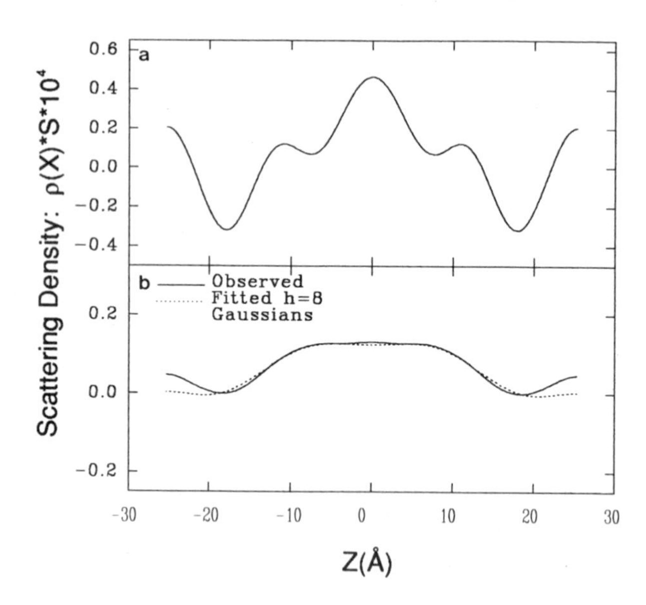

Figure 6. (a) Difference profiles calculated by subtracting the scattering due to DOPC alone from DOPC with 1 mol% influenza A M2 peptide This profile may be taken as representing the distribution of M2 peptide within the DOPC bilayer, with any changes in lipid or water distribution superimposed upon it. Profile (b) is a difference scattering profile which describes the distribution of deuterium on ($^2H_8$-Val 27)-M2 peptide in DOPC bilayers. The dotted line is the sum of two pairs of Gaussians fitted to the difference in diffraction space to eliminate the effects of termination error. The major Gaussians lie in the fatty-acyl region of the bilayer at $Z_V = \pm 7.75 \pm 0.3$Å, $A_V = 10.24 \pm 0.3$Å. The minor peaks, in the water layer ($Z_V = \pm 24.34 \pm 0.5$Å, $A_V = 5.65 \pm 0.2$Å), represent only 9 % of the total deuterium.

position. $^2H_{15}$-amantadine was synthesised by Dr. M. R. Alecio (Shell Research Centre, Sittingbourne, Kent, UK). Different combinations of these molecules were incorporated into stacked DOPC bilayers and studied by neutron diffraction (Duff *et al.,* 1994) using the instrument H3B at the HFBR (Brookhaven National Laboratory, New York, USA).

Difference profiles (Figure 6a), suggested that the M2 peptide was indeed transbilayer. The dominant feature of the peptide distribution revealed by these plots is a broad peak of neutron scattering density which spans the bilayer. The density peaks at the centre of the bilayer. Such a distribution would be difficult to achieve if the peptide were not transbilayer. This picture is confirmed by the observation of a continuous band of deuterons spreading right across the bilayer in difference profiles which describe the water distribution within the system (Figure 4c).

Figure 6b shows the intrabilayer distribution of deuterium on M2 peptide synthesised with $^2H_8$-valine at residue 6. The structure factors for this reconstruction were calculated by subtracting structure factors for bilayers containing unlabeled peptide from those for bilayers with the deuterated peptide. The profile may be defined as two Gaussians located at 8Å from the centre of the bilayer. The approach here is to fit the Gaussians in reciprocal space (ie to the structure factors) in order to eliminate the effects of Fourier series termination error.

Figure 7 shows profiles describing the bilayer distribution of $^2H_{15}$-amantadine hydrochloride in (a) pure DOPC, (b) DOPC with 1% M2 peptide and (c) DOPC with 1% mutant M2 peptide. The profiles were calculated in a manner corresponding to the $^2H_8$-valine profile in Figure 6. Each of the three amantadine profiles is unique, indicating some specificity of interaction with the peptide. Profile (b) contains four major peaks, two of which

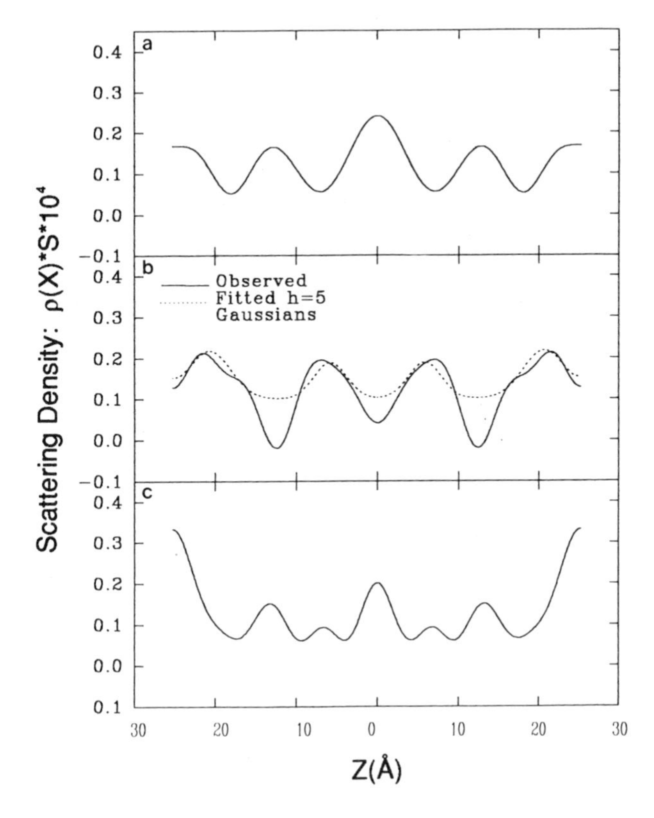

Figure 7. Difference neutron scattering density profiles, calculated using eight orders of diffraction, showing the distribution of $^2H_{15}$-amantadine HCl in bilayers of: (a) pure DOPC; (b) DOPC containing 1 mol% influenza A M2 peptide; (c) DOPC containing 1 mol% (Ala 27)-M2 peptide. Plot (b) also shows two pairs of Gaussians fitted to the corresponding difference structure factors in diffraction space. $Z_A = \pm 5.81 \pm 0.1$Å, $A_A = 2.88 \pm 0.1$Å, 37 % of total and $Z_A = \pm 20.65 \pm 0.1$Å, $A_A = 3.74 \pm 0.1$Å, 63 % of total.

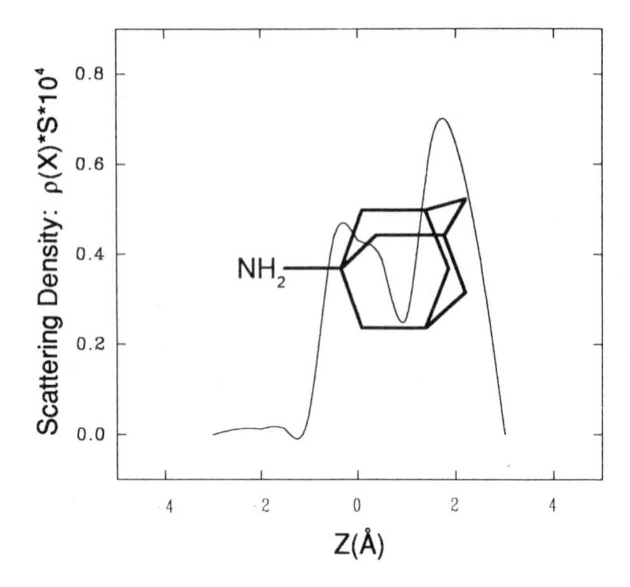

Figure 8. The structure of 1-amino-($^{2}H_{15}$)-adamantane hydrochloride superimposed on its calculated neutron scattering density profile at 0.02M concentration. All of the hydrogens have been replaced with deuterons, with the exception of those on the amine group. The profile is smeared horizontally by ±1Å.

occur within the bilayer, two in the water layer. Model-fitting shows that the inner peaks may be described, at low resolution, by a pair of Gaussians at 6Å from the centre of the bilayer.

Can these data be interpreted? In a sense the hard work was done for us by Sansom and Kerr (1993) who used computer modeling to predict the location of a possible interaction between amantadine and the transbilayer domain of influenza M2. Their findings suggested that the amine group of amantadine could interact with serine 31, with the protuberances of the fused ring lying in pockets formed by valine 27 (corresponding to residue 6 of the peptide). This would locate the centre of amantadine 5Å from the middle of the bilayer. The neutron measurements place the amantadine at 6Å, is this compatible with the computer predictions? The answer is yes, they agree extremely closely, especially when we take into account the fact that the centre of neutron scattering density and the centre of mass of the amantadine do not coincide, as shown in Figure 8. The scattering centre is displaced from the geometric centre by 1Å away from the amine group, as a result of the uneven deuteration. Therefore the neutron and computer studies tell exactly the same story (see Figure 9). It is tempting to believe that here we are observing the molecular mechanism of amantadine sensitivity and resistance in influenza A.

CONCLUSION

Used correctly, neutron diffraction is a powerful technique for the extraction of positional information from phospholipid bilayers and membrane-active molecules within them. Selectively deuterated groups can be localised to better than Ångstrom precision along the bilayer normal, thereby giving information about conformation and potential for interaction of membrane components. The three peptides studied by us to date, namely melittin, calcitonin and influenza A virus M2 peptide, have demonstrated the potential of this technique, which it is hoped will see much wider application in years to come.

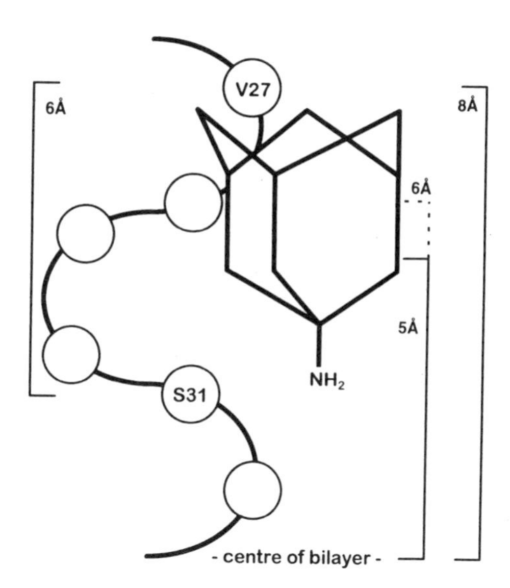

Figure 9. A possible molecular mechanism for the blockade of influenza A M2 proton channels by amantadine. For clarity, only one of the monomers of a tetrameric channel protein is shown. The cationic amine group interacts with the oxygens on the four serine residues at position 31 whilst the tricyclohexane ring interdigitates with pockets formed by valine 27. The model was first proposed by Sansom and Kerr (1993) and later confirmed by neutron diffraction (Duff *et al.,* 1994).

ACKNOWLEDGMENTS

We thank A. Watts (University of Oxford), A.J.C. Cudmore (University of Edinburgh), X. Bai (Brookhaven National Laboratory) and E. Pebay (Institut Laue Langevin) for assistance with various aspects of the work described. Gratitude is also extended towards Dr M.R. Alecio (Shell Research Centre), for preparing the deuterated amantadine. Data collection and initial reduction were carried out on the instruments D16 at the ILL, and H3B at Brookhaven. The work was supported by the Office of Health and Environment of the U. S. Department of Energy, the Institut Laue Langevin and the U. K. Science and Engineering Research Council.

REFERENCES

Belshe, R.B., & Hay, A.J., (1989). Genetic basis of resistance to rimantadine emerging during treatment of influenza virus infection. *J. Resp. Dis.* (suppl.) 52–61.

Bradshaw, J.P., Dempsey, C.E., & Watts, A., (1994). A combined X-ray and neutron diffraction study of selectively deuterated melittin in phospholipid bilayers: the effect of pH. *Molecular Membrane Biology,* 11:79–86.

Duff, K.C., & Ashley, R.H., (1992). The transmembrane domain of influenza A M2 protein forms amantadine-sensitive proton channels in planar lipid bilayers. *Virology,* 190:485–489.

Duff, K.C., Cudmore, A.J., & Bradshaw, J.P., (1993). The location of amantadine hydrochloride and free base within phospholipid multilayers : a neutron and X-ray diffraction study. *Biochim. Biophys. Acta,* 1145:149–156.

Duff, K.C., Gilchrist, P.J., Saxena, A. M., & Bradshaw, J.P., (1994). Neutron diffraction reveals the site of amantadine blockade in the influenza A M2 ion channel. *Virology,* 202:287–293.

Duff, K.C., Kelly, S.M., Price, N.C., & Bradshaw, J.P., (1992). The secondary structure of influenza A M2 transmembrane domain : a circular dichroism study. *FEBS,* 311:256–258.

Hay, A.J., (1989). In *Concepts in Viral Pathogenesis III* (A.L. Notkins and M.B.A. Oldstone, editors) pp561–567, Springer-Verlag, New York.

Hay, A.J., Wolstenholme, A.J., Skehel, J.J., & Smith, M.H., (1985). The molecular basis of the specific anti-influenza action of amantadine. *EMBO J.,* 4:3021–3024.

Martin, K., & Helenius, A., (1991). Nuclear transport of influenza virus ribonucleoproteins: the viral matrix protein (M1) promotes export and inhibits import. *Cell,* 67:117–130.

Sansom, M.S.P., & Kerr, I.D., (1993). Influenza virus M2 protein: a molecular modelling study of the ion channel. *Protein Engineering,* 6:65–74.

Segrest, J.P., Jackson, R.L., Morrisett, J.D., & Gotto, A.M. Jr., (1974). A molecular theory for protein-lipid interactions in plasma lipoproteins. *FEBS Lett.,* 38:247–253.

Sugrue, R.J., Bahadur, G., Zambon, M.C., Hall-Smith, M., Douglas, A.R., & Hay, A.J., (1990). Specific structural alteration of the influenza haemagglutinin by amantadine. *EMBO J.,* 9:3469–3476.

Itzstein, M.v., Wu, W.Y., Kok, G.B., Pegg, M.S., Dyason, J.C., Jin, B., Phan, T.V., Smythe, M.L., White, H.F., Oliver, S.W., Colman, P.M., Varghese, J.N., Ryan, D.M., Woods, J.M., Bethell, R.C., Hotham, V.J., Cameron, J.M., & Penn, C.R., (1993). Rational design of potent sialidase-based inhibitors of influenza virus replication. *Nature,* 363:418–423.

Wiener, M.C., King, G.I., & White, S.H., (1991). Structure of fluid DOPC bilayer determined by joint refinement of X-ray and neutron data. *Biophys. J.,* 60:568–576.

Zebedee, S.L., & Lamb, R.A., (1988). Influenza A virus M2 protein: monoclonal antibody restriction of virus growth and detection of M2 in virions. *J. Virol.,* 62:2762–2772.

17

FLUID BILAYER STRUCTURE DETERMINATION

Joint Refinement in 'Composition Space'

S. H. White[1] and M. C. Wiener[2]

[1]Department of Physiology and Biophysics
University of California
Irvine, California 92717-4560
[2] Department of Biochemistry and Biophysics
University of California
San Francisco, California 94143-0448

ABSTRACT

Experimentally-determined structural models of fluid lipid bilayers are essential for verifying molecular dynamics simulations of bilayers and for understanding the structural consequences of peptide interactions. The extreme thermal motion of bilayers precludes the possibility of atomic-level structural models. Defining 'the structure' of a bilayer as the time-averaged transbilayer distribution of the water and the principal lipid structural groups such as the carbonyls and double-bonds (quasimolecular fragments), one can represent the bilayer structure as a sum of Gaussian functions referred to collectively as the quasimolecular structure. One method of determining the structure is by neutron diffraction combined with exhaustive specific deuteration. This method is impractical because of the expense of the chemical syntheses and the limited amount of neutron beam time currently available. We have therefore developed the 'composition space' refinement method for combining X-ray and minimal neutron diffraction data to arrive at remarkably detailed and accurate structures of fluid bilayers. The composition space representation of the bilayer describes the probability of occupancy per unit length across the width of the bilayer of each quasimolecular component and permits the joint refinement of X-ray and neutron lamellar diffraction data by means of a single quasimolecular structure that is fitted simultaneously to both data sets. Scaling of each component by the appropriate neutron or X-ray scattering length maps the composition-space profile to the appropriate scattering length space for comparison to experimental data. The difficulty with the method is that fluid bilayer structures are generally only marginally determined by the experimental data. This means that the space of possible solutions must be extensively explored in conjunction with a thorough analysis of errors. The composition-space refinement method will be discussed in detail using the results of measurements made on dioleoylphosphatidylcholine (DOPC) bilayers at low water contents. (This work supported by grants from the NIH [GM-37291, GM-46823] and the NSF [DMB-8807431].)

NEUTRON REFLECTIVITY STUDIES OF SINGLE LIPID BILAYERS SUPPORTED ON PLANAR SUBSTRATES

S. Krueger,[1] B. W. Koenig,[2] W. J. Orts,[1] N. F. Berk,[1] C. F. Majkrzak,[1] and K. Gawrisch[2]

[1] National Institute of Standards and Technology
Gaithersburg, Maryland 20899
[2] National Institutes of Health
Bethesda, Maryland 20892

ABSTRACT

Neutron reflectivity was used to probe the structure of single phosphatidylcholine (PC) lipid bilayers adsorbed onto a planar silicon surface in an aqueous environment. Fluctuations in the neutron scattering length density profiles perpendicular to the silicon/water interface were determined for different lipids as a function of the hydrocarbon chain length. The lipids were studied in both the gel and liquid crystalline phases by monitoring changes in the specularly-reflected neutron intensity as a function of temperature. Contrast variation of the neutron scattering length density was applied to both the lipid and the solvent. Scattering length density profiles were determined using both model-independent and model-dependent fitting methods. During the reflectivity measurements, a novel experimental set-up was implemented to decrease the incoherent background scattering due to the solvent. Thus, the reflectivity was measured to $Q \sim 0.3Å^{-1}$, covering up to seven orders of magnitude in reflected intensity, for PC bilayers in D_2O and silicon-matched (38% D_2O/62% H_2O) water. The kinetics of lipid adsorption at the silicon/water interface were also explored by observing changes in the reflectivity at low Q values under silicon-matched water conditions.

INTRODUCTION

The neutron reflectivity technique (see, for example, Majkrzak & Felcher, 1990) is well-suited to the study of structural changes in lamellar systems such as lipid monolayers, bilayers or multilayers and biological membranes. It is important to be able to measure a single lipid bilayer in aqueous solution since it represents a working model for a

Neutrons in Biology, edited by Schoenborn and Knott
Plenum Press, New York, 1996

biological membrane. Parameters of interest include the thickness of the bilayer and its individual components, penetration of water into the bilayer, asymmetry of the lipid distribution and changes in membrane properties due to interactions with small ions, peptides, proteins and other lipids. In addition, the depth of penetration of protein amino acid side chains and the localization of small molecules could be investigated.

Recently, the technique has come into wider use in the measurement of lipid monolayer and bilayer systems in aqueous environments. The structure of dimyristoylphosphatidlycholine (DMPC) bilayers adsorbed onto planar substrates has been studied by Johnson *et al.* (1991). Reinl *et al.* (1992) used neutron reflectivity to measure changes in the thickness of single dipalmitoylphosphatidlycholine (DPPC) bilayers on quartz surfaces after incorporation of cholesterol. In addition, several experiments have focussed on the study of streptavidin binding to biotinylated lipid monolayers, which are either formed at the air/water interface (Vaknin *et al.,* 1991; Lösche *et al.,* 1993) or adsorbed onto solid substrates (Schmidt *et al.,* 1992).

In this work, neutron reflectivity has been used to probe the structure of single phosphatidylcholine (PC) bilayers adsorbed onto a planar silicon surface in an aqueous environment. Fluctuations in the neutron scattering length density profiles perpendicular to the silicon/water interface were determined as a function of the hydrocarbon chain length and the phase state of the lipid. A novel experimental set-up was implemented to decrease the incoherent background scattering from the water, thereby extending the angular range and increasing the resolution of the measurements (Krueger *et al.,* 1995). For different PC bilayers in D_2O and silicon-matched water (SMW), reflectivity was obtained to $Q \sim 0.3 Å^{-1}$, which covered up to seven orders of magnitude in reflected intensity.

Contrast variation experiments were used to aid in the accurate modeling of the scattering data. Neutron scattering length density profiles were obtained from the data using two different fitting methods (Ankner & Majkrzak, 1992; Berk & Majkrzak, 1995). The use of the two fitting methods greatly aided in the understanding of the sensitivity of the measured reflected intensities to changes in the bilayer structure. It was found that the technique is highly-sensitive to small variations (on the order of 1–2Å) in overall thickness of the lipid bilayer and, consequently, to changes in overall hydrocarbon chain length. However, regions of different scattering length density within the bilayer must be on the order of 10Å in thickness to be located accurately under the present experimental conditions. Thus, efforts to improve the instrumental setup in order to increase this sensitivity are currently underway.

MATERIALS AND METHODS

Two different phosphatidylcholines, DPPC and distearoylphosphatidlycholine (DSPC), were measured at two different temperatures. A schematic of a PC lipid is shown in Figure 1. The only difference between the two lipids is the length of the hydrocarbon chains, which are 16 and 18 hydrocarbons for DPPC and DSPC, respectively. Lipids with hydrogenated chains (CH_2) were prepared in D_2O and lipids with deuterated chains (CD_2) were prepared in SMW. The measurement temperatures were chosen for each lipid based on their phase transition temperatures (Cevc, 1993) between liquid crystalline and gel phases. The DSPC samples were measured at 75°C and 30°C and the DPPC samples were measured at 60°C and 20°C.

Neutron reflectivity measurements were performed at the BT7 spectrometer at the National Institute of Standards and Technology (Majkrzak, 1991). Filtered, monochro-

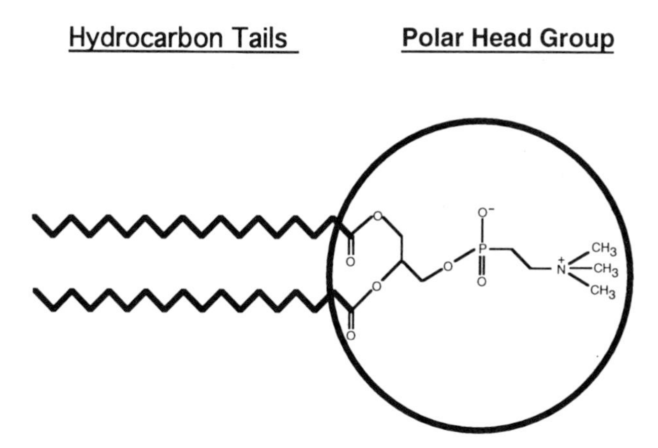

Figure 1. Schematic drawing of a phosphatidylcholine lipid molecule. A simple model, in which all portions of the lipid which are not explicitly part of the CH_2 chains have been designated collectively as the 'lipid head group', has been used to calculate the average neutron scattering length density.

matic (2.35Å) neutrons are collimated by two slits, defined by absorbing masks, located before the sample position. Neutrons specularly reflected from the sample, the plane of which is oriented vertically, are detected by a highly (>90%) efficient ^3He detector. The reflected intensity is measured as a function of glancing angle of incidence, θ, by performing standard θ-2θ scans, where 2θ is the angle of the detector relative to the incident beam direction.

The lipid bilayers were formed by allowing a suspension of sonicated vesicles to come into contact with two planar single-crystal silicon substrates, separated by several centimeters, in a teflon sample cell. Each silicon substrate contains a thin (~10Å) silicon oxide layer at the surface. The adsorption of the lipid onto one substrate surface can be monitored in situ by measuring the reflected intensity from CD_2 lipids in SMW. The reflected intensity was measured at a single incident angle ($\theta = 0.1°$) as a function of time. Within 5 minutes, the intensity reaches a plateau and further adsorption takes place at a much lower rate. After approximately 1 hour, the sample cell is flushed several times with water so that only a single bilayer remains on each substrate. Finally, the two silicon substrates are moved close to one another so that only a thin (~10μm) water layer remains between them.

The final sample geometry is shown schematically in Figure 2. Note that the neutron beam enters and exits through one silicon plate, reflecting from the bilayers adsorbed at the two surfaces. The incident beam of wavevector k_i strikes the first surface at the angle, θ, and the reflected beam of wavevector k_f exits at the same angle. The reflectivity is measured as a function of wavevector transfer, $Q = |k_f - k_i|$ in the direction normal to the bilayer surface (Z). The intensity of specularly reflected neutrons was measured for wavevector transfers up to $Q \sim 0.3Å^{-1}$. Beyond this value, the signal became noisy due to incoherent scattering from the water. The reflectivity curves were corrected for background, slit opening size and finite sample size and then converted to a $\log_{10}$(reflectivity) vs Q scale. The thin (~10μm) water layer which existed between the two silicon plates allowed for a greatly-reduced incoherent background, making measurements out to this high Q value possible. However, reflections from the bilayer on the surface of the back silicon substrate make a significant contribution to the total reflectivity and must be taken into account when analyzing the reflectivity data.

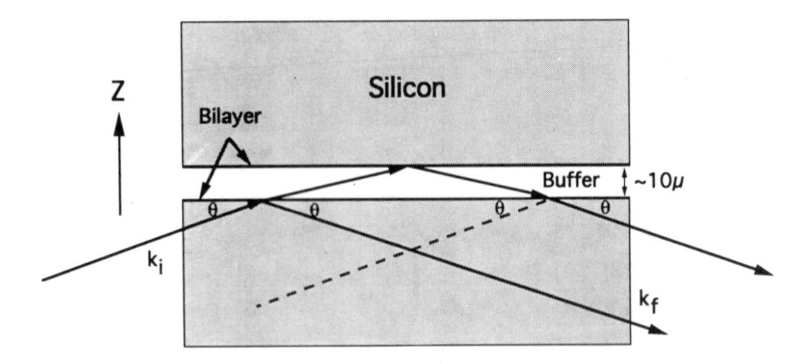

Figure 2. Schematic of the experimental setup. The neutron beam enters and exits through one silicon substrate and reflects at the two silicon/water interfaces containing lipid bilayers. The small gap between the silicon substrates results in a greatly-reduced incoherent background scattering from the water.

The lipid bilayer can be modeled as series of three layers, with an average neutron scattering length density for each layer representing the lipid head group or the hydrocarbon chain region. The program TMLAYER (Ankner & Majkrzak, 1992), in which the scattering length density profiles are composed of histogram functions based on the presumed lipid composition, was used to perform model-dependent fits to the data. Here, all portions of the lipid which are not explicitly part of the CH_2 chains have been considered collectively as the 'lipid head group' (see Figure 1). Model-independent fits were performed using the program PBS (Berk & Majkrzak, 1995), in which the scattering length density profiles are composed of randomly-initialized smooth functions represented by parametric B-splines. In general, the PBS fitting method leads to one or more families of solutions which fit to the reflectivity data equally well and which thus allow a determination of the shape uncertainty in the scattering length density profiles. Both fitting procedures explicitly take into account the reflections from the bilayer on the surface of the back silicon substrate.

RESULTS AND DISCUSSION

The reflected intensities from DSPC-CH_2 and DPPC-CH_2 bilayers in D_2O in the gel phase are plotted in Figure 3. The (~5Å) difference in bilayer thickness is readily apparent. The intensity from the thicker DSPC bilayer shows a minimum at $Q = 0.14Å^{-1}$, whereas that from the DPPC bilayer shows a minimum at $Q = 0.17Å^{-1}$. Figure 4 shows the reflected intensity from DPPC-CH_2 in D_2O in both the gel and liquid crystalline phases. The minimum in the reflected intensity occurs at a higher Q value for the liquid crystalline phase, indicating that the bilayer is thinner under these conditions.

Figure 5 shows a comparison between the model-dependent (TMLAYER) and model-independent (PBS) fits to the data from DPPC in the gel phase. Figure 5a shows DPPC-CH_2 in D_2O while Figure 5b shows DPPC-CD_2 in SMW. It is evident that satisfactory fits can be obtained from both methods. Neutron scattering length density profiles, $\rho(z)$, in the direction perpendicular to the bilayer, obtained from the model-independent (PBS) fits, are shown for DPPC in the gel phase in Figure 6. Figures 6a and 6b show the profiles for DPPC-CH_2 in D_2O and Figures 6c and 6d show the profiles for DPPC-CD_2 in

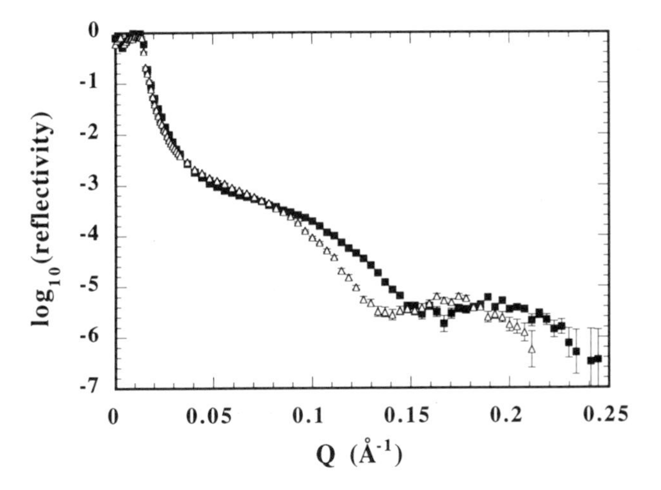

Figure 3. Measured specular reflectivity from DSPC-CH_2 (Δ) and DPPC-CH_2 (■) in D_2O in the gel phase.

SMW. In both cases, two distinct families of curves, which can be related by symmetry operations, were found to fit equally well to the reflectivity data.

The DPPC samples in D_2O contain CH_2 chains which have a lower scattering length density than the D_2O. Thus, Figure 6b can be ruled out as unphysical since the hydrocarbon chains appear as a high scattering length density region. On the other hand, the DPPC samples in SMW contain CD_2 chains, which have a higher scattering length density than the SMW. Since it is also known that an oxide layer exists at the surface of the silicon substrate, it must appear as a region of scattering length density higher than that of the substrate and the water, but still lower than that of the hydrocarbon chains. Thus, Figure 6d can be eliminated since no silicon oxide layer is evident at the surface of the substrate. In this manner, the PBS fitting method was used in order to narrow down the possible scattering length density profiles. The resultant profiles (Figure 6a and 6c) were then used

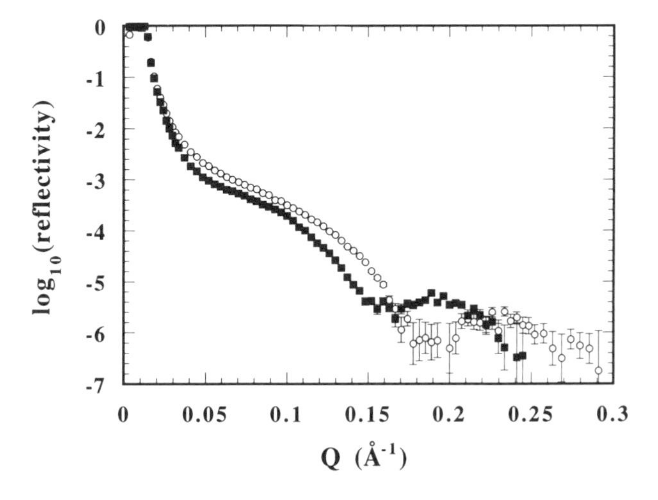

Figure 4. Measured specular reflectivity from DPPC-CH_2 in D_2O, in both the gel (■) and the liquid crystalline (O) phases.

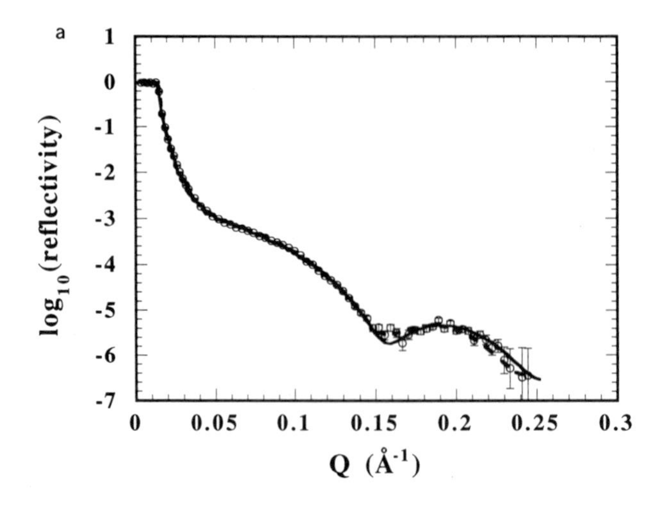

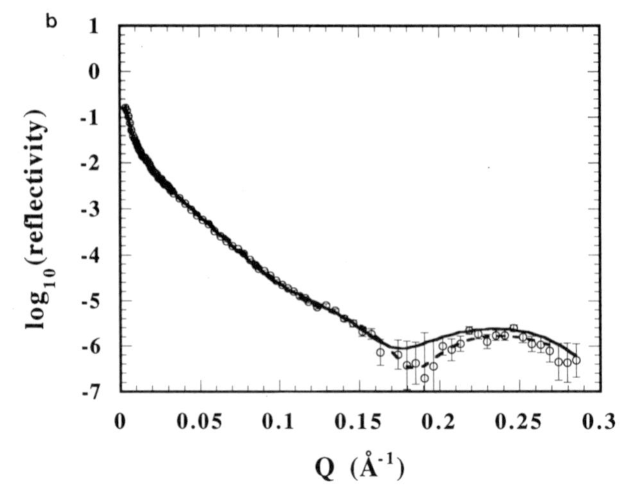

Figure 5. Fits to the measured reflectivity from: (a) DPPC-CH_2 in D_2O and (b) DPPC-CD_2 in silicon-matched water (SMW), in the gel phase (○). In both cases, the solid line represents the fit obtained from the model-dependent (TMLAYER) method and the dashed line represents equally good fits obtained from the model-independent (PBS) method.

as starting points for refining the model-dependent fits made with TMLAYER (Koenig *et al.,* 1995). In this case, the data were fit using the smallest possible number of steps in the model function. The parameter space was systematically explored starting with a coarse grid size, with the choice of initial parameters guided by the PBS profiles. The fits were compared with the data by calculating the root mean square deviation (χ^2). The grid size was decreased in the vicinity of local minima until the minimum χ^2 value was verified. Since the model parameters are interdependent, the global minima of χ^2 proved to be shallow in each case, giving rise to errors on the order of 10% in the thickness of the hydrocarbon chain regions.

In addition to the silicon oxide layer at the surface of the silicon substrate, a ~10Å-thick water layer is clearly present between the oxide and the lipid head group layers under SMW conditions (Figure 6c). This water layer is also evident in D_2O (Figure 6a), but its scattering length density is somewhat lower than that of the D_2O due, at least in part, to penetration of water into the lipid head group layer. A schematic of the lipid bilayer shows the approximate location of the lipid head group and hydrocarbon chain layers for Figures 6a and 6c. The thickness of the hydrocarbon chain region for the DPPC-CH_2 bilayers in

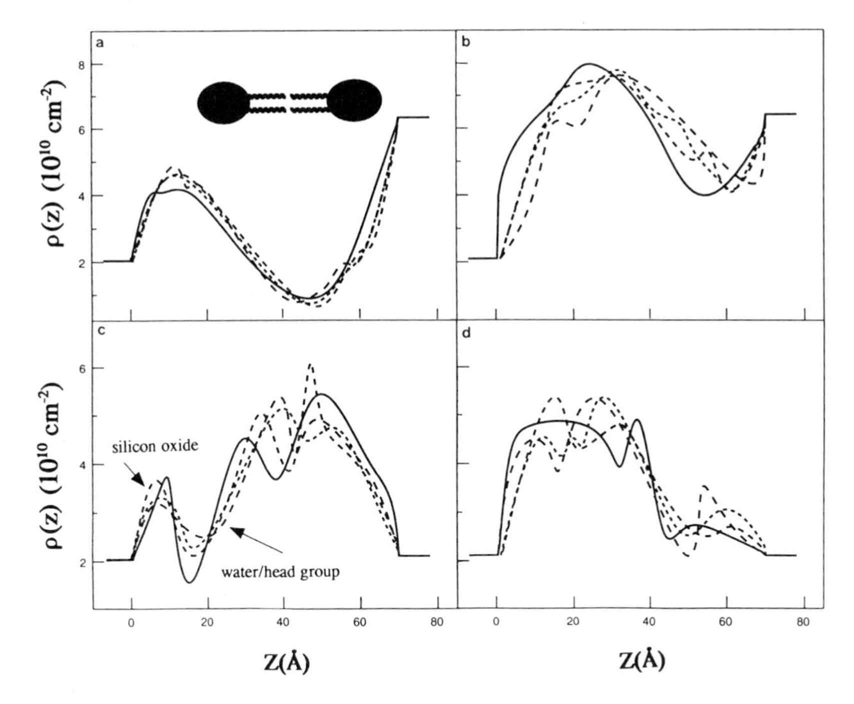

Figure 6. The families of neutron scattering length density profiles corresponding to the PBS fits to the reflectivity data from DPPC in the gel phase. (a) and (b) are for DPPC-CH_2 in D_2O while (c) and (d) are for DPPC-CD_2 in SMW. The silicon substrate is located at $Z \leq 0$. A schematic of the lipid bilayer in (a) indicates the approximate location of the lipid head group and hydrocarbon chain layers. In SMW, the silicon oxide layer at the surface of the silicon substrate is readily apparent and a ~10Å-thick water layer exists between the oxide layer and the lipid bilayer, as indicated in (c).

D_2O was found to be 32 ± 2Å in the gel phase and 28 ± 2Å in the liquid crystalline phase. Similarly, the values for DPPC-CD_2 in SMW were found to be 36 ± 2Å in the gel phase and 29 ± 2Å in the liquid crystalline phase. These values are in agreement with those obtained for X-ray diffraction from fully-hydrated DPPC multilamellar membrane stacks in the gel phase (Wiener *et al.*, 1989) and from unilamellar DPPC vesicles in the liquid crystalline phase (Lewis & Engelman, 1983). For DSPC-CH_2 bilayers in D_2O in the gel phase, the hydrocarbon chain region thickness was found to be 38 ± 2Å.

An examination of the families of scattering length density profiles obtained from the PBS fitting method for any of the data sets leads to the conclusion that the measurements are sensitive to changes on the order of 1–2Å in overall bilayer or hydrocarbon chain region thickness. However, structural details within the bilayer cannot be determined under the present experimental conditions. This is further illustrated in Figure 7, where the inset shows a model scattering length density profile which has been increased in thickness by 2Å and to which has been added a sinusoidal ripple of width 2Å. The corresponding calculated reflectivity shows that the change in overall thickness can be seen very clearly below $Q = 0.3\text{Å}^{-1}$. However, changes in the reflected intensity due to the ripple in the profile cannot be discerned below $Q = 0.6\text{Å}^{-1}$. Currently, improvements are being made to the instrument which should allow measurements down to 10^{-8} in reflected intensity. Therefore, it may be possible to obtain reflected intensity curves beyond Q =

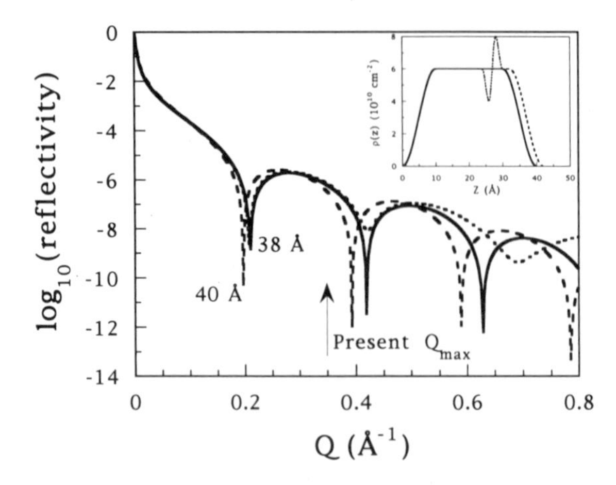

Figure 7. Model neutron reflectivity curves corresponding to the scattering length density profiles shown in the inset. The dashed profile is 2Å thicker than the solid profile while the dotted profile contains a sinusoidal density fluctuation with a thickness of 2Å. The present experimental maximum Q value is indicated by the arrow.

0.6Å^{-1}, thus gaining the desired increase in sensitivity to small changes within the lipid bilayers.

CONCLUDING REMARKS

The structures of planar lipid bilayers on silicon substrates in aqueous solution have been described using the neutron reflectivity technique. Reflected intensities were obtained out to Q ~ 0.3Å^{-1} and over seven orders of magnitude by reducing the incoherent scattering from water significantly by allowing only a thin (~10μm) gap between two silicon blocks. Good fits to the reflectivity data could be found using model-dependent and model-independent fitting methods and good agreement was found between the resultant scattering length density profiles, within the given spatial resolution. The results indicated that a thin (~10Å) water layer exists between the silicon oxide layer at the surface of the substrate and the lipid bilayer. Values obtained for the hydrocarbon chain region thickness are reasonable compared with those obtained from other methods.

ACKNOWLEDGMENTS

We thank Dr. J.F. Ankner (University of Missouri) for providing the TMLAYER program. BWK was partially supported by a post-doctoral fellowship from the German Academic Exchange Service. SK, WJO and CFM were partially supported by the National Science Foundation under Agreement No. DMR-9122444.

REFERENCES

Ankner, J.F., & Majkrzak, C.F., (1992). Subsurface profile refinement for neutron specular reflectivity. In *S.P.I.E. Conference Proceedings*, Vol. 1738. C.F. Majkrzak and J.L. Wood, editors. S.P.I.E., Bellingham, Wash.

Berk, N.F., & Majkrzak, C.F., (1995). Using parametric B-splines to fit specular reflectivities. *Phys. Rev.* B, 51:11296–11309.

Cevc, G., (1993). Appendix B: Thermodynamic Parameters of Phospholipids. In *Phospholipids Handbook*. G. Cevc, editor. Marcel Dekker, New York, NY, pp939–956.

Johnson, S.J., Bayerl, T.M., McDermott, D.C., Adam, G.W., Rennie, A.R., Thomas, R.K., & Sackmann, E., (1991). Structure of an adsorbed dimyristoylphosphatidylcholine bilayer measured with specular reflection of neutrons. *Biophys. J.*, 59:289–294.

Koenig, B.W., Krueger, S., Orts, W.J., Majkrzak, C.F., Berk, N.F., Silverton, J.V., & Gawrisch, K., (1995). Neutron reflectivity and atomic force microscopy studies of a lipid bilayer in water adsorbed to the surface of a silicon single crystal. *Langmuir,* (submitted).

Krueger, S., Ankner, J.F., Satija, S.K., Majkrzak, C.F., Gurley, D., & Colombini, M., (1995). Extending the angular range of neutron reflectivity measurements from planar lipid bilayers: Application to a model biological membrane. *Langmuir* (in press).

Lewis, B.A., & Engelman, D.M., (1983). Lipid bilayer thickness varies linearly with acyl chain length in fluid phosphatidylcholine vesicles. *J. Mol. Biol.,* 166:211–217.

Lösche, M., Piepenstock, M., Diederich, A., Grünewald, T., Kjaer, K., & Vaknin, D., (1993). Influence of surface chemistry on the structural organization of monomolecular protein layers adsorbed to functionalized aqueous interfaces. *Biophys. J.*, 65:2160–2177.

Majkrzak, C.F., & Felcher, G.P., (1990). Neutron scattering studies of surfaces and interfaces. *Mater. Res. Soc. Bull.,* 15:65–72.

Majkrzak, C.F., (1991). Polarized neutron reflectometry. *Physica B,* 173:75–88.

Reinl, H., Brumm, T., & Bayerl, T.M., (1992). Changes of the physical properties of the liquid-ordered phase with temperature in binary mixtures of DPPC with cholesterol. *Biophys. J.*, 61:1025–1035.

Schmidt, A., Spinke, J., Bayerl, T.M., & Knoll, W., (1992). Streptavidin binding to biotinylated lipid layers on solid supports. *Biophys. J.*, 63:1185–1192.

Vaknin, D., Als-Nielsen, J., Piepenstock, M., & Lösche, M., (1991). Recognition processes at a functionalized lipid surface observed with molecular resolution. *Biophys. J.*, 60:1545–1552.

Wiener, M.C., Suter, R.M., & Nagle, J.F., (1989). Structure of the fully hydrated gel phase of dipalmitoylphosphatidylcholine. *Biophys. J.*, 55:315–325.

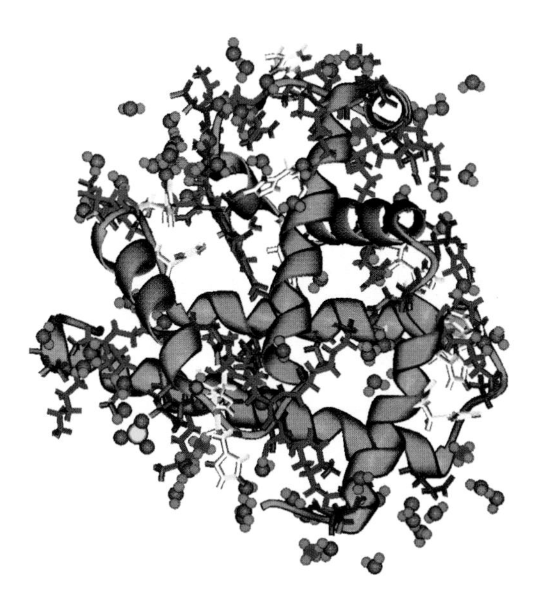

Introduction, Figure 1. The myoglobin structure with water molecules as determined by neutron diffraction studies. Water is shown as dotted clouds on the surface of the protein (after Cheng & Schoenborn, 1990). Observed bound water molecules and 5 solvent ions identified in this study are depicted. Of the total of 89, 39 water molecules are bound to protein side chain atoms; 16 are bound to main chain atoms; and 12 are in bridges between protein atoms - 10 are intramolecular and 1 intermolecular. The 22 remaining water molecules are bound only to other water molecules. All water molecules bound to the protein are hydrogen bonded to polar or charged side chains that are depicted in the diagram.

Introduction, Figure 2. Three orthogonal views of the low resolution structure of the 30S ribosomal subunit of *E.coli* as determined by small angle neutron scattering techniques using specifically deuterated proteins. This structure was constructed by positioning the 21 proteins of known molecular weight by measuring the distance between each pair of proteins. The complete set of distances was determined by an extensive series of neutron scattering experiments. Literally hundreds of ribosome samples were required with one or two proteins deuterated according to a strict protocol (after Capel *et al.*, 1987). The maximum linear dimension of the array is about 190Å. The size of the spheres is proportional to their molecular weight. The numbering of the proteins follows the standard nomenclature for ribosomal proteins. The two views are front and back views of the model.

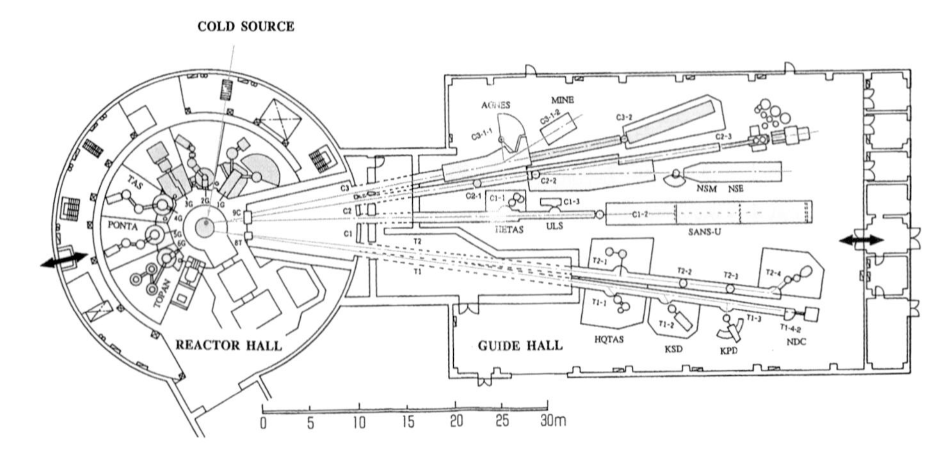

Figure 10.1. The whole layout of JRR-3M.

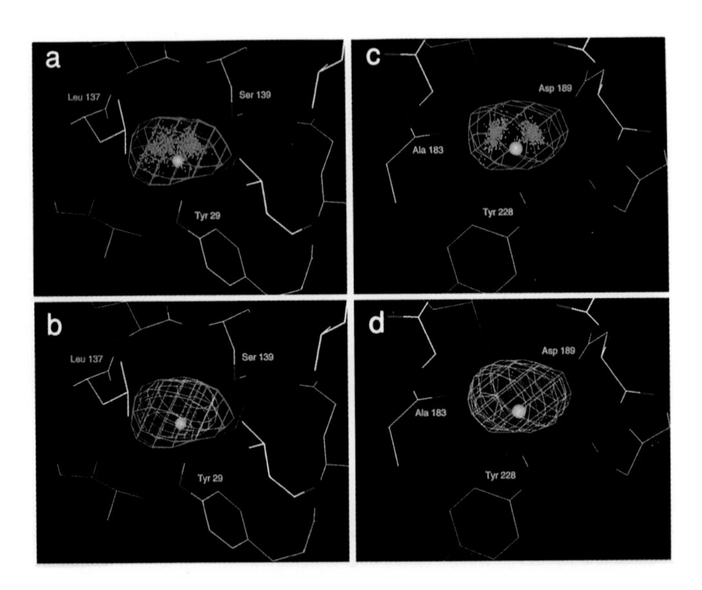

Figure 23.4. Neutron density and contoured hydrogen distributions for internal waters-314 (a-b) and 347 (c-d). Depicted in panels (a) and (c) are the sampled hydrogen coordinates from the dynamics trajectory (light blue) superimposed on the neutron density (green contour). A 'time-averaged' depiction of these trajectories (pink contours), obtained by summing gaussian functions representing the instantaneous hydrogen coordinates, is compared to the neutron density in panels (b) and (d). X-ray water positions are illustrated as yellow spheres.

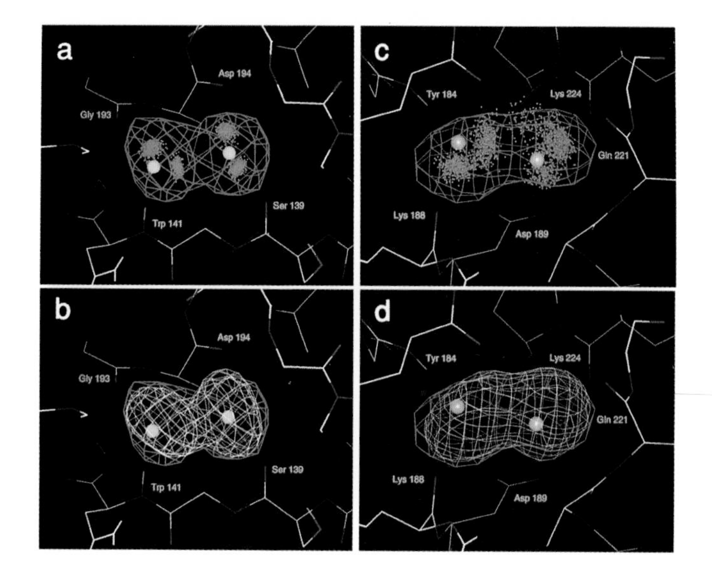

Figure 23.5. Primary hydrogen-bonding configurations of internal water clusters 275/326 (a-b) and 269/346 (c-d). The representations are identical to those used in Figure 4. In panels (a)-(b), water-275 is on the left side; in panels (c)-(d) water-346 is on the left.

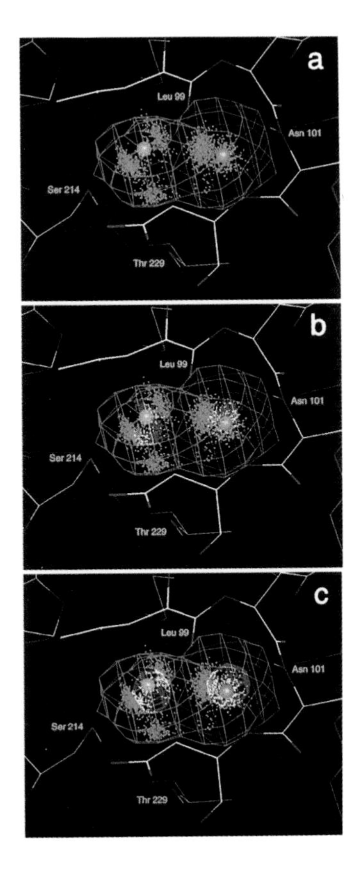

Figure 23.6. Primary hydrogen-bonding configurations of internal water cluster 352/353. In panel (a) the hydrogen coordinates for waters-352 (left) and 353 are superimposed on the neutron density as in Figures 4 and 5. Scatter plots of the oxygen coordinates are illustrated in red in panel (b). Note that water-352 has a trimodal hydrogen distribution and a bimodal oxygen distribution, indicating static disorder between two stable hydrogen-bonding configurations. If the oxygen distributions are contoured to produce a 'time averaged' representation, the loci of the peaks correspond well with the crystallographic positions assigned to the waters (red contours; panel c).

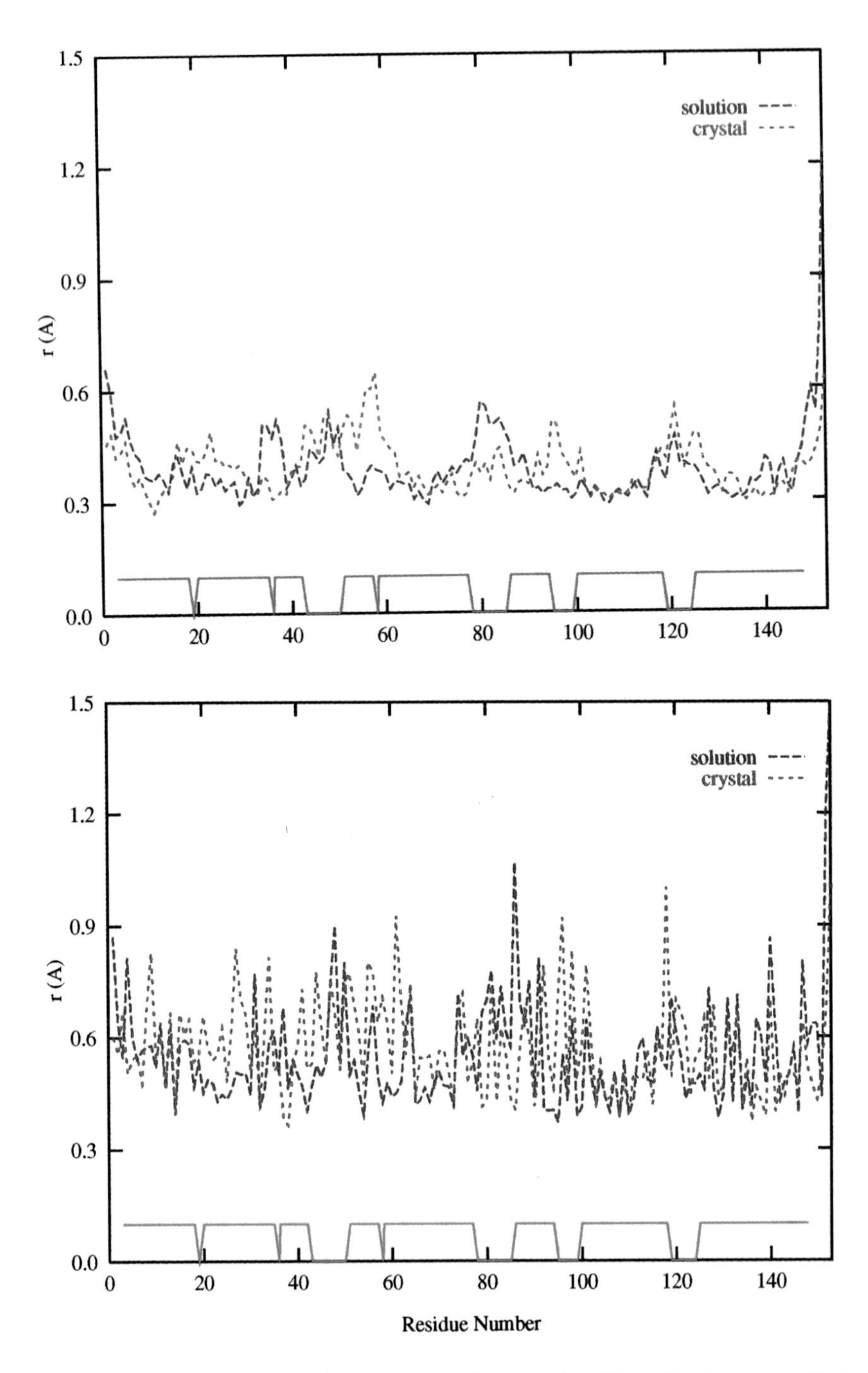

Figure 24.1. The rms fluctuations of protein atoms, averaged over each residue and trajectory coordinates of the simulations: (a) backbone atoms; (b) side-chain atoms.

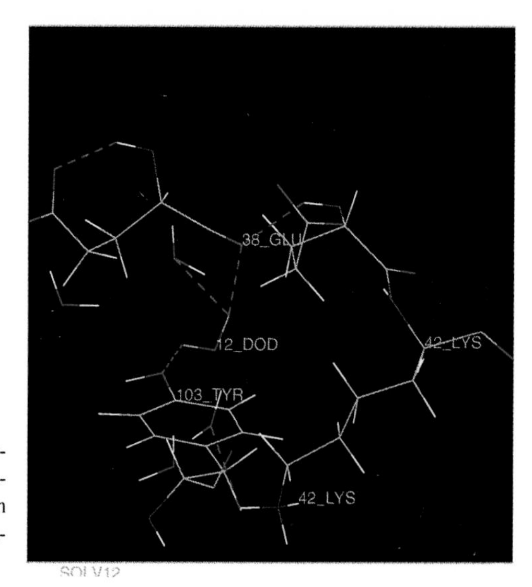

Figure 24.5. The hydrogen-bonding geometry for water-12 as observed in the neutron map. Hydrogen bonds are depicted in red dashed lines; oxygen atoms in red; nitrogen in blue; hydrogen white; carbon and deuterium atoms are displayed in green.

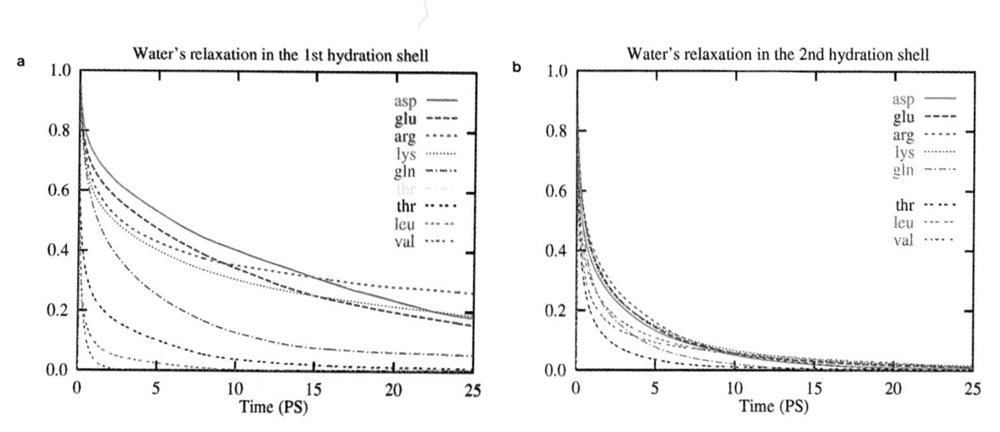

Figure 24.6. The relaxation times of water molecules adjacent to different side-chains in the solution system: (a) in the first hydration shell; (b) in the second hydration shell.

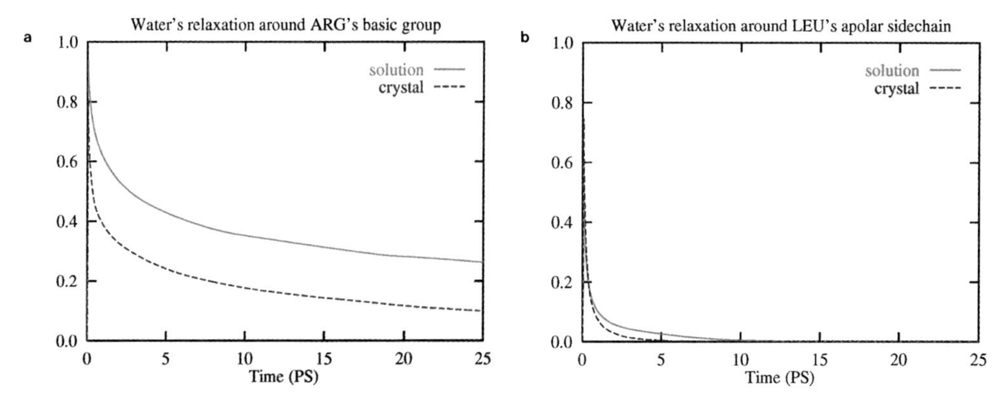

Figure 24.7. Comparison of the relaxation times of water molecules near (a) arginine side-chains in solution and crystal form: (b) leucine side-chains.

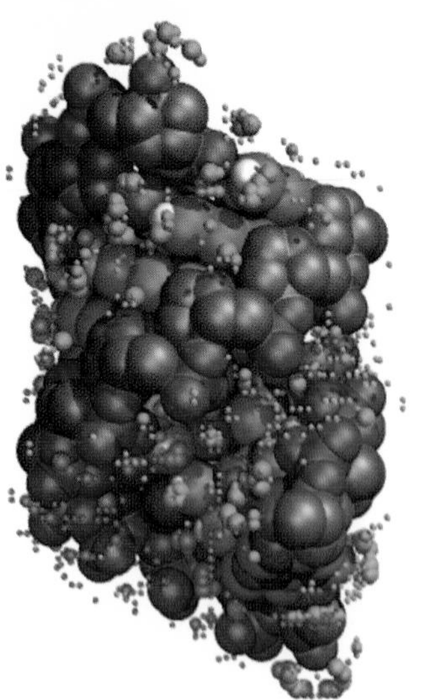

Figure 25.1. Minor- and major-groove view of d(GGGGGCCCCC)$_2$ in the A-DNA conformation. The high-density points of the local water density calculated in a regular grid are represented by cyan spheres of radii: 0.6Å for ρ larger than 7.0ρ_0, 0.4Å for ρ between 5 and 7ρ_0, and 0.2Å for ρ between 3 and 5ρ_0. DNA atoms are color coded by N=blue, C=grey, H=white, O=red and P=dark blue.

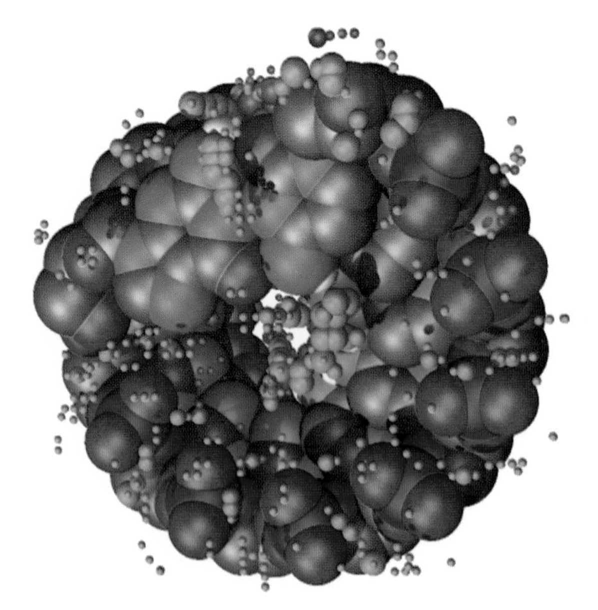

Figure 25.2. Top view of d(GGGGGCCCCC)$_2$ in the A-DNA conformation. Details as in Figure 1.

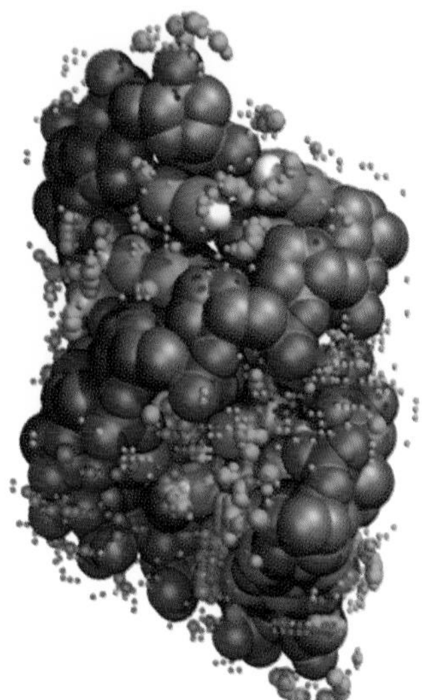

Figure 25.3. Minor- and major-groove view of d(GCGCGCGCGC)$_2$ in the A-DNA conformation. Details as in Figure 1.

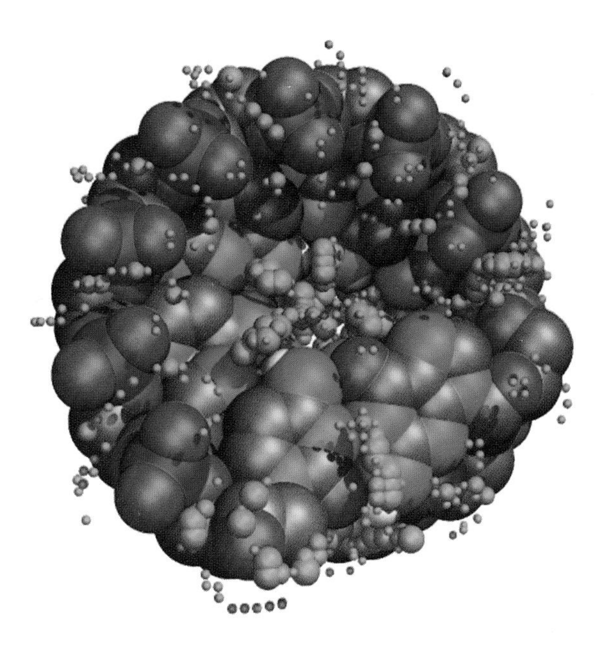

Figure 25.4. Top view of d(GCGCGCGCGC)$_2$ in the A-DNA conformation. Details as in Figure 1.

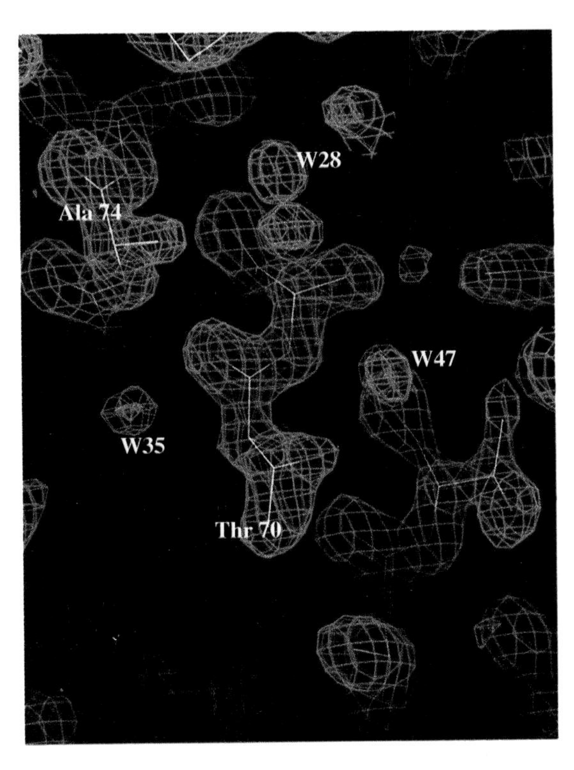

Figure 26.5. Comparison of $(2F_o\text{-}F_t,\phi_t)$ and $(2F_n\text{-}F_p,\phi_p)$ maps. Both maps are drawn at 1.5σ ($0.72e^-/\text{Å}^3$), where σ is the rms deviation from the mean electron-density in each map calculation. The $(2F_o\text{-}F_t,\phi_t)$ map is in blue; the $(2F_n\text{-}F_p,\phi_p)$ map is in magenta. Stronger electron densities at water sites W28, W35, and W47 in the $(2F_o\text{-}F_t,\phi_t)$ map, but similar at protein region, indicate that solvent features are revealed better after solvent phases are included in the map calculation.

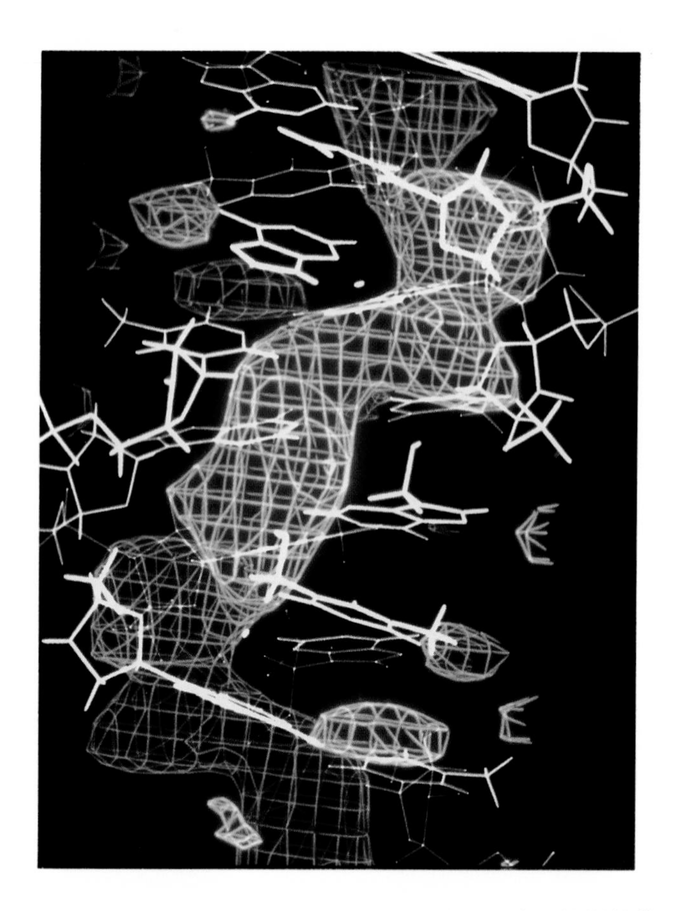

Figure 29.4. Difference Fourier map generated from the D_2O and H_2O datasets from D-DNA illustrated in Figures 2 and 3. Particularly noticeable in this image is the network of hydration which runs down the narrow minor groove of the DNA double helix.

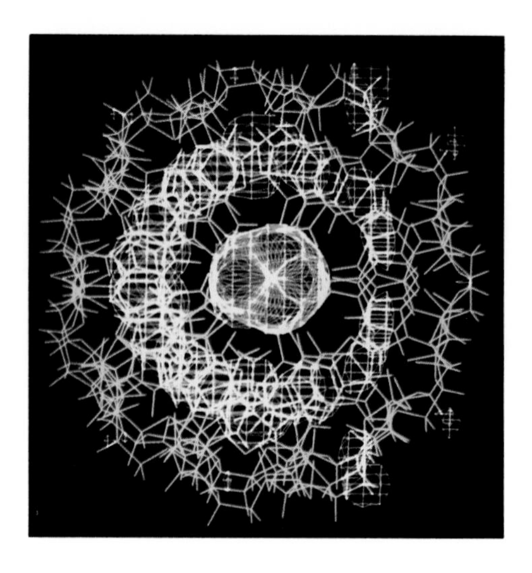

Figure 29.7. A projection down the helix axis of the Fourier difference map generated using the D_2O and H_2O datasets from A-DNA illustrated in Figure 5.

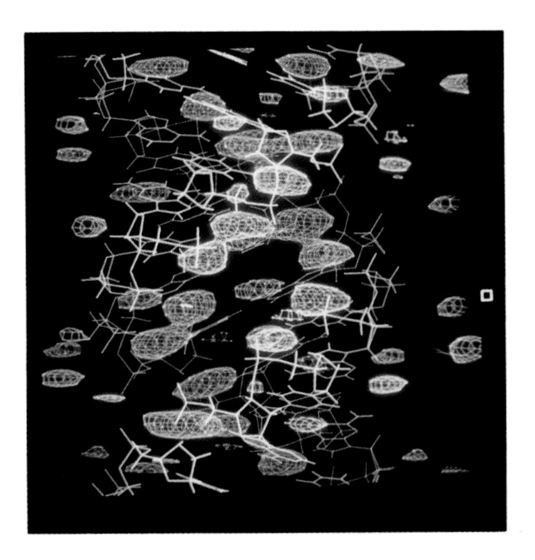

Figure 29.8. Water peaks associated with the A-form of DNA viewed perpendicular to the helix axis in the Fourier difference map generated using the D_2O and H_2O datasets illustrated in Figure 5.

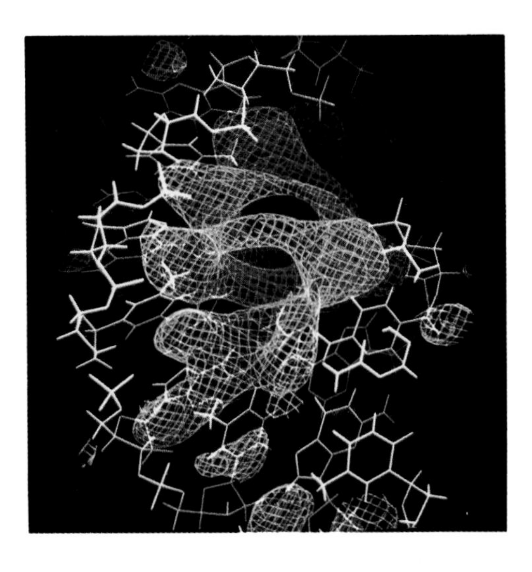

Figure 29.9. An off-helix axis view of the location of water in the wide groove of the A-form of the DNA double helix.

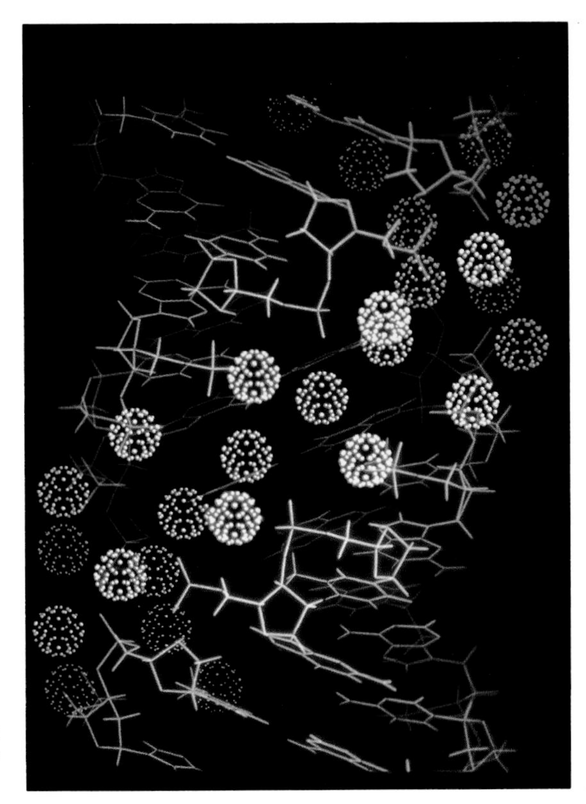

Figure 29.10. The distribution of ions (shown as stippled spheres) with respect to the major groove of the A form of the DNA double helix.

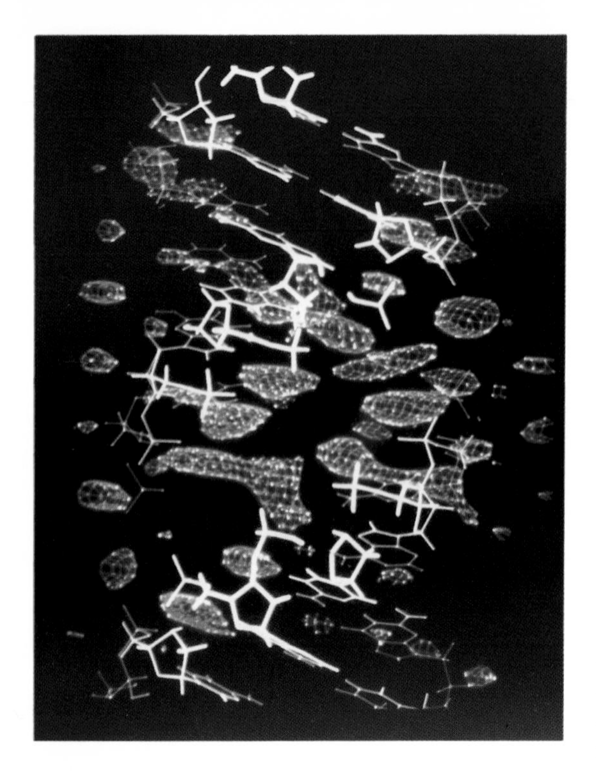

Figure 30.5. Difference Fourier map calculated using data recorded at SXD. Peaks associated with water located between successive O1 phosphate oxygen atoms are clearly visible.

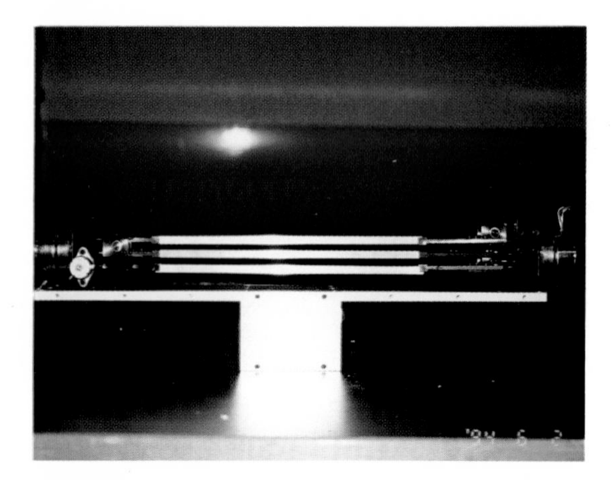

Figure 34.1. Monochromator system of the BIX.

Figure 34.2. The photograph of the BIX. (a) incident beam flight tube, (b) beam monitor, (c) sample and (d) detectors and their shielding house.

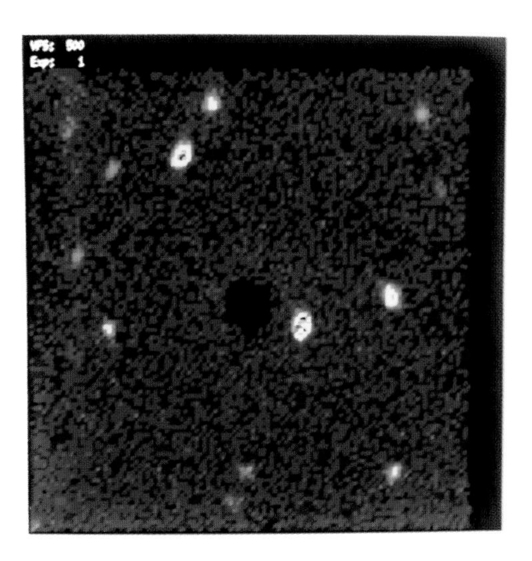

Figure 34.3. The diffraction patterns of hen egg white lysozyme single crystal recorded on the gas-filled position sensitive detector.

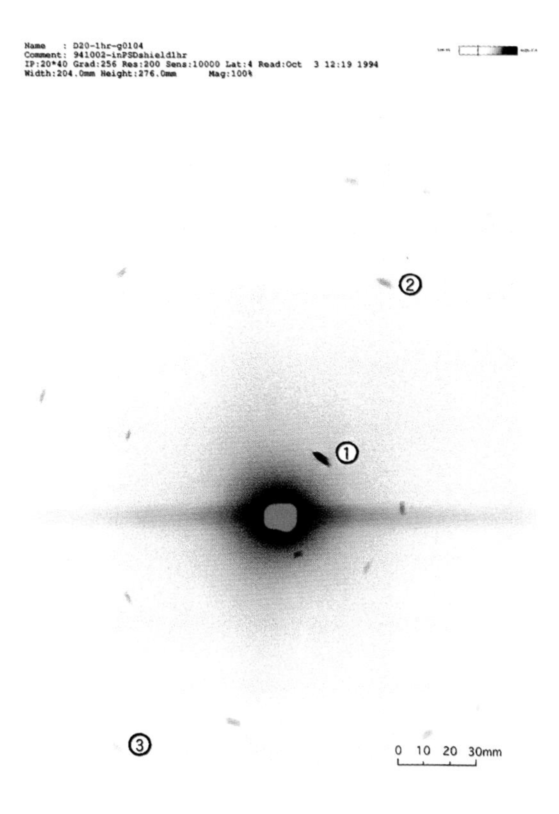

Figure 34.4. The diffraction patterns of hen egg white lysozyme single crystal recorded on the imaging plate neutron detector.

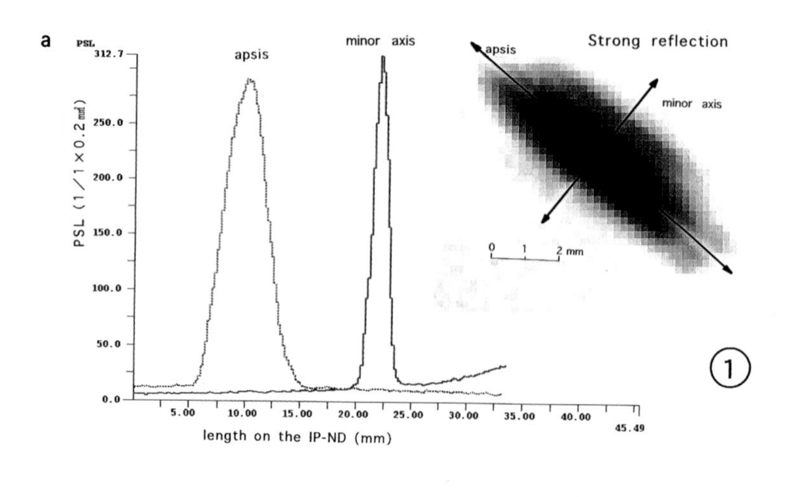

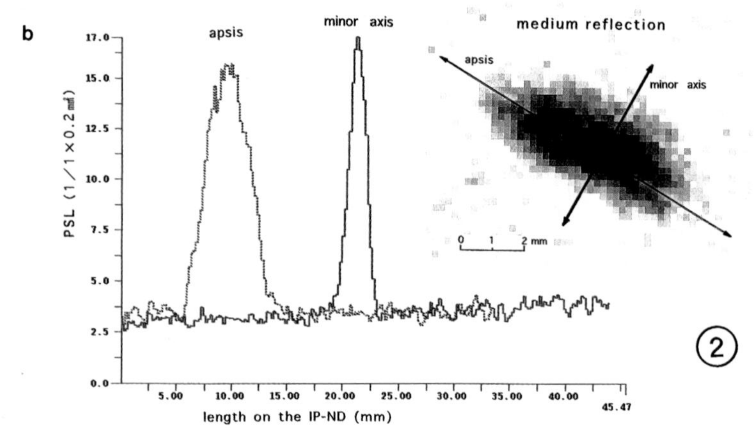

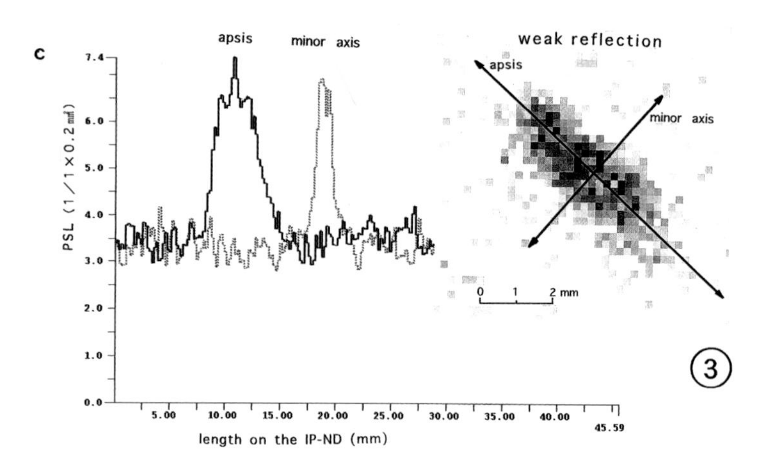

Figure 34.5. The enlarged spot and the profile along the apsis and the minor axis of the spot. (a), (b) and (c) correspond to the reflection 1, 2, and 3, respectively, in Figure 4. The ordinate value is the PSL per 5 × 1 pixels (1mm × 0.2mm).

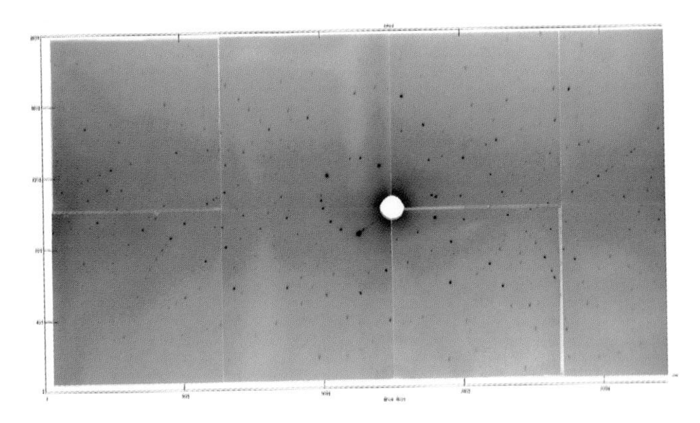

Figure 35.4. Laue diagram from a crystal of triclinic hen egg-white lysozyme.

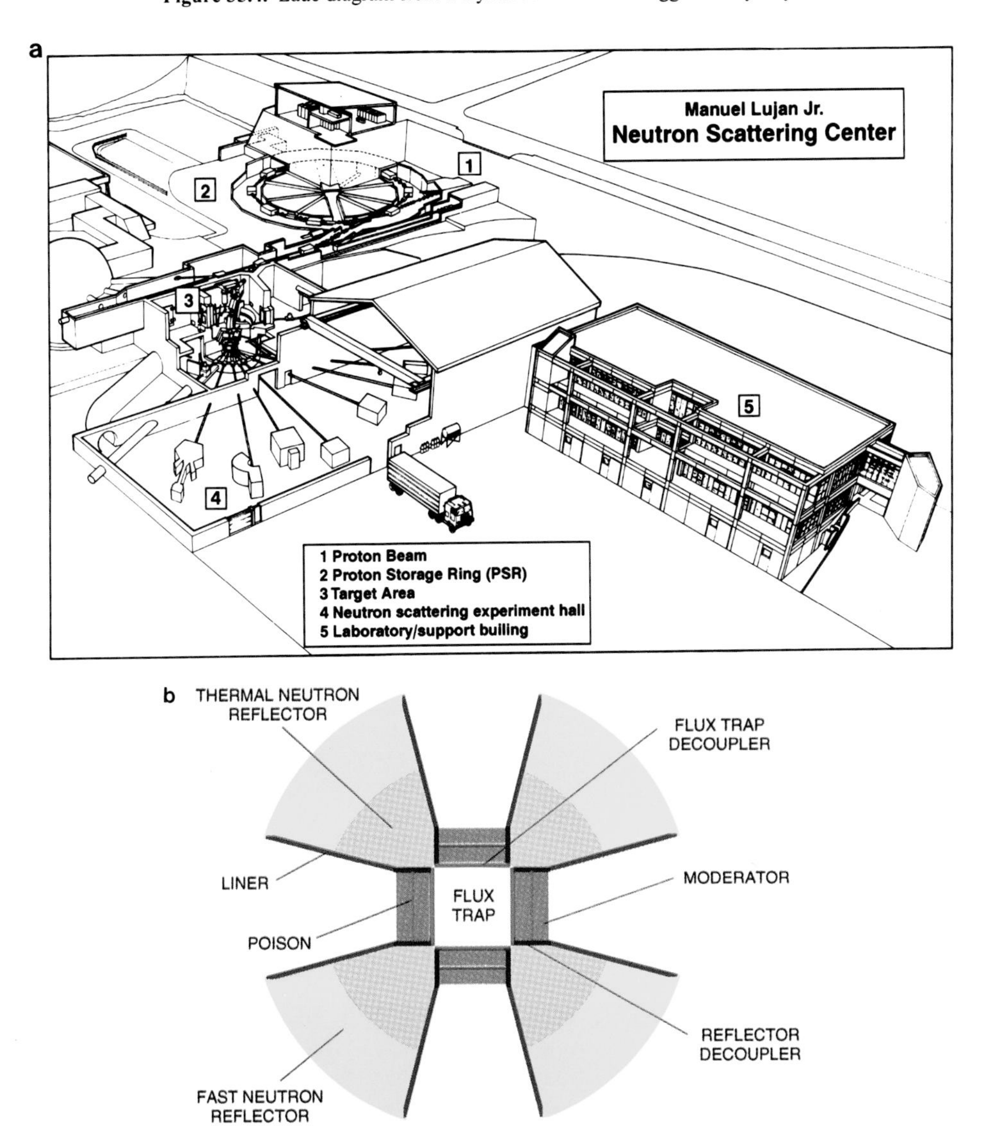

Figure 36.1. (a) Overview of the Los Alamos neutron scattering facilities. (b) Horizontal cut-away view of the target moderator area showing the liners (decouplers) and poison layers used to maintain the pulsed nature of the neutrons by eliminating 'stray neutrons' that would broaden the time/energy band width. The poison layer (Gd) in the moderator itself is used to limit the volume of moderator medium 'viewed' by the instrument.

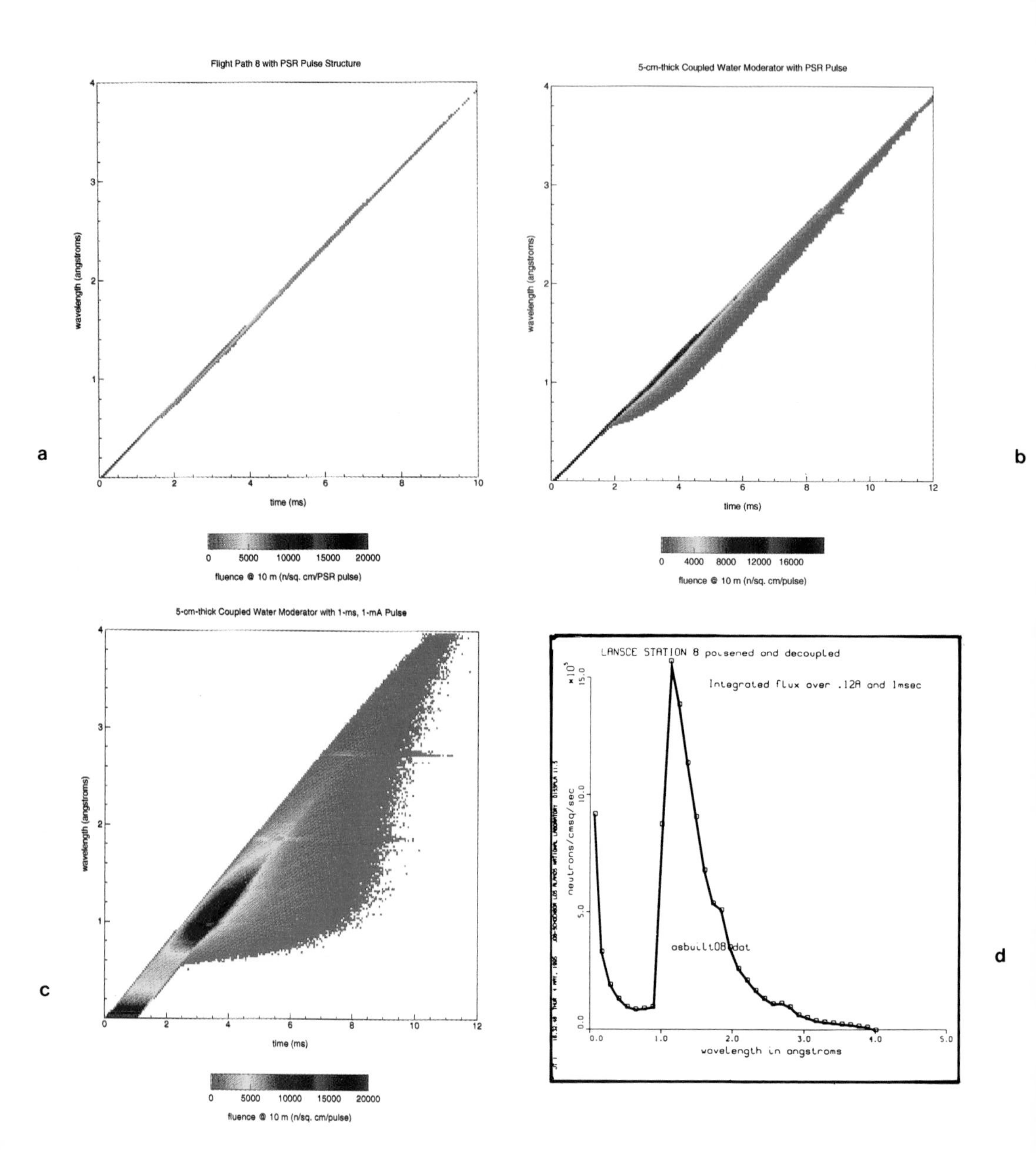

Figure 36.2. (a) Observed and calculated neutron flux at a sample position 10m from the neutron source of Station 8 at LANSCE using a 0.1mA pulse with a pulse duration less than 1μsec from the PSR (proton storage ring). Beam-line 8 uses a water moderator. Use of a colder moderating medium, such as methane will shift the spectrum toward longer wavelength, as is desirable for protein and membrane crystallography where maximum flux between 1 and 5Å is most effective. (b) The flux distribution with the same pulse structure but using a fully coupled and unpoisoned water moderator. (c) A similar spectral representation but with a 1msec proton pulse (LPSS) with 1mA current at the end of the LAMPF Linac. Neutrons are moderated with a 50mm water moderator at ambient temperature. (d) Integrated flux over 1msec and 0.12Å as a function of wavelength for the case illustrated in Figure 1a; this is equivalent to the flux on the crystal for one snapshot of the timed Laue data collection. Departures from the smooth curve are thought to be artefact of the water scattering kernel used in the calculation, and are not expected in reality. Calculated flux profiles for the 'as built' conditions agree well with measured flux.

19

INTERCALATION OF SMALL HYDROPHOBIC MOLECULES IN LIPID BILAYERS CONTAINING CHOLESTEROL

D. L. Worcester,[2] K. Hamacher,[1] H. Kaiser,[1] R. Kulasekere,[1] and J. Torbet[3]

[1] Physics Department and Research Reactor
[2] Biology Division
University of Missouri
Columbia, Missouri 65211
[3] Institut Laue Langevin, 156X
38042 Grenoble Cedex 9, France

ABSTRACT

Partitioning of small hydrophobic molecules into lipid bilayers containing cholesterol has been studied using the 2XC diffractometer at the University of Missouri Research Reactor. Locations of the compounds were determined by Fourier difference methods with data from both deuterated and undeuterated compounds introduced into the bilayers from the vapor phase. Data fitting procedures were developed for determining how well the compounds were localized. The compounds were found to be localized in a narrow region at the center of the hydrophobic layer, between the two halves of the bilayer. The structures are therefore intercalated structures with the long axis of the molecules in the plane of the bilayer.

INTRODUCTION

Lecithin and cholesterol mixtures form bilayer membranes that are good models for studying molecular arrangements and dynamics of membranes. Such model membrane studies may have small polypeptides or other components incorporated (Gogol *et al.*, 1983; Duff *et al.*, 1994). Cholesterol is particularly prevalent in eukaryotic plasma membranes and is therefore needed in model studies of structure and function of these membranes. The properties cholesterol confers to the plasma membrane bilayer are important for many membrane processes, some of which are poorly understood.

In this work, partitioning of small hydrophobic molecules into lipid bilayers containing cholesterol has been studied using the 2XC diffractometer at the University of Mis-

Neutrons in Biology, edited by Schoenborn and Knott
Plenum Press, New York, 1996

souri Research Reactor. Bilayer uptake of decane, hexane, and toluene has been studied, with the compounds introduced into the bilayer from the vapor phase. The bilayers consisted of diacyl phosphatidylcholines (PC) and cholesterol, with cholesterol content ranging from 0.3 to 0.5 mole percent. Locations of the small hydrophobic molecules were determined by Fourier difference profiles with data from both deuterated and undeuterated compounds.

At low bilayer contents, the compounds were found to be localized in a narrow region at the center of the hydrophobic layer, between the two halves of the bilayer. The structures formed are therefore intercalated structures and the long axis of intercalated molecules such as hexane and decane must be in the plane of the bilayer. Fourier difference profiles show very narrow peaks, comparable in width to that of deuterium label on the cholesterol steroid ring. For a fixed vapor pressure of the hydrophobic compounds, the amount of compound taken up by the bilayer depended on the fatty acyl chain length of the phosphatidylcholine. The results suggest a new concept (intercalation) for the partitioning of small hydrophobic molecules into bilayer membranes containing cholesterol. This concept may also be pertinent to the interaction of side chains of transmembrane alpha helices with bilayers containing cholesterol.

MATERIALS AND DIFFRACTION METHODS

Lecithins and cholesterol were obtained from Sigma Chemical Co., St. Louis, and from Fluka, Switzerland. Cholesterol with five deuterium atoms near the hydroxyl group was prepared from cholestenone by enolization and reduction of the enol acetate derivative with sodium borodeuteride (Figure 1). The reduction product contains a few percent epicholesterol which is removed by cholesterol purification as the digitonide and recrystallization from ethanol before use. We have used this deuterium labeled cholesterol to enhance the internal neutron scattering contrast of the bilayers, producing better neutron diffraction data, and to obtain accurate measurements of the position, disorder and dynamics of cholesterol in bilayers.

Highly oriented bilayers were prepared on glass or quartz substrates by slow evaporation of solvent from solutions in 80% ethanol at 40°C. Mosaic spreads of the samples were about 0.5° full width at half maximum. The samples are best described as smectic liquid crystals. The layers are highly ordered in stacks and the molecules are well positioned in the direction perpendicular to the layers, producing up to 12 orders of lamellar Bragg neutron diffraction. There is liquid-like disorder in the bilayer plane. In this work we have concentrated on using the lamellar Bragg diffraction and deuterium labeled compounds to obtain accurate positions and displacement (thermal) parameters for the labeled molecules.

Introduction of hexane and decane into the bilayers was made from the vapor phase (White *et al.*, 1981; King *et al.*, 1985). Decane was introduced into the sample chamber as pure decane. Hexane was introduced as a mixture with heptadecane (1/1 by volume) to control the amount of hexane in the bilayer. Measurements were made at 23°C. The bilayers were hydrated at controled relative humidities using saturated salt solutions.

The bilayer smectic liquid crystals have some properties that make accurate measurements of structure factors difficult. We have investigated these properties extensively in order to reduce possible systematic errors. Although the samples show narrow mosaic spreads, as measured by the full width at half maximum of rocking curves, there is usually a small amount of the sample that is less oriented. As a result, double diffraction of the first Bragg order can occur. The effect requires a large first order intensity, as with in-

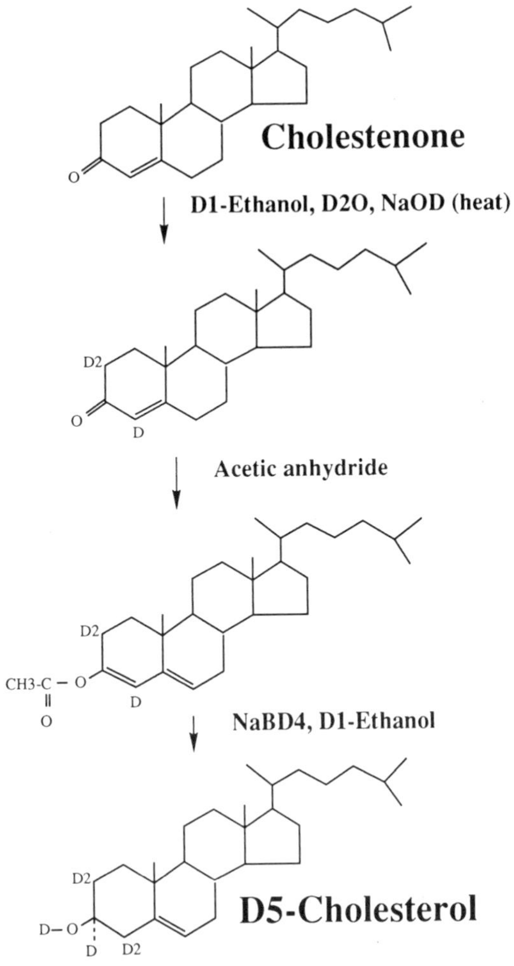

Cholestenone

D1-Ethanol, D2O, NaOD (heat)

Acetic anhydride

NaBD4, D1-Ethanol

D5-Cholesterol

Figure 1. Cholesterol labeled with five deuterium atoms covalently bound to carbon atoms 2, 3 and 4 (adjacent to the hydroxyl group) was prepared by the steps shown here. The hydroxyl deuterium is labile and readily exchanges with water so is not present unless D_2O hydration is provided. Projected onto the long axis of the cholesterol molecule, the five deuterium are within 1Å of each other and therefore provide a highly localized label for studying cholesterol location and dynamics across the bilayer.

creasing D_2O hydration. The double diffraction is demonstrated by a rocking curve for the second order, as shown in Figure 2. The two smaller peaks result from double diffraction of the first order. For the peak at 2.25°, intense first order diffraction is diffracted again by less oriented material. At 6.75°, first order diffraction is produced by the less oriented material, and then is diffracted again by the well oriented part of the sample. Only the middle peak is the true second order intensity. The double diffraction results in reduced intensity of the first order. The intensity of the second order is increased if the detector accepts diffraction at the second order scattering angle over the range of θ that includes the double diffraction. Even small samples of only a few μm thickness show this double diffraction, so scanning strategy is important, especially for the first two Bragg orders.

Another feature is that the first order diffraction is often broadened by small angle scattering since the path of the neutrons through the sample is large and there are apparently large scale density fluctuations in the sample. Lipid mixtures often show non-ideal mixing, resulting in excess volumes and density variations that produce small angle neutron scattering (Knoll *et al.*, 1981). Also, sample morphology is probably more complex than just lamellar. The small angle scattering effect is prominent with D_2O hydration be-

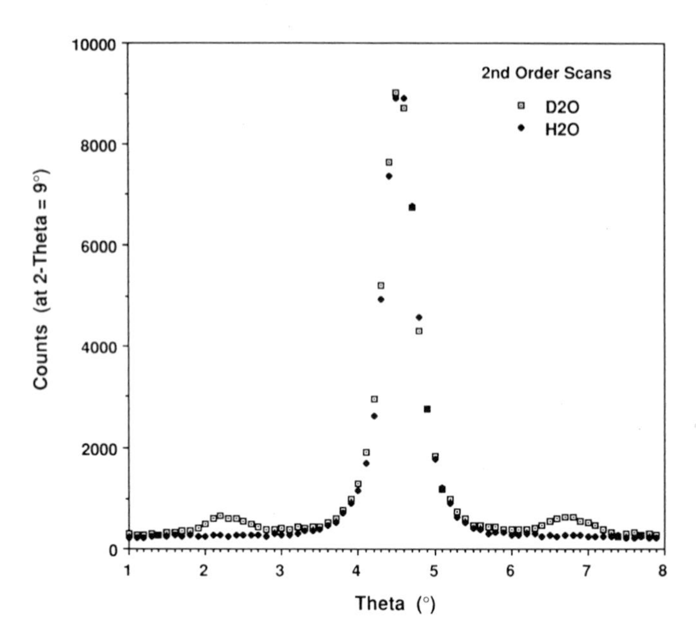

Figure 2. Second order θ scan for a sample of dimyristoyl phosphatidylcholine (DMPC) + 0.3M cholesterol showing double diffraction of the first order at 2.2 and 6.8° when the sample is hydrated with D_2O at 86% relative humidity. The detector is positioned at 9° with an aperture subtending 2.4° in 2θ during sample rotation. The sample is 5μm thick and well oriented, but some unoriented material is also present producing the double diffraction as described in the text.

cause of high contrast, and is demonstrated by broader rocking curves with D_2O hydration than with H_2O. Care is needed to obtain the full integrated intensity. The effect is troublesome because the contrast dependence maintains linearity of first order structure factor with H_2O/D_2O, though intensity is lost. Because of these features, and to ensure accurate absorption corrections, we have mainly used thin samples (about 10μm thickness), consisting of only a few thousand bilayers with good orientation. With these thin samples the broadening is still measurable, but is small. Good measurements of higher order diffraction usually required larger samples (up to 50μm thickness).

Diffraction measurements were made using the 2XC diffractometer at the University of Missouri Research Reactor. The instrument uses a beryllium filtered 4.36Å beam from a vertically focussing pyrolytic graphite monochromator. Aperture collimation was used throughout since soller collimators can cause shadowing effects. The detector was a single 2" diameter counter with acceptance mask subtending 2.4° in 2θ, as viewed from the sample. Data was collected primarily using θ/2θ scans. Additional θ scans were often made for the low orders, to monitor the effects described above. The incident beam was monitored after the beryllium filter for normalizing the counting at each angle. An example of lamellar diffraction data is shown in Figure 3.

Integrated intensities were corrected for absorption according to the formula:

$$I_0 = I_{meas.}(z/(1-\exp(-z))) \tag{1}$$

where $z = (2wN_A / A\sin\theta)\Sigma n_i\sigma_i / \Sigma m_i$, I_0 is the true intensity, θ is the Bragg angle, w is the weight of lipid covering area A, N_A is Avogadro's number, and n_i is the relative

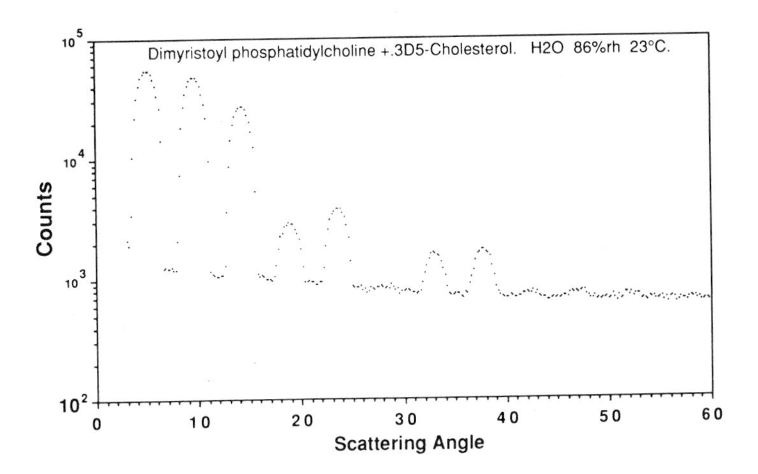

Figure 3. Neutron diffraction pattern for DMPC and 0.3M D5-Cholesterol hydrated with H_2O at 86% relative humidity. The sample is 20mg lipid and about 40μm thick. Diffraction through the 8th order is strong, with the 9th, 10th and 11th orders also present.

number of atoms of total neutron cross section σ_i and mass m_i in the lipid mixture. This correction is rigorous for samples of uniform thickness and uniform diffracting power throughout the thickness. These requirements may not be met by real samples, but the above correction factor is a good approximation if the correction is kept small. For many samples, the correction increased intensities by 10% or less.

We found that some of the smectic liquid crystal samples were not stable in diffracting power over the course of measurements. Many changes of humidity and other conditions were made during the course of a study taking several days, and apparently can produce changes in the amount of less oriented material in a sample. The intensity changes we observed were only a few percent, but in consequence, a scaling factor was used as a free parameter for deuterium difference analysis.

DATA ANALYSIS

In this work we have concentrated on H/D isotopic substitution methods for obtaining detailed information from limited diffraction data, and on the data fitting methods needed for providing and evaluating that information. Structure factors for both H and D samples were obtained from absorption corrected intensities according to:

$$|F(n)| = \sqrt{(I_0(n)\sin 2\theta_n)} \tag{2}$$

where $1/\sin 2\theta_n$ is the Lorentz factor for the nth order Bragg diffraction. Phase assignments were made by swelling methods (Worcester & Franks, 1976; Worcester *et al.*, 1992) and by using the perdeuterated hexane or decane as a heavy atom (Kaiser *et al.*, 1994). Structure factors for samples with H/D isotopes were analyzed using:

$$aF_D(n) = F_H(n) + c \cdot \cos(2\pi n X_0 / d)\exp(-n^2 B / 4d^2) \tag{3}$$

where X_0 and B are the main parameters to be determined. $F_D(n)$ are the structure factors for the bilayers with deuterium at position X_0, $F_H(n)$ are the structure factors without deuterium, a and c are scale factors, d is the bilayer repeat spacing and B is the exponent of the Debye Waller factor, which is used here as a general displacement parameter that includes both disorder and dynamics. The single cosine term requires that the deuterium label be localized at a single position. In real space the deuterium label is a Gaussian distribution centered at X_0 with full width at half maximum given by $(1/\pi)\sqrt{(B\,ln2)}$. Thus, in isotopic substitution studies such as this, B is often a more precise measure of resolution than d_{min} since it is directly related to an observed linewidth, as in spectroscopy.

The value of B is determined by the disorder and the thermal dynamics of the deuterium label. Separating the dynamics from the disorder can be difficult, but measurements at different temperatures can be made as a start. B is related to deuterium displacements by $B = 8\pi^2\mu^2$, where μ is the root mean square displacement in the X direction, perpendicular to the bilayer plane. Neutron diffraction, with isotopic substitution for emphasis, should be able to provide accurate B values for evaluating molecular dynamics, especially because the nuclei are point scatterers (Willis & Pryor, 1975). These B values can complement quasi-elastic neutron scattering studies of bilayer dynamics (Pfeiffer *et al.*, 1989). We have concentrated on methods for determining accurate B values from bilayer studies. The methods and results may be of value in neutron diffraction studies of other macromolecules, such as proteins and nucleic acids.

Two fitting procedures have been used. In one procedure, X_0 and B are specified and best fit values for a and c were calculated by the linear least squares method. This was repeated with different values for X_0 and B until an overall least squares difference between measured and fitted terms was achieved. This can be done with or without variance weighting. Initial fitting of the parameters was fairly quickly done this way.

A second fitting procedure was used for best evaluation of B with associated uncertainties. It was particularly useful for data with deuterated hydrocarbons introduced into the bilayer, since these concentrate at the center of the bilayer where $X_0 = 0$ for appropriate signs of the structure factors. In this case, the cosine term becomes 1 and Equation 3 can be written as:

$$ln(aF_D - F_H) = ln(c) - n^2B/4d^2 \tag{4}$$

This relation allows easy determination of B and propagated uncertainties using variance weighted least squares fitting. Values for 'a' can be tried until the reduced χ^2 is minimized. For stable samples with hydrocarbons introduced from the vapor phase, 'a' is very close to 1, since F_D and F_H are measured for the same sample.

RESULTS

A Fourier difference profile for decane introduced into a bilayer of DMPC+0.5M D5-Cholesterol is shown in Figure 4. There is little or no change in the repeat spacing upon introduction of the decane so it must occupy void spaces, as previously described for hexane in dioleoyl phosphatidylcholine bilayers (King *et al.*, 1985).

The degree of localization of the hydrocarbons is best evaluated by fitting the structure factors in reciprocal space to obtain B values complete with uncertainties and some measure of the goodness of fit. To do this we used Equation 4 and variance weighted lin-

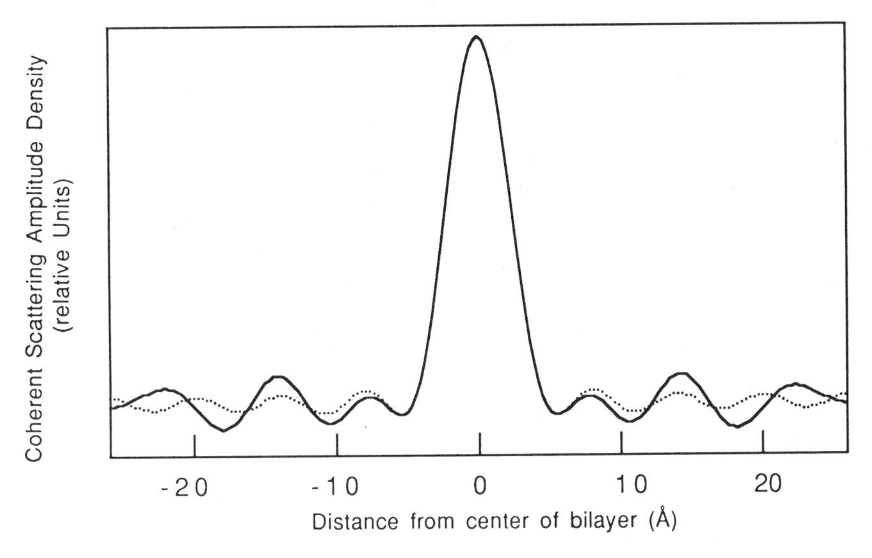

Figure 4. The Fourier difference profile for decane introduced from the vapor phase into bilayers of DMPC + 0.5M D5-Cholesterol hydrated at 86% relative humidity (solid line). Structure factors for the sample exposed to H-decane were subtracted from structure factors for the sample exposed to perdeuterated decane, with scaling, for the profile. Structure factors (aF_D - F_H) are 7.47, 7.33, 6.31, 6.23, 4.04, 2.27, 3.65, 1.45, with a = 0.952. The repeat spacing is 52.0Å. The dotted line is an unweighted best fit using Equation 3 with X_0 = 0 and B = 255Å^2. The decane is clearly localized in a very narrow region at the center of the bilayer. Note that all the structure factors are positive, so the perdeuterated decane can serve as a heavy atom at a center of symmetry for phase retrieval (Kaiser *et al.*, 1994). This sample was 28μm thick. See Kaiser *et al.* (1994) for a diffraction pattern of this sample.

ear least squares fitting methods. Figure 5 shows the linear plots of data obtained for hexane and decane introduced into DMPC + 0.5M D5-cholesterol bilayers. The B value for hexane, introduced as a 1/1 (vol.) mixture in heptadecane, is significantly larger than for decane. Hexane is therefore less localized than the decane, but is nevertheless narrower by a factor of 2 in this cholesterol containing bilayer than in dioleoyl phosphatidylcholine bilayers (King *et al.*, 1985). The higher accuracy for B with hexane compared to decane, is due to more deuterium (as perdeuterated hexane) in the bilayer, hence larger structure factor differences.

Systematic sources of error, as discussed above in the Materials and Diffraction Methods Section, frequently caused problems for the first and second orders. Other sources of systematic error that affect all diffraction orders are the stability of the counting electronics and changes in background. These may be important because nearly all the fits have reduced χ-squared values greater than one, so counting statistics alone do not represent all the uncertainties in the structure factors. Small differences in repeat spacings can cause errors, especially for diffraction orders where the bilayer transform is changing rapidly. If the different repeat spacings are accurately known, corrections can be made using the Shannon sampling theorem (Büldt *et al.*, 1979), provided that it is also known whether the changes are in the hydrocarbon region or the hydrated region of the bilayer. Because of uncertainties in making corrections, it is important to keep the repeat spacings as constant as possible for the two data sets used in the difference analysis. In addition to humidity control, temperature control is important since the bilayers have coefficients of thermal expansion in the direction across the bilayer of -0.1Å/°C.

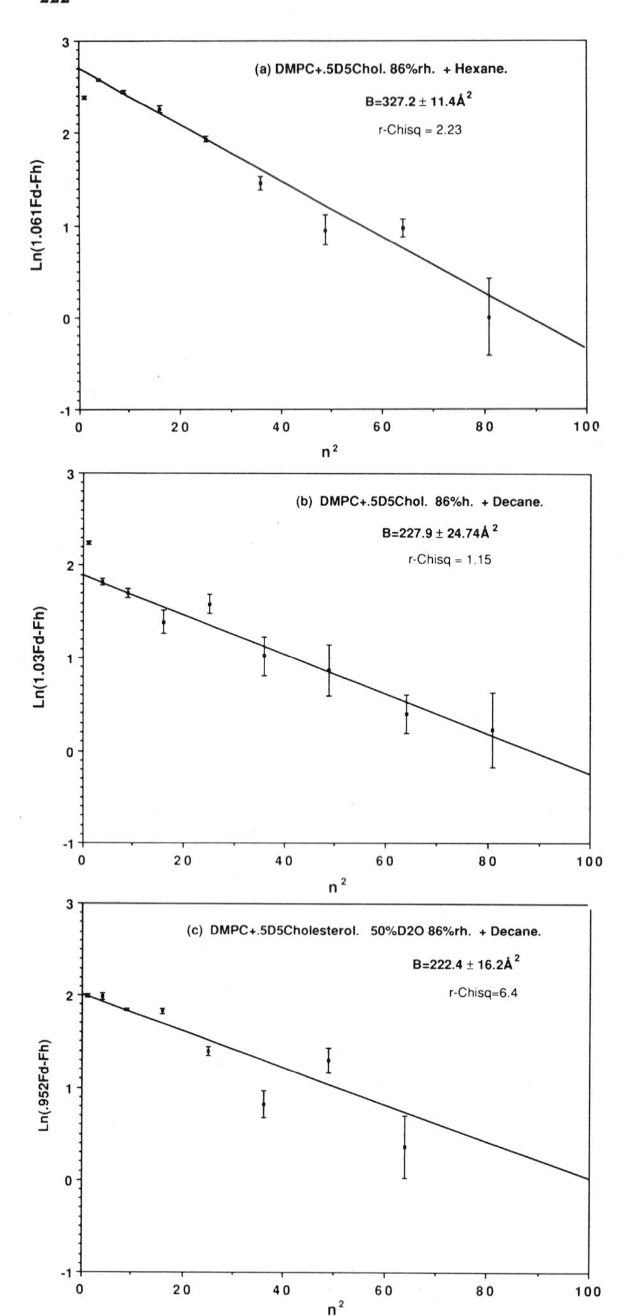

Figure 5. Plots of H/D difference structure factors for hexane (a) and decane (b) and (c) introduced into bilayers of DMPC + 0.5M D5-cholesterol at 86%rh. The error bars represent only counting statistics. The data for Figure 4 are evaluated again in (c). Data are fit according to Equation 4 to obtain the values for B, which are obtained from the slope of the variance weighted best fit line. Note that for (a) and (b), the first orders are far from the fitted line compared to the uncertainties from counting statistics, so were omitted from the fits.

To understand the significance of the small values for B obtained for decane in the bilayers, comparison can be made to difference data for the five deuterium atoms on the cholesterol. In this case, separate samples are needed for the deuterium labeled and unlabeled bilayers. The analysis is done as before, but the cosine term of Equation 3 is not equal to 1 and so must be included in Equation 4. The position of the label (X_0) is an additional parameter in the fitting. Some results are shown in Figure 6. Data for Figure 6a are

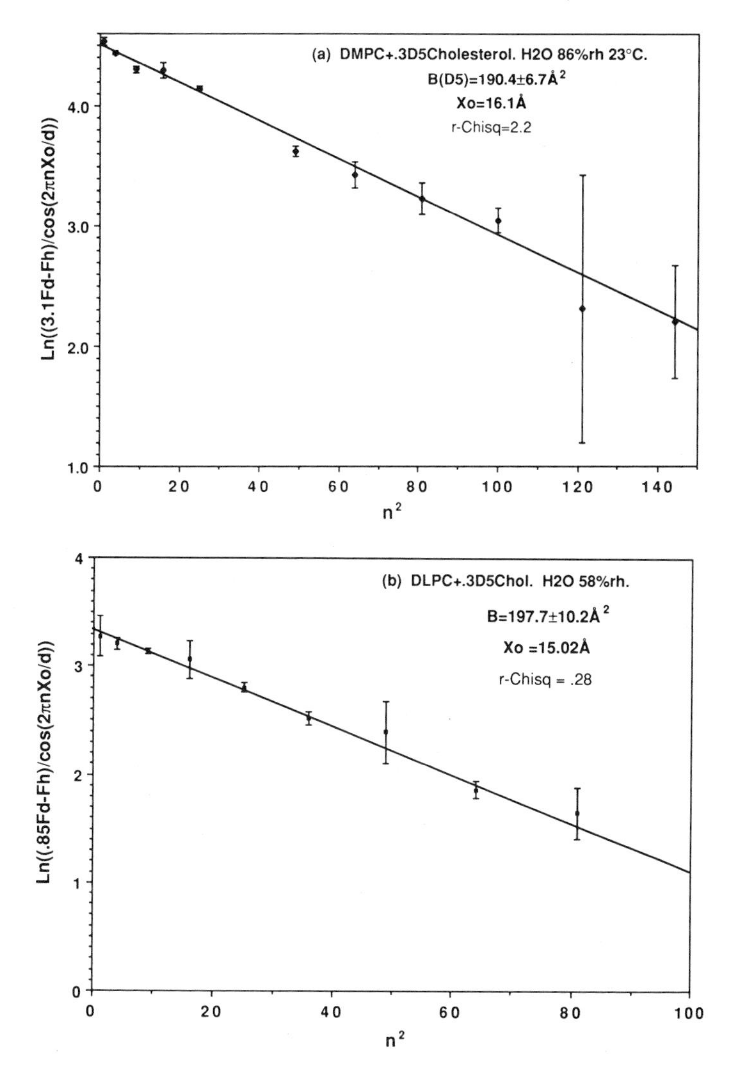

Figure 6. Variance weighted least squares fits with structure factors for bilayers with deuterium labeled cholesterol and with unlabeled cholesterol are shown for bilayers of dimyristoyl phosphatidylcholine (a) and dilauroyl phosphatidylcholine (b). In these cases the deuterium label is not at $X_0 = 0$ and so the cosine term of Equation 3 needs to be incorporated into Equation 4. As a result, uncertainty in the repeat spacing d needs to be included with the uncertainties from counting statistics. An uncertainty in d of ±0.25Å was included for these fits. The cosine term can cause fitting problems when it is near zero for a particular Bragg order. Structure factor differences are usually small in these cases. The data for that value of n are best omitted from the fit, as is the case in (a) for n = 6 where both F_D and F_H were very small. Also, uncertainties can be greatly increased by the cosine term, as occurs for n = 11 in (a). The fit is unusually good in (b).

from the diffraction pattern of Figure 3. The values of B obtained for the deuterium label on cholesterol are very similar to the values of B obtained for decane in the bilayers. This is remarkable because cholesterol is much larger than decane and therefore less mobile, and the deuterium label is on the rigid steroid part of the cholesterol molecule, whereas decane is highly flexible. Nevertheless the labels are about equally localized. It must be concluded that the decane is oriented with its long axis in the plane of the bilayer and is intercalated between the two halves of the bilayer. The decane may be largely in the all-trans conformation so that flexibility is reduced.

CONCLUSIONS

In this work we have shown that hydrophobic molecules such as hexane and decane are highly localized at the center of lipid bilayers containing cholesterol. The molecules must therefore be intercalated between the two halves of the bilayer with long axis in the plane of the membrane, as shown schematically in Figure 7. Decane is more localized than hexane. Both decane and hexane are at least a factor of two more localized in bilayers

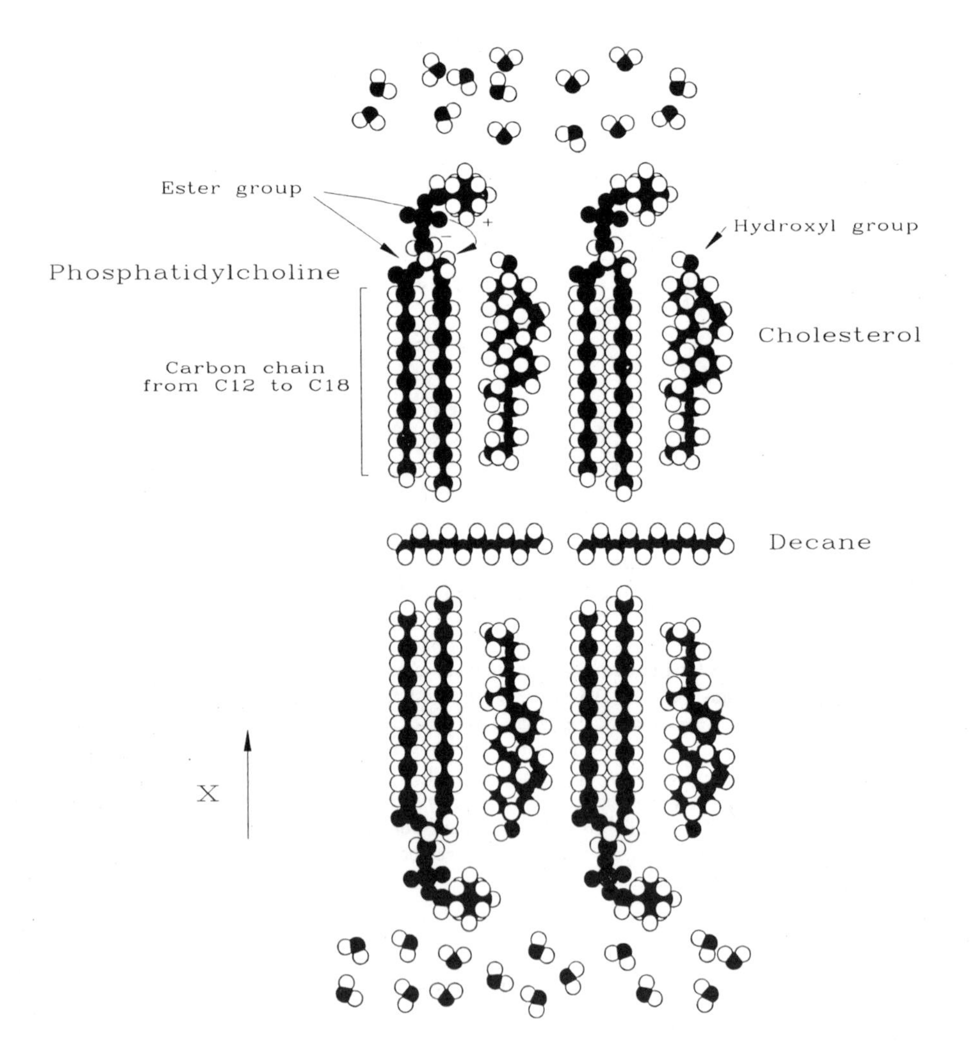

Figure 7. Schematic illustration of the molecular arrangement of decane in the lecithin/cholesterol model membranes studied here. The small B values obtained for decane require that the decane molecules be aligned in the plane of the membrane, and so are perpendicular to the hydrocarbon chains of the lipids, which are not depicted in the liquid crystalline state for this illustration.

containing cholesterol than was found for hexane in dioleoyl phosphatidylcholine bilayers. It is probable that this difference is caused by the additional molecular order conferred on the bilayer by cholesterol.

The demonstration of intercalated structures suggests a new concept for how small hydrophobic molecules interact with bilayers containing cholesterol, and how they partition into the bilayers. Clearly the bilayer hydrophobic region is not a uniform solvent to these small hydrophobic molecules. The center of the bilayer is a highly preferred location. Orientation is in the plane of the membrane rather than parallel to lipid hydrocarbon chains at this preferred location. Transport across the membrane must therefore occur in several discrete steps: partitioning into one half of the bilayer, diffusion to the center and orientation in the bilayer plane at the center where residence time is relatively large, partition into the other half of the bilayer and diffusion to the opposite surface.

Large hydrophobic side chains of amino acids in transmembrane alpha helices may also form intercalated structures with the lipid bilayer if the large side chains are predominantly at the middle of the transmembrane alpha helix. Volume profiles for the side chains of these helices show larger volume at the middle, which must be accommodated by the bilayer. The results for hexane and decane suggest that this could be done by intercalation.

This work has also demonstrated methods for obtaining accurate displacement parameters in neutron diffraction studies of lipid bilayers and model membranes. The main previous work of this type was on phosphatidylcholine model membranes without cholesterol (Büldt *et al.*, 1979; Zaccai *et al.*, 1979). In these early studies the displacement parameter used was the Gaussian half-width at 1/e height, notated by ν, where $B = 4\pi^2\nu^2$. In terms of B, the results in these papers ranged from $B = 67 \pm 51 Å^2$ to $B = 456 \pm 161 Å^2$ (Büldt *et al.*, 1979; Zaccai *et al.*, 1979). The analysis methods presented here build on this early work and demonstrate that increased accuracy in values for B can be obtained. The methods need further testing and applications to problems in structure and dynamics of model membranes, particularly with comparisons of B values at different temperatures.

ACKNOWLEDGMENTS

The 2XC diffractometer facility was developed by the University of Missouri Research Reactor (MURR) and the Physics Department. We thank Professors S.A. Werner and H. Taub, as well as colleagues at MURR for major contributions to providing and maintaining this instrument. We also gratefully acknowledge that this work is based on and is a direct continuation of earlier work by DLW at AERE, Harwell, U.K. which was also supported by the Science Research Council.

REFERENCES

Büldt, G., Gally, H.U., Seelig, J., & Zaccai, G., (1979). Neutron diffraction studies on phosphatidylcholine model membranes. I. Head group conformation. *J. Mol. Biol.*, 134:673–691.

Duff, K.C., Gilchrist, P.J., Saxena, A.M., & Bradshaw, J.P., (1994). Neutron diffraction reveals the site of adamantane blockade in the influenza A M2 ion channel. *Virology*, 202:287–293.

Gogol, E.P., Engelman, D.M., & Zaccai, G., (1983). Neutron diffraction analysis of cytochrome b_5 reconstituted in deuterated lipid multilayers. *Biophys. J.*, 43:285–292.

Kaiser, H., Hamacher, K., Kulasekere, R., Lee, W.-T., Ankner, J.F., DeFacio, B., Miceli, P., & Worcester, D.L., (1994). Neutron inverse optics in layered materials. *Proc. SPIE* 2241:78–89.

King, G.I., Jacobs, R.E., & White, S.H., (1985). Hexane dissolved in dioleoyllecithin bilayers has a partial molar volume of approximately zero. *Biochemistry,* 24:4637–4645.

Knoll, W., Haas, J., Stuhrmann, H.B., Füldner, H.-H., Vogel, H., & Sackmann, E., (1981). Small-angle neutron scattering of aqueous dispersions of lipids and lipid mixtures. A contrast variation study. *J. Appl. Cryst.,* 14:191–202.

Pfeiffer, W., Henkel, Th., Sackmann, E., Knoll, W., & Richter, D., (1989). Local dynamics of lipid bilayers studied by incoherent quasi-elastic neutron scattering. *Europhys. Lett.,* 8:201–206.

White, S.H., King, G.I., & Cain, J.E., (1981). Location of hexane in lipid bilayers determined by neutron diffraction. *Nature,* 290:161–163.

Willis, B.T.M., & Pryor, A.W., (1975). *Thermal Vibrations in Crystallography*, Cambridge University Press.

Worcester, D.L., & Franks, N.P., (1976). Structural analysis of hydrated egg lecithin and cholesterol bilayers II. Neutron diffraction. *J. Mol. Biol.,* 100:359–378.

Worcester, D.L., Kaiser, H., Kulasekere, R., & Torbet, J., (1992). Phase determination using transform and contrast variation methods in neutron diffraction studies of biological lipids. *Proc. SPIE,* 1767:451–456.

Zaccai, G., Büldt, G., Seelig, A., & Seelig, J., (1979). Neutron diffraction studies on phosphatidylcholine model membranes II. Chain conformation and segmental disorder. *J. Mol. Biol.,* 134:693–706.

20

CYLINDRICAL AGGREGATES OF CHLOROPHYLLS STUDIED BY SMALL-ANGLE NEUTRON SCATTERING

D. L. Worcester[1] and J. J. Katz[2]

[1] Biology Division
University of Missouri
Columbia, Missouri 65211
[2] Chemistry Division
Argonne National Laboratory
Argonne, Illinois 60439

ABSTRACT

Neutron small-angle scattering has demonstrated tubular chlorophyll aggregates are formed by self-assembly of a variety of chlorophyll types in nonpolar solvents. The size and other properties of the tubular aggregates can be accounted for by stereochemical properties of the chlorophyll molecules. Features of some of the structures are remarkably similar to light harvesting chlorophyll complexes *in vivo*, particularly for photosynthetic bacteria. These nanotube chlorophyll structures may have applications as light harvesting biomaterials where efficient energy transfer occurs from an excited state which is highly delocalized.

INTRODUCTION

Tubular aggregates of chlorophylls self-assemble in nonpolar solvents such as octane or toluene upon addition of small amounts of water to dry, concentrated solutions. The hollow cylinders have a variety of diameters, optical and electronic properties depending on chlorophyll type. We have continued to study these interesting structures since their discovery (Worcester *et al.,* 1986) and the finding that the different diameters of the hollow cylinders are approximately quantized. These 'nanotubes' of chlorophylls may have applications as biomaterials. Structures formed by chlorophyll-a, chlorophyll-a/b mixtures, pyrochlorophyll-a, acetyl chlorophyll-a, bacteriochlorophyll-a and bacteriochlorophyll-c have been characterized. Current work is on the properties of bacteriochlorophyll-c cylinders, pyrochlorophyll-a cylinders and the possible similarity of these artificial structures to the cylindrical structures seen by freeze fracture electron microscopy in chlorosomes of the photosynthetic bacteria *Chlorobium* and *Chloroflexus*. A variety of molecular structures serve for photosynthetic light harvesting *in vivo*. Comparisons of the natural light harvesting systems with properties of the artificial chlorophyll aggregates should help to clarify the properties and molecular arrangements that provide the most efficient light harvesting and energy transfer.

Chlorophyll molecular arrangements in some of the artificial aggregates have been proposed, and specific stereochemical features can account for the approximate quantization of diameters (Worcester *et al.*, 1990). The chlorophyll must be in highly ordered arrangements in the hollow cylinders. Electron spin resonance spectroscopy shows that electrons undergo rapid, one-dimensional transport along the cylinder surface for some chlorophylls, resulting in narrow electron spin resonance signals (Bowman *et al.*, 1988). Also, the long wavelength optical absorption spectra of the aggregates are considerably red shifted (Katz *et al.*, 1991), suggesting that the chlorophyll macrocycles are in contact by their π electron orbitals (Thompson & Zerner, 1988). Chlorophylls which differ slightly because of different substituent groups around the macrocycle show characteristic diameters, as determined by neutron scattering, so specific self-assembly processes must be responsible for cylinder formation and the resulting properties.

CHLOROPHYLL TYPES

The molecular structures of the six chlorophyll types that have been studied are shown in Figure 1. The main groups involved in the formation of hollow cylinder aggregates are probably the ring V keto oxygen which is present in all chlorophyll types and the carbonyl oxygens present in ring I or II of some chlorophylls. The hydroxyl group present in ring I of bacteriochlorophyll-c may also be involved. The ring V carbomethoxy group is important as a steric blocking group that restricts the ways in which the chlorophyll macrocycles can stack, since it projects above the macrocylce plane. Bacteriochlorophyll-c and pyrochlorophyll-a lack this carbomethoxy group so additional molecular arrangements may be possible for these chlorophylls. Support for the involvement of some of these groups in forming chlorophyll stacks is provided by X-ray structures obtained with single crystals of chlorophyll derivatives (Strouse, 1973; Chow *et al.*, 1975; Kratky & Dunitz, 1977). In these structures, a water molecule is coordinated to the chlorophyll magnesium atom and forms a hydrogen bond to the ring V keto oxygen of another chlorophyll molecule to form long chains, or stacks. In organic solvents, such chains could form cylinders because of curvature induced in the chlorophyll macrocycles by the coordination of water on only one side (Worcester *et al.*, 1986).

The hollow cylinder aggregates are prepared by addition of microliter amounts of water to carefully dried chlorophyll dissolved in dry octane, toluene, or a mixture of these at about 10mg/ml concentration. Dry starting material is particularly important because non-cylindrical aggregates containing water that are already present do not readily reform into cylinders.

NEUTRON SCATTERING RESULTS

Small angle neutron scattering is the only technique that has been able to demonstrate that the chlorophyll aggregates are long, hollow cylinders. Good contrast of scatter-

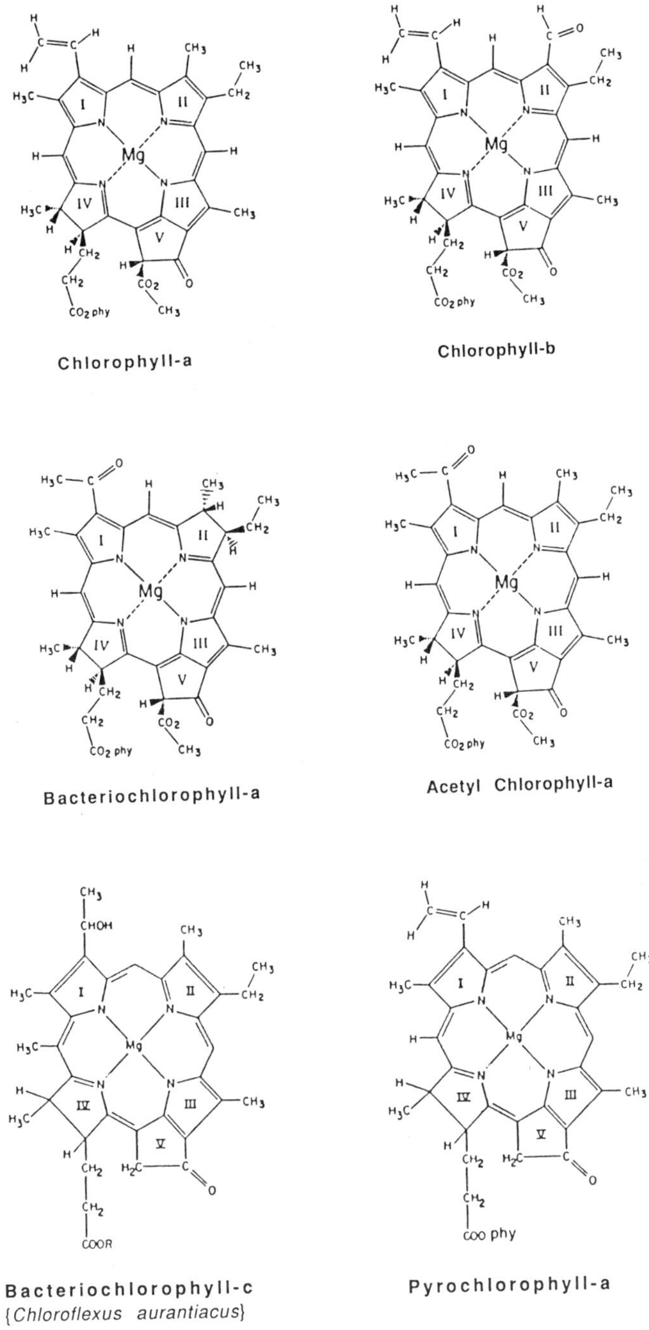

Chlorophyll-a

Chlorophyll-b

Bacteriochlorophyll-a

Acetyl Chlorophyll-a

Bacteriochlorophyll-c
{*Chloroflexus aurantiacus*}

Pyrochlorophyll-a

Figure 1. Molecular structures of the different chlorophyll types. Phy is the 20 carbon phytyl chain. In bacteriochlorophyll-c more than one type of chain (mainly farnesyl) is present, and is indicated by 'R'.

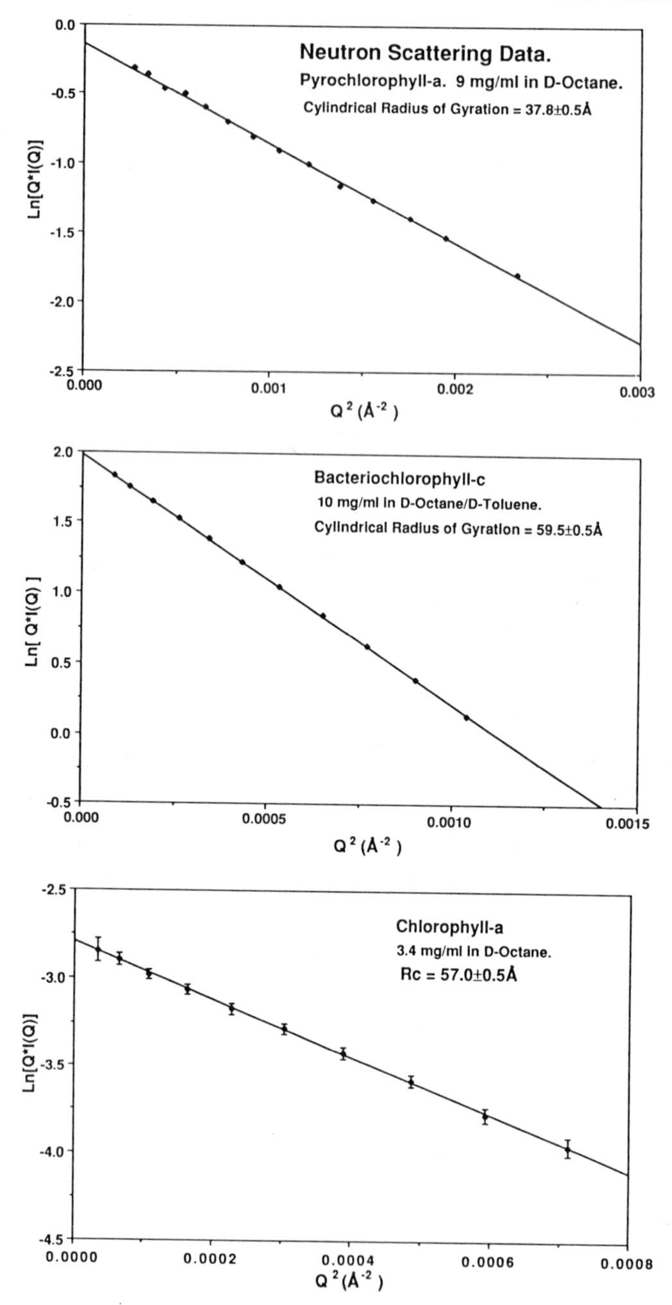

Figure 2. Guinier type plots of the neutron scattering data for determining cross-sectional radii of gyration.

ing density between the chlorophylls and the nonpolar solvent is provided by using perdeuterated solvents. Samples are about 1% chlorophyll of natural isotopic composition contained in quartz cells of 1 or 2mm thickness. Such samples give good neutron scattering signals and allow measurement of the Guinier regions for cylindrical structures at low Q as well as the scattering at higher Q which shows secondary maxima arising from the

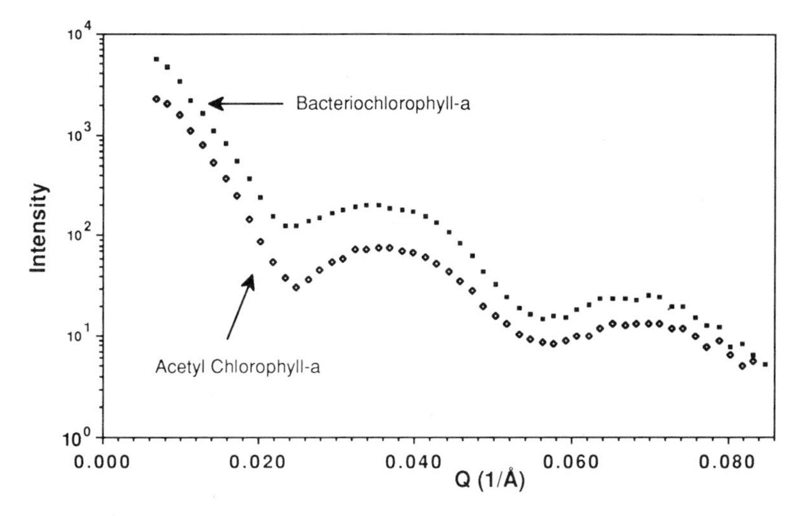

Figure 3. Neutron scattering curves for bacteriochlorophyll-a hydrated in perdeuterated octane/toluene (3/1 by volume) and acetyl chlorophyll-a hydrated in perdeuterated toluene. The different solvents are needed for solubility. The similar positions of the secondary maxima demonstrate that these two chlorophyll types form cylinders of the same size. Polydispersity of diameters is small.

Bessel function transform of the hollow cylinders (Worcester *et al.,* 1989). Cylinder diameters can be determined from Guinier type plots for cylindrical structures, from the positions of the secondary maxima, or by modelling the complete scattering curve. Figure 2 shows Guinier type plots for a number of chlorophylls. The cross-sectional radius of gyration is a good measure of cylinder diameter when the wall thickness of the hollow cylinder is small compared to the radius. This is the case for many of the cylinder types, as established by comparing the full scattering curve to the calculated scattering for a cylinder of specified radius and wall thickness.

Small diameter cylinders, and samples with appreciable polydispersity in diameters were best evaluated over a wider Q range than just the Guinier region, by comparing the measured scattering with the calculated scattering for cylinders of specified wall thickness and diameter polydispersity. In many cases polydispersity was small, as evidenced by the deep minima between the scattering maxima such as shown in Figure 3. For cylinders of pyrochlorophyll-a, polydispersity was appreciable.

An interesting feature of the cylinders for the two chlorophyll types shown in Figure 3 is that their optical absorption spectra are qualitatively different even though the cylinder diameters and neutron scattering curves are essentially the same. Figure 4 shows the optical absorption spectra for the cylinders of bacteriochlorophyll-a and of acetyl chlorophyll-a. Red shift of the long wavelength absorption peak upon formation of the hydrated cylinders is present in both cases, but bacteriochlorophyll-a gives a double peak whereas only one red shifted peak is present for acetyl chlorophyll-a. The additional peak for bacteriochlorophyll-a is the peak at about 800nm, which is much less red shifted. A reason for this difference has not been established, but may result from some bacteriochlorophyll-a molecules having less π orbital overlap in bacteriochlorophyll-a aggregates (Thompson & Zerner, 1988) due to the fact that ring II is reduced. Reduction of ring II is the only difference between the molecular structures of bacteriochlorophyll-a and acetyl chlorophyll-a (see Figure 1). It is of interest that the optical spectrum of the bacteriochlorophyll-a cylin-

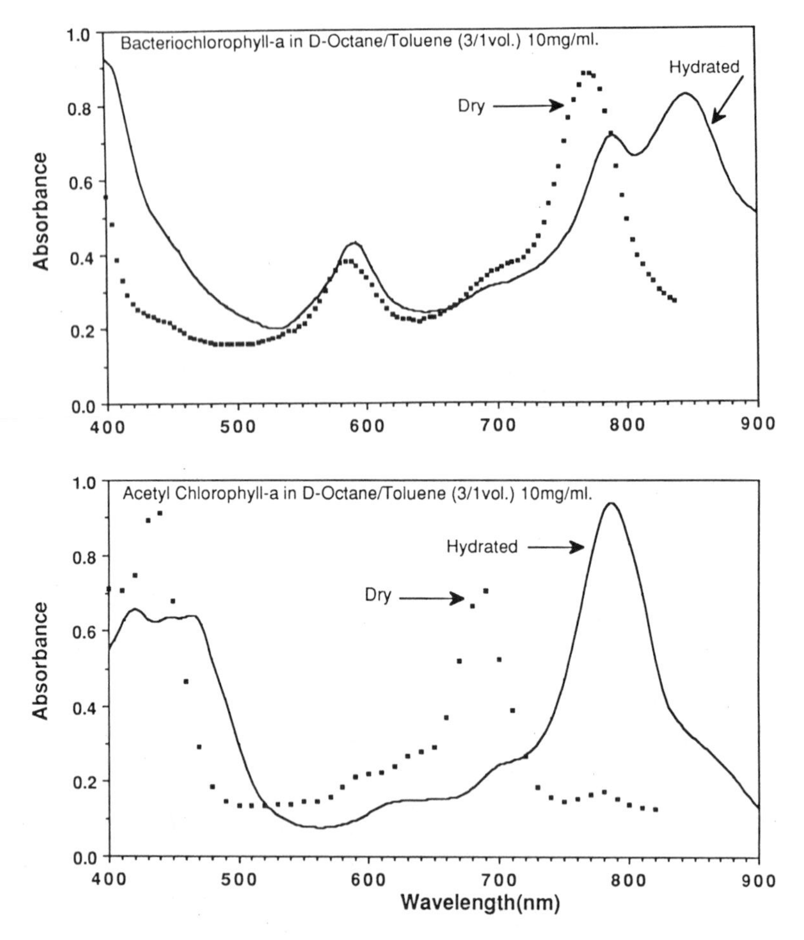

Figure 4. Optical absorption spectra for samples of bacteriochlorophyll-a and acetyl chlorophyll-a which in hydrated form give the scattering patterns for cylinders shown in Figure 3. Note the substantial difference in the long wavelength absorptions for the hydrated cylinders of these two types of chlorophylls.

ders is very similar to that of antenna systems of photosynthetic bacteria which contain predominantly bacteriochlorophyll-a.

Many results for cylinders of the six chlorophyll types are summarized in Table I. An interesting feature is that, with the exception of pyrochlorophyll-a, the diameters are approximately quantized, as they are roughly in the ratio of 100Å/200Å/400Å.

QUANTIZED DIAMETERS

The apparent quantization of chlorophyll diameters can be accounted for in terms of the different molecular features of the chlorophylls and a stereochemical mechanism (Worcester *et al.*, 1990). A key feature is that most of the chlorophylls have two distinct sides with respect to the approximate plane of the macrocycle. On one side, the ring V carbomethoxy group extends above the macrocycle and provides a major steric obstruction to overlap of macrocycle rings I and III in chlorophyll stacks. The only possible

Table I. Summary of the cylinder diameters found for the six different chlorophyll types by small angle neutron scattering

Chlorophyll Type	Cylinder Diameter (Å)
Pyrochlorophyll-a	60–80
Chlorophyll-a Bacteriochlorophyll-c	100–120
Bacteriochlorophyll-a Actylchlorophyll-a	200–230
Chlorophyll-a + Chlorophyll-b (1/1) Bacteriochlorophyll-a	380–440

stacking is by contact of the ring III surface below the carbomethoxy group. This steric requirement, together with the possibility of water coordination to the magnesium atom on either side of the macrocycle, are sufficient to account for diameter quantization, as will be shown below. It explains why the chlorophylls we have studied, which all have the ring V carbomethoxy group, form cylinders which are either about 100Å, 200Å or 400Å diameter (Table I).

To see how quantization arises from chlorophyll stereochemistry, a simple notation is useful for the main features of the chlorophyll macrocycle. This notation represents the macrocycle as a line when viewed from the side, perpendicular to the ring I to ring III axis. The line is bent at the middle when water is coordinated to the magnesium atom, to represent the distortion that is produced by adding water as a fifth ligand to the originally planar macrocycle and four coordinated magnesium. The bend is exaggerated here for demonstration purposes. In reality it is only a few degrees. A short line is added as a barb at one end of the line to represent the ring V carbomethoxy group and the steric obstruction that it provides. Water coordination on either side of the macrocycle produces two distinctly different structures, as shown in Figure 5. The two positions of water coordination are 'cis' and 'trans' to the carbomethoxy group, so the two structures are notated 'C' and 'T' respectively. Stacks of all C type macrocycles will form a circle and hence cylinders, whereas stacks of all 'T' type macrocycles are linear, as shown in Figure 5. Stacks containing both C and T type macrocycles can have specific diameters if specific combinations of these two form a repeat unit. The important feature is that the direction of bend changes after every T, regardless of the type of the subsequent macrocycle. With repeat units of the type (C_xT_y), where x and y are integers, all cases of odd values for y have no net bend. Values for y that are even numbers result in continuous curvature, and cylinders can form. The diameter is increased compared to the diameter of the all C stack by the factor (x+y)/x.

Many stacks made from repeat units of the type (C_xT_y) are possible, but so far the only observed diameters are increased ×2 and ×4. These are accounted for by (C_2T_2) and (C_2T_6) repeat units respectively. Diameters increased by ×3 would be formed by (C_1T_2) repeat units, but have not been observed. This suggests a special stability to pairs of chlorophyll molecules. More complex repeat units are also possible, such as $(C_1T_2C_1T_4)$ with diameter also increased by ×4, so that the precise type of aggregate may not be unique for particular diameters. In addition, certain repeat units may not be exclusively used for particular cylinder types, resulting in some polydispersity and variation in cylinder diameter. However, the clear secondary maxima in the neutron scattering for many of the samples demonstrate that such structural heterogeneity cannot be very substantial.

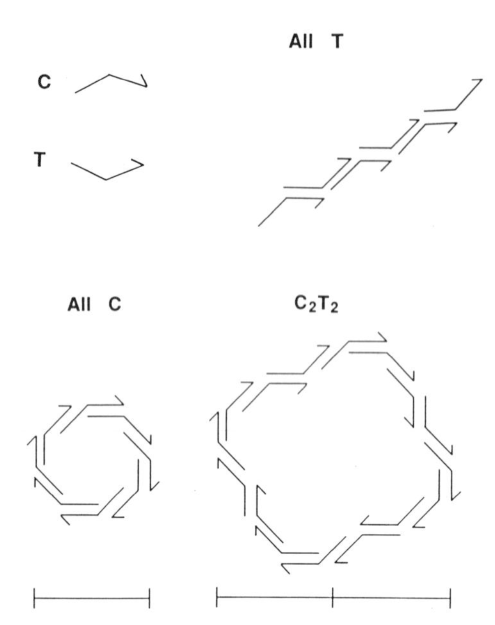

Figure 5. The two types of water coordinated chlorophyll macrocycles are represented in side view as a line. The line is bent, or curved when water is coordinated to the magnesium atom. The water can be on either side of the macrocycle with respect to the ring V carbomethoxy group which is represented as a barb. The two types form very different stacks and diameters can be quantized if cylinders are made of specific repeat units containing both types. The doubled diameter of a stack of (C_2T_2) repeat units compared to the diameter of an all C stack is shown. Note that the increased diameter is produced by the presence of macrocycles that bend in the opposite direction to others. (From Worcester *et al.*, 1990).

Diameter quantization can thus be accounted for by the stereochemistry of the chlorophyll macrocycle, but many specific structures are possible based on stereochemistry alone. Also, an acetyl group on ring I , as found in bacteriochlorophyll-a and acetyl chlorophyll-a apparently makes hydrogen bonding possible that gives more stability to the larger diameter stacks compared to stacks of chlorophyll-a. There is no apparent stereochemical reason which prevents bacteriochlorophyll-a and acetyl chlorophyll-a from forming cylinders of the same diameter as chlorophyll-a, but these haven't been observed. For the larger diameters which are observed, pairs of chlorophylls are required, suggesting special stability to dimers of these chlorophylls. Pairs of T type macrocycles are especially important for cylinder formation because groups of odd numbers of T macrocycles change the direction of curvature and therefore greatly inhibit cylinder formation. Details of hydrogen bonding and/or π orbital overlap may be needed for a complete account of diameter quantization.

BACTERIOCHLOROPHYLL-C AND PYROCHLOROPHYLL-A

The ring V carbomethoxy group is absent in bacteriochlorophyll-c and pyrochlorophyll-a (see Figure 1). Additional types of aggregate formation are therefore possible. Our early studies of bacteriochlorophyll-c demonstrated cylinders of about 100Å diameter (Worcester *et al.*, 1986), the same as for chlorophyll-a (Figure 2 and Table I). An important difference however, was that cylinders of bacteriochlorophyll-c were not disaggregated by drying, as were cylinders of other chlorophylls, suggesting that the ring I hydroxyethyl group of bacteriochlorophyll-c, rather than water, was coordinated to the magnesium atom for cylinder formation (Worcester *et al.*, 1986; see also Smith *et al.*, 1983). These findings for bacteriochlorophyll-c are especially interesting because freeze fracture electron microscopy of the bacteriochlorophyll-c rich chlorosomes present in the green photosynthetic bacteria *Chlorobium* and *Chloroflexus* reported cylindrical structures. In *Chlorobium* chlorosomes, the cylinders were reported to be nearly the same size (Staehlin *et al.*, 1980) as the artificial aggregates we have made using bacteriochlorophyll-c (100Å diameter). Chlorosomes contain some protein as well as the large amounts of bacteriochlorophyll-c which serves to collect light energy and transfer it to the photosynthetic reaction centers. The arrangement of protein and bacteriochlorophyll-c in chlorosomes is controversial. Some groups report that proteins are involved in the organization of bacteriochlorophyll-c in *Chloroflexus* chlorosomes (Niedermeier *et al.*, 1992), whereas other studies have reported that the interior of chlorosomes in *Chloroflexus*, where the cylinders are found, does not contain protein, (Holzwarth *et al.*, 1990) and that protein-free cylinders can be isolated. It is possible therefore, that the artificial bacteriochlorophyll-c cylinders we have been studying are much the same as those found in chlorosomes of green photosynthetic bacteria.

In *Chloroflexus*, the cylinders are reported to be about 52Å diameter (Staehlin *et al.*, 1978), which is smaller than the 100Å cylinders found in *Chlorobium* and the artificial cylinders we obtained with bacteriochlorophyll-c. To determine how the smaller cylinders could be formed, we investigated what structures might be formed by bacteriochlorophyll-c if the ring I hydroxyethyl group was not used for coordination to magnesium. Thus, if the two cylinder sizes found in chlorosomes arise only from bacteriochlorophyll-c aggregation, then bacteriochlorophyll-c must be able to aggregate in two different ways. To test if aggregation without involvement of the hydroxyl group produces smaller cylinders we studied pyrochlorophyll-a. Like bacteriochlorophyll-c, this chlorophyll lacks the ring V carbomethoxy group, but it also lacks the ring I hydroxyethyl group and therefore water molecules and molecular stacks (Kratky *et al.,* 1977) similar to chlorophyll-a stacks are needed for cylinder formation.

Neutron scattering data for hydrated aggregates of pyrochlorophyll-a are shown in Figure 6. A Guinier type plot is shown in Figure 2. These data clearly demonstrate smaller cylinders, which are comparable to the cylinder sizes reported in *Chloroflexus* chlorosomes. The secondary maximum in the scattering data of Figure 6 is not very clear due to the lack of distinct minima. This is characteristic of polydispersity in cylinder diameters. The radius of gyration in such cases is a weighted average of the different radii, so that evaluation of the full scattering curve gives a much better description of the cylinder sizes. This is done by comparing the measured scattering to the calculated scattering for a distribution of sizes, as shown in Figure 6. The range of diameters for pyrochlorophyll-a is fairly large but the results clearly demonstrate that artificial cylinder diameters of about 52Å are possible. These may be good models for the cylinders in *Chloroflexus* chlorosomes.

CONCLUSION

In conclusion, the results for bacteriochlorophyll-c and pyrochlorophyll-a suggest that the smaller diameter cylinders found in *Chloroflexus* could be due to bacteriochlorophyll-c stacks which are formed with a water molecule coordinated to magnesium of one chlorophyll and hydrogen bonded to the ring V carbonyl oxygen of another chlorophyll, as in chlorophyll-a stacks, but with different hydrogen bonding than in chlorophyll-a stacks because of the absence of the ring V carbomethoxy group. The ring I hydroxyl group

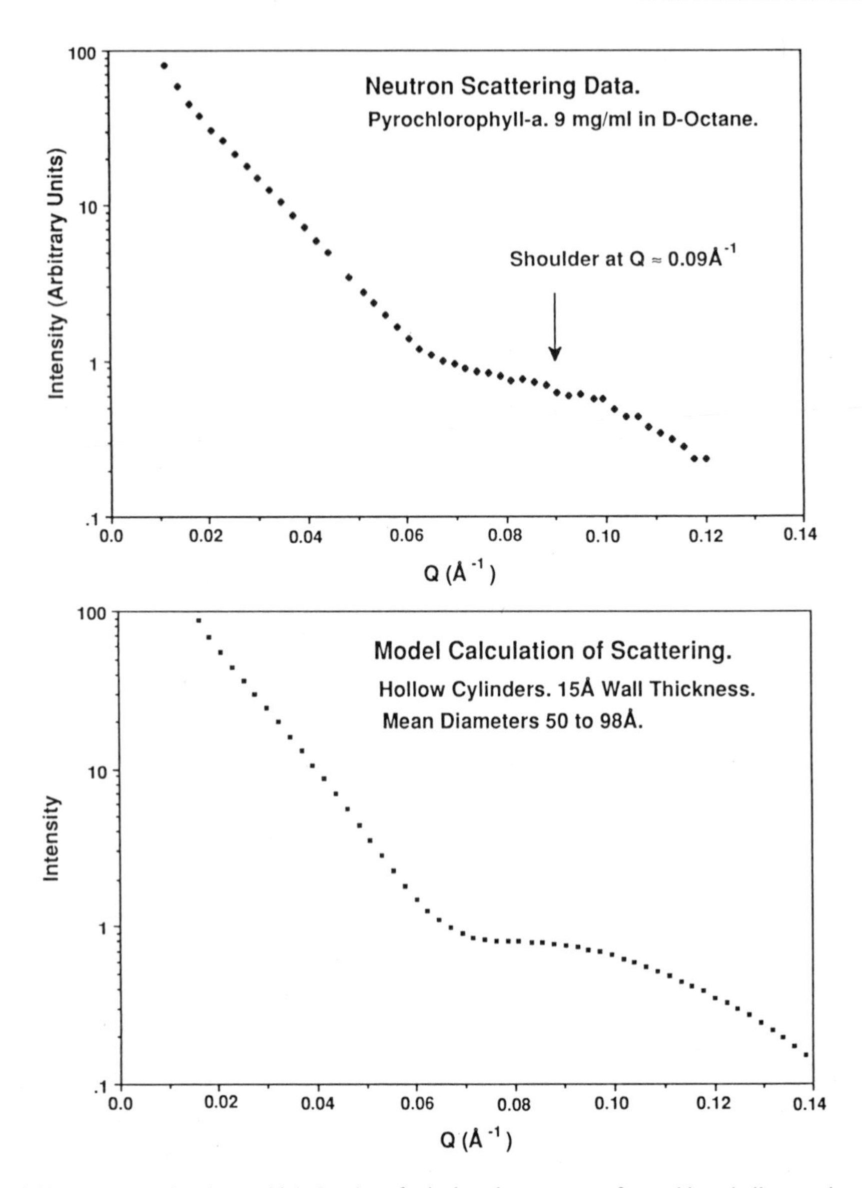

Figure 6. Neutron scattering data to high Q values for hydrated aggregates of pyrochlorophyll-a are shown and compared with model calculations for polydisperse hollow cylinders with 15Å wall thickness and mean diameters from 50 to 98Å. Such polydispersity clearly accounts for the shoulder being present in the scattering data, rather than a distinct secondary maximum.

would not be coordinated to magnesium in these smaller cylinders, but would be involved in hydrogen bonding. This different hydrogen bonding could produce the smaller cylinders if it results in macrocycles that are much more tilted with respect to the cylinder axis. In bacteriochlorophyll-c cylinders of 100–120Å diameter, the ring I hydroxyethyl group would be coordinated to magnesium. Self-assembly of bacteriochlorophyll-c has only produced the 100–120Å cylinders so the reason for smaller cylinders *in vivo* might be found in assembly processes that involve other molecules.

Note Added in Proofs

Rings of closely interacting, red-shifted bacteriochlorophyll-a molecules have been demonstrated in light harvesting antenna systems LH1 and LH2 of purple bacteria (McDermott *et al.,* 1995. *Nature* 374:517–521; Karrasch *et al.,* 1995. *EMBO J.* 14:631–638). Circular structures of chlorophylls thus appear to be common both *in vitro* and *in vivo*.

ACKNOWLEDGMENTS

Neutron scattering measurements were made at the Intense Pulsed Neutron Source (IPNS), Argonne, at the University of Missouri Research Reactor and at the Institut Laue Langevin, Grenoble, France. Work at IPNS, Argonne was supported by the Chemistry Branch of Basic Energy Sciences, U.S. Dept. of Energy , under contract W-31–109-Eng-38.

REFERENCES

Bowman, M.K., Michalski, T.J., Tyson, R.L., Worcester, D.L., & Katz, J.J., (1988). Electron spin resonance of charge carriers in chlorophyll-a/water micelles. *Proc. Natl. Acad. Sci. USA,* 85:1498–1502.

Chow, H.-C., Serlin, R., & Strauss, C.E., (1975). The crystal and molecular structure and absolute configuration of ethyl chlorophylide a dihydrate. A model for the different spectral forms of chlorophyll-a. *J. Am. Chem. Soc.,* 97:7230–7237.

Holzwarth, A.R., Griebenow, K., & Schaffner, K., (1990). A photosynthetic antenna system which contains a protein-free chromophore aggregate. *Z. Naturforsch.,* 45c:203–206.

Katz, J.J., Bowman, M.K., Michalski, T.J., & Worcester, D.L., (1991). Chlorophyll aggregation: Chlorophyll/water micelles as models for *in vivo* long-wavelength chlorophyll. In *Chlorophylls*, H. Scheer, editor. CRC Press: 211–235.

Kratky, C. & Dunitz, J.D., (1977). Ordered aggregation states of chlorophyll-a and some derivatives. *J. Mol. Biol.,* 113:431–442.

Kratky, C., Isenring, H.P., & Dunitz, J.D. (1977). Methylpyrochlorophyllide-a monohydrate monoetherate, *Acta Cryst.,* B33:547–549.

Niedermier, G., Scheer, H., & Feick, R.G., (1992). The functional role of protein in the organization of bacteriochlorophyll-c in chlorosomes of *Chloroflexus aurantiacus. Eur. J. Biochem.,* 204:685–692.

Smith, K.M., Kehres, L.A., & Fajer, J., (1983). Aggregation of bacteriochlorophylls c, d, and e. Models for the antenna chlorophylls of green and brown photosynthetic bacteria. *J. Am. Chem. Soc.,* 105:1387–1389.

Staehelin, L.A., Golecki, J.R., & Drews, G., (1980). Supramolecular organization of chlorosomes (chlorobium vesicles) and their membrane attachment sites in *Chlorobium limicola. Biochim. Biophys. Acta,* 589:30–45.

Staehelin, L.A., Golecki, J.R., Fuller, R.C., & Drews, G., (1978). Visualization of the supramolecular architecture of chlorosomes (Chlorobium type vesicles) in freeze fractured cells of *Chloroflexus aurantiacus. Arch. Microbiol.,* 119:269–277.

Strouse, C.E., (1973). The crystal and molecular structure of ethyl chlorophylide-a.2H_2O and its relationship to the structure and aggregation of chlorophyll-a. *Proc. Natl. Acad. Sci. USA,* 71:325–328.

Thompson, M.A., & Zerner, M.C., (1988). On the red shift of the bacteriochlorophyll-b dimer spectra. *J. Am. Chem. Soc.,* 110:606–607.

Worcester, D.L., Michalski, T.J., & Katz, J.J., (1986). Small-angle neutron scattering studies of chlorophyll micelles: Models for bacterial antenna chlorophyll. *Proc. Natl. Acad. Sci. USA,* 83:3791–3795.

Worcester, D.L., Michalski, T.J., Tyson, R.L., Bowman, M.K., & Katz, J.J., (1989). Structure, red-shifted absorption and electron transport properties of specific aggregates of chlorophylls, *Physica B,* 156&157:502–504.

Worcester, D.L., Michalski, T.J., Bowman, M.K., & Katz, J.J., (1990). Quantized diameters in self-assembled cylindrical aggregates of chlorophylls. *Materials Research Society Symposium Proceedings,* 174:157–162.

21

NEUTRON SCATTERING STUDIES OF THE DYNAMICS OF BIOPOLYMER-WATER SYSTEMS USING PULSED-SOURCE SPECTROMETERS

H. D. Middendorf[1] and A. Miller[2]

[1] Clarendon Laboratory
University of Oxford
Oxford OX13PU, United Kingdom
[2] The Principal's Office
Stirling University
Stirling FK94LA, United Kingdom

ABSTRACT

Energy-resolving neutron scattering techniques provide spatiotemporal data suitable for testing and refining analytical models or computer simulations of a variety of dynamical processes in biomolecular systems. This paper reviews experimental work on hydrated biopolymers at ISIS, the UK Pulsed Neutron Facility. Following an outline of basic concepts and a summary of the new instrumental capabilities, the progress made is illustrated by results from recent experiments in two areas: quasi-elastic scattering from highly hydrated polysaccharide gels (agarose and hyaluronate), and inelastic scattering from vibrational modes of slightly hydrated collagen fibers.

INTRODUCTION

In the past, neutron scattering has contributed to our understanding of the structure and dynamics of biomolecules primarily through time-averaged information from diffraction techniques (Stuhrmann & Miller, 1978; Schoenborn, 1984; Zaccaï, 1994). The large scattering contrast between protons and deuterons makes it easy to locate hydrogen atoms and hydrogenous groups, and this unique feature of neutron scattering has been exploited in numerous studies. Because of the low energies of neutrons with de Broglie wavelengths in the 1 to 20Å range, it is possible to go beyond diffraction and to analyse the quasi-elastic and inelastic components of the scattering which are due to thermal or subthermal

Neutrons in Biology, edited by Schoenborn and Knott
Plenum Press, New York, 1996

atomic and molecular motions. Energy-resolving neutron scattering techniques, by virtue of the parameter domain covered and the potential for hydrogen/deuterium contrast variation, are clearly able to contribute substantially to the study of biomolecular dynamics (Middendorf, 1984a; 1992; Middendorf & Randall, 1985; Martel, 1992).

Although neutron spectroscopy was first applied to biomolecular problems around 1970, this area of neutron scattering developed very slowly owing to the small number of suitable instruments and severe technical as well as access limitations. In recent years, fresh impetus for experimental work has come from a new generation of neutron spectrometers employing pulsed beams rather than continuous beams from reactor sources (Windsor, 1981; Newport *et al.*, 1988). Pulsed-beam spectrometers have reached a high level of development now. At ISIS, the UK pulsed neutron facility, they are being used in several projects on the dynamics of biomolecular systems, their building blocks, and certain model compounds. The purpose of this paper is to discuss the instrumental advances in the light of current work on problems of biophysical and biochemical relevance, and to illustrate the progress made by reviewing selected results from recent experiments on polysaccharides and polypeptides.

DYNAMIC STRUCTURE FACTORS AND CORRELATION FUNCTIONS

A scattered radiation field, in general, carries information on both the structure and the dynamics of the sample. Conventional diffraction theory can serve as a point of departure for introducing the basic notions of energy-resolving neutron scattering techniques. For unpolarized neutrons, each neutron-nucleus collision is governed by two conservation equations defining momentum and energy transfer:

$$\hbar\mathbf{Q} = \hbar(\mathbf{k}_f - \mathbf{k}_o) \qquad \text{(momentum)} \tag{1}$$

$$\hbar\omega = E_f - E_o = 1/2\, m(v_f^2 - v_o^2) = (\hbar^2/2m)(k_f^2 - k_o^2) \qquad \text{(energy)} \tag{2}$$

Here E, v and $|\mathbf{k}| = k = 2\pi/\lambda$ refer to the energy, velocity, wavenumber and wavelength of incident neutrons (subscript o), or neutrons scattered into solid angle element $d\Omega$ around a scattering angle 2θ (subscript f). From diffraction theory it is well known that momentum transfer $\hbar Q$ and the real-space radius vector r are Fourier conjugates; now an additional pair of Fourier transform variables—energy transfer $\hbar\omega$ and time t—is necessary to describe scattering with energy changes. The significance of Q is that motions are probed over scale lengths of $d = 2\pi/Q$.

With the exception of spin-echo instruments (Richter, 1992), all neutron spectrometers* provide raw data in the form of a double differential cross-section, $d^2\sigma/d\Omega dE$, which represents the intensity scattered into solid angle element $d\Omega$ within the energy interval

* In the literature, the terms 'spectrometer' and 'diffractometer' are used rather loosely. In particular, instruments producing S(Q) data for diffraction analyses are sometimes called 'spectrometers'. While this usage may be justified in the case of Laue time-of-flight instruments, it can be confusing. In this paper, instruments used primarily for S(Q) data collection and therefore structural studies are called 'diffractometers'; the term 'spectrometer' is reserved for instruments providing momentum *and* energy resolved data sets for dynamical studies.

dE. In the simplest case of a monatomic assembly of N nuclei this is (Windsor, 1981; Bée, 1988):

$$d^2\sigma / d\Omega dE \approx N(k_f / k_o)[\sigma_{inc} S_{inc}(\mathbf{Q},\omega) + \sigma_{coh} S_{coh}(\mathbf{Q},\omega)] \tag{3}$$

where σ_{inc} and σ_{coh} are incoherent and coherent scattering cross-sections.

The aim of analyzing $d^2\sigma/d\Omega dE$ data from a quasi-elastic or inelastic scattering experiment is to extract the dynamic structure factors, or scattering laws, $S_{inc}(\mathbf{Q}, \omega)$ and $S_{coh}(\mathbf{Q}, \omega)$, for incoherent and coherent scattering, respectively (Lovesey, 1984; Bée, 1988). These functions are generalisations of the static structure factors $S_{inc}(\mathbf{Q}) \approx 1$ and $S_{coh}(\mathbf{Q}) \equiv S(\mathbf{Q})$ measured in neutron diffraction experiments, such that the energy transfer enters as a new and to some extent independent variable. In diffraction experiments, the incoherent scattering is not normally of interest since it contributes only a diffuse, essentially Q-independent background. Analysed spectrally, however, this 'background' is a rich source of information on atomic and molecular motions.

The quantitative connection between $S_{inc}(\mathbf{Q}, \omega)$ or $S_{coh}(\mathbf{Q}, \omega)$ and the real space dynamics is established by appropriate Fourier transform. For an ensemble of nuclear trajectories $\mathbf{R}_i(t)$, the $\omega \rightarrow t$ transform first gives Q-dependent time correlation functions, the so-called 'intermediate' scattering functions:

$$F_{coh}(\mathbf{Q},t) = \boldsymbol{FT}\{S_{coh}(Q,\omega)\} = (1/N)\sum_{i,j}\langle \exp\{-i\mathbf{Q}\cdot\mathbf{R}_i(0)\}\exp\{i\mathbf{Q}\cdot\mathbf{R}_j(t)\}\rangle \tag{4a}$$

$$F_{inc}(\mathbf{Q},t) = \boldsymbol{FT}\{S_{inc}(Q,\omega)\} = (1/N)\sum_{i}\langle \exp\{-i\mathbf{Q}\cdot\mathbf{R}_i(0)\}\exp\{i\mathbf{Q}\cdot\mathbf{R}_i(t)\}\rangle \tag{4b}$$

(i,j = 1...N; angular brackets denote thermal ensemble averages). Equation 4a relates to coherent scattering since it covers all pair correlations including the 'self' terms i = j, whereas Equation 4b describes single-particle correlations for which only the diagonal terms i = j are needed. A further Fourier transformation, now with respect to the conjugate variables $\mathbf{Q}$ and $\mathbf{r}$, leads from $F_{inc}(\mathbf{Q},t)$ and $F_{coh}(\mathbf{Q},t)$ to the van Hove space-time correlation functions $G_s(\mathbf{r},t)$ and $G(\mathbf{r},t)$, respectively. These relate to experimentally accessible quantities to a well-established body of theoretical work on the time-dependent statistical mechanics of systems of interacting particles, thus providing a basis for interpretation that is often more 'direct' than that of other experimental techniques.

Since protons dominate the scattering from hydrogenous molecules, $S_{inc}(\mathbf{Q}, \omega)$ is of central interest for biomolecular applications (Middendorf, 1984a; 1984b). It is related by double Fourier transformation to $G_s(\mathbf{r},t)$, the diagonal or 'self' part of G(r,t). The simplicity of point-like nuclear scattering facilitates the interpretation of $d^2\sigma/d\Omega dE$ data in terms of frequency distributions and correlation functions, and this is exploited increasingly by large-scale computer simulations of atomic and molecular motions in crystallographically known systems (Karplus & Petsko, 1990; Goodfellow, 1990; Smith, 1991).

The intensities $d^2\sigma/d\Omega dE$ observed in energy-resolved neutron experiments are not subject to symmetry selection rules as in infrared or Raman spectroscopy. Neutrons 'see everything that moves', in the sense that the elemental intensity δI contributed by nucleus i is governed essentially by its cross-section σ_i and by $\langle u_i^2\rangle Q^2$, the mean-square displacement along Q multiplied by the square of the momentum transfer. The differential intensity $d^2\sigma/d\Omega dE$ scattered into solid angle element $d\Omega$ within the energy interval dE is expressed by appropriate correlation and thermal averages over $\mathbf{R}_i(t)$. In any somewhat more complex sample, these N nuclei will consist of two or more species with different cross-sections, and each of these will in general comprise dynamically non-equivalent groups of nuclei. For incoherent scatterers, these summations are fairly straightforward; for samples that scatter coherently the resulting expressions contain interference terms and are much more difficult to evaluate (Egelstaff *et al.*, 1975).

The theory and practice of hydrogen/deuterium contrast variation, well developed in diffraction work with neutrons, carry over to spectral analyses of both the incoherent and coherent scattering. By exchange with H_2O/D_2O buffers and/or covalent deuteration, it is possible to create a wide range of contrast and to accentuate or 'fade out' the scattering due to particular constituents.

PULSED-SOURCE VERSUS REACTOR BASED INSTRUMENTS

Neutron scattering experiments using pulsed beams are not in principle different from those using continuous beams from fission reactors, but there are significant differences in instrument design and experimental strategy (Windsor, 1981; Newport *et al.*, 1988). These are dictated by the need to utilize the time structure and polychromaticity of neutron pulses emitted by a spallation source with a frequency between 20 and 50Hz. With reference to structural studies, it must be pointed out first that small-angle neutron scattering (SANS) from larger biomolecules is of little interest because diffraction applications of this kind depend on both a high time-averaged flux and the ability to reach very low momentum transfers $\hbar Q$ (down to $Q \approx 5 \times 10^{-4}$Å^{-1}). Existing pulsed-source facilities and their SANS instruments cannot meet these requirements. There is more scope for structural work at moderately low momentum transfers ($0.01 \leq Q \leq 0.1$Å^{-1}) and beyond the SANS regime up to values of the order of 10Å^{-1}; here the ability of time-of-flight Laue diffractometers to provide simultaneous S(Q) patterns over a wide Q-range with excellent resolution is a unique asset. In the context of this paper, work at ISIS on developing the application of pulsed neutron techniques to fiber diffraction problems is of special interest (Langan *et al.*, 1995).

By and large, however, the potential for innovative biomolecular work using pulsed neutron sources is greatest in the area of quasi-elastic and inelastic scattering. Apart from economical considerations, the basic advantages are, first, the high gain factors that can be achieved by pulsed-beam operation relative to instruments on continuous beams; and second, the downscattering mode of data collection that makes efficient use of the wider incident spectrum and gives access to a larger momentum-energy domain. In addition, pulsed-beam spectrometers are easy to equip with auxiliary time-of-flight detectors which record Laue diffraction patterns monitoring structural properties of the sample simultaneous with the acquisition of $d^2\sigma/d\Omega dE$ data by the main detector array.

Spectrometers on pulsed neutron sources often work in 'inverse' scattering geometry: the beam incident on a sample is not monochromatized before scattering (as on reactor sources), but consists of a train of intense, short pulses with a nearly 'white' energy distribution up to the epithermal region. During the dead times between pulses, energy analysis is performed on the scattered fraction of neutrons by some combination of time-of-flight, crystal reflection, crystal filtering, or resonance absorption techniques. Spectrometers on steady-state reactor sources also need to create time windows for spectral analysis by means of mechanical devices (Fermi choppers) producing neutron pulses, or alternatively, they employ finely energy-collimated beams to scan a spectral window peri-

Table I. Principal neutron research centers operating pulsed sources

Country	Name	Institution	Current* (μA)	Frequency* (Hz)
UK	ISIS	Rutherford Appleton Laboratory, Chilton, Oxon.	180	50
USA	LANSCE	Los Alamos National Laboratory, Los Alamos, NM	70	20
USA	IPNS	Argonne National Laboratory, Argonne, IL	15	30
Japan	KEK	Nat'l Laboratory High Energy Physics, Tsukuba	5	30

*Proton synchrotron current and pulse frequency

odically. Both techniques require discarding ≥99% of the 'raw' neutrons, and they are generally less efficient than well-designed spectrometers on sources delivering polychromatic pulses.

There are currently four research centers that operate pulsed neutron sources (see Table I). The most powerful of these is ISIS at the Rutherford Appleton Laboratory near Oxford. With its suite of 7 state-of-the-art neutron spectrometers commissioned between 1987 and 1993 (Boland & Whapham, 1993), ISIS offers excellent opportunities for work on a wide range of biomolecular problems. At present, however, exploitation of this unique resource is limited by the small fraction of instrument time (around 5%) allocated to biophysics and biochemistry projects. Also, despite the instrumental progress made and the demand for S(Q, ω) data created by the growth in computer simulations of biomolecular dynamics, so far only two ISIS spectrometers—IRIS and TFXA—have been used in a somewhat more systematic manner for work in this area. Two further ISIS instruments, MARI and eVS, have begun to be used very recently in efforts to widen the Q, ω-domain studied. Biochemically and biophysically oriented work at ISIS has largely been concerned with fibrous biopolymers, polypeptides, and some of their building blocks such as amino acids, DNA bases, and certain hydrogen-bonded model compounds. As yet, globular proteins have received little attention, in spite of the fact that pulsed-source instruments cover the Q, ω-domain of computer simulations much better than instruments at reactor sources, and that realistic simulations (including the water of hydration) are best developed for globular proteins.

The present generation of neutron spectrometers, ie those on pulsed as well as steady-state sources, can probe molecular motions with energies from the neV to the eV region, or 10^{-5} to 10^{5}cm^{-1} in terms of optical wavenumbers (1cm^{-1} = 123.98μeV). This corresponds to characteristic times that range from the Brownian dynamics of macromolecules (microseconds) well into the quantum realm of femtosecond processes. The regions covered by instruments in routine operation at the two leading research centres are shown in outline in Figure 1. Neutron scattering thus gives access, at least nominally, to substantial portions of the large space-time domain characteristic of biomolecular interaction processes. This statement must be qualified as follows:

i. At very low $\hbar\omega$, between ~5 and 500 × 10^{-5}cm^{-1} (6–600neV), there are only two instruments—the spin-echo spectrometers IN11 at Grenoble (Institut Laue-Langevin) and MESS at Saclay (Laboratoire Léon Brillouin). The spin-echo technique has been used sporadically in experiments demonstrating its potential for quantifying domain motions in proteins, but access limitations have not yet allowed more comprehensive studies.

ii. At very high $\hbar\omega$, in the 10^{4} to 5 × 10^{5}cm^{-1} (1.2–60eV) region of Neutron Compton Scattering (NCS), there is a single recently commissioned instrument, eVS

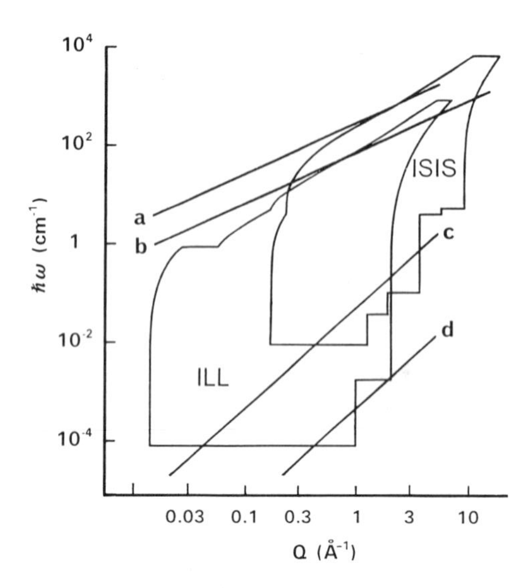

Figure 1. Envelopes of the Q, ω-domains of commonly used spectrometers at ILL (IN5, IN6, IN10, IN11) and at ISIS (IRIS, TFXA, MARI, HET), for standard operating parameters. Straight lines: momentum-energy relations for acoustic phonons [(a) and (b), V_{ac} = 3000 and 1000m/sec] and for diffusion [(c) and (d), $D_s = 10^{-6}$ and 10^{-8}cm^2/sec].

at ISIS (Mayers & Evans, 1991). Pilot experiments on DNA, collagen and agarose have been carried out during 1994, but the data are difficult to analyse and it is not yet clear in what way and to what extent NCS techniques can provide data on hydrogen-bond potentials and related parameters in complex biomolecular systems.

An important experimental parameter in this context is the momentum transfer Q_{min} at $2\theta_{min}$, the smallest scattering angle for which $d^2\sigma/d\Omega dE$ spectra can be measured with reasonable intensity. Low Q_{min} values are essential for many biophysical studies since they determine the largest scale length probed. At present, $2\pi/Q_{min}$ values of 100Å or larger can be achieved only by instruments on reactor sources, ie by spin-echo spectrometers and by cold-neutron time-of-flight spectrometers equipped with good low-angle detector arrays. The only really useful instrument in the latter category is IN5 at ILL Grenoble), but a new multi-chopper spectrometer on a smaller reactor at Berlin (Hahn Meitner Institut) promises to give good coverage of the low-Q region.

FIBROUS BIOPOLYMERS: HYDROGEN-BONDING, WATER DYNAMICS, MODE COUPLING

Neutron S(Q) and S(Q, ω) data from experiments at ISIS on fibrous biopolymers and their building blocks are contributing in a number of ways to our understanding of hydrogen-bonding, hydration processes and associated dynamical phenomena (Deriu *et al.*, 1993b; Middendorf, 1994; Middendorf *et al.,* 1995). Quantitative comparisons with simulations are beginning to be made here.

The remarkable polymorphism and hydrogen-bonding properties of fibrous biopolymers give rise to a rich diversity of secondary and tertiary structures, among which the helix is a common motif. Single-helical (α-helical polypeptides, polysaccharides), double-helical (DNA, polysaccharides) and triple-helical (collagen, related polypeptides) biopolymers aggregate with water and salt ions to form hydrated fibrils that are both internally

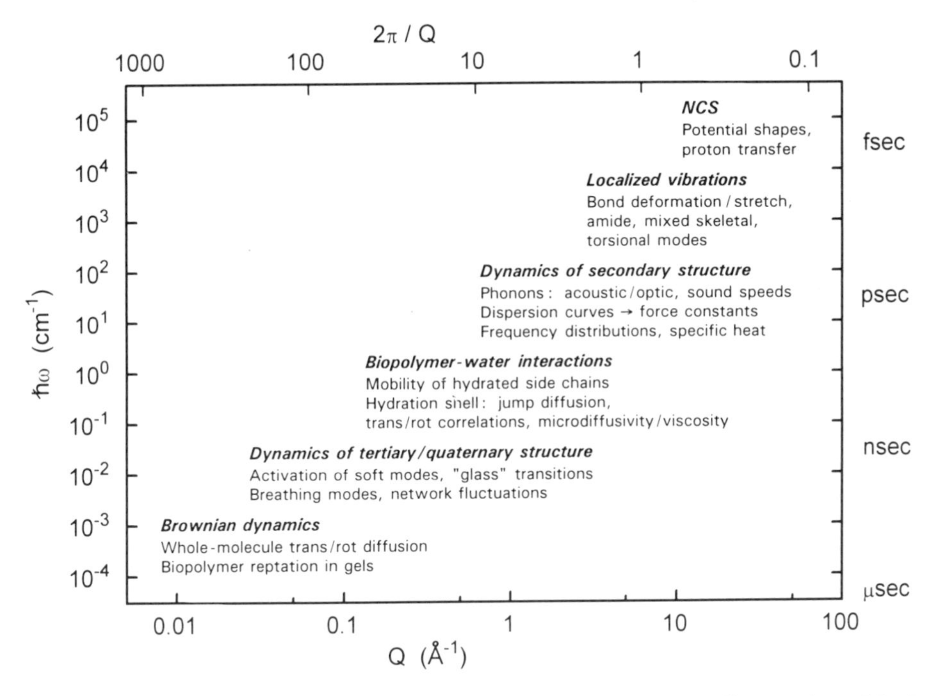

Figure 2. Synopsis of effects observed and information derived from experiments in different regions of the Q, ω-plane (NCS = Neutron Compton Scattering).

and externally hydrogen-bonded. Internally, elements of conjugate helix strands may be linked directly by hydrogen bonds, as in agarose and hyaluronate, two important polysaccharides. In DNA, on the other hand, the sugar-phosphate backbones are hydrogen-bonded indirectly *via* 2 or 3 O···H-N and N···H-N bonds between the complementary base pairs. In collagen, internal linkages comprise direct interchain hydrogen bonds as well as indirect bonds *via* water bridges that are an integral part of the structure (Bella *et al.*, 1994). The network of hydrogen bonds external to a fibrous biomolecule involves mainly the closely associated water molecules of the primary hydration shell. Some of these waters may be well-ordered and irrotationally 'bound' at the biopolymer surface, especially in the helix grooves. Others will be part of an extended network of hydrogen-bonded water bridges to neighboring helix strands, or they will be in direct contact with the more mobile hydration shell. Depending on hydration level, ionic milieu and temperature, a smaller or larger fraction of these waters will only be weakly 'bound' and thus be fairly mobile, although often anisotropically (Finney, 1986; Jeffrey & Saenger, 1991).

This structural complexity of fibrous biopolymers is reflected in a wide range of functionally important dynamical processes, of a collective as well as diffusive nature (Figure 2). Processes that are individually well known from simpler systems are not only present simultaneously, but may interact to give novel effects. A detailed characterisation of the dynamics of hydrogen-bonding, of the water mobility and of how water motions couple to the chain dynamics is essential for a molecular understanding of the functional role of these biomaterials and for their applications in biotechnology. Currently, the region of greatest interest for neutron work encompasses the transition from vibrational excitations at intermediate frequencies to those at lower frequencies and further to predominantly diffusive interactions, with scale lengths from a few Å (dimensions of backbone 'beads', side chains, water clusters) to lengths of the order of 100Å (phonon wavelengths, correlation lengths of aggregates).

QUASI-ELASTIC SCATTERING

The IRIS Spectrometer

The backscattering spectrometer IRIS, by virtue of its high Q-resolution and wide dynamic range, is a versatile instrument for studying motions of protons and protonated groups with characteristic times in the nsec to psec range (Carlile & Adams, 1992). IRIS is being used in a number of projects aiming to quantify biomolecular sorption processes, the mobility of water in biopolymer gels and biomembranes, the excitation of 'soft' degrees of freedom in proteins at temperatures between 150 and 270K, and base pair fluctuations in DNA on approach to melting.

A scattering diagram of IRIS is shown in Figure 3. Cold neutron pulses from the liquid-H_2 moderator on the ISIS target travel through a 36m long guide tube; its final 2.5m section consists of a focussing supermirror guide. In the 'inverted' backscattering geometry employed here, the energy scan is achieved by the spread in velocity as neutron pulses travel towards the 20 × 35mm slab sample. The scattered neutrons are energy-analyzed by 175°-backscattering from two large, semi-circular crystal arrays close to the instrument axis: (i) a 51-element mica array between $2\theta = 18°$ and $162°$ ($0.3 < Q < 1.9Å^{-1}$ for 006 reflection, resolution $0.09cm^{-1}$), and (ii) a similar graphite array on the opposite semi-circle between $2\theta = 200°$ and $340°$ ($0.3 < Q < 1.9Å^{-1}$ for 002 reflection, resolution $0.12cm^{-1}$). Neutrons backscattered from these arrays are detected by 2 × 51 scintillator counters each producing a 2000-point spectrum covering the energy window $-1 \leq \hbar\omega \leq +10cm^{-1}$. These 102 spectra are recorded simultaneously. By rephasing a Fermi chopper in the guide tube, without changing scattering geometry or disturbing samples, it is possible to record a third set of 51 wide-window spectra ($-15 \leq \hbar\omega \leq +60cm^{-1}$) for the 004 reflection of the graphite array ($0.9 < Q < 3.8Å^{-1}$, resolution $0.4cm^{-1}$). The Q-resolution $\delta(Q)/\delta(2\theta) = (2\pi/\lambda)\cos\theta$ varies from $0.045Å^{-1}$ (lowest Q) to $0.01Å^{-1}$ (high Q); these values are smaller by factors of

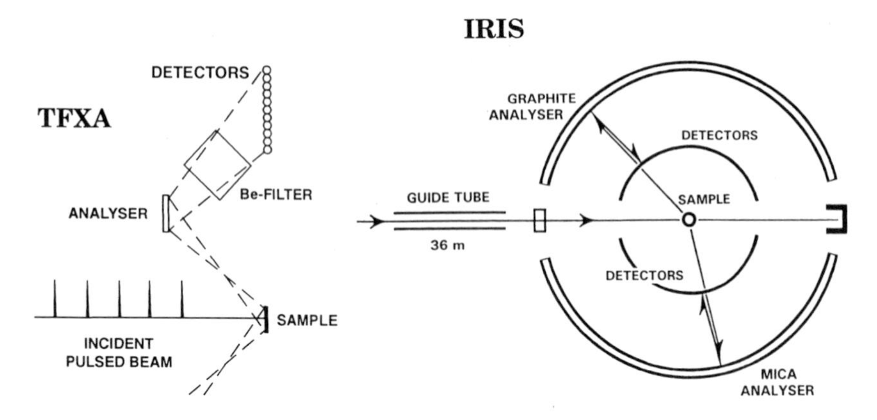

Figure 3. Scattering geometry (schematic) of the backscattering spectrometer IRIS and the time-focusing crystal analyser spectrometer TFXA at the Pulsed Neutron Facility ISIS (Rutherford Appleton Laboratory, Chilton, Oxon.). For TFXA, only one limb of the symmetrical scattering configuration is shown.

5 to 20 than those of reactor instruments. Overall, for comparable acquisition times (5–10 hours), the quality of the spectra is much better than that of reactor-based backscattering instruments. Longer run times are needed for spectra with energy resolutions of 0.01 or $0.05cm^{-1}$ (002 or 004 mica analyser reflections); these reach lower Q_{min} and allow measurements of translational diffusion coefficients down to a few $10^{-8}cm^2/sec$.

Scattering from Hydrated Systems

The majority of neutron studies performed so far has been concerned with hydration problems, and those at ISIS have concentrated on fibrous biomolecules. Quasi-elastic scattering from hydrated biopolymers at $T \geq 250K$ is dominated by diffusive processes (Middendorf, 1984a; Bée, 1988; Deriu *et al.*, 1993b). Small, Doppler-like velocity changes upon scattering lead to nearly symmetric broadenings of the elastic lines centered on $\hbar\omega = 0$, for each $Q \approx const$ spectrum (Figure 4). In hydrogenous systems of complex structure, these broadenings are due to incoherent scattering events involving a variety of proton motions with $|\hbar\omega| << k_BT$. Qualitatively, the broadenings observed tend to increase with Q and with temperature T, and to decrease as the polymer concentration C in a hydrated sample is raised.

There are, broadly, two ways of analysing sets of measured lineshapes in terms of molecular parameters. The first, conventional one is to construct a plausible dynamical model on the basis of whatever qualitative or semi-quantitative information is available, and to express it analytically. The second approach, suitable for crystallographically known systems, is based on detailed solutions of $\mathbf{R}_i(t)$ using molecular dynamics packages such as CHARMM or GROMOS; it aims to compare the resulting Fourier-transformed correlation functions directly with $S(Q,\omega)$ or $F(Q,t)$ data from experiments.

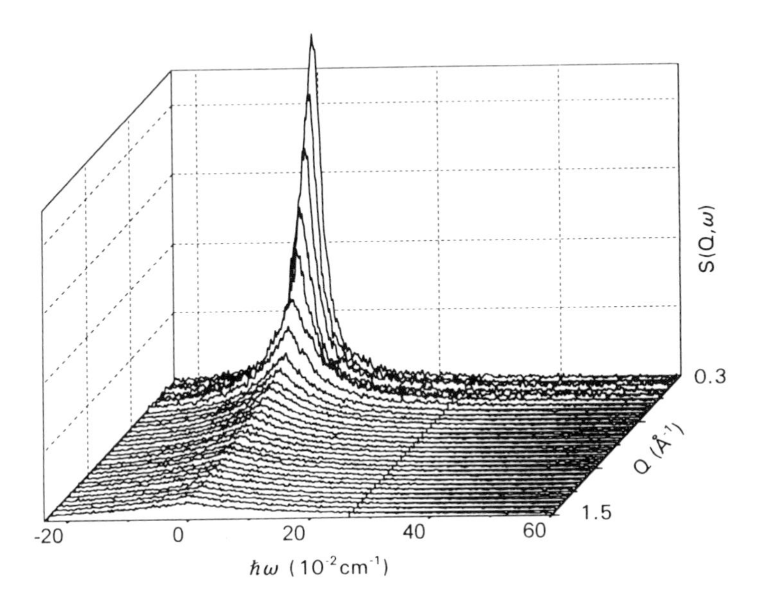

Figure 4. A set of quasi-elastic neutron spectra for an 8% hyaluronate-H_2O gel at 277K, measured at ISIS using the backscattering spectrometer IRIS (002 reflection of graphite analyser array, resolution $0.13cm^{-1}$ FWHM from spectra for a 2mm vanadium plate).

The object of lineshape analyses based on explicit models is to decompose the sets of spectra from each run (ie for a fixed pair of T,C values) into two or more components corresponding to distinct classes of mobile protons, and to extract from least-mean-square fits the dependence of model parameters on momentum transfer $\hbar Q$. For the case of dominant incoherent scattering, lineshape decomposition is usually based on expressions of the form:

$$S_{inc}(Q,\omega) = \exp(-Q^2\langle u_p^2\rangle)\left[A_o(Q)\delta(\omega) + (1 - A_o(Q))S_{eq}(Q,\omega)\right] \quad (5)$$

where S_{qe} is the scattering law describing line broadening.

In the $Q \rightarrow 0$ limit, the line widths Γ(HWHM) are dominated by translational diffusion which may proceed by discrete steps (jump diffusion on random or regular sites) or according to a more complex stochastic process (Bée, 1988). Self-diffusion coefficients D_s derived from lineshape analyses are always 'microscopic' in character, ie they reflect the mobility of protons and protonated groups over distances $d = 2\pi/Q$ from 2 or 3Å to around 50Å. From theoretical considerations and simulation studies it is well known that the time dependence of the mean-square displacement $\langle r^2\rangle$ of an atom or small molecule is often bimodal in the sense that at short times there is an approximately linear region with a steeper slope, followed by an asymptotic region with a slope proportional to the 'macroscopic' self-diffusion coefficient. For this reason the values of D_s derived from neutron $\Gamma(Q)$ are broadly comparable with, but in general not equal to, the self-diffusion coefficients measured by techniques probing longer distances and times. High-resolution data covering a wide Q-range may allow to extract two D_s values, one characterizing the short-time dynamics ($0.5 \leq Q \leq 5$Å^{-1}, identified with D_t below) and one that is closer to a 'macroscopic' D_s ($Q \leq 0.1$Å^{-1}). Clearly, the space-time domain covered makes data of this kind more directly comparable to simulation results emphasizing relatively short times and distances.

At intermediate and higher momentum transfers, librational and rotational motions contribute broader lineshape components. Assuming separable rotational and translational correlation functions, it is possible to write explicit models for $S_{qe}(Q, \omega)$ in terms of translational and rotational diffusivities and correlation times. The rotational part is frequently approximated by a Sears expansion for continuous rotational diffusion (Bée, 1988). Structural information is contained in $A_o(Q) \equiv F_{inc}(Q,\infty)$, the elastic incoherent structure factor or EISF (Bée, 1988). The EISF is the square of the Fourier transform of the spatial domains swept out by protons and protonated groups as $t \rightarrow \infty$. High-frequency amplitude information enters into the Debye-Waller factor, $\exp(-\langle u_p^2\rangle Q^2)$, where $\langle u_p^2\rangle$ is an average mean-square displacement of protons due to all vibrational motions with energies outside the quasi-elastic window.

IRIS is being used for experiments on fibrous biopolymers in two ways:

i. Sorption experiments for $0.05 \leq h \leq 0.8$ (specific hydration h = gr water/gr dry biopolymer). Samples consisting of sheets of parallel fibers are hydrated in steps from the dry or nearly dry state up to levels of water content sufficient for the primary and secondary hydration shells. By analysing sets of $S(Q, \omega)$ data for a number of h values at T = *const*, ie along a sorption isotherm, it is possible to gain insight into the structural and dynamical changes in response to partial hydration, and to quantify the dynamics of water closely associated with the biopolymer surface (Rupley & Careri, 1991; Schreiner *et al.,* 1988).

ii. Fully hydrated systems for which $1 < h \leq 100$, in the form of more or less concentrated solutions or gels. Here the biopolymer components are completely solvated by a spatially contiguous aqueous phase that may comprise bulk-like regions.

Next to hydration, temperature is the second important variable in experiments on the molecular dynamics of such systems. Temperature variation can be combined very effectively with H/D contrast variation to sort out the main contributions to S(Q, ω). By raising the temperature in steps from ~20 to 350K, it is possible to follow the gradual activation or 'unfreezing' of different degrees of freedom associated with interactions between elements of the secondary, tertiary, and quaternary structure. Temperature regions of particular interest are those associated with glass-like transitions (150–250K), with the restructuring and melting of hydration water (250–275K), and with thermally or biochemically activated fluctuations leading to transient, localized chain unfolding ($320 \leq T \leq 350K$).

Structure and Dynamics of Gels

The aqueous gels formed by biopolymers possess physicochemical properties that are unique among two-component polymeric systems. Despite the importance of biopolymer gels as constituents of living systems and their widespread use for electrophoresis and in medical applications, the molecular structure and dynamics of gels remain relatively unexplored (Burchard & Ross-Murphy, 1990; De Rossi *et al.*, 1991).

Biopolymer gels are soft nanostructured materials with unique properties. They behave as viscoelastic materials over scale lengths longer than several microns (ie lengths large compared with the maximum polymer network dimension), but microscopically they are fluids except for the 'molecular scaffolding' of the polymer network or matrix. The structure and rigidity of this 'scaffolding' can vary substantially, and Coulomb interactions may play an important role. From X-ray and neutron fiber diffraction work (Arnott *et al.*, 1983; Miller, 1984; Langan *et al.*, 1995), we know in some detail the primary and secondary structure of slightly hydrated biopolymer fibers, principally collagen, agarose, hyaluronate, and DNA. Our knowledge of the highly hydrated gel state was for a long time limited to results from techniques probing distances d ≥ 1000Å (optical techniques) and motions with frequencies ≤ 10MHz (rheology). In recent years, electron microscopy as well as X-ray and neutron diffraction studies of such gels have provided data on their molecular structure from ~10Å (diffraction) to values of the order of microns (EM). Neutron diffraction results have been reported mainly for agarose gels (Deriu *et al.*, 1993a; Krueger *et al.*, 1994); for dilute gels the logS(Q) *vs.* logQ plots are linear over 1–2 decades with fractal exponents between 2 and 2.5 (Middendorf *et al.*, 1990; Middendorf & Hotz de Baar, 1996). To understand quantitatively how the electrophoretic and rheological behavior of aqueous gels is determined by their microscopic properties, it is essential to obtain data linking the space-time domains covered so far, and to examine in particular the transition region from long-range diffusive processes to more localized interactions at the biopolymer-water interface.

Recent gel experiments using IRIS have focused on agarose and hyaluronic acid (HA) (Middendorf *et al.*, 1989; Deriu *et al.*, 1993a; Middendorf *et al.*, 1994). Both are linear polydisaccharides forming helical structures that have been characterized extensively by X-ray fiber diffraction and model building. Sulfur-free agarose is uncharged, whereas HA molecules carry a high density of negative charge. Like all sugars, they are strongly hydrophilic due mainly to OH groups that mesh easily with the transient hydrogen-bond network of surrounding water molecules, and additionally (in the case of HA) due to COO^- groups on glucuronic acid rings.

In the gel state, the interplay between structure and dynamics is quite different for these two polysaccharides, and the macroscopic properties differ accordingly. Agarose molecules associate by hydrogen-bonding to give long rod-like bundles of 10 to 30 double helices (Djabourov *et al.,* 1989; Dormoy & Candau, 1991). These bundles can aggregate into more or less random 3D networks that are relatively stiff, with interstitial spaces large enough ($\leq$ 3000Å in low concentration gels) to allow diffusion of globular proteins or rotation of sizeable DNA fragments. The network nodes are somewhat tangled 'junction zones' where bundles join or split up and pass over into adjacent rods.

In sodium or potassium HA gels, on the other hand, the long, highly hydrated molecules appear to adopt extended and possibly entangled random-coil conformations dominated by a dynamic equilibrium between electrostatic interactions and osmotic forces involving Na^+ or K^+ counterions (McDonald, 1988). Macroscopically, the resulting swollen gels possess unusual rheological properties and have been described as a 'viscoelastic putty' (Atkins & Sheehan,1973). The ability to 'bind' large amounts of water already at low concentrations is important physiologically in tissues where hydrated HA facilitates cell migration, and in synovial fluid where HA is a key determinant of the rheological properties.

A variety of experimental and simulation studies support the notion that only a few layers of the water of hydration around a biopolymer fibril or a globular protein are strongly affected by the surface properties, and the translational and rotational mobilities of water molecules approach their bulk values fairly quickly with increasing distance from the surface. Experimentally, however, it is quite difficult to quantify these properties at and near a structurally well-defined biomolecular surface. In terms of a simple two-phase model of hydration, a distinction may be made between an interface region extending over a relatively small number of water layers, and adjoining bulk-like 'pools' of solvent filling the interstitial spaces (Figure 5). Protein crystals are by far the best characterized systems here, with interstitial water volumes of 10^4 to 10^5Å^3 for which Fourier maps from X-ray

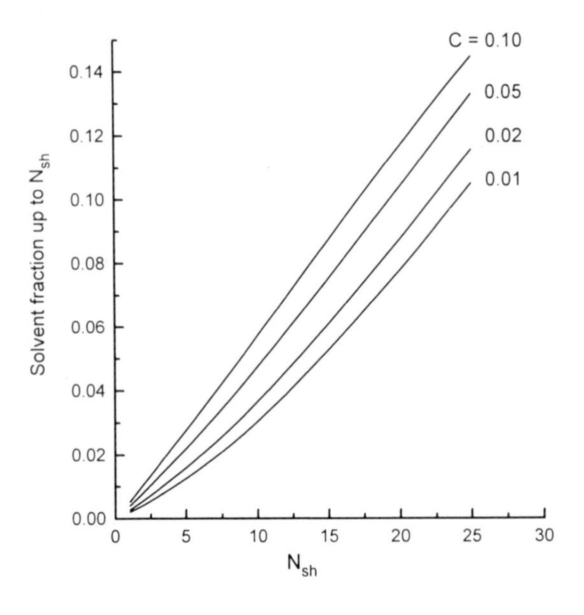

Figure 5. Fraction of solvent contained in hydration shells of thickness N_{sh} (multiples of 2.95Å water layers) around the cylindrical rods of a regular 3D network (cubic close-packed structure) modelling an agarose gel. The rods are bundles of polysaccharide double helices with diameters 20, 29, 47, 69Å for the concentrations shown (C defined as biopolymer weight divided by total weight). See Middendorf & Hotz de Baar (1996) for data on random network models.

and neutron crystallography provide detailed time-averaged data on the hydration of surface groups and the location of closely associated water molecules. By comparison, the interstitial solvent volumes in agarose gels are highly irregular geometrically, and are larger by factors of 10^2 to 10^4.

Quasi-Elastic Scattering from Polysaccharide Gels

The IRIS data sets obtained so far have all been analysed in terms of explicit models based on the convolution approximation discussed in the section *Scattering from hydrated systems*. High-resolution quasi-elastic spectra from gels typically consist of three components: a narrow, nearly elastic component of low intensity, a broader line whose width depends markedly on momentum transfer $\hbar Q$, and a third component in the form of a very broad underlying feature which at all Q-values fills the entire $\hbar\omega$ window accessible through the 002 analyser option of IRIS. For a series of gels with water contents between 80% and 99.5%, it is natural to proceed from the theoretical framework used in most earlier neutron studies of water, and to interpret the data in terms of the classical Chudley-Elliott (CE) model of proton jumps to a random, radially symmetric distribution of sites (Teixeira *et al.,* 1985; Teixeira, 1993). In analyses such as these, the prominent Q-dependent component is attributed to translational jump diffusion accompanied by the breaking of hydrogen bonds, whereas the broader, nearly Q-independent part of the spectrum is regarded as due to small-amplitude rotational motions. Considering the difference in broadening between these two components, there is some justification in adopting the usual *faute de mieux* assumption of decoupled translational and rotational degrees of freedom. In bulk water, the relevant characteristic times are separated by factors of 5 to 10 in the supercooled regime, but they become comparable at ordinary temperatures (Teixeira *et al.,* 1985) and a factorization of the intermediate scattering functions $F_{inc}(Q,\omega)$ may be a poor approximation.

In a heterogeneous system, explicit models for $S_{inc}(Q,\omega)$ require a partitioning of the sample volume into regions whose proton populations possess qualitatively different mobility characteristics. There are, first, protons that do not move appreciably over the longest time scale probed by IRIS ($\sim 5 \times 10^{-9}$sec): these belong to the polymer backbone and to a small fraction of very closely associated, essentially irrotationally 'bound' water molecules. A second regime contains protons site-bound for times of the order of 10^{-9}sec and capable of performing only hindered rotations in a restricted space. Finally, the protons in the interstitial aqueous phase proper will have translational and rotational degrees of freedom similar to bulk water protons, but the corresponding relaxation times are likely to be longer. Complementary information from other techniques, in particular NMR and RSMR (Rayleigh Scattering of Mössbauer Radiation, Albanese & Deriu, 1987), may be incorporated in the analysis to extend the time scale to values below 10^{-9}sec with the aim of resolving the scattering from slowly moving protons that contribute an effectively elastic component to IRIS spectra.

Least-mean-square fits of measured lineshapes to models formulated along these lines provide data on four parameters characterising the dynamics of water protons: (i) a mean-square vibrational displacement $\langle u_p^2 \rangle$ derived from Debye-Waller factors; (ii) a jump diffusion residence time τ_0; (iii) a translational diffusivity D_t; and (iv) rotational relaxation time τ_r (from a Sears model for continuous rotational diffusion). Detailed results for agarose gels with concentrations between 1% and 20% and temperatures in the 267 to 325K range have been reported by Deriu *et al.* (1993a) and by Middendorf *et al.* (1994).

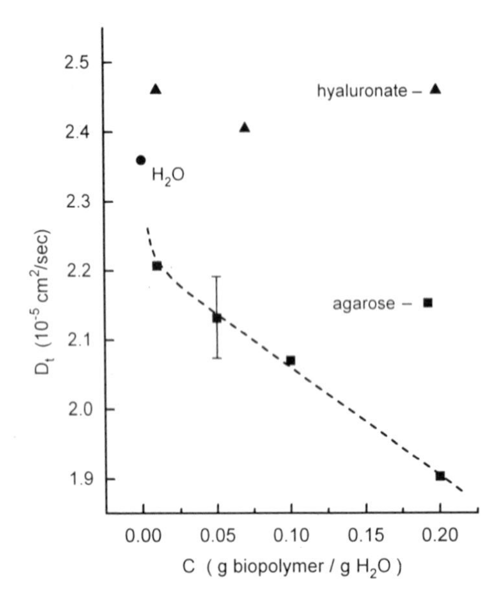

Figure 6. Microscopic translational diffusivity D_t at 294K as a function of polysaccharide concentration C for agarose (■) and hyaluronate (▲) gel water. Error bar refers to agaros gel data; the corresponding errors for HA data (from work in progress) are larger by ~50%. Bulk H_2O value (●) from IRIS measurements by Cavatorta *et al.* (1994).

The corresponding data set for hyaluronate gels is incomplete; only two concentrations have been examined so far. The principal results may be summarized as follows:

- Agarose gel water is less mobile than bulk H_2O at the same temperature, while the properties of HA gel water are fairly close to those of H_2O. On the basis of limited data for HA, the concentration dependence of the microscopic water diffusivity D_t for agarose and HA appears to differ sharply: for HA gels, $D_t(C)$ is essentially equal to the bulk value of $2.43 \times 10^{-5} cm^2 sec^{-1}$ and does not change much on going from a 1% to an 8% gel. For agarose, on the other hand, D_t first drops from the C = 0 value (bulk $H_2 0$) to the 1% gel value rather abruptly and then continues to decrease (Figure 6).
- The residence times τ_0 for translational jump diffusion (CE model) are very similar in HA and agarose gels. The temperature dependence of τ_0 suggests a lowered activation energy for translational jump diffusion relative to bulk water values; as a function of inverse temperature, τ_0 passes through a 'knee' at $T \approx 274K$ and then rises steeply (Figure 7). The thermodynamic implications of this behaviour are intriguing and need to be examined. For $T > 270K$, the temperature dependence of τ_0 in both gels resembles that of pure supercooled water at some corresponding temperature between 250 and 270K. It is possible, therefore, to rescale $\tau_0(T)$ to a common curve by assigning to gel water a 'structural temperature' lower than the actual thermodynamic temperature. Similar concepts have been put forward to describe the properties of hydration water of proteins and around phospholipid bilayers (Nimtz *et al.*, 1988).
- The rotational relaxation time τ_r assumes values of 0.3–0.4psec at 293K for agarose water whereas for HA $\tau_r \approx$ 1psec, close to the bulk H_2O value of about 1.2psec.
- The overall Q-dependence of translational broadenings Γ_t extracted from lineshape analyses for agarose gels is well described by a CE model with a random distribution of proton jumps to a shell of nearest-neighbor sites centered on 1.1Å.

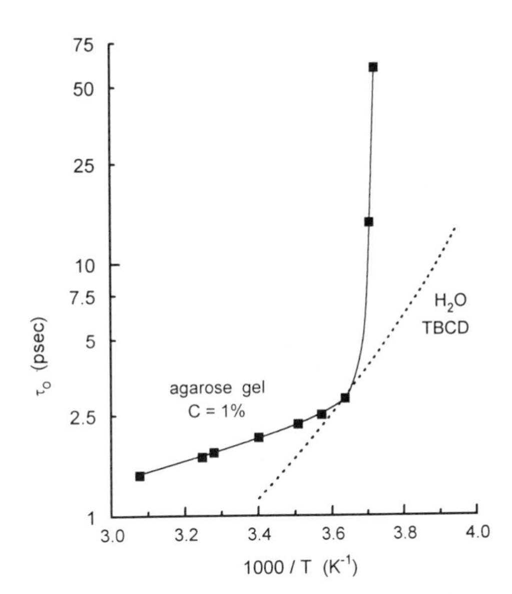

Figure 7. Residence times τ_0 for translational jump diffusion (Chudley-Elliott model) as a function of inverse temperature for water in a 1% agarose gel (■), compared with pure H_2O data measured at ILL Grenoble (instrument IN6) by Teixeira *et al.* (1985).

For agarose gels with $C \geq 0.1$, however, the high Q resolution and good statistics of IRIS spectra also reveal a small oscillatory component modulating the CE curve and its Fourier transform. These oscillations could be due to collective motions involving 10–40 water molecules, an interpretation suggested by molecular dynamics simulations (Ohmine *et al.*, 1988). The transient water clusters in models of this kind are characterized by correlation lengths of a few molecular diameters (6–8Å), ie values close to the wavelength of the oscillatory component observed in gel line broadenings. Such oscillations are less conspicuous in hyaluronic acid, and more data on HA are needed to substantiate this point.

INELASTIC SCATTERING

Molecular Spectroscopy Instruments

At higher energy transfers, overlapping with IRIS up to about 60cm^{-1}, the molecular spectroscopy instruments TFXA and MARI cover almost 2½ decades in $\hbar\omega$, from the low-frequency acoustic phonon region (~10cm^{-1}) to the highest bond-stretching frequencies and their overtones ($\leq 4000\text{cm}^{-1}$) (Penfold & Tomkinson, 1986; Eckert & Kearley, 1992). While the $\hbar\omega$-ranges of these two instruments are similar, their momentum transfer characteristics are quite different:

TFXA is a crystal analyser instrument with a special time-focusing scattering geometry shown in Figure 3. For all pencils of radiation backscattered from the sample, the effective scattering angle 2θ is 135° so that $\cos 2\theta = \mathbf{k}_f \cdot \mathbf{k}_o / k_f k_o = -1/\sqrt{2}$. For this condition the conservation equations for momentum and energy lead to:

$$Q^2 = k_f^{\,2}\left[2 + \varpi + (1+\varpi)^{1/2}\sqrt{2}\right]; \qquad \varpi = \hbar\omega / E_f \tag{6}$$

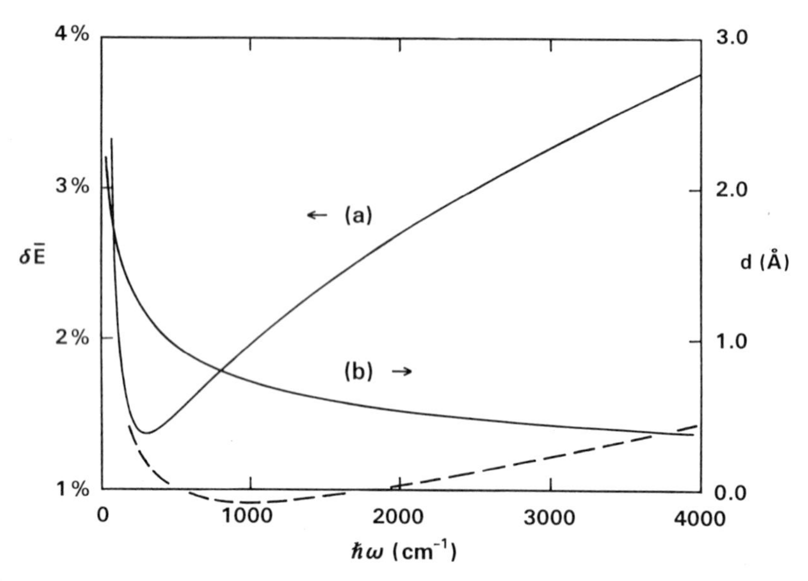

Figure 8. (a) Left-hand scale: Energy resolution δE/E of TFXA as a function of energy transfer $\hbar\omega$ (solid curve); theoretical δE/E of a projected spectrometer expected to replace TFXA during 1996 (dashed curve). (b) Right-hand scale: Distances $d = 2\pi/Q(\text{Å})$ probed by TFXA as a function of $\hbar\omega$.

where $E_f = 31.86\text{cm}^{-1}$ is the fixed analyser energy. TFXA thus produces a single spectrum per run, with a nonlinear Q(ω) relationship according to which Q increases from 2–3Å^{-1} at low $\hbar\omega$ to around 18Å^{-1} at the high-energy end. The spectrum obtained corresponds to a nonlinear 'slice' through the Q, ω-domain shown in Figure 1. At larger energy transfers the Q-dependence approaches the asymptote $Q^2(\text{Å}^{-2}) = 0.06\ \hbar\omega(\text{cm}^{-1})$. The scale lengths probed decrease from 2.2 to 0.4Å as $\hbar\omega$ increases from 15 to 4000cm^{-1}. This dependence is shown in Figure 8 together with the energy resolution δE/E.

MARI, by contrast, is a direct-geometry chopper spectrometer with a large array of detectors distributed over scattering angles 2θ from 3° to 135°. For each run, more than 100 spectra of ~1000 points each are recorded simultaneously. This instrument is slower than TFXA because it has no focusing capability and the data are not constrained to a single Q(ω) curve. However, the spectra obtained cover the entire Q, ω-domain accessible for any given incident energy (Figure 9), and the S(Q, ω) functions derived from them can be visualized as 3D intensity surfaces over the Q, ω-plane.

Both these instruments give energy-loss or 'downscattering' spectra, corresponding to Stokes scattering in optical terminology. The principal factors determining the differential intensity due to a vibrational mode with frequency ω_j are the following: (i) the coherent or incoherent cross-sections, (ii) the Debye-Waller factor, (iii) a 'polarization' term $(\mathbf{Q}.\mathbf{C}_{pj})^2$ where $\mathbf{C}_{pj}$ is the normalized amplitude vector of nucleus p vibrating in mode j, (iv) a thermal population factor (Bose-Einstein factor), and (v) a product of δ-functions ensuring conservation of energy and momentum (Lovesey, 1984; Middendorf, 1984b). For energy-loss spectra, the population factor assumes the form:

$$n_j + \tfrac{1}{2} + \tfrac{1}{2} = \left[\exp(\hbar\omega_j / k_B T) - 1\right]^{-1} + 1 \rightarrow 1 + \varepsilon \ \text{ for } \hbar\omega_j \geq 4k_B T \tag{7a}$$

whereas for energy-gain or 'upscattering' spectra:

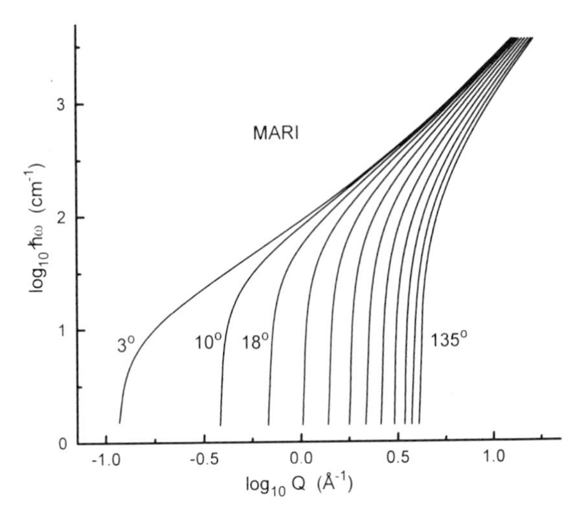

Figure 9. Family of curves showing the Q, ω-dependence of time-of-flight spectra measured for an array of $2\theta = const$ detectors ($2\theta = 3°, 10°, 18°, ..., 135°$). Each curve corresponds to a single nonlinear 'slice' through the Q, ω-domains shown in Figure 1, such that $Q \approx const$ up to $\hbar\omega$ of the order of $1cm^{-1}$ whence it gradually veers towards higher Q and approaches the asymptote Q^2 (Å^{-2}) = 0.06 $\hbar\omega$(cm^{-1}).

$$n_j + \tfrac{1}{2} - \tfrac{1}{2} = \left[\exp(\hbar\omega_j / k_B T) - 1\right]^{-1} \rightarrow \exp(-\hbar\omega_j / k_B T) \leq 0.02 \text{ for } \hbar\omega_j \geq 4k_B T \qquad (7b)$$

This basic fact, combined with a more or less uniform energy resolution of 1–2% over a wide $\hbar\omega$ window, gives pulsed-source instruments such as TFXA and MARI a decisive advantage over reactor-based instruments for neutron studies of molecular vibrations in complex systems.

Since protons dominate the scattering from hydrogenous biopolymers, the frequencies and amplitudes of modes involving protons are of central interest in experiments of this kind. Hydrogen/deuterium substitution is a powerful tool here for differentiating between contributions due to polymer backbone groups and side chains. At temperatures below about 100K, where Doppler broadenings and other degrading effects are small, neutron spectra from pulsed-source instruments are highly resolved and can be comparable in quality to data from routine IR or Raman work. TFXA and MARI spectra are thus opening up a new 'window' on proton dynamics of biomolecular relevance, complementing the large body of optical work this area but also challenging certain long-standing assumptions and results based on IR and Raman data (Kearley *et al.*, 1994).

Vibrational Spectroscopy of Biomolecules

Progress in understanding the vibrational spectra of biomolecules is currently being made on two fronts: On the one hand, several small component or model molecules (such as N-methylacetamide, acetanilide, amino acids, dipeptides, DNA bases) are being studied in depth with the aim of accounting for their molecular dynamics and hydrogen-bonding properties in considerable detail, ideally from first principles (ie quantum chemistry). Experiments here typically involve anhydrous powder or crystal samples at T < 100K, conditions that are far from the 'working environment' of biomolecules. On the other hand, experimental and simulation work on the dynamics of larger biomolecular systems (especially proteins and nucleic acids) increasingly aims to study more realistic conditions, ie physiological temperatures and hydration levels sufficient for functional interactions. There is at present little common ground between these two broad areas of work, the main connection being that force constants and potential parameters derived from low-temperature work on smaller molecules are needed for simulating the dynamics of larger systems at ordinary temperatures. Neutron data and S(Q, ω) simulations on oligopeptides and oligonucleotides provide important links between these areas of work.

Spectra from biopolymers in the Q, ω-domain covered by TFXA and MARI may be interpreted with reference to two well-established areas of work: the theory of neutron scattering from phonons propagating in polymers and hydrogen-bonded molecular crystals, and that from localized oscillators in a molecular environment modeled as a 'thermal bath' (Lovesey, 1984). For spectra from biopolymers which cannot easily be obtained in selectively deuterated form, the assignment of spectral features to particular modes requires a good deal of supplementary information from experiments on structurally similar polymers and their building blocks.

A useful starting point is to distinguish between low-frequency cooperative modes (phonon bands) and more localized group excitations at higher $\hbar\omega$. Quantitatively this amounts to a separation of time scales, ie a factorization of $F_{inc}(Q,t)$ (see Equation 5) into two or possibly three components (Warner *et al.,* 1983). The biopolymer is regarded as a structured chain of 'beads' supporting low-frequency skeletal modes (translational, torsional, accordion, etc.), with a number of independent localized oscillators 'embedded' in each chain element. This factorization of $F_{inc}(Q,t)$ requires the oscillator frequencies to be large relative to the frequency of the highest phonon band. Detailed analysis has established that this is a reasonable approximation already for frequency ratios of around 2. In the form of 'extended atom' approximations, this approach is commonly used in normal-mode analyses or full MD simulations in order to reduce the number of degrees of freedom, in particular by grouping the hydrogens together with the heavy atoms to which they are attached (Fanconi, 1980; Smith, 1991). The polypeptide and collagen spectra measured on TFXA suggest a time scale separation corresponding to intensity minima between 350 and $400 cm^{-1}$. An important aspect is that the low-frequency dynamics is largely determined by the heavy atoms (mainly C, O, N), whereas the incoherent scattering comes almost exclusively from hydrogen atoms. Insofar as these can be regarded as rigidly attached to the former, and this is a reasonable approximation at lower $\hbar\omega$, they serve as probes of the dynamics of smaller and larger chain elements.

Neutron Spectroscopy of Collagen

The only natural biopolymer for which vibrational neutron spectra are being studied is collagen, the principal protein component of connective tissues in vertebrates (Fraser *et al.*, 1987; Miller, 1984; Kadler, 1994). It consists of hydrogen-bonded bundles of triple-helical polypeptide chains with a characteristic Gly-X-Y repeat. Gly is the amino acid glycine, and X and Y can be any other amino acid, but the imino acids proline or hydroxyproline occur frequently in these positions. The structural properties of collagen have been investigated extensively by X-ray and neutron diffraction. To understand the mechanical properties of collagen at the molecular level, it is essential to be able to interpret stress-strain curves and elastic moduli in terms of the underlying bonding patterns, force constants, and chain dynamics. Sound velocities derived from Brillouin line shifts have already been used in attempts to relate the macroscopic elasticity parameters to structural detail and hydration (Harley *et al.*, 1977; Randall & Vaughan, 1979; Cusack & Lees, 1984), but spatiotemporal data over a wider parameter domain are required for a comprehensive characterisation of the molecular dynamics.

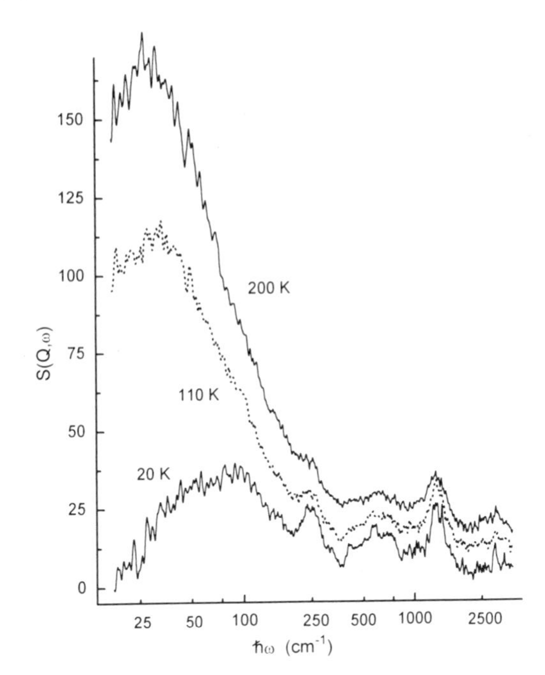

Figure 10. Neutron spectra of slightly D_2O-hydrated collagen fibers at 20, 110 and 200K (Type I collagen from rat tail tendon, fibers oriented at right angles to scattering plane). Not corrected for thermal population factor n_{BE}, to show the strong temperature dependence of low-frequency modes up to a few $100cm^{-1}$ (shifted by $\Delta S = 0.1$ and 0.2 relative to the 20K curve).

The TFXA spectra obtained in recent work at ISIS provide the first high-resolution neutron views of the vibrational excitations of collagen and related triple-helical polypeptides (Middendorf, 1994). Low-temperature spectra from aligned, slightly hydrated arrays of collagen fibers all show several bands with considerable fine structure (Figure 10). At temperatures below 100K, the spectra display the nearly harmonic excitations of the collagen triple helix. At higher temperatures (150 to 250K), they give insight into the broad glass-like transition and the onset of mode-coupling around 200K as the result of the 'unfreezing' of low-frequency degrees of freedom. Parallel IRIS experiments reveal that up to ~190K there is essentially no quasielastic broadening at any Q; above 200K there are small intensity increases in the wings which become much more intense and broad towards 270K (Middendorf *et al.*, 1996). The Q-dependence of the difference intensities observed on IRIS, analyzed in conjunction with TFXA data covering the 140 to 270K temperature region, can give important information on the way in which the unfreezing of sidechain motions is correlated with changes in high-frequency excitations, in particular amide modes.

Low-Frequency Region: $\hbar\omega \leq 350cm^{-1}$

As expected for a polymer with a complex repeat unit of variable structure, the low-frequency region in collagen spectra consists of two very broad bands each carrying a number of distinct lines: the first one centred on $\hbar\omega \approx 80cm^{-1}$, and the second on $\hbar\omega \approx 240cm^{-1}$. The TFXA spectra shown in Figure 10 also reveal a strong temperature dependence of the intensity of the lowest lying band. A few of the peaks seen in earlier work on dry, unoriented collagen (Berney *et al.*, 1987) can be identified in these spectra, but the high-resolution TFXA spectra from oriented samples discussed here show much more detail.

The quantitative interpretation of collagen S(Q, ω) data requires comparison with neutron spectra, optical spectra, and normal-mode calculations for triple-helical polypeptides, in particular polyglycine II, polyproline II, and the oligopeptide $(\text{Pro-Pro-Gly})_{10}$ (Krimm & Bandekar, 1986; Fanconi, 1980; Baron *et al.*, 1989). In the following, these three systems will be abbreviated as PGII, PPII, and PPG10, respectively. The qualitative features of the low-temperature density of states calculated for PGII (Fanconi & Finegold, 1975) are reflected in the 20K neutron spectrum of collagen, although for a heterogeneous polypeptide the band gap from 130 to 200cm^{-1} is of course 'filled in' to a considerable extent. Collagen, PPG10 and PPII all display a similar overall structure in the low-frequency region, but there are significant differences at intermediate and high frequencies (Figure 11). The spectral signatures of polypeptides for $\hbar\omega \geq 100\text{cm}^{-1}$ are due to complex interactions between skeletal and deformation modes which reflect properties of the secondary and tertiary structure and become more localized at higher $\hbar\omega$.

The general features of the $\hbar\omega \leq 400\text{cm}^{-1}$ region are similar for collagen, PPG10 and PGII. The first band is due to dispersive phonon modes and the second one to weakly dispersive skeletal deformation modes. The structure of the first band in dry collagen and in PPG10 are essentially identical, with submaxima at 45 and 110cm^{-1}. Theoretical results for PGII give a maximum due to longitudinal modes at 13cm^{-1} in the isolated molecule, but this is shifted upward to 40cm^{-1} when hydrogen-bonding to neighboring chains is taken into account. These calculations are in broad agreement with neutron spectra for PGII (Baron *et al.*, 1989). The maximum in the scattering intensity appears to be due to an inflection in the longitudinal mode branch, induced by softening of modes with wavevectors matching the pitch of the crystallographic helix. The 45cm^{-1} peak in collagen and PPG10 spectra can be identified with this maximum for longitudinal acoustic modes in PGII-like chains that are hydrogen-bonded and supercoiled to form the collagen triple helix. The longitudinal acoustic phonon frequency in the polypeptide matches that of its associated interhelical ice shell, suggesting efficient coupling between modes in the hydration shell and the polypeptide backbone (White, 1976). Calculations for PGII have revealed torsional modes between 50 and 120cm^{-1} (Krimm & Bandekar, 1986) that can be expected to lead to a neutron intensity maximum near 110cm^{-1}; a broad maximum around this frequency is clearly seen in collagen and PPG10 spectra.

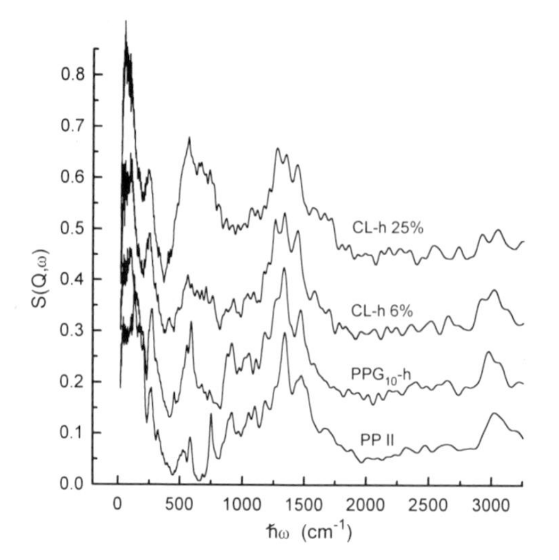

Figure 11. Comparison of neutron spectra at intermediate and high frequencies (all at 30K, n_{BE}-corrected): PPII = polyproline II (dry, 0% H_2O); PPG10-h = hydrogenous $(\text{Pro-Pro-Gly})_{10}$ ('dry', 6% H_2O); CL-h = collagen fibers oriented at right angles to the scattering plane, 'dry' (6% H_2O) and 25% H_2O-hydrated. Spectra shifted by $\Delta S = 0.12$, 0.24 and 0.36 relative to PPII.

Density of States and Mean-square Displacements

A basic quantity characterising the dynamics of polymers is the distribution of vibrational frequencies $Z(\omega)$ (Peticolas, 1979; Fanconi, 1980). A proton-weighted function $Z_p(\omega)$, which at low frequencies closely approximates $Z(\omega)$, can be extracted from predominantly incoherent $d^2\sigma/d\Omega dE$ spectra measured on TFXA or MARI. Apart from a uniform average weighting by the square of the atomic polarization vectors, this involves extracting a one-phonon scattering function and an evaluation of the Debye-Waller factor (Lovesey, 1984). Appropriately normalized $Z_p(\omega)$ functions may be used in three ways:

i. to estimate an effective mass for the 'proton-labeled' chain elements determining the low-frequency modes, in conjunction with independent measurements or calculations of Debye-Waller factors;
ii. to test and refine the parameters entering into MD simulations;
iii. to evaluate the low-temperature specific heat C_v for comparison with calorimetric data.

In proton-dominated spectra, the Debye-Waller factor is essentially given by exp(-$\langle u_p^2 \rangle Q^2$) where $\langle u_p^2 \rangle$ is the average mean-square displacement of the hydrogens. At low temperatures $\langle u_p^2 \rangle$ mainly reflects the low-frequency modes. For a harmonic system at temperature T, it can be expressed in terms of $Z_p(\omega)$ by the integral (Lovesey, 1984):

$$\langle u_p^2 \rangle = \int_0^{\omega_c} Z_p(\omega)[\hbar / 2\omega m(\omega)] \coth\{\hbar\omega / 2k_B T\} d\omega \tag{8}$$

where $m(\omega)$ is a frequency-dependent effective mass and ω_c a cut-off frequency. The low-temperature $Z_p(\omega)$ function for minimally hydrated Type I collagen is shown in Figure 12 together with the integrand of Equation 8 to exhibit the extent to which modes below 50 to 100cm^{-1} determine the $\langle u_p^2 \rangle$ values. The specific heat, on the other hand, is also dominated by the lowest acoustic modes for $T \leq 50$K, but with increasing temperature the integrand in the corresponding expression for C_v samples much more rapidly the entire low-frequency region up to a few 100cm^{-1} (Figure 12).

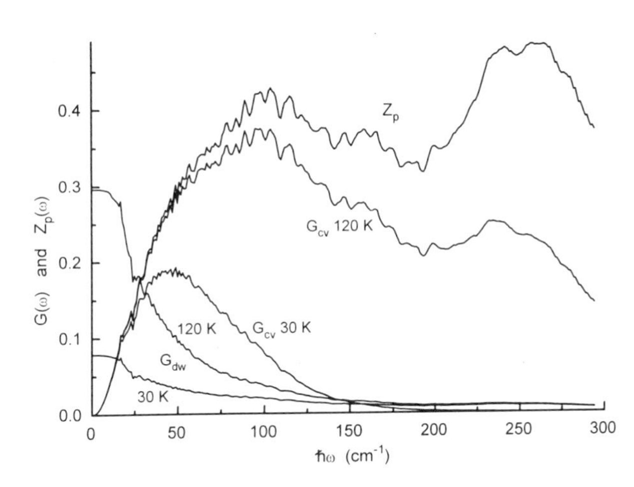

Figure 12. Low-temperature vibrational density of states Z_p for nearly dry, H_2O-exchanged collagen fibers at 30K, with integrand functions F_{dw} and F_{cv} determining the Debye-Waller factor and specific heat in the harmonic regime.

To extract an average value for the effective mass of oscillators dominating the low-frequency dynamics, it may be assumed that m is approximately constant and equal to M^* for $0 < \omega < \omega_c$. If the Debye-Waller factor is evaluated from quasi-elastic scattering data for identical samples in the same temperature region, M^* may be determined from Equation 8. For collagen, the $\langle u_p^2 \rangle$ values obtained from parallel IRIS experiments lie between 0.02 to 0.03Å^2 (Middendorf *et al.*, 1996); a range that agrees fairly well with neutron results for globular proteins and ordinary polymers.

For collagen, analyses along these lines give M^* values in the 10 to 20 dalton range, at temperatures where anharmonic effects are likely to be small or negligible. These M^* values are consistent with an idealized model consisting of extended atom groups as chain elements, in which librational motions can have an effective mass comparable with or less than the atomic mass. Models and approximations of this kind are important for extending the long time limit of MD simulations into the nsec to μsec domain.

A comparison of spectra from dry H_2O-exchanged collagen (6% H_2O) and 25% H_2O-hydrated collagen reveals excess intensity in the region where pure H_2O ice shows an acoustic phonon band. In the 25% hydrated sample, all the water is closely associated with protein so that there is no pure ice. Water molecules in the interstices between collagen helices form extended, hydrogen-bonded water chains and clusters, packed in quasi-hexagonal arrangement (Grigera & Berendsen, 1979). These can support collective low-frequency modes, and the 50cm^{-1} peak in difference spectra must be assigned to such modes propagating in interhelical water structures. This assignment is consistent with the absence of a similar 50cm^{-1} peak in the difference spectrum between hydrogenous and amide-deuterated PPG10. The 6% water present in PPG10 samples is all tightly 'bound' in water bridge sites along the peptide backbone; these water molecules cannot form an extended hydrogen-bonded network and there are no ice-like modes. This is borne out by the observation that H_2O/D_2O exchange does not lead to spectral changes.

Intermediate and High Frequencies: Amide Modes and Triple-Helical Structure

At energy transfers intermediate between the intensity minimum at 370cm^{-1} and that near 900cm^{-1}, low-temperature TFXA spectra from oriented collagen fibers show bands with a level of fine structure not much inferior to that of optical spectra. The Amide V band is one focus of attention here; comparisons with the corresponding IR and Raman bands give new insight into proton-related features.

All neutron spectra from collagen are very complex at intermediate and higher frequencies. The modes observed involve predominantly stretching or bending of C-H, N-H and O-H bonds, together with torsional deformations. Examples are the various amide modes between 700 and 1550cm^{-1}, the in-plane and out-of-plane bending modes of aromatic ring hydrogens between 600 and 1600cm^{-1}, and the methylene and methyl deformation modes between 950 and 1450cm^{-1}. Analysis of TFXA spectra from polyglycine I has led to the conclusion recently that at low temperatures the amide proton moves in an asymmetric double-well potential and that its modes are more localized than was thought previously (Kearley *et al.*, 1994). The assignments of several bands and peaks in collagen spectra on the basis of optical data, normal-mode calculations and results for triple-helical polypeptides have been discussed in some detail by Middendorf *et al.* (1995).

In the 500 to 900cm^{-1} region, several features may be ascribed to hydrogen-bonded amide hydrogens. In PGII, polyalanine, and the hydrogen-bonded crystals of the peptide models N-methylacetamide (NMA) and acetanilide (ACN), the NH out-of-plane mode,

Amide V, mixed with C=O in-plane bend, appears at 750cm^{-1}. A mode involving C=O in-plane bend mixed with skeletal deformation is expected at 707cm^{-1}; PGII spectra show a strong peak in this region. Triple-helical supercoiling leads to a downward shift of the Amide V mode of the Gly-Pro linkage, together with skeletal deformation and C=O in-plane bending modes of the Pro-Pro linkages. Because of the background due to water libration, the NH out-of-plane bending mode is not clearly identified by H/D exchange in the triple-helical peptides. Distortions of the intra-helical hydrogen bonds and the backbone dihedral angles due to supercoiling appear to play a role here. The equivalent hydrogen-bond and backbone geometries in models of collagen vary with the details of the molecular structure. Normal-mode calculations have predicted significant effects of hydrogen bonding and bond angle variation on Amide V modes (Krimm & Bandekar, 1986). Relative to the triple-helical polypeptides investigated, neutron spectra for dry collagen give much more intensity between 600 and 800cm^{-1}. Contributions from amino acid side chains, in particular out-of-plane bending modes of side chain NH groups, are expected here, on top of a broad background of librational modes from intra-chain H_2O molecules. The intensity from phenylalanine and tyrosine ring modes should be small because of the low abundance of these residues in collagen. By contrast, the high abundance of imino acid residues results in fairly strong bands at 1320 and 1450cm^{-1} due to CH_2 twist and wag modes of pyrrolidine rings. A moderately intense peak at 1320cm^{-1}, assigned to CH_2 twist motion, has been observed in IR and Raman spectra of polypeptides containing proline.

The imide II band between 1445 and 1485cm^{-1} has been the subject of resonance Raman studies of X-Pro or Pro-Pro cis/trans isomerization (X = any a.a. residue) and imino hydrogen bonding. This is a high-frequency mode sensitive to changes in backbone conformation. In neutron spectra it seems to be masked by the strong pyrrolidine ring CH_2 deformation band at 1450cm^{-1}. Excess intensity around 1435cm^{-1} in collagen spectra is likely to come from methyl group deformations, and methyl modes may also contribute intensity near 1060cm^{-1}. Neutron data and spectral simulations for the model compounds NMA and ACN (Barthés *et al.*, 1992; Fillaux *et al.*, 1993; Hayward *et al.*, 1995) are consistent with these assignments.

Some features in neutron spectra from collagen and related triple-helical polypeptides depend sensitively on deuteration of the amide hydrogen. The most conspicuous change is a loss of intensity at 1250cm^{-1}, with lesser changes at 1500 and 1550cm^{-1}, and around 930 and 1020cm^{-1} in the PPG10-d spectra. The peaks at 1250 and 1550cm^{-1} are absent from the PPII spectra since no amide hydrogens are present. As expected from the higher amino/imino ratio in collagen, both Amide II and Amide III modes are stronger in dry collagen spectra than in PPG10. The spectral changes observed near 925 and 1025cm^{-1} upon deuteration of PPG10 are difficult to interpret as there are no amide modes expected in this region. It may be that these changes are related to excitations involving tightly bound water molecules. Amide I should be weak in neutron spectra if it involves little hydrogen motion. There is little intensity at 1650cm^{-1} where Raman and IR spectra have revealed the Amide I peak in collagen and PPG10. This supports the assignment of Amide I to C=O stretch, a mode involving little hydrogen motion.

CONCLUSIONS

The results reviewed in this paper demonstrate that pulsed-beam spectrometers have an important role to play in experimental work on biopolymer dynamics. Outstanding assets of neutron techniques in this context are the closeness of measured $S(Q, \omega)$ functions to dynamic structure factors derived from simulations, and the complementarity of the information obtained to that from optical spectroscopy at intermediate and high frequencies, and to that from nuclear magnetic resonance techniques in the quasi-elastic regime.

The interpretation of neutron spectra from biomolecular systems, and the exploitation of the knowledge gained, poses challenging questions. In recent years substantial progress has been made in simulating the molecular dynamics of proteins, nucleic acids and polysaccharides for increasingly realistic microenvironments and interactions with smaller molecules. The total effort going into simulation studies has grown enormously compared to that expended on developing experimental techniques providing spatiotemporal data over the parameter domain of interest. As a result, current quasi-elastic and inelastic neutron scattering work is largely 'theory-driven', at two closely related but in practice somewhat distinct levels: first, low-temperature experiments provide force constants and potential parameters quantifying in detail the harmonic and weakly anharmonic dynamics in minimally hydrated biopolymers or their anhydrous component molecules. At a second level of analysis and experiment, parameters derived from the data obtained are used in conjunction with other basic input data for large-scale MD simulations of fully hydrated systems at temperatures $\geq$ 270K. In this latter category, there is a dearth of $S(Q, \omega)$ data against which simulation results can be tested and refined.

Neutron spectrometers such as IRIS, TFXA and MARI provide information on biomolecular dynamics and interaction processes that cannot be obtained in any other way. At present, however, the fraction of instrument time available for experiments of this kind is inadequate to sustain a rate of development commensurate with the general growth in the use of advanced spectroscopic and simulation techniques in the life sciences. Also, the fragmental system of allocating time on different instruments makes no allowance for the large Q,ω-domain of biomolecular dynamics, and is ill-adapted to the fact that MD simulations routinely cover the $0.1 \leq t \leq 100$psec region ($300 \leq \hbar\omega \leq 0.3\text{cm}^{-1}$). Relative to reactor-based facilities of comparable size, the research environment at ISIS is narrower and support for work on biomolecular problems is very limited. It has been difficult in particular to initiate neutron projects of biomedical and biotechnological relevance. With reference to the scientific programs motivating the construction of new instruments at ISIS and the planning for a European Spallation Source, we may hope to convince the 'neutrocrats' that the applications discussed in this paper hold much promise for the future and are well worth more attention and resources.

ACKNOWLEDGMENTS

We gratefully acknowledge the continued collaboration of our colleagues in Parma, Messina and Edinburgh, in particular Prof. Deriu, Prof. Wanderlingh, and Drs Bradshaw, Cavatorta, DiCola and Hayward. We thank the ISIS Pulsed Neutron Facility for the use of facilities, and the IRIS, TFXA and MARI teams for much help and advice on instrumental aspects. This work was supported by the UK Biotechnology and Biological Sciences Research Council.

REFERENCES

Albanese, G., & Deriu, A., (1979). High energy resolution X-ray spectroscopy. *Rivista del Nuovo Cimento,* 2:1–40.

Arnott, S., Mitra, A.K., & Raghunathan, S., (1983). Hyaluronic acid double helix. *J. Mol. Biol.*, 169:861–872.

Atkins, E.D.T., & Sheehan, J.K., (1973). Hyaluronic acid: A novel, double helical molecule. *Science*, 179:560–563.

Baron, M.H., Fillaux, F., & Tomkinson, J., (1989). Inelastic neutron scattering study of the proton dynamics in polyglycine I and II. *Third European Conference on Spectroscopy of Biological Molecules*. Bologna.

Barthés, M., Eckert, J., Johnson, S.W., Moret, J., Swanson, B.I., & Unkefer, C.J., (1992). Anomalous vibrational modes in acetanilide as studied by inelastic neutron scattering. *J. de Physique I*, 2:1929–1939.

Bella, J., Eaton, M., Brodsky, B., & Berman, H.M., (1994). Crystal and molecular structure of a collagen-like peptide at 1.9Å resolution. *Science*, 266:75–79.

Bée, M., (1988). *Quasi-elastic Neutron Scattering*. Adam Hilger, Bristol.

Berney, C.V., Renugopalakrishnan, V., & Bhatnagar, R.S., (1987). Collagen: an inelastic neutron scattering study of low frequency vibrational modes. *Biophys. J.*, 52:343–345.

Boland, B., & Whapham, S., (editors) (1993). *ISIS Experimental Facilities*. SERC, Rutherford Appleton Laboratory, Chilton, UK.

Burchard, W., & Ross-Murphy, S.B., (editors) (1990). *Physical Networks: Polymers and Gels*. Elsevier, Amsterdam.

Carlile, C.J., & Adams, M.A., (1992). The design of the IRIS inelastic neutron spectrometer and improvements to its analysers. *Physica B*, 182:431–440.

Cavatorta, F., Deriu, A., DiCola, D., & Middendorf, H.D., (1994). Diffusive properties of water studied by incoherent quasi-elastic neutron scattering. *J. Phys.: Condensed Matter*, 6:A113-A117.

Cusack, S., & Lees, S., (1984). Variation of longitudinal acoustic velocity at gigahertz frequencies with water content in rat-tail tendon fibers. *Biopolymers*, 23:337–351.

Deriu, A., Cavatorta, F., DiCola, D., & Middendorf, H.D., (1993a). Large-scale structure and dynamics of polysaccharide gels. *J. de Physique IV-C1*, 3:237–247.

Deriu, A., Cavatorta, F., Cabrini, D., Carlile, C.J., & Middendorf, H.D., (1993b). Water dynamics in biopolymer gels by quasi-elastic neutron scattering. *Europhys. Lett.*, 24:351–357.

De Rossi, D., Kajiwara, K., Osada, Y., & Yamauchi, A., (editors) (1991). *Polymer Gels: Fundamentals and Biomedical Applications*. Plenum Publishing Corporation, New York.

Djabourov, M., Clark, A.H., Rowlands, D.W., & Ross-Murphy, S.B., (1989). Small-angle X-ray scattering characterization of agarose sols and gels. *Macromolecules*, 22:180–188.

Dormoy, Y., & Candau, J., (1991). Transient electric birefringence study of highly dilute agarose solutions. *Biopolymers*, 31:109–117.

Eckert, J., & Kearley, G.J., (editors) (1992). Spectroscopic applications of inelastic neutron scattering: Theory and practice. *Spectrochimica Acta*, 48A:269–476.

Egelstaff, P.A., Gray, C.G., Gubbins, K.E., & Mo, K.C., (1975). Theory of inelastic neutron scattering from molecular fluids. *J. Statist. Phys.*, 13:315–330.

Fanconi, B., (1980). Molecular vibrations of polymers. *Ann. Rev. Phys. Chem.*, 31:265–291.

Fanconi, B., & Finegold, L., (1975). Vibrational states of the biopolymer polyglycine II: theory and experiment. *Science*, 190:458–459.

Finney, J.L., (1986). The role of water perturbations in biological processes. In *Water and Aqueous Solutions*. (G.W. Neilson and J.E. Enderby, editors). pp227–244. Adam Hilger, Bristol.

Fillaux, F., Fontaine, J.P., Baron, M.-H., Kearley, G.J., & Tomkinson, J., (1993). Inelastic neutron-scattering study of the proton dynamics in N-methyl acetamide at 20K. *Chem. Phys.*, 176:249–278.

Fraser, R.D.B., MacRae, T.P., & Miller, A., (1987). Molecular packing in type I collagen fibrils. *J. Mol. Biol.*, 193:115–125.

Goodfellow, J.M., (editor) (1990). *Molecular Dynamics: Applications in Molecular Biology*. Macmillan Press, Basingstoke.

Grigera, J.R., & Berendsen, H.J.C., (1979). The molecular details of collagen hydration. *Biopolymers*, 18:47–57.

Harley, R., James, D., Miller, A., & White, J.W., (1977). Phonons and the elastic moduli of collagen and muscle. *Nature*, 267:285–287.

Hayward, R.L., Middendorf, H.D., Wanderlingh, U., & Smith, J.C., (1995). Normal mode and anharmonic dynamics in crystalline acetanilide: a combined computer simulation and inelastic neutron scattering analysis. *J. Chem. Phys.*, (in press).

Jeffrey, G.A., & Saenger, W., (1991). *Hydrogen Bonding in Biological Structures*. Springer-Verlag, Berlin.

Kadler, K., (1994). Extracellular matrix 1: fibril-forming collagens. *Protein Profile*, 1:519–612.

Karplus, M., & Petsko, G., (1990). Molecular dynamics simulations in biology. *Nature*, 347:631–639.

Kearley, G.J., Fillaux, F., Baron, M.-H., Bennington, S., & Tomkinson, J., (1994). A new look at proton transfer dynamics along hydrogen bonds in amides and peptides. *Science*, 264:1285–1289.

Krimm, S., & Bandekar, J., (1986). Vibrational spectroscopy and conformation of peptides, polypeptides, and proteins. *Adv. Prot. Chem.*, 38:181–364.

Krueger, S., Andrews, A.P., & Nossal, R., (1994). Small angle neutron scattering studies of structural characteristics of agarose gels. *Biophys. Chem.*, 53:85–94.

Langan, P., Forsyth, V.T., Mahendrasingam, A., Dauvergne, M.T., Mason, S.A., Wilson, C.C., & Fuller, W., (1995). Neutron fibre diffraction studies of DNA hydration. *Physica* B (in press).

Lovesey, S.W., (1984). *Theory of Neutron Scattering from Condensed Matter*, Vol.1. Clarendon Press, Oxford.

Martel, P., (1992). Biophysical aspects of neutron scattering of vibrational modes in proteins. *Prog. Biophys. Mol. Biol.*, 57:129–179.

Mayers, J., & Evans, A.C., (1991). Measurement of atomic momentum distribution functions by neutron Compton scattering. *Report RAL-91–048*. Rutherford Appleton Laboratory, Chilton, U.K.

McDonald, J.A., (1988). Extracellular matrix assembly. *Ann. Rev. Cell Biol.*, 4:183–208.

Middendorf, H.D., (1984a). Biophysical applications of quasi-elastic and inelastic neutron scattering. *Ann. Rev. Biophys. Bioeng.*, 13:425–451.

Middendorf, H.D., (1984b). Inelastic scattering from biomolecules: Principles and prospects. In *Neutrons in Biology.* (B.P. Schoenborn, editor). pp401–436. Plenum Publishing Corporation, New York.

Middendorf, H.D., (1992). Neutron studies of the dynamics of globular proteins. *Physica B,* 182:415–420.

Middendorf, H.D., (1994). Fibrous biopolymers: New experimental approaches using pulsed-source neutron techniques. In *Hydrogen Bond Networks.* (M.-C. Bellissent-Funel and J.C. Dore, editors). pp529–532. Kluwer, Dordrecht.

Middendorf, H.D., & Hotz de Baar, O.F.A., (1996). Simulation of small-angle scattering from complex, hierarchically structured biopolymer networks. *Molec. Phys.,* (in press).

Middendorf, H.D., & Randall, J.T., (1985). Neutron spectroscopy and protein dynamics. In *Structure and Motion: Membranes, Nucleic Acids and Proteins.* (E. Clementi, G. Corogiu, M.H. Sarma and R.H. Sarma, editors). pp219–241. Adenine Press, New York.

Middendorf, H.D., Cavatorta, F., & Deriu, A., (1990). Small-angle neutron scattering from polysaccharide gels. *Progr. Colloid Polym. Sci.,* 81:275–278.

Middendorf, H.D., Cavatorta, F., Deriu, A., & Steigenberger, U., (1989). Quasi-elastic neutron scattering from polysaccharide gels. *Physica B,* 156&157:456–460.

Middendorf, H.D., DiCola, D., Cavatorta, F., Deriu, A., & Carlile, C.J., (1994). Water dynamics in charged and uncharged polysaccharides gels by quasi-elastic neutron scattering. *Biophys. Chem.,* 47:145–153.

Middendorf, H.D., Hayward, R.L., Parker, S.F., Bradshaw, J., & Miller, A., (1995). Neutron spectroscopy of collagen and model polypeptides. *Biophys. J.,* 69:660–673.

Miller, A., (1984). Collagen: the organic matrix of bone. *Phil. Trans. R. Soc. Lond.,* B304:455–477.

Newport, R.J., Rainford, B.D., & Cywinski, R., (editors) (1988). *Neutron Scattering at a Pulsed Source.* Adam Hilger, Bristol.

Nimtz, G., Marquardt, P., Stauffer, D., & Weiss. W., (1988). Raoult's law and the melting point depression in mesoscopic systems. *Science,* 242:1671–1675.

Ohmine, I., Tanaka, H., & Wolynes, P.G., (1988). Large local energy fluctuations in water. II. Cooperative motions and fluctuations. *J. Chem. Phys.,* 89:5852–5860.

Penfold, J., & Tomkinson, J., (1986). The ISIS time-focussed crystal analyser spectrometer, TFXA. *Report RAL-86–019.* Rutherford Appleton Laboratory, Chilton, U.K.

Peticolas, W.L., (1979). Low frequency vibrations and the dynamics of proteins and polypeptides. *Methods Enzymol.,* 61:425–458.

Randall, J.T., & Vaughan, J.M., (1979). Brillouin scattering in systems of biological significance. *Phil. Trans. R. Soc. Lond.,* A293:341–348.

Richter, D., (1992). Neutron spin-echo investigations on molecular motion in polymers. *Physica B*, 182:7–14.

Rupley, J.A., & Careri, G., (1991). Protein hydration and function. *Adv. Protein Chem.,* 41:38–129.

Schoenborn, B.P., (editor) (1984). *Neutrons in Biology*. Plenum Press, New York.

Schreiner, L.J., Pintar, M.M., Dianoux, A.J., Volino, F., & Rupprecht, A., (1988). Hydration of sodium DNA by neutron quasi-elastic scattering. *Biophys. J.,* 53:119–122.

Smith, J., (1991). Protein dynamics: comparison of simulations with inelastic scattering experiments. *Q. Rev. Biophys.,* 24:227–291.

Stuhrmann, H.B., & Miller, A., (1978). Small-angle scattering of biological structures. *J. Appl. Cryst.,* 11:325–340.

Teixeira, J., Bellissent-Funel, M.-C., Chen, S.H., & Dianoux, A.J., (1985). Experimental determination of the nature of diffusive motions of water molecules at low temperatures. *Phys. Rev. B,* 31:1913–1917.

Teixeira, J., (1993). The physics of liquid water. *J. de Physique IV, Coll.,* C1–3:163–169.

Warner, M., Lovesey, S.W., & Smith, J., (1983). The theory of neutron scattering from mixed harmonic solids. *Z. Phys. B—Condensed Matter,* 51:109–126.

White, J.W., (1976). Inelastic neutron scattering from synthetic and biological polymers. In *Neutron Scattering for the Analysis of Biological Structures*, VI, pp3–26. Brookhaven Symp. Biol. Vol.27. (B.P. Schoenborn, editor). Brookhaven National Laboratory, Upton, New York.

Windsor, C.G., (1981). *Pulsed Neutron Scattering*. Taylor and Francis, London.

Zaccaï, G., (editor) (1994). Special issue on neutrons in biology. *Biophys. Chem.*, 47:1–189.

22

PROTEIN-DETERGENT INTERACTIONS IN SINGLE CRYSTALS OF MEMBRANE PROTEINS STUDIED BY NEUTRON CRYSTALLOGRAPHY

P. A. Timmins[1] and E. Pebay-Peyroula[2]

[1] ILL
BP 156, 38042 Grenoble Cedex 9, France
[2] IBS-UJF
41 Avenue des Martyrs, 38027 Grenoble Cedex 1, France

ABSTRACT

The detergent micelles surrounding membrane protein molecules in single crystals can be investigated using neutron crystallography combined with H_2O/D_2O contrast variation. If the protein structure is known then the contrast variation method allows phases to be determined at a contrast where the detergent dominates the scattering. The application of various constraints allows the resulting scattering length density map to be realistically modeled. The method has been applied to two different forms of the membrane protein porin. In one case both hydrogenated and partially deuterated protein were used, allowing the head group and tail to be distinguished.

INTRODUCTION

The high resolution structures of a number of membrane proteins have been solved by high resolution X-ray crystallography. The first of these were the photo-synthetic reaction centers from *Rhodopseudomonas viridis* (Deisenhofer *et al.*, 1985) and *Rhodobacter sphaeroides* (Arnoux *et al.*, 1989); followed by a number of porins from *Rhodobacter capsulatus* (Weiss *et al.*, 1991), the OmpF porin from *E.Coli* in a trigonal crystal form (Cowan *et al.*, 1992), PhoE by molecular replacement based on the trigonal form of OmpF (Cowan *et al.*, 1992), lamB (maltoporin) (Schirmer *et al.*, 1995) and recently a porin from *Rhodopseudomonas blastica* (Kreusch & Schultz, 1995). The structure of the *E.Coli* porin (OmpF) in a tetragonal form solved using molecular replacement techniques based on the trigonal form has also been completed (Cowan *et al.*, 1995). This crystal form was in fact the first membrane protein crystal produced which diffracted X-rays to high resolution (Garavito & Rosenbusch, 1980).

Neutrons in Biology, edited by Schoenborn and Knott
Plenum Press, New York, 1996

Each of these crystals was formed from detergent solubilised membrane protein but in no case could a significant amount of electron density attributable to detergent be identified in the crystal. In the case of the tetragonal porin (Garavito *et al.*, 1983) density equivalent to about one detergent molecule per protein monomer can be seen. A similar observation was made in crystals of the photosynthetic reaction center from *Rh. viridis* (Deisenhofer *et al.*, 1985).

The structure of the detergent in membrane protein crystals is of interest for two principal reasons: (i) it can be an element in the further understanding of the process of crystallization, and (ii) protein-detergent interactions may, at least partially, mimic protein-lipid interactions.

The only technique for studying the structure of disordered detergent in membrane protein crystals is neutron diffraction (Timmins *et al.*, 1994) where the substitution of heavy water (D_2O) for light water can increase the contrast between water, detergent and protein. By judicious choice of H_2O/D_2O mixtures the contrast can be manipulated such as to enhance the contrast of either protein or detergent, a technique commonly used in small-angle scattering (Timmins & Zaccai, 1988). The contrast can also be manipulated by specific deuteration of either or both of the other components, protein or detergent. Here we describe the application of this technique to two different porins and compare the results with those obtained for two photosynthetic reaction centers.

NEUTRON CRYSTALLOGRAPHY AND CONTRAST VARIATION

Protein/detergent complexes are a class of objects particularly suited to H_2O/D_2O contrast variation. Figure 1 shows how the contrast for a typical protein and the headgroups and hydrophobic tails of a hydrogenated and a tail deuterated detergent vary as a function of the deuterium content of the aqueous environment.

The application of contrast variation to single crystals was first described for X-rays by Bragg and Perutz (1952) and has been developed for neutrons over the past 15 years. The method is based essentially upon the fact that for a system in which the scattering length density of the macromolecule is a linear function of the deuterium content of the solvent water then the complex structure factor is also a linear function of the deuterium content of the solvent:

$$\underline{F}(hkl) = \underline{F}_O(hkl) + X\underline{F}_{HD}(hkl)$$

where $\underline{F}_O(hkl)$ is the structure factor in H_2O, and $\underline{F}_{HD}(hkl)$ is the structure factor difference between H_2O and D_2O.

This relationship allows one to determine the phase difference (with an uncertainty in sign) between one contrast and another (Roth *et al.*, 1984). If, as is often the case, the structure at one contrast is known then the absolute phase at any other contrast may be determined (with an ambiguity due to the uncertainty in sign). Thus, for example, in a protein/detergent complex the protein structure may be known from X-ray crystallography and thus the phases are known at the contrast at which the detergent is invisible. Using the contrast variation relationship we may then calculate the phases at the contrast where the protein is invisible (~40% D_2O/H_2O). This gives a scattering density map of the detergent alone. Such a map contains, of course, errors due to the phase uncertainties mentioned above and is, as such, analogous to a single isomorphous replacement map in X-ray crystallography.

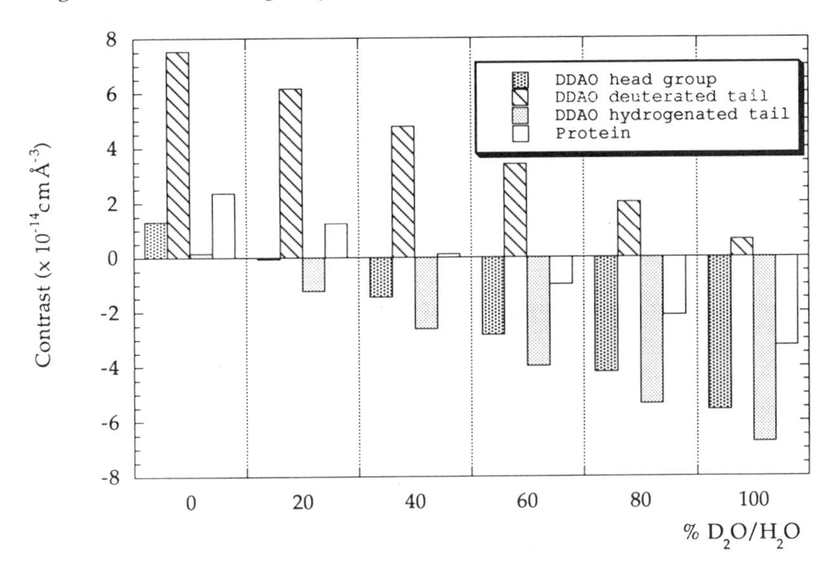

Figure 1. The variation of contrast with H_2O/D_2O content of the water for protein and for the detergent decyl-N,N'-dimethyl amine oxide (DDAO). The contrast for the detergent is calculated separately for the head group and for the hydrophobic tail. In the latter case it is also shown for the fully deuterated tail.

In order to determine the true detergent structure it is necessary to improve the maps using other constraints which allow a realistic modeling of the detergent.

Some of the constraints which may be applied are as follow:

- The volume occupied by the protein is invariant as a function of contrast and is usually known from X-ray studies.
- The fraction of the unit cell occupied by detergent may be known or realistic limits may be placed upon it.
- The aqueous solvent is a flat continuum of density and so 'solvent flattening' can be applied.
- The variation in scattering density of the protein (internal structure) is also known from the X-ray structure.

The way in which this modeling may be carried out is illustrated by the example of the detergent structure in tetragonal crystals of the Ompf porin from *E.Coli* (Pebay-Peyroula *et al.,* 1995). Here, data were measured both for crystals containing β-octyl glucoside and fully tail deuterated decyl N,N', dimethyl amine oxide. In the case of the latter detergent the head group is calculated to be contrast matched in 19% D_2O/H_2O and the deuterated decyl chain in 110% D_2O/H_2O, (a value which is, of course, experimentally unattainable but for which data may be calculated by extrapolation). Diffraction data were measured from crystals soaked in 0, 35, 70 and 100% D_2O/H_2O. Using the protein coordinates obtained from the high resolution structure determined by X-ray crystallography (Cowan *et al.,* 1995) and the phase difference information obtained from the neutron diffraction data scattering length density maps were calculated at 19, 14 and 100% D_2O/H_2O. Each of these maps, after appropriate smoothing (Roth, 1991), was modeled according to the criteria described above. The perdeuterated hydrophobic tails were modeled from the 19% D_2O/H_2O map (where the heads are invisible) and the 40% D_2O/H_2O map (where the protein is invisible). The headgroups were modeled from the 110% D_2O/H_2O map where the decyl chain is invisible. The detergent content of the unit cell was not known very precisely and the modeling was therefore carried out for a range of different possible detergent contents. For each model, structure factors were calculated and compared with the experimental structure factors via a conventional R-value. The final protein-detergent model had R-factors varying between 28 and 34% for the different contrasts compared with 41 to 66% for the protein alone.

The modeling of the hydrophobic tails was rather easier than that of the heads for a number of probable reasons. In this particular case the volume of the headgroup is rather small compared with that of the hydrophobic tail ($81Å^3$ *versus* $296Å^3$). The contrast of the hydrophobic tail (Figure 1) is very different from that of the protein whereas that of the headgroup is not so different. Finally the modeling was done using the assumption that both tails and headgroups constitute well defined condensed localized phases. Whilst this assumption probably holds rather well for the tails it is most probable that the headgroups are hydrated. Thus instead of having a scattering length difference distributed over a volume corresponding to the dry volume of the head, they may effectively occupy a volume considerably greater with a lower mean scattering length density. It has been demonstrated, for example, that in free LDAO micelles the headgroup is associated with about three water molecules (Timmins *et al.,* 1988). In this case the effective volume of the headgroup would increase from $81Å^3$ to $81 + 3\times30 = 171Å^3$ with a scattering length density $0.354 \times 10^{-14}cmÅ^{-3}$ instead of $0.748 \times 10^{-14}cmÅ^{-3}$ calculated from the unhydrated head group (Timmins *et al.,* 1994).

DETERGENT STRUCTURE IN PORINS AND REACTION CENTERS

Although the detergent structures are known in only two reaction centers (Roth *et al.,* 1989; Roth *et al.,* 1991) and two porins (Pebay-Peyroula *et al.,* 1995; Timmins, Pebay-Peyroula, Penel, Welte & Wacker, in preparation) some preliminary comparisons can be made. In the two reaction centers the detergents used were LDAO in the case of the protein from *Rhodopseudomonas viridis* and β-octyl glucoside for that from *Rhodobacter sphaeroides*. In both cases the detergent forms a ring around the trans-membrane part of the molecule although the ways in which different protein detergent complexes interact is different. The thickness of the ring is about 25Å in a direction perpendicular to the supposed membrane plane. This is very close to twice the length of the LDAO molecule and implies that the detergent molecules lie with their long axes parallel to and in contact with the hydrophobic surface of the protein. The zone of the protein which is covered by detergent defined by a band of hydrophobic residues which are delimited at the upper and lower surfaces by a region rich in arginine, histidine and tyrosine residues.

The porin structures are also characterized by detergent belts surrounding the proteins which, in this case, present a β-barrel as the hydrophobic membrane-contacting surfaces. The case of the porin from *E.Coli* is particularly interesting as the structure is known from two crystals of the same form, and hence same molecular packing, but with different detergents. In the one case the detergent was β-octyl glucoside and the other perdeuterated (d10) decyl N,N′ dimethyl amine oxide as described above. The detergent's rings around the porin trimers have almost identical geometry in the two cases (Pebay-Peyroula *et al.,* 1995), each ring being delimited by rings of tyrosine and phenyl alanine residues. Views of the detergent belts parallel and perpendicular to the three-fold axes of the porin trimer are shown in Figure 2.

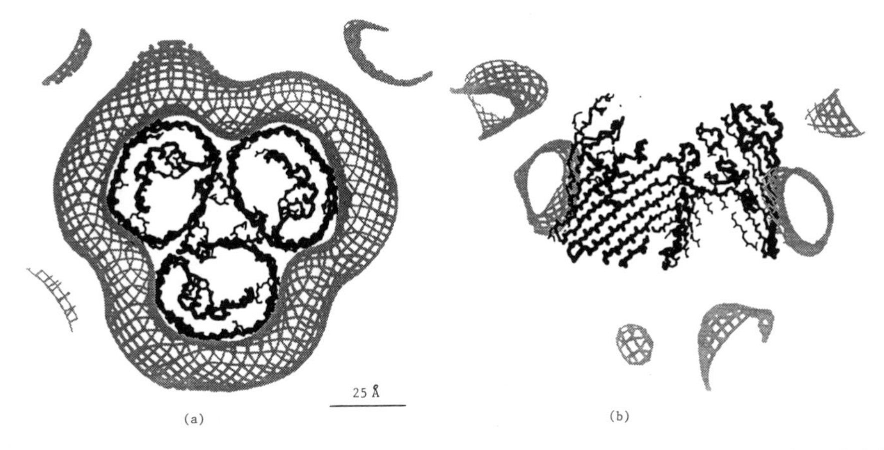

Figure 2. The distribution of detergent (DDAO or β-OG) around a trimer of the OmpF porin from *E.Coli.* (a) view parallel to the trimer three-fold symmetry axis, (b) view perpendicular to the trimer three-fold symmetry axis. The detergent represents 20% of the unit cell volume.

The packing of the protein-detergent complexes in the crystal lattice is of particular interest. If we consider the protein structure alone the crystal comprises two independent interpenetrating networks of tetrahedra of porin trimers. There are no protein-protein contacts between the two interpenetrating networks. The complete structure of the protein/detergent complex shows that these networks are joined by detergent/detergent contacts, probably via the headgroups. It appears therefore that contacts between disordered components (detergent) are necessary to maintain the atomic resolution order of the protein.

Crystals of the porin from *Rhodobacter capsulatus* have a very different packing from the three mentioned above. Moreover, the detergent used was C8E4 (Weiss *et al.,* 1991). The detergent forms rings about the porin trimers in a way very similar to that of the *E.Coli* porin. The interactions between the different components is, however, very different. The protein molecules in this crystal form are arranged in layers with all molecules pointing in the same direction. Although there are no side-to-side interactions between molecules there are sufficient protein-protein interactions between molecules in successive layers to form a three-dimensional lattice without the presence of the detergent. The neutron diffraction results (Timmins, Pebay-Peyroula, Penel, Wacker and Welte, in preparation) show that the role of the detergents is to cover the hydrophobic surfaces of the protein in the same way as for the *E.Coli* protein and presumably in this way to stabilise the crystal. The surface covered by the detergent is delimited again by two rings of aromatic residues.

CONCLUSIONS

Neutron diffraction is the only way to directly visualize detergent in crystals of membrane proteins. In the four cases studied to date the detergent forms clearly defined belts around the trans-membrane part of the protein, the limits of which are demarcated by more or less well defined rings of aromatic residues which presumably determine the thickness of the lipid at the point of insertion in the real membrane. The role of the detergent in the growth of crystals is somewhat different from one form to another. In the extreme case the crystals of *E.Coli* the three-dimensional integrity of the crystal depends on detergent-detergent interactions whereas in the *Rh. capsulatus* protein the crystal can be formed from protein-protein contacts alone.

ACKNOWLEDGMENTS

We would like to thank all our colleagues who have participated in the two porin projects, Wolfram Welte and Thomas Wacker (University of Freiburg), Michael Garavito (University of Chicago), Jurg Rosenbusch (Biozentrum, University of Basel) and Martin Zulauf (Hoffmann-La Roche, Basel). We are also grateful to Michel Roth (IBS, Grenoble) for discussions and help in the use of his data reduction and analysis programs.

REFERENCES

Arnoux, B., Ducruix, A., Reiss-Husson, F., Lutz, M., Norris, J., Schiffer, M., & Chang, C.H., (1989). Structure of spheroidene in the photosynthetic reaction center from Y *Rhodobacter sphaeroides*. *FEBS Lett.*, 258:47–50.

Cowan, S.W., Garavito, R.M., Jansonius, J.N., Jenkins, J., Karlsson, R., Koenig, N., Pai, E., Pauptit, R.A.,Rizkallah, P.J., Rosenbusch, J.P., Rummel, G., & Schirmer, T., (1995). The structure of the outer membrane protein porin in a tetragonal crystal form. *Structure* (submitted).

Cowan, S.W., Schirmer, T., Rummel, G., Steiert, M., Ghosh, R., Pauptit, R.A., Jansonius, J.N., & Rosenbusch, J.P., (1992). Crystal structures explain functional properties of two *E.Coli* porins. *Nature*, 358:727–733.

Deisenhofer, J., Epp, O., Miki, K., Huber, R., & Michel., H., (1985). Structure of the protein sub-units in the photosynthetic reaction centre of *Rhodopseudomonas viridis* at 3Å resolution. *Nature*, 318:618–624.

Garavito, R.M., & Rosenbusch, J.P., (1980). Three-dimensional crystals of an integral membrane protein: an initial X-ray analysis. *J. Cell Biol.*, 86:327–329.

Garavito, R.M., Jenkins, J., Jansonius, J.N., Karlsson, R., & Rosenbusch, J.P., (1983). Solubilization, characterization and reconstitution of membrane proteins X. An X-ray diffraction analysis of an integral membrane protein: porin from *E.Coli*. *J. Mol. Biol.*, 164:312–327.

Kreusch, A., & Schultz, G.E., (1995). Refined structure of the porin from *Rhodopseudomonas blastica*. Comparison with the form from *Rhodobacter capsulatus*. *J. Mol. Biol.*, 243:891–905.

Pebay-Peyroula, E., Garavito, M., Rosenbusch, J.P., Zulauf, M., & Timmins, P., (1995). Detergent structure in tetragonal crystals of porin from the outer membrane of *E.Coli*. *Structure* (submitted).

Roth, M., Lewit-Bentley, A., & Bentley, G.A., (1984). Scaling and phase difference determination in solvent-contrast variation experiments. *J. Appl. Cryst.*, 17:77–84.

Roth, M., Lewit-Bentley, A., Michel, H., Deisenhofer, J., Huber, R., & Oesterhelt, D., (1989). Detergent structure in crystals of a bacterial photosynthetic reaction centre. *Nature*, 340:659–662.

Roth, M., (1991). Phasing at low resolution. In: *Crystallographic Computing 5*, D.M. Moras, A.D. Podjarny and J.C. Thierry, editors. (Oxford University Press, Oxford) 229.

Roth, M., Arnoux, B., Ducruix, A., & Reiss-Husson, F., (1991). Structure of the detergent phase and protein-detergent interactions in crystals of the wild-type (Strain Y) *Rhodobacter sphaeroides* photochemical reaction center. *Biochemistry*, 30:9403–9413.

Schirmer, T., Keller, T.A., Wang, Y-F., & Rosenbusch, J.P., (1995). Structural basis for sugar translocation through maltoporin channels at 3.1Å resolution. *Science*, 267:512–514.

Timmins, P.A., & Zaccai, J., (1988). Low resolution structures of biological complexes studied by neutron scattering. *Eur. Biophys. J.*, 15:257–268.

Timmins, P.A., Leonhard, M., Weltzien, H.U., Wacker, T., & Welte, W., (1988). A physical characterization of some detergents of potential use for membrane protein crystallization. *FEBS Lett.*, 238:361–368.

Timmins, P.A., Pebay-Peyroula, E., & Welte, W., (1994). Detergent organisation in solution and in crystals of membrane proteins. *Biophys. Chem.*, 53(1–2):27–36.

Weiss, M.S., Kreusch, A., Schultz, E., Nestel, U., Welte, W., Weckesser, J., & Schultz, G.E., (1991). The structure of porin from *Rhodobacter capsulatus* at 1.8Å resolution. *FEBS Lett.*, 280:379–382.

23

HYDROXYL AND WATER MOLECULE ORIENTATIONS IN TRYPSIN

Comparison to Molecular Dynamics Structures

Robert S. McDowell[1] and Anthony A. Kossiakoff[2*]

[1] Department of Bioorganic Chemistry
[2] Department of Protein Engineering
Genentech Inc.
460 Pt. San Bruno Blvd., South San Francisco, California 94080

ABSTRACT

A comparison is presented of experimentally observed hydroxyl and water hydrogens in trypsin determined from neutron density maps with the results of a 140ps molecular dynamics (MD) simulation. Experimental determination of hydrogen and deuterium atom positions in molecules as large as proteins is a unique capability of neutron diffraction. The comparison addresses the degree to which a standard force-field approach can adequately describe the local electrostatic and van der Waals forces that determine the orientations of these hydrogens. The molecular dynamics simulation, based on the all-atom AMBER force-field, allowed free rotation of all hydroxyl groups and movement of water molecules making up a bath surrounding the protein. The neutron densities, derived from 2.1Å D_2O-H_2O difference Fourier maps, provide a database of 27 well-ordered hydroxyl hydrogens. Virtually all of the simulated hydroxyl orientations are within a standard deviation of the experimentally-observed positions, including several examples in which both the simulation and the neutron density indicate that a hydroxyl group is shifted from a 'standard' rotamer. For the most highly ordered water molecules, the hydrogen distributions calculated from the trajectory were in good agreement with neutron density; simulated water molecules that displayed multiple hydrogen bonding networks had correspondingly broadened neutron density profiles. This comparison was facilitated by development of a method to construct a pseudo 2Å density map based on the hydrogen atom distributions from the simulation. The degree of disorder of internal water molecules

* To whom correspondence should be addressed.

Neutrons in Biology, edited by Schoenborn and Knott
Plenum Press, New York, 1996

is shown to result primarily from the electrostatic environment surrounding that water molecule as opposed to the cavity size available to the molecule. A method is presented for comparing the discrete observations sampled in a dynamics trajectory with the time-averaged data obtained from X-ray or neutron diffraction studies. This method is particularly useful for statically-disordered water molecules, in which the average location assigned from a trajectory may represent a site of relatively low occupancy.

INTRODUCTION

X-ray protein structures provide a wealth of information about stereochemical details of the time-averaged structure; however, it is generally difficult to sort out the relative importance of steric and electrostatic effects in defining certain conformations. An important adjunct in studying these effects is molecular mechanics and thus, the application of force-field potential calculations has become an important tool in describing a wide range of biophysical and biochemical properties of proteins. A primary strength of the approach is that it can provide a window on certain facets of protein structure and activity that are not easily accessible by direct experimentation.

The force-field potential functions forming the basis for the calculations are in a continuing state of evolution, being improved when better correlations between theory and experiment are established. At present the components that make up the energy expression used in the calculations vary in their quantitative reliability. Thus, it is crucial to develop some experimental touchstone to evaluate the accuracy of the individual energy parameters. Although very few conformations within a polypeptide chain are isolated from some long-range factors, certain categories of interactions exist that can be separated and analyzed for direct short range effects. Particularly informative groups in this type of assessment are hydroxyl hydrogens (-O-H) and water molecules (H-O-H). These groups act as electrostatic probes having degrees of freedom that allow them to orient themselves optimally within the local electrostatic environment. In X-ray diffraction, hydrogen positions are not observable, therefore assigning an orientation to these groups is not possible. On the other hand, in neutron diffraction the scattering properties of hydrogen and deuterium are sufficient to allow them to be located in density maps, facilitating the determination of the orientation of hydroxyls and waters when they are well ordered and the analysis is done at high resolution.

We report here a comparison of the positions of the exchangeable hydrogens in trypsin determined from neutron diffraction studies with the results predicted by molecular dynamics calculations based on a 140 picosecond (ps) simulation. The positions of the hydroxyl and water hydrogens have been determined using a D_2O-H_2O solvent difference map that by its nature has significantly more interpretable neutron density than conventional maps (Kossiakoff *et al.*, 1992; Shpungin & Kossiakoff, 1986). The purpose of this study is twofold: to assess the degree to which a molecular mechanics approach can adequately describe the non-covalent forces that determine hydroxyl and water orientations, and to examine the local environments within the protein that are responsible for those forces.

MATERIALS AND METHODS

Neutron Density Maps

D_2O-H_2O Difference Map Calculations. The difference map peaks used in the assignments of hydroxyl hydrogen and water molecule orientations were obtained from a

2.1Å refinement of the D_2O-H_2O trypsin data (Kossiakoff *et al.*, 1992; Shpungin & Kossiakoff, 1986). This approach has been shown to significantly improve the signal-noise ratio in Fourier maps, allowing for a more accurate assignment of labile hydrogen positions. The final R value of the trypsin structure was 0.141 (Kossiakoff *et al.*, 1992). Not all hydroxyl hydrogens were observed in these maps, because in some cases the hydroxyl oxygen itself was too disordered to be unambiguously assigned. In all, out of 52 total, 27 hydroxyl orientations could be defined with high confidence and these were used as the basis for the analysis described here.

About 300 water sites were assigned from the refined solvent density; 140 of these sites were defined in the maps as discrete peaks, while the remaining were found within less-ordered channels of density. Twelve internal waters were sufficiently highly ordered to assign hydrogen orientations to them.

Hydroxyl Hydrogen Densities. The breadth of a neutron density peak profile is determined by both the positional probability of the hydrogen and by the resolution of the data; in most cases, however, the extent of the peak is predominantly influenced by the resolution. At 2.1Å resolution, the error in assigning torsion angles from the density map has been assessed to be about 15° - 20° for the well-ordered hydroxyl hydrogens (Kossiakoff *et al.*, 1990). Water orientations were determined by using the X-ray defined position of the oxygen as a fixed pivot point and orienting the hydrogen atoms to best fit the difference density. Although a quantitative assessment of errors in orientational assignment is difficult, a reasonable estimate based on empirical fitting of the density is about 20° - 30°.

Calculation of the Molecular Dynamics Trajectory

Energy Calculations. All energy calculations were conducted using the all-atom AMBER force-field (Weiner *et al.*, 1984; Weiner *et al.*, 1986) as implemented in a modified version of the Discover program (Biosym, San Diego). Nonbonded interactions were calculated with a cutoff of 11.5Å; nonbonded interactions between 11.5Å and 13.0Å were attenuated using a 5th-order polynomial switching function (Kitchen *et al.*, 1990). A fixed dielectric constant of 2.0 was used in the electrostatic calculations.

Model Construction. The initial model of the trypsin/water system was constructed using the crystallographic coordinates of bovine trypsin (4PTP) (Chambers & Stroud, 1979) deposited in the Brookhaven Protein Data Bank (Bernstein *et al.*, 1977). The model contains 168 water molecules; 68 additional surface water molecules were added to the system as suggested by the neutron density; internal waters were numbered according to the scheme used in the X-ray structure. The resulting model was energy-minimized in two stages: initially, non-hydrogen atoms were kept rigid, while all hydrogen positions were optimized by the sequential use of steepest descents and conjugate gradient minimization to maximum gradient thresholds of 0.1kcal/Å and 0.01kcal/Å, respectively. Water oxygen atoms were subsequently allowed to move, and the system was further energy-minimized using conjugate gradients until the maximum gradient was less than 0.001kcal/Å.

Rigid Rotor Calculation. For internal hydroxyl groups not associated with water molecules, approximate torsional energy surfaces were calculated by sampling the static potential energy of the system as the hydroxyl hydrogen was rotated in 10° intervals about the $C\text{-}O_{(hydroxyl)}$ bond. All other atoms were fixed in their energy-minimized coordinates.

The relative energies at each rotor angle thus reflected the combined torsional, electrostatic, van der Waals, and H-bonding components of the AMBER force-field.

Dynamics Simulation. The dynamic properties of internal water molecules and hydroxyl groups within trypsin were studied using a constrained molecular dynamics simulation. Heavy atoms of the protein were fixed at their crystallographic coordinates, while the oxygen atoms of surface water molecules were tethered to their minimized coordinates using a one-sided square well potential. A tethering potential simultaneously was used allowing surface water molecules free mobility within a 6.0Å-diameter sphere but prevented them from 'boiling away'. Internal water oxygens and all hydrogen atoms were allowed to move freely.

Starting from the minimized structure, the system was allowed to 'warm' from 0K to 300K over a 20ps interval, and maintained at 300K for 100ps. An integration step of 0.001ps was used. Structures were sampled during the post-equilibration phase of the calculation from 60 to 140ps every 0.2ps. Convergence was tested by monitoring the system energy and the relative contribution of the tethering potential during the sampling period.

RESULTS

Construction of Water Model

Following energy minimization, the average displacement of crystallographic water molecules from their starting positions was 0.52 ± 0.31Å. Surface molecules were displaced an average of 0.56Å, while internal molecules were displaced an average of 0.33Å. The most significant displacements, observed in surface waters-325 (1.95Å) and 363 (1.96Å), removed unfavorable van der Waals contacts present in the 4PTP model. Following minimization, all protein residues and water molecules had net negative nonbonded energies.

Distribution of Hydroxyl Orientations

The well-ordered hydroxyl groups in trypsin were divided into two categories: groups that are internal to the molecule and interact either with the protein itself or with buried water molecules, and groups that are external and exposed to surface waters.

Trypsin has only four internal hydroxyl groups that are not associated with a water molecule: Ser-37, Ser-54, Thr-241, and Tyr-172. For these hydroxyl groups, it was possible to directly compare the neutron densities and static energy surfaces obtained from a rigid-rotor scan about the $C\text{-}O_{(hydroxyl)}$ bond with the dihedral angle distributions obtained from the dynamics simulation. The torsion angles resulting from the simulation generally cluster around the minimum on the potential energy surface and are near the maxima in the observed neutron density (Figure 1). In the case of Ser-54, the hydroxyl rotamer is shifted from a staggered value in order to optimize the hydrogen-bonding distance with a backbone oxygen.

The hydroxyl orientations determined by sampling the neutron density were graphically compared to the results of the dynamics simulation using a polar scatter plot, in which variations in a dihedral angle over the trajectory are mapped to a circle whose radius is proportional to the simulation time (Swaminathan *et al.,* 1990). Shown in Figure 2 are polar plots for the internal hydroxyl hydrogens in trypsin, including the four residues

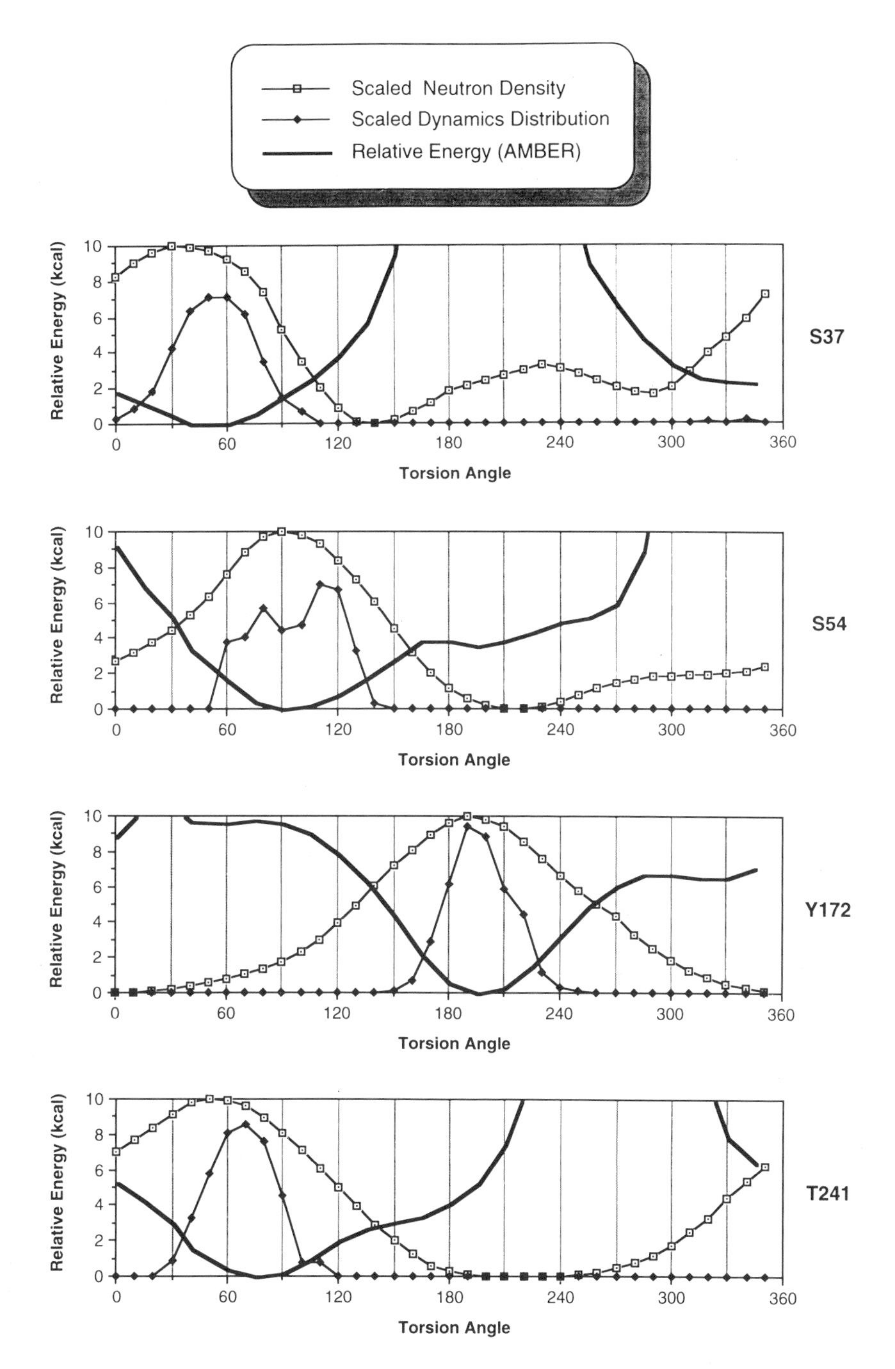

Figure 1. A comparison of neutron density peak profiles, dynamics distributions, and rotational energy surfaces for internal hydroxyl groups in trypsin. For internal hydroxyl groups that are not coordinated to water molecules, the neutron-density peak profiles (□) are compared with the dihedral angle distributions observed in the dynamics simulation (■), and the relative potential energy calculated by a rigid-rotor sampling (thick solid line). The local minima in the potential surface correlate well with the maxima observed in the neutron density, and are also indicative of the dominant conformations produced by molecular dynamics.

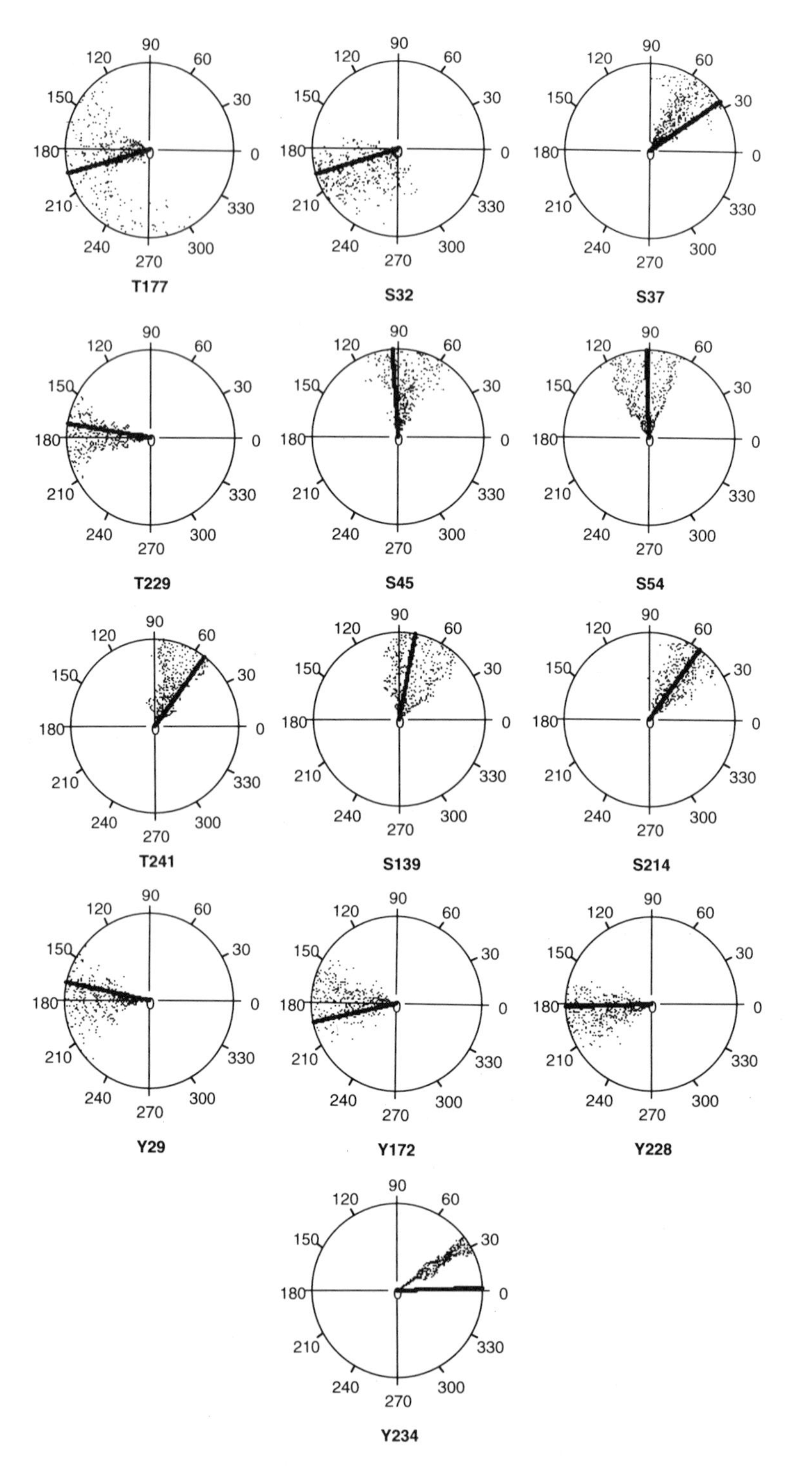

Figure 2. Polar scatter plots illustrating hydroxyl torsion angles. For internal hydroxyl groups within trypsin, the dihedral angles observed in the dynamics trajectory are mapped to a polar scatter plot whose radius is proportional to the simulation time following equilibration. The dihedral angle inferred from the neutron density maps is indicated as a thick line. The dihedral angles generated by the simulation do not display a strong time-dependence, further indicating an equilibrated system.

Table I. Dihedral angles and hydrogen-bonding geometries of internal hydroxyl groups. 'Distance' refers to the OH-acceptor distance, while 'Angle' refers to the OH-acceptor angle

Hydroxyl Group	Neutron χ	Simulation c		H-bond Acceptor	Distance (Å)		Angle (°)	
		Avg.	σ		Avg.	σ	Avg.	σ
Ser-32	−163.	−143.	28.	Wat-344 O	2.58	0.80	147.	22.
Ser-37	35.	54.	13.	Ser-37 O	1.92	0.09	131.	8.
Ser-45	94.	82.	14.	Wat-358 O	1.91	0.09	162.	10.
Ser-54	92.	97.	20.	Gly-43 O	2.24	0.06	156.	12.
Ser-139	79.	73.	21.	Gln-30 $O_{\varepsilon 1}$	1.83	0.08	137.	10.
Ser-214	54.	54.	12.	Asp-102 $O_{\delta 2}$	1.74	0.04	164.	10.
Thr-177	−164.	−169.	30.	Asn-100 $O_{\delta 1}$	1.85	0.12	147.	15.
Thr-229	171.	178.	12.	Asp-102 O	1.82	0.08	156.	14.
Thr-241	53.	68.	15.	Trp-237 O	1.88	0.11	145.	13.
Tyr-29	168.	178.	14.	Wat-314 O	2.16	0.29	160.	11.
Tyr-172	−167.	177.	15.	Pro-225 O	1.76	0.08	147.	12.
Tyr-228	−178.	−171.	14.	Wat-347 O	1.96	0.15	155.	10.
Tyr-234	2.	33.	4.	Asn-101 O	1.76	0.03	75.	4.

not coordinated to water. Also shown for comparison is the analogous plot for a disordered surface residue. Superimposed on each graph is a solid line indicating the torsion angle at which the maximum value of the neutron density is observed. The distribution of the points in the plots indicates that none of the hydroxyl rotors displays a significant time-dependence of its orientation, further suggesting an equilibrated system.

Table I lists a comparison of the mean and standard deviation of each internal hydroxyl torsion angle with the optimum angle inferred from the neutron density; also listed are statistics describing the donor-acceptor geometries that determine the hydroxyl orientations. For most of the internal hydroxyl groups, the difference between the mean dihedral angle observed in the dynamics simulation and the optimum angle observed in the neutron density is within experimental error (± 20°). The standard deviation of each angle's fluctuation during the trajectory is likewise of a similar order as the experimental error.

The hydroxyl hydrogens of Ser-32 and Thr-177 display the largest degree of fluctuation throughout the trajectory (Table I); although these hydrogens interact with protein oxygens, they are also coordinated to disordered water molecules that are exposed to the surface of the protein through water networks. This fluctuation may largely be due to the treatment of surface water as a flexible 'shell', which does not adequately reproduce the surface environment found in the crystal. There is also evidence in the density map for partial protonation of the $N_{\delta 1}$ nitrogen of His-40, which could also affect the hydrogen-bonding mode of Ser-32.

Four of the surface residues, Ser-146, Ser-150, Thr-21, and Tyr-39 are ordered primarily through hydrogen bonding to neighboring donor atoms since they lack consistent acceptor atoms that associate with the hydroxyl hydrogen, yet show reasonable agreement with the neutron density. Three of these residues, Ser-146, Thr-21, and Tyr-39, display hydroxyl conformations that maximize the distance between the hydroxyl hydrogen and a neighboring donor hydrogen that coordinates to the hydroxyl oxygen atom. The hydroxyl conformation of Ser-150 is likely determined by electrostatic repulsion with the nearby side chain of Lys-145. Both Ser-150 and Tyr-39 have sets of hydrogen-bonding partners that only coordinate with the hydroxyl group during approximately 50% of the trajectory, yet these residues maintain a fairly well-defined conformation. Four other surface residues, Ser-170, Thr-26, Thr-134, and Tyr-20, lack corresponding donors that associate with the hydroxyl oxygen. The conformations simulated for these residues are likewise close to the experimentally-observed values. Tyr-151, which lacks potential hydrogen-bonding partners, maintains its conformation primarily for steric reasons: the alternate conformation (χ = 180°) is calculated to be > 1.0kcal/mol higher in energy due to crowding from the side chain methylene group of Gln-192.

Internal Water Orientations

Unlike the hydroxyl hydrogens that were anchored by the fixed parent oxygen atom, the internal water molecules in the simulation were allowed to move freely, subject only to the nonbonded restraints imposed by the protein structure. The distances between the crystallographic locations of internal water molecules and the calculated centers of mass from the trajectory were used to assess the adequacy with which the simulation described those nonbonded interactions. These distances are listed in Table II, along with the standard deviations of motion, cavity sizes, and crystallographic temperature factors. Because environmental differences are expected to produce variations in the degree of disorder of internal water molecules, comparisons of crystallographic and calculated water positions should take into account the relative disorder of the molecules, which is given by their standard deviations. As shown in Figure 3, the correlation between the two is nearly linear: the distance between a water's crystallographic position and its calculated center of mass is within one standard deviation of the molecule's motion. No such correlation is observed between the crystallographic temperature factors and the standard deviations of motion. Although some of the discrepancy in the temperature factors may reflect regions of flexibility of the protein that are not reproduced in this simulation, these effects are expected to be small.

Table II. Dynamics profiles of internal water molecules

	Distance		
Water ID	X-ray-data (Å)	σ	Cavity Volume (Å^3)
314	0.35	0.53	14.52
328	0.13	0.33	10.18
347	0.10	0.31	15.90
358	0.06	0.20	10.75
269	0.59	0.56	11.95
346	0.33	0.56	11.95
275	0.10	0.16	8.55
326	0.07	0.20	8.55
352	0.30	0.56	11.95
353	0.20	0.43	11.95
307	0.77	0.96	N.D.
330	0.56	0.76	N.D.
331	1.15	1.17	N.D.
334	1.46	1.46	N.D.
339	0.56	0.57	N.D.
340	0.68	1.43	N.D.

Single Internal Waters. Trypsin has four single internal waters: 314, 328, 347, and 358 that interact solely with protein groups. The most disordered of these molecules, water-314, cannot simultaneously form strong hydrogen bonds with the acceptor atoms of Ser-139 and Leu-137, which are 5.45Å apart. During the simulation, the hydrogens of water-314 rapidly exchanged hydrogen-bonding partners, as reflected qualitatively by the somewhat diffuse envelope for both the calculated and experimental map contours shown in Figure 4 (a,b). By contrast, water-328 is more ordered, and is capable of simultaneously forming strong hydrogen-bonds with the N-terminus of the protein (2.1Å) and with the carbonyl oxygen of Gly-142 (2.0Å). The mean hydrogen-bonding distance to Gly-140 (2.2Å) indicates a weaker hydrogen-bond. Water-347 (Figure 4 (c,d)) is ordered, although it occupies the largest cavity (15Å^3). The acceptor atoms of Asp-189 and Ala-183 are sufficiently close (4.71Å) that hydrogen-bonds between water-347 and its partners are retained throughout the trajectory. Although water-358 is the most highly-ordered single water in the simulation, and displays the closest agreement between the averaged center-of-mass location and the crystallographic coordinates (0.06Å), it shows the poorest agreement between the simulated and neutron density contours.

Internal Water Clusters. Trypsin contains three sets of internal 2-water clusters that can be treated using a similar analysis (Figure 5). In the simulation waters-275 and 326 are highly-ordered and have well-defined hydrogen positions, as evidenced by the contours illustrated in Figure 5 (a,b). They display consistently strong hydrogen-bonding networks and occupy a compact cavity (8.5Å^3/water). Waters-269 and 346, which have a more diffuse neutron density (Figure 5 (c,d)), have fairly disordered trajectories. Their center of mass locations taken from the trajectory differ from the X-ray coordinates by

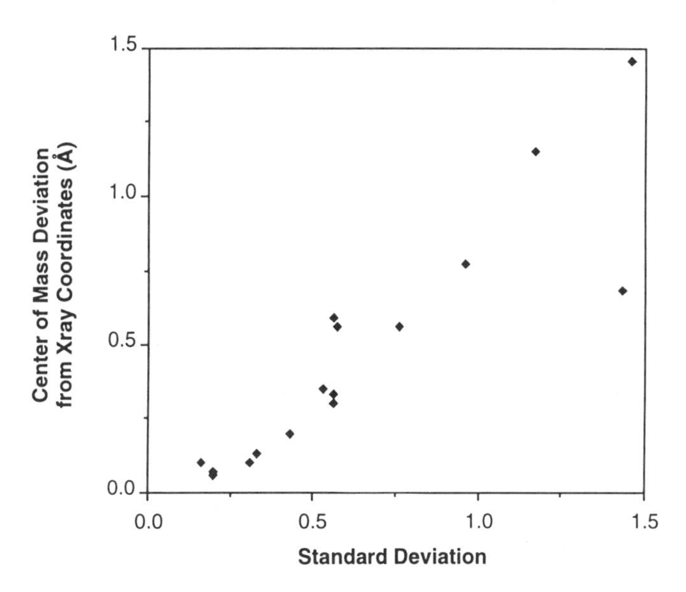

Figure 3. Deviation from X-ray coordinates versus degree of disorder for internal water molecules within trypsin. For each internal water molecule in trypsin, the deviation between the time-averaged center-of-mass determined from the dynamics trajectory and the location of the water molecule determined from X-ray crystallography is compared to the standard deviation of that molecule's motion during the simulation. The comparison is nearly linear, indicating that the discrepancy between the crystallographic and simulated water positions is within one standard deviation of the molecule's motion.

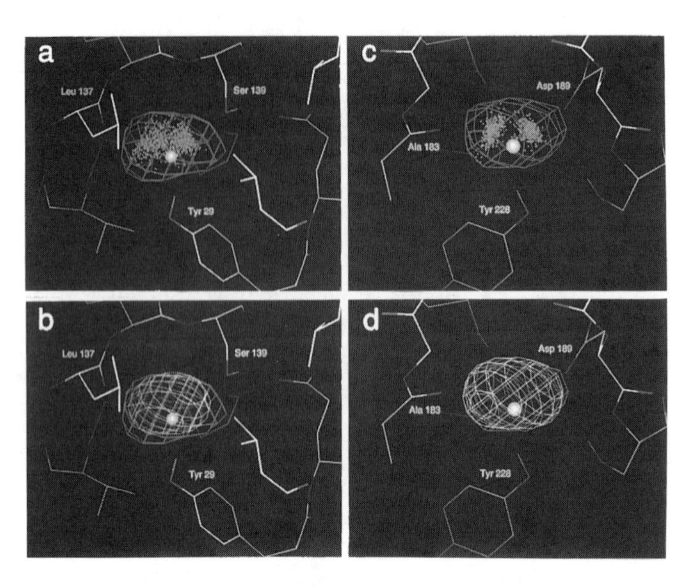

Figure 4. Neutron density and contoured hydrogen distributions for internal waters-314 (a-b) and 347 (c-d). Depicted in panels (a) and (c) are the sampled hydrogen coordinates from the dynamics trajectory (light blue) superimposed on the neutron density (green contour). A 'time-averaged' depiction of these trajectories (pink contours), obtained by summing Gaussian functions representing the instantaneous hydrogen coordinates, is compared to the neutron density in panels (b) and (d). X-ray water positions are illustrated as yellow spheres. (A color version of this figure appears in the color insert following p. 214.)

0.59Å and 0.33Å, respectively. Water-269 can potentially form hydrogen-bonds with several acceptors in the cavity: the backbone oxygens of Tyr-184, Gln-221, and Lys-224, and the $O_{\delta 1}$ oxygen of Asp-189. If the distances between each of these acceptors and the closest hydrogen on water-269 are used as a metric for cluster analysis, 88% of the sampled structures can be represented by one of two major configurations. A total of 12 structures from the trajectory were minimized; all converged to one of these two major configurations. Two minor configurations account for the remaining 12% of the trajectory.

Waters-352 and 353 display a more striking example of multiple hydrogen-bonding configurations, which is shown in Figure 6. Cluster analysis using distances between the hydrogens of water-352 and the potential acceptor atoms in the cavity (the amide oxygen of Leu-99 and the side chain oxygens of Ser-214 and Thr-229) indicates that two hydrogen-bonding configurations account for over 95% of the structures sampled in the trajectory.

Both the 269/346 the 352/353 clusters are moderately disordered due to multiple hydrogen-bonding configurations; however they convey distinctly different behaviors during the trajectory. The rapid exchange of hydrogen-bonding partners accounts for the seemingly random hydrogen positions of water-269, which switches between interactions to Gln-221 O and Asp-189 $O_{\delta 1}$. Water-346 retains the same hydrogen-bonding partners and has localized (though diffuse) hydrogen coordinates. Similar to water-269, water-352 is statically disordered; however, the water oxygen and hydrogens alternate between two more well-defined states than was the case for water-269. Water-353 maintains a consistent hydrogen-bonding network. The static disorder of the water-352 oxygen between hydrogen bonding environments is seen in the trimodal hydrogen distributions and bimodal oxygen distributions illustrated in Figure 6. There are two distinct positions, neither of which coincide with the X-ray determined position (yellow sphere). However, the calcu-

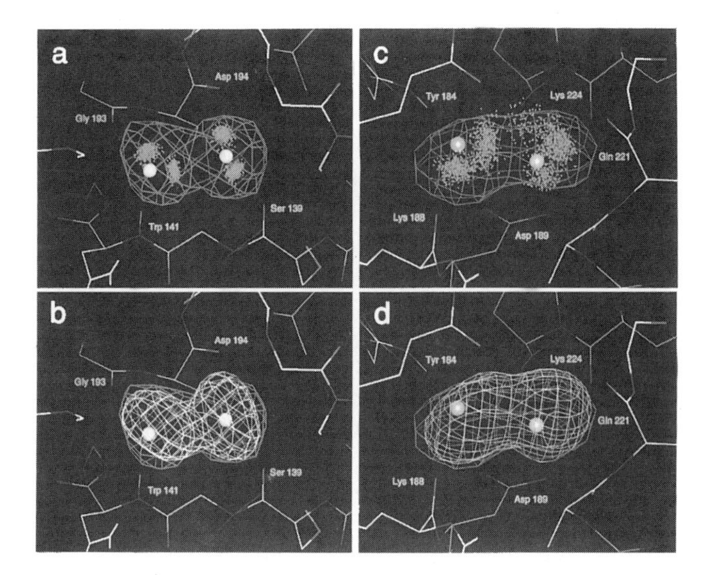

Figure 5. Primary hydrogen-bonding configurations of internal water clusters 275/326 (a-b) and 269/346 (c-d). The representations are identical to those used in Figure 4. In panels (a)-(b), water-275 is on the left side; in panels (c)-(d) water-346 is on the left. (A color version of this figure appears in the color insert following p. 214.)

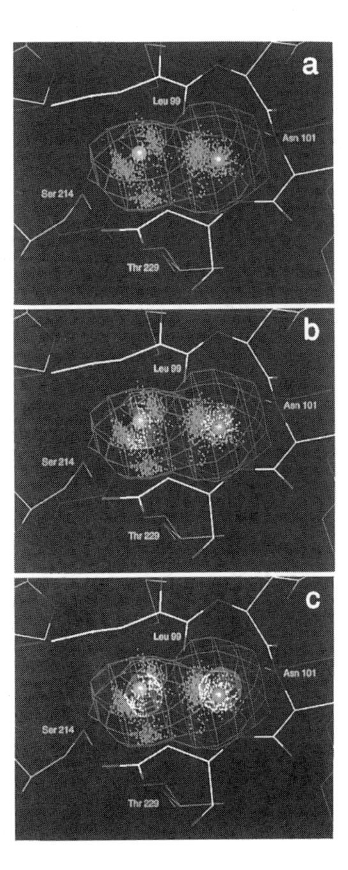

Figure 6. Primary hydrogen-bonding configurations of internal water cluster 352/353. In panel (a) the hydrogen coordinates for waters-352 (left) and 353 are superimposed on the neutron density as in Figures 4 and 5. Scatter plots of the oxygen coordinates are illustrated in red in panel (b). Note that water-352 has a trimodal hydrogen distribution and a bimodal oxygen distribution, indicating static disorder between two stable hydrogen-bonding configurations. If the oxygen distributions are contoured to produce a 'time averaged' representation, the loci of the peaks correspond well with the crystallographic positions assigned to the waters (red contours; panel c). (A color version of this figure appears in the color insert following p. 214.)

lated time-averaged gaussian contour (panel c) shows a good correspondence to the X-ray position.

DISCUSSION

The experimental determination of hydroxyl and water molecule orientations using neutron solvent difference maps provides a unique database to evaluate the accuracy to which force-field dependent calculations can reproduce experimental observations. Overall, the agreement between the observed and calculated orientations is good indicating that in most cases the calculated short range energies are accurate to within experimental error. We note that several other groups likewise have recently attempted to correlate calculated hydroxyl positions with those observed in neutron diffraction studies of several proteins (Bass *et al.,* 1992; Brünger & Karplus, 1988). These studies had mixed results. There were, on average, more instances of disagreement between calculated and observed orientations than were reported for trypsin and subtilisin by this laboratory (Kossiakoff *et al.*, 1990; Kossiakoff *et al.,* 1991). The study here differs in several key aspects. Firstly, the database is derived from assignments determined from D_2O-H_2O neutron difference maps, which effectively increases the signal to noise ratio twofold compared to conventional refinement results. The refinement method used to determine the hydroxyl orientations was wholly unbiased in that it involved no restrictions concerning the distance or orientation of the hydroxyl hydrogen relative to the parent oxygen (Kossiakoff *et al.,* 1992). Secondly, only well-ordered groups are included in the database; while this decreases the size of the database slightly (about 20%), it greatly improves the reliability of the correlations. Thirdly, the calculated hydrogen positions are obtained using a molecular dynamics simulation, which reduces the problem of the hydrogen orientation being trapped in a false minimum and better reflects the time-averaged positions observed in the neutron difference maps.

Hydroxyl Rotor Orientations; Observations and Conclusions

The hydroxyl rotor analysis showed that the well ordered hydroxyl groups have characteristically diverse energy profiles - some having quite deep and narrow energy wells, others having broad minima. The diffuseness of the neutron peaks generally tracks with the character of the energy profiles. Overall, the shapes of the profiles are a good indicator of the character of local forces influencing the hydroxyl group's orientation. In most instances (ie Ser-54) where the potential energy surface predicts the hydroxyl orientation to deviate from a staggered conformation, the neutron density supports the assignment. With the exception of Tyr-234, the dihedral angles of internal hydroxyl groups generated by the simulation tend to cluster around the locus of maximum neutron density. For the hydroxyl groups not coordinated to water molecules, the noncovalent forces exerted by the rigid protein structure are primarily responsible for the unique local minima observed in the rigid-rotor energy surfaces. These local minima are sufficiently well described by the force-field approach that both simple scanning calculations and molecular dynamics simulations reproduce the experimental hydroxyl group orientations.

The orientations of hydroxyl hydrogens that are coordinated to water molecules are a product of forces exerted by both the static protein structure and the adjacent water molecules. The accurate prediction of torsional angles for solvated hydroxyl groups therefore implies a reasonably balanced description of the dynamic noncovalent forces that simultaneously determine both the alignment of bound water molecules and the orientations of side chain hydroxyl groups that interact with those water molecules. A significant number of surface hydroxyl groups were ordered in both the neutron density and in the simulation. The analysis suggests hydroxyl groups that lack specific acceptor atoms may still be ordered due to electrostatic repulsion from nearby donor atoms, while hydroxyl groups lacking rigid donors and acceptors (eg Tyr-20) may still be ordered, if they interact with an acceptor atom that is fairly localized.

An energy scanning analysis of the hydroxyl rotors in the subtilisin neutron structure likewise showed a strong correlation between the observed and calculated orientations (Kossiakoff *et al.,* 1991). In subtilisin there was a set of five well ordered hydroxyls that indicated somewhat high-energy orientations. In each case, the orientations were predicted to within 10° - 15° by the energetic criteria in the force-field calculation. Combining the trypsin and subtilisin results for hydroxyl groups principally affected by other protein groups several general conclusions are drawn: (i) non-bonded energy terms have the largest influence on the orientations; (ii) in situations where the H-bonding energy opposes the energy for the inherent barrier to rotation, the H-bonding effect dominates; and (iii) the orientation of a hydroxyl is more highly influenced when the hydroxyl acts as a donor than as a H-bond acceptor.

Water Molecule Orientations

The model of protein hydration determined both experimentally (Otting *et al.,* 1992; Otting & Wüthrich, 1989) and by other MD simulations (Ahlström *et al.,* 1988; Brunne *et al.,* 1993; Brooks & Karplus, 1989; Gunsteren *et al.,* 1983) is one where the lifetime of any particular water at a specific site is short, on the order of picoseconds. This is even the case for surface waters involved with extensive hydrogen bonds. What is observed is that while one water is displaced from/or moves along the surface, another water immediately takes its place; a hydrogen bonding site is rarely unoccupied. This description corresponds to what is observed in the simulation reported here. There is also agreement in the observation that the 'buried' waters are much slower to exchange and do not do so during the limited duration of the trajectory.

To introduce the discussion that follows we emphasize the difference between time-averaged and dynamic disorder. The 'time-averaged' disorder of a water determined from a diffraction experiment is quite distinct from the 'lifetime' of a water at a particular site as assessed by NMR measurements or MD simulations. The diffraction experiment is the measure of the occupancy of water binding at a site: not necessarily the same water molecule, but any water molecule. Thus, if a water is exchanged out of a site after a 10ps lifetime and is replaced with another water in 1ps (a situation that describes a 'dynamic' disorder event by the time-averaged definition), the site has essentially a full occupancy. In the following discussion we have chosen to follow convention and use the term 'disordered water molecules' to mean 'disordered binding site'.

For most of the highly ordered waters, the water hydrogen distributions calculated from the trajectory were in qualitative agreement with the neutron density. The primary discrepancies between the simulation and experimental observation were attributable to differences between the surface environment in the crystal and the flexible shell model used in the simulation. An important observation picked out from the simulation and corroborated by inspection of the neutron maps is that there are water molecules whose oxygen atom keeps a constant hydrogen bonding interaction to a particular donor, while the hydrogens flip between different acceptors. Even in a high resolution X-ray structure analysis, the water, as described by the density, will appear highly ordered because only the oxygen position is observed. In actual fact, the water orientation within the hydrogen bonding environment is highly variable.

CONCLUDING REMARK

By most measures, the comparisons noted in this study between theory and experiment are impressive. In almost every instance, the prediction of hydroxyl rotamer orientations are within experimental error. This is the case whether the hydroxyl functions as both a hydrogen bond donor and acceptor, as an acceptor alone or as a donor alone. In instances where the calculation shows evidence that the orientation is perturbed out of the staggered conformer, the neutron density corroborates the assignment. The calculated orientations of most the highly ordered internal water molecules agreed with the shapes of their respective neutron density peaks. The correspondence between the hydrogen distributions in the simulation and the neutron density peaks was sufficient to assign confidently those waters that have multiple hydrogen bonding orientations. As stated in the INTRODUCTION the goal of the study presented here was to provide new and unique experimental data to judge the accuracy of several important parameters used in the forcefield. We feel to a large extent that this goal was achieved; however, the database needs to be further expanded. A limitation is the availability of high quality neutron diffraction structural data. In this regard the future is brighter than the past considering the planned improvements in the few available neutron sources. It is hoped that neutron structure information to complement that presented here will be available in the future.

REFERENCES

Ahlström, P., Teleman, O., & Jöhsson, B., (1988). Molecular dynamics simulation of interfacial water structure and dynamics in a parvalbumin solution. *J. Am. Chem. Soc.*, 110:4198–4203.

Bass, M.B., Hopkins, D.F., Jaquysh, W.A., & Ornstein, R.L., (1992). A method for determining the positions of polar hydrogens added to a protein structure that maximizes protein hydrogen bonding. *Proteins: Struct. Funct. Genet.*, 12:266–277.

Bernstein, F.C., Koetzle, T.F., Williams, G.J.B., Meyer, E.F., Brice, M.D., Rodgers, J.R., Kennard, O., Shimanouchi, T., & Tasumi, M., (1977). The protein data bank: A computer-based archival file for macromolecular structures. *J. Mol. Biol.*, 112:535–542.

Brünger, A.T., & Karplus, M., (1988). Polar hydrogen positions in proteins: empirical energy placement and neutron diffraction comparison. *Proteins: Struct. Funct. Genet.*, 4:148–156.

Brunne, R.M., Liepinsh, E., Otting, G., Wüthrich, K., & Gunsteren, W.F.v., (1993). Hydration of proteins. A comparison of experimental residence times of water molecules solvating the bovine pancreatic trypsin inhibitor with theoretical model calculations. *J. Mol. Biol.*, 231:1040–1048.

Brooks, C.L., & Karplus, M., (1989). Solvent effects on protein motion and protein effects on solvent motion. Dynamics of the active site region of lysozyme. *J. Mol. Biol.*, 208:159–181.

Chambers, J.L., & Stroud, R.M., (1979). The accuracy of refined protein structures, a comparison of two independently refined models of bovine trypsin. *Acta Cryst.*, B35:1861–1874.

Gunsteren, W.F.v., Berendsen, H.J.C., Hermans, J., Hol, W.G.J., & Postma, J.P.M., (1983). Computer simulation of the dynamics of hydrated protein crystals and its comparison with X-ray data. *Proc. Natl. Acad. Sci. USA*, 80:4315–4319.

Kitchen, D.B., Hirata, F., Westbrook, J.D., Levy, R., Kofke, D., & Yarmush, M., (1990). Conserving energy during molecular-dynamics simulations of water, proteins, and proteins in water. *J. Comp. Chem.*, 11:1169–1180.

Kossiakoff, A.A., Shpungin, J., & Sintchak, M.D., (1990). Hydroxyl hydrogen conformations in trypsin determined by the neutron diffraction solvent difference map method: relative importance of steric and electrostatic factors in defining hydrogen-bonding geometries. *Proc. Natl. Acad. Sci. USA*, 87(12):4468–4472.

Kossiakoff, A.A., Sintchak, M.D., Shpungin, J., & Presta, L.G., (1992). Analysis of solvent structure using neutron D_2O-H_2O solvent maps: pattern of primary and secondary hydration of trypsin. *PROTEINS: Struct. Funct. Genetics,* 12(3):223–236.

Kossiakoff, A.A., Ultsch, M., White, S., & Eigenbrot, C., (1991). Neutron structure of subtilisin BPN: effects of chemical environment on hydrogen-bonding geometries and the pattern of hydrogen-deuterium exchange in secondary elements. *Biochemistry,* 30(5):1211–1221.

Otting, G., Liepinsh, E., & Wüthrich, K., (1992). Polypeptide hydration in mixed solvents at low temperatures. *J. Am. Chem. Soc.,* 114:7093–7095.

Otting, G., & Wüthrich, K., (1989). Studies of protein hydration in aqueous solution by direct NMR observation of individual protein-bound water molecules. *J. Am. Chem. Soc.,* 111:1871–1875.

Shpungin, J., & Kossiakoff, A.A., (1986). A method of solvent structure analysis for proteins using D_2O-H_2O neutron difference maps. *Methods Enzymol.,* 127:329–342.

Swaminathan, S., Ravishanker, G., Beveridge, D.L., Lavery, R., Etchebest, C., & Sklenar, H., (1990). Conformational and helicoidal analysis of the molecular dynamics of proteins: 'curves', dials and windows for a 50 psec dynamic trajectory of BPTI. *PROTEINS: Struct. Funct. Genetics,* 8:179–193.

Weiner, S.J., Kollman, P.A., Case, D.A., Singh, U.C., Ghio, C., Alagona, G., Profeta, S.P., & Weiner, P., (1984). A new force field for molecular mechanics simulation of nucleic acids and proteins. *J. Am. Chem. Soc.,* 106:765–784.

Weiner, S.J., Kollman, P.A., Nguyen, D.T., & Case, D.A., (1986). An all-atom forcefield for simulations of proteins and nucleic acids. *J. Comp. Chem.,* 7:230–252.

24

UNDERSTANDING WATER

Molecular Dynamics Simulations of Myoglobin

Wei Gu, Angel E. Garcia, and Benno P. Schoenborn

Los Alamos National Laboratory
Los Alamos, New Mexico 87545

ABSTRACT

Molecular dynamics simulations were performed on CO myoglobin to evaluate the stability of the bound water molecules as determined in a neutron diffraction analysis. The myoglobin structure derived from the neutron analysis provided the starting coordinate set used in the simulations. The simulations show that only a few water molecules are tightly bound to protein atoms, while most solvent molecules are labile, breaking and reforming hydrogen bonds. Comparison between myoglobin in solution and in a single crystal highlighted some of the packing effects on the solvent structure and shows that water solvent plays an indispensable role in protein dynamics and structural stability. The described observations explain some of the differences in the experimental results of protein hydration as observed in NMR, neutron and X-ray diffraction studies.

INTRODUCTION

The interaction of water molecules with proteins determine the structure, stability and function of proteins. In protein crystals, nearly half of the unit cell content is solvent. The structure and dynamics of water molecules are correlated with the protein environment associated with them, and are affected by the polarity, dynamics, and the geometric arrangement of protein atoms. Water could be tightly bound to the flexible side-chains and fluctuate strongly, or could be in contact with hydrophobic groups and interact weakly by means of van der Waals interactions. There are four questions normally asked when dealing with water of hydration:

- Where are these water molecules?
- How long do they stay at specific sites?
- How strongly do they bind to protein atoms?
- How does water influence the structure and dynamics of proteins?

Neutrons in Biology, edited by Schoenborn and Knott
Plenum Press, New York, 1996

Hydrogen bonding is a very important factor when dealing with protein-solvent interactions. Hydrogen atoms in proteins cannot directly be determined by X-ray crystallography. Neutrons interact however much stronger with hydrogen and deuterium atoms, and have been used to locate hydrogen atoms in a number of protein structures. For example, in carbonmonoxymyoglobin, 89 solvent molecules with their hydrogen bonding have been determined (Cheng & Schoenborn, 1991). By contrast, NMR studies of proteins observe only very few hydration sites (Koenig *et al.*, 1992). Water molecules observed by neutron and X-ray diffraction techniques are found at sites with high occupancy, while NMR techniques observe only water molecules with long (> 300ps) residence times (Otting *et al.*, 1991).

In order to study this phenomena and evaluate the effect of crystal packing, simulations were carried out with myoglobin in the crystalline and solution state. The myoglobin coordinates derived from the neutron structure analysis provided the starting conditions in the simulations. The simulations show that only a few water molecules are continuously 'bound' to protein atoms during the whole time period of the simulation, while most solvent molecules are labile, breaking and reforming hydrogen bonds.

METHODS

The molecular dynamics simulations of myoglobin were performed using the program CHARMM (Brooks *et al.*, 1983) with its standard potential energy functions. All hydrogens atoms were explicitly included in the calculations. The TIP3 water model (Jorgensen *et al.*, 1983) was used in these calculations to model the solvent. The starting structure for the simulation, included 89 water molecules, 5 ammonia ions, 1 sulfate ion and a CO ligand as described in the neutron diffraction analysis (Cheng & Schoenborn, 1991). The coordinates of the starting structure were augmented by a surrounding, pre-equilibrated water bath at 300K and a density of 1g/cc. Water molecules overlapping with protein atoms were removed from the water bath if the oxygen atoms were closer than 2.7Å.

Simulations were performed in crystal and in solution environments. For the simulation in solution, solvent was constrained within a 9.0Å shell from the protein surface. This energy-minimized (Gu & Schoenborn, 1995) protein water system contained 2242 water molecules. The rms coordinate change for the whole protein was 0.37Å, and 0.26Å for backbone atoms. Stochastic boundary conditions were applied to the system (Brooks *et al.*, 1985) and water molecules were constrained in 2.0Å shells by harmonic forces, using 0.0kcal/mol for the innermost water molecules; to 7.0kcal/mol for the outer shell. A long cut-off distance of 14.5Å was applied in calculating all non-bonding interactions. The total simulation time for the solution case was 80ps.

For the simulation in the crystal environment, a periodic boundary condition with $P2_1$ symmetry was applied. The crystal parameters of myoglobin reported by Cheng and Schoenborn (1991) were used. The space between the two protein molecules in the unit cell was filled with 868 water molecules. After energy minimization of this system, the rms differences from the neutron structure was 0.29Å for the whole protein and 0.20Å for backbone atoms. A cut-off distance of 14.0Å was applied in calculating all non-bonding interactions. The total simulation time for crystal system was 150ps.

Both simulations were performed by numerically integrating Newton's equations of motion for all the atoms, with a step size of 0.001ps. A constant dielectric was used to calculate electrostatic interactions. A *Switching* function was used to calculate the potential energy for van der Waals interactions with a range between 9.0Å and 13.5Å (Brooks *et al.*, 1988). The potential energy function used to calculate coulomb interactions between atoms was in the form of a *Shifting* function (Brooks *et al.*, 1988). All covalent bonds involving hydrogen atoms were constrained by the SHAKE algorithm (van Gunsteren & Berendsen, 1977). Both simulations included 15ps of heating to bring the system to room temperature and 15ps to equilibrate the system. These 30ps are defined as the initial period.

RESULTS AND DISCUSSION

The averaged dynamic structure calculated from the last 50ps of the simulation for all myoglobin atoms in the solution case shows an rms deviation of 1.34Å from the neutron structure with 0.87Å for backbone atoms only. For the system in the crystal form, the averaged dynamic structure was calculated from the last 120ps simulation with an rms deviations of 1.27Å for the whole protein and 0.81Å for main chain atoms, as compared to coordinates in the original neutron structure. The averaged rms fluctuations of atoms in the backbone and in side-chains are shown in Figure 1. The fluctuations in helical regions are smaller than those in bends. As expected, fluctuations of side-chain atoms are larger than those observed for the backbone.

The radial distribution function of water molecules in the equilibrated myoglobin systems are shown in Figure 2. This distribution function is plotted as a function of the approach distance of water molecules from their interacting protein atoms, including hydrogen atoms. The peaks below 2Å correspond to the close contacts between oxygen atoms of water molecules and hydrogen atoms in polar groups of the protein. The broadness of these peaks are caused by van der Waals contacts. The second peak at 2.74Å shows the binding interactions between water molecules and non hydrogen atoms of the protein. The large peak is indicative of hydrogen bonding interactions, while the width of the peak reflects the van der Waals interactions between water molecules and non-polar groups. The first hydration layer is distinguishable in both solution and crystal systems. The higher peaks in the radial distribution function for the crystal system is related to the confinement for the water molecules caused by crystal packing and the different number of water molecules in the two cases.

Out of the 89 water molecules clearly localized in the neutron structure, 67 are in the first hydration shell. These water molecules interact with protein atoms while the other 22 water molecules are in the second hydration shell and interact directly only with other water molecules. Hydrogen bonding energies (Schoenborn *et al.*, 1994; Boobbyer *et al.*, 1989) were calculated for water molecules in the first hydration shell and are displayed in Figure 3. In these calculations, only contributions between protein and water atoms were considered. Most of these water sites have multiple interactions with protein atoms and therefore show large hydrogen bonding energies.

To compare water sites between the crystal (neutron) structure and the simulations, the position of the water molecules were evaluated relative to their relevant protein acceptor/donor atoms. To determine the position of these sites in the dynamically changing protein structure, a water molecule was considered to be occupying a site if it remained within a distance of 4.7Å from these atoms. Figure 4 shows the number of water molecules localized in the first hydration shell of the crystal structure (ie 67) that remain at the same site for a time, t, during the simulation. The number of sites occupied by water molecules in the neutron structure that show water molecules, but not always the same water

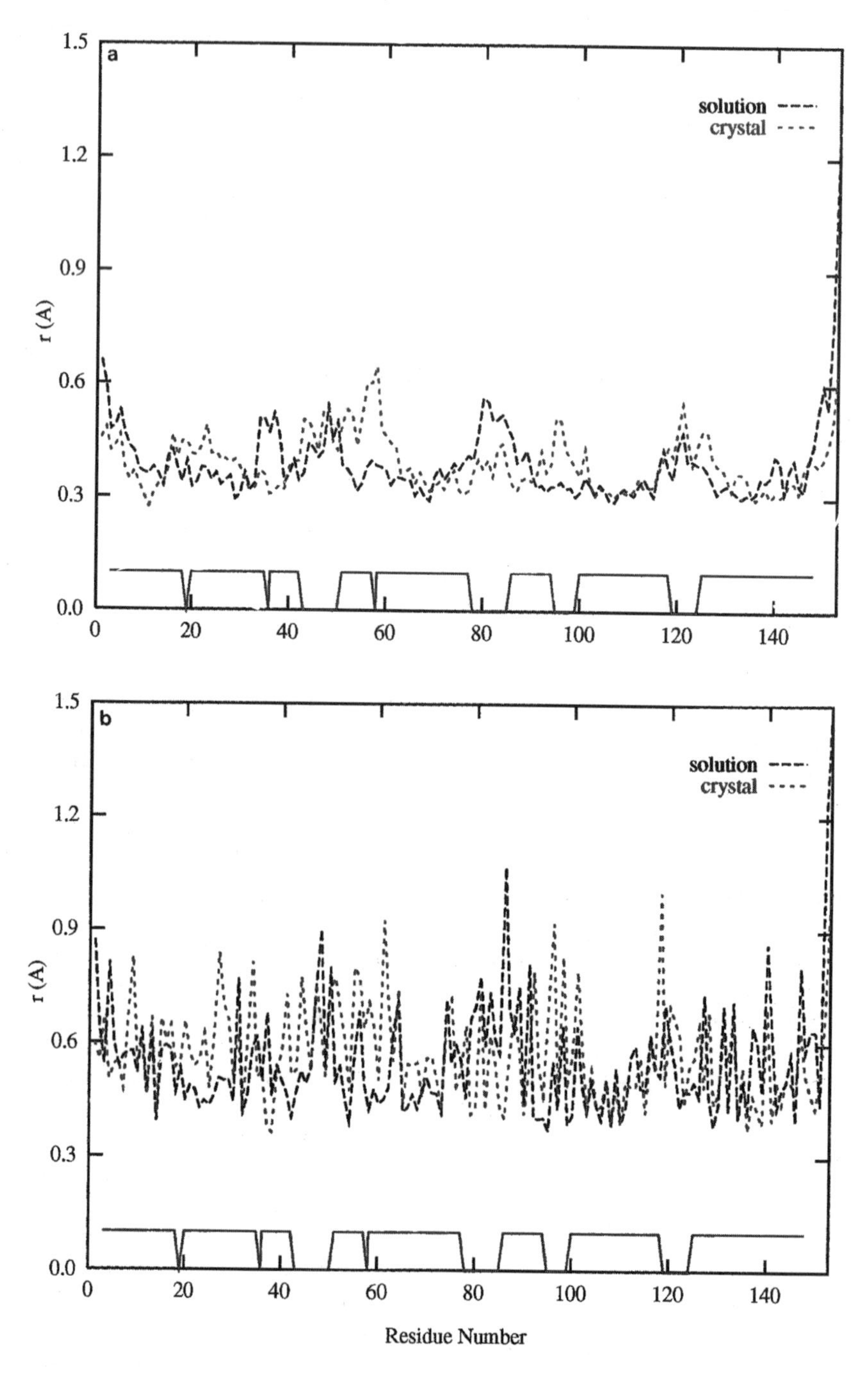

Figure 1. The rms fluctuations of protein atoms, averaged over each residue and trajectory coordinates of the simulations: (a) myoglobin backbone atoms; (b) myoglobin side-chain atoms. (A color version of this figure appears in the color insert following p. 214.)

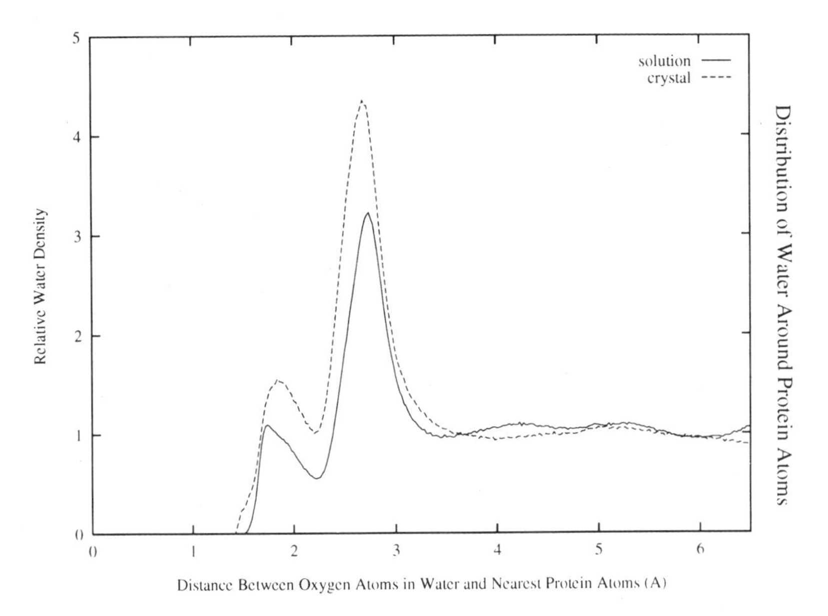

Figure 2. The radial distribution function of water around protein atoms calculated by using trajectory coordinates at an interval of 0.02ps.

molecule, at a given time, t, during the simulation are also shown in Figure 4. Fifty eight (58) of these sites are occupied by water molecules at any instant, but not necessarily by the same water molecule. Four (4) of these sites are, however, permanently occupied by the same water molecule. Most bound water molecules exist in an equilibrium state where hydrogen bonds are continuously broken and made in such a way that a hydrogen bonding loci is occupied for most of the time by a water molecule - but not necessarily the same molecule. The number of hydration sites occupied by water is much larger than the number of permanently bound sites during the 50ps simulation period. In addition to the

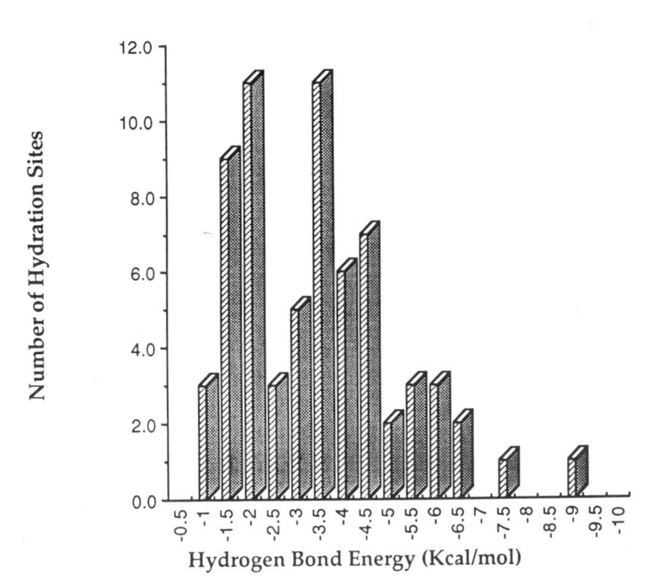

Figure 3. Hydrogen bond energies for the 67 bound water molecules as observed in the first hydration shell in the neutron diffraction analysis. The hydrogen bond energies are calculated according to Boobbyer (Schoenborn *et al.*, 1994; Boobbyer *et al.*, 1989).

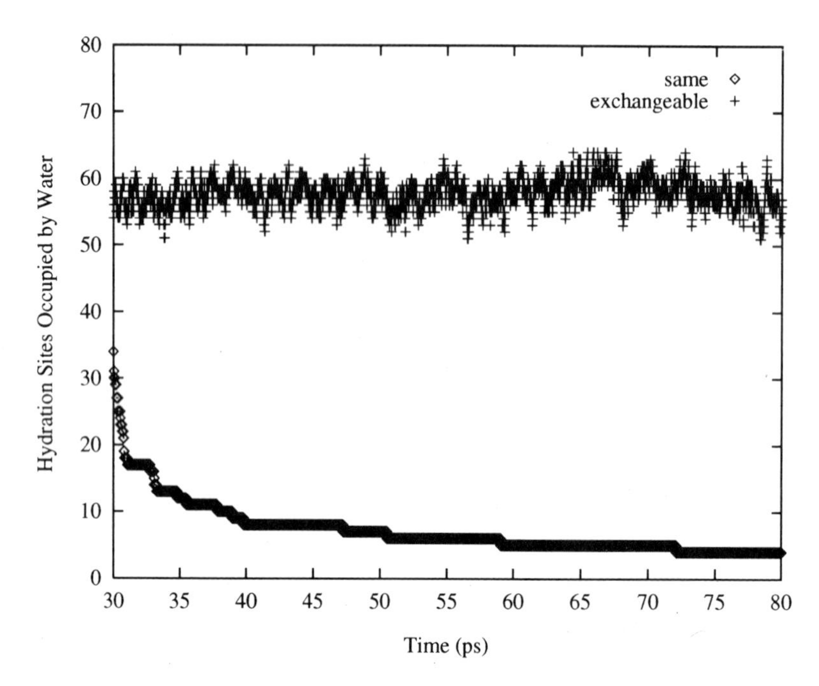

Figure 4. The number of hydration sites observed in the simulation (solution) compared to the 67 'bound' water molecules as determined in the first hydration shell of the neutron analysis. Sites depicting water that are located within 2.0Å from the original sites are included. The bottom curve shows the number of hydration sites occupied by the same water at each time step of 0.02ps. The top curve shows the number of hydration sites occupied by water (not always the same water molecule) at a given time step.

four tightly bound sites mentioned above, there are six other sites that show unexchangeable water sites but they are displaced relative to the sites found in the neutron map. We also calculated the time-averaged occupancies for each hydration site in the solution system and found that among the 67 sites in the first hydration shell, 57 sites have occupancies larger than 70%. In the crystal system there are only two unexchanged water sites remaining for the whole 120ps of the simulation. Among these two sites, one (water-23) is equivalent to that found in the solution system. The other one (water-17) binds to protein atoms but is displaced from the site occupied in solution. In the protein crystal, water-17 is bound between the two symmetry related proteins, stabilizing this configuration.

Table I shows hydrogen bond energies (kcal/mol) calculated according to Boobbyer (Boobbyer *et al.*, 1989) for the stable water molecules observed in the dynamic simulations and in the neutron map. The energy calculated includes only water to protein terms and excludes any water to water interactions. The target atoms listed are the protein atoms with the strongest interactions with the given water molecule. Hydrogen bonds are only included with energies lower than -0.5kcal/mol. Water molecules are identified according to the nomenclature used by Cheng & Schoenborn (1991).

The hydrogen bonding energies of water molecules tightly bound to sites identified in the neutron structure were calculated and averaged over 5ps. Averaged energies for selected sites are shown in Table I. Water molecules bound to these sites show large hydrogen bonding energies during the full course of the simulations as well as for the neutron structure. The average binding configurations in the simulation are similar to the neutron

Table I. Hydrogen bond energies (in kcal/mol) between selected protein side-chains and 'bound' water molecules in the MD simulation and the neutron diffraction structure

		Dynamic simulation		Neutron structure	
Water #*	Target	Energy	# H-bonds	Energy	# H-bonds
12^S	TYR-103 OH	−5.08	2	−4.65	2
16^S	ILE-75 O	−6.96	3	−6.10	3
17^C	GLU-136 $O_{\varepsilon 2}$	−7.15	3	−6.53	5
23^{SC}	GLU-4 O	-5.29^{SC}	2^{SC}	−3.93	2
33^S	HEM-154 O_{B31}	−6.49	3	−5.47	3

* ^{S} solution, ^{C} crystal

Note: The energies are calculated following Boobbyer (Boobbyer *et al.*, 1989). Water molecules are identified following the nomenclature used by Cheng and Schoenborn (1991).

structure. All water molecules bound at these sites have multiple energetically favorable interactions with protein atoms. One water molecule with multiple hydrogen bonding interactions to protein atoms is shown in Figure 5. The average binding configuration of this water molecule is the same as in the neutron structure.

An interesting event occurring during the simulation was the interchange of two bound water molecules (Gu & Schoenborn, 1995). In the neutron map, water-19 is hydrogen-bonded to aspartic acid-44, histidine-48, and another water molecule (water-35) with an energy of -8.0kcal/mol. This hydration site showed full occupancy in the simulations. After 15ps, water-19 and water-35 exchanged positions with each other. The region from residue-44 to residue-48 in the protein molecule is in a highly polar loop-region with large B-factors, suggesting large mobility. The water molecules in this region balance the local electrostatic force field of this polar loop, and the large mobility of the side chains permits the exchange of water molecules. All hydration sites in the first hydration shell show multiple bonds to protein atoms with high occupancy in the simulations as well as the neutron derived structure. The occupancy of the hydration sites is correlated with the number of hydrogen bonds to protein atoms.

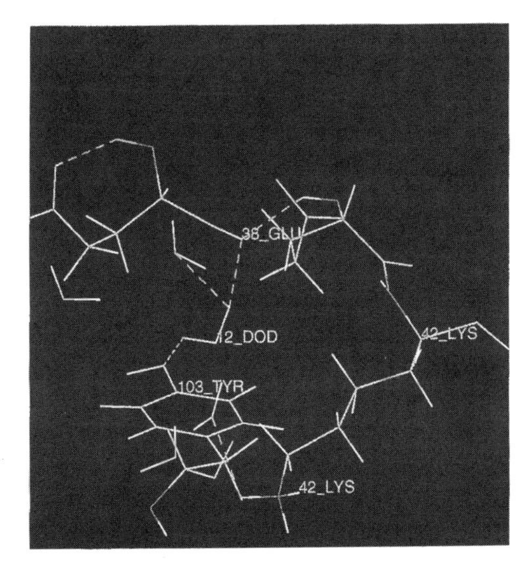

Figure 5. The hydrogen-bonding geometry for water-12 as observed in the neutron map. Hydrogen bonds are depicted in red dashed lines; oxygen atoms in red; nitrogen in blue; hydrogen white; carbon and deuterium atoms are displayed in green. (A color version of this figure appears in the color insert following p. 214.)

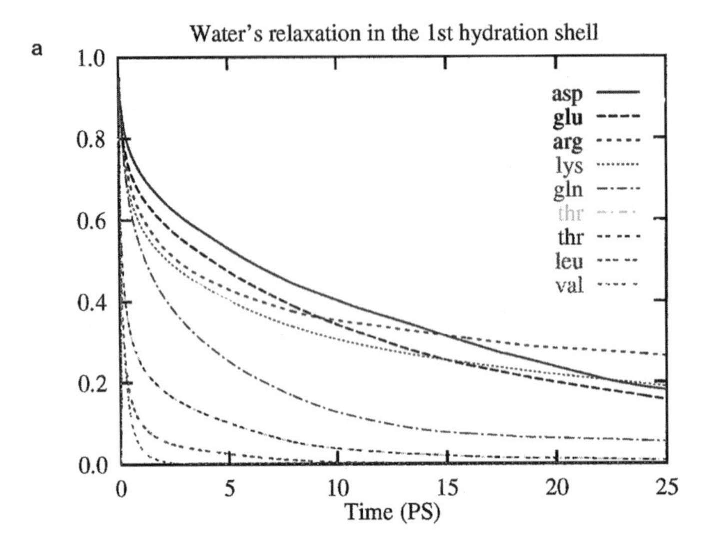

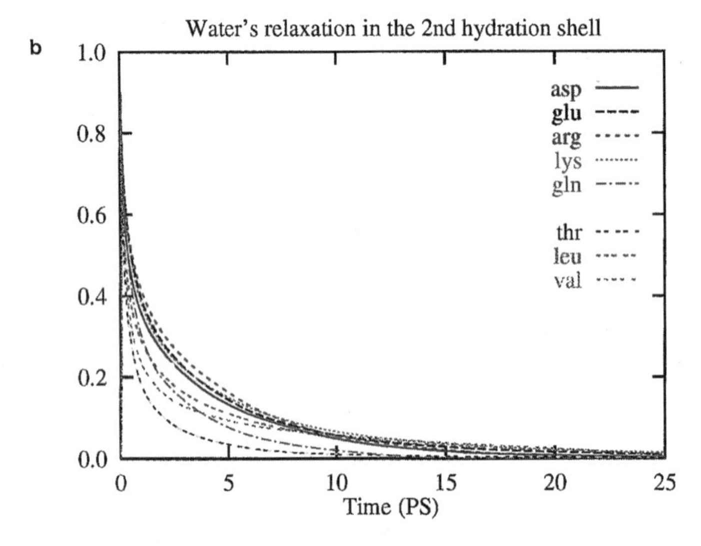

Figure 6. The relaxation times of water molecules adjacent to different side-chains in the solution system: (a) in the first hydration shell; (b) in the second hydration shell. (A color version of this figure appears in the color insert following p. 214.)

The dynamics of bound water is well described by their relaxation rates, which can be defined as the average surviving time of the original bound water (Garcia & Stiller, 1993). The calculation of water's relaxation rates show that in the first hydration shell, the relaxation rates is dependent on the character of the side-chains (Figure 6a). On average, water's relaxation rates around charged side-chains are among the slowest (> 10ps), water's relaxation rates around apolar side-chains are among the fastest (~1ps), and the waters' relaxations around polar side-chains are intermediate (a few ps). As expected, solvent in the second hydration shell show smaller differences (Figure 6b). Calculations of such relaxation times in the crystal system show similar trends. Surprisingly however, the crystal case shows a faster exchange rate for all charged and polar groups (Figure 7). A possible explanation for this observation may be the reduced electrostatic interactions due to neighboring charges and polar groups in a symmetry related molecule. For non-polar side-chains, the relaxation rates are similar in both cases.

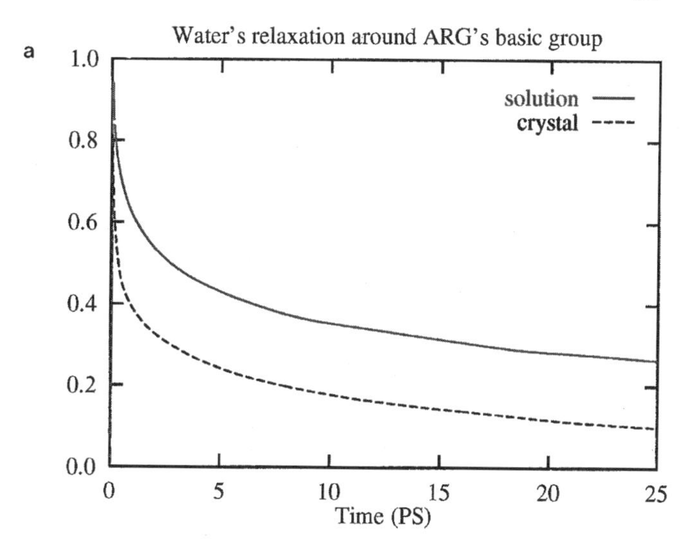

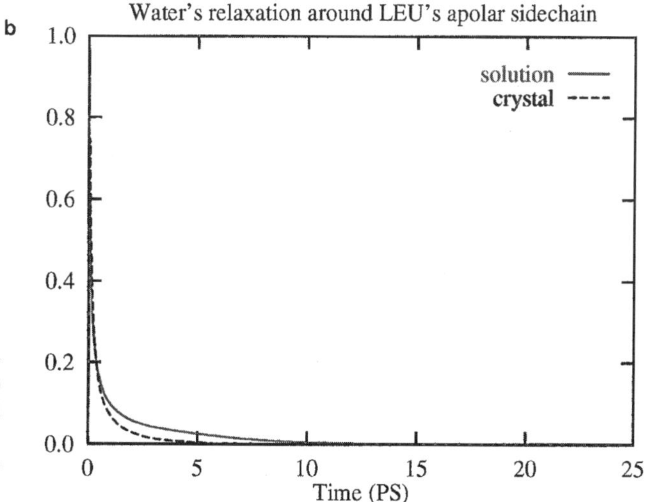

Figure 7. Comparison of the relaxation times of water molecules near (a) arginine side-chains in solution and crystal form: (b) leucine side-chains. (A color version of this figure appears in the color insert following p. 214.)

CONCLUSIONS

The molecular dynamics simulations described here show that most bound water sites observed in the neutron study show similarly high occupancies of water in both the solution and crystal form of myoglobin. Only a few sites are tightly bound to protein atoms during the whole time frame of the simulations. Therefore, most of bound water sites observed in the neutron maps do not have residence times long enough to be detected by NMR experiments. NMR experiment and neutron diffraction give us very important but different information regarding protein hydration. Neutron diffraction provides an average picture of the interactions between water and protein atoms and measures the average occupancies while NMR experiments are sensitive to the dynamic stability of hydration and measures only cases where water molecules are tightly bound to protein atoms and do not tumble for a relatively long time.

From calculations of hydrogen bond energies, it can be shown that hydrogen bonding is a significant factor to the stability of the water structure and that at least two hydrogen-bonds with a total energy much larger than kT are needed to 'bind' a water molecule. The occupancy and residence times of water molecules at hydration sites are clearly correlated with the number of hydrogen bonds that a water molecule shares with the protein atoms. Therefore, the positions of hydrogen atoms in polar groups provide very important information for understanding protein hydration.

By comparing the dynamics of hydration water between solution and crystal systems, it could be seen that the differences in the environments around the protein-solvent interface have large effects on the interactions within the protein-solvent interface.

ACKNOWLEDGMENTS

This research was done under the auspices of the Office of Health and Environmental Research of the U.S. Department of Energy. The authors thank Fong Shu for calculating the hydrogen bonding energies reported here.

REFERENCES

Boobbyer, N.A., Goodford, J.P., McWhinnie, P.M., & Wade, R.C., (1989). New hydrogen-bond potentials for use in determining energetically favorable binding sites on molecules of known structure. *J. Med. Chem.*, 32:1083–1094.

Brooks, B.R., Bruccoleri, R.E., Olafson, B.D., States, D.J., Swaminathan, S., & Karplus, M., (1983). CHARMM: A program for macromolecular energy, minimization, and dynamics calculations. *J. Comp. Chem.*, 4:187–217.

Brooks III, C.L., Brünger, A., & Karplus, M., (1985). Active site dynamics in protein molecules: A Stochastic boundary molecular-dynamics approach. *Biopolymers*, 24:843–865.

Brooks III, C.L., Karplus, M., & Pettitt, B.M., (1988). *Proteins: A theoretical perspective of dynamics, structure and thermodynamics*. John Wiley & Sons, New York.

Cheng, X., & Schoenborn, B.P., (1991). Neutron diffraction study of carbonmonoxymyoglobin. *J. Mol. Biol.*, 220:381–399.

Garcia, A.E., & Stiller, L., (1993). Computation of the mean residence time of water in the hydration shells of biomolecules. *J. Comp. Chem.*, 14:1396–1406.

Gu, W., & Schoenborn, B.P., (1995). Molecular dynamics simulation of hydration in myoglobin. *Proteins*, 22:20–26.

van Gunsteren, W.F., & Berendsen, H.J.C., (1977). Algorithms for macromolecular dynamics and constraint dynamics. *Mol. Phys.*, 34:1311–1327.

Jorgensen, W.L., Chandrasekhar, J., & Madura, J.D., (1983). Comparison of simple potential functions for simulating liquid water. *J. Chem. Phys.*, 79(2):926–935.

Koenig, S.H., Brown III, R.D., & Ugolini, R., (1992). A unified view of relaxation in protein solutions and tissue, including hydration and magnetization transfer. *Magnetic Resonance in Medicine*, 193:77–83.

Otting, G., Liepinsch, E., & Wüthrich, K., (1991). Protein hydration in aqueous solutions. *Science*, 254:974–980.

Schoenborn, B.P., Gu, W., & Shu, F., (1994). Hydrogen bonding and solvent in proteins. In *Synchrotron Radiation in the Biosciences*. Oxford Univ. Press.

THEORETICAL DESCRIPTION OF BIOMOLECULAR HYDRATION

Application to A-DNA

A. E. García,[1] G. Hummer,[1] and D. M. Soumpasis[2]

[1] Theoretical Biology and Biophysics Group, T10, MS K710
Los Alamos National Laboratory
Los Alamos, New Mexico 87545
[2] Department of Molecular Biology
Max Planck Institute for Biophysical Chemistry
PO Box 2841 D-37018 Göttingen, Germany.

ABSTRACT

The local density of water molecules around a biomolecule is constructed from calculated two- and three-points correlation functions of polar solvents in water using a Potential-of-Mean-Force (PMF) expansion. As a simple approximation, the hydration of all polar (including charged) groups in a biomolecule is represented by the hydration of water oxygen in bulk water, and the effect of non-polar groups on hydration are neglected, except for excluded volume effects. Pair and triplet correlation functions are calculated by molecular dynamics simulations. We present calculations of the structural hydration for ideal A-DNA molecules with sequences $[d(CG)_5]_2$ and $[d(C_5G_5)]_2$. We find that this method can accurately reproduce the hydration patterns of A-DNA observed in neutron diffraction experiments on oriented DNA fibers (P. Langan *et al.* J. Biomol. Struct. Dyn., 10, 489 (1992)).

INTRODUCTION

The interaction of biomolecules with water plays an important role in the structure, dynamics and stability of biomolecules. Binding of water molecules at specific sites has a determinant influence on specificity and function of biomolecules. There has been a continuing interest in studying the structure and dynamics of water molecules solvating biomolecules (Nemethy & Scheraga, 1962). Raman (Tominaga *et al.*, 1985), Brillouin (Tao, 1988; Tao *et al.*, 1987; Lindsay *et al.*, 1984), infrared (Austin *et al.*, 1989), NMR (Otting & Wüthrich,

1989), elastic and quasi elastic neutron scattering techniques (Grimm *et al.*, 1987; Doster *et al.*, 1989) and calorimetric studies (Marky & Kupke, 1989) give information about the dynamics and stability of water molecules in the first and second hydration shells around biomolecules. High-resolution neutron (Schoenborn, 1988; Teeter, 1991) and X-ray diffraction (Berman, 1991) on crystals and fibers (Langan *et al.*, 1992; Forsyth *et al.*, 1992) give detailed information about the position of water surrounding a biomolecule. Theoretical modeling mostly relies on the use of computer simulation methods [Monte Carlo (MC) (Clementi, 1976) and molecular dynamics (MD) (Brooks *et al.*, 1988; Levitt, 1989)]. Computer simulations of large solvated biomolecules require enormous computing times, with the relaxation times of the solution covering a wide range of time scales. In addition, the rather open liquid structure of water with a low particle density ($\rho_0 \approx 0.033\text{Å}^{-3}$) results in poor density statistics that limit the studies of the structural hydration of biomolecules.

In this paper we calculate the local density of water around a biomolecule (or any inhomogeneity) based on a Potential-of-Mean-Force (PMF) expansion of the N-points correlation function in terms of a hierarchy of approximations involving pair, triplet and higher-order correlation functions (Soumpasis, 1993; Hummer & Soumpasis, 1994a; 1994b; 1994c; Hummer *et al.*, 1995a; 1995b). This method is orders of magnitude faster than computer simulations and allows the study of solvated molecules of almost arbitrary size. This formalism has been used to describe the hydration of nucleic acids in solution and in the crystal environment. In this report, we calculate the structural hydration around A-DNA molecules.

THEORY

The PMF expansion formalism has been described in previous articles. Here we will briefly describe the main results. To define the system we employ $\vec{r}_{i_\alpha}$ to represent the positions of N_α atoms of M different types α that compose the solvated biomolecule. The position of all atoms, denoted by $\{\vec{r}_{i_\alpha}\}$, is presumed fixed. The conditional water density at position $\vec{r}$ can be expressed as a configuration space integral, where the solvated macromolecule appears as an external field. This conditional probability can, in turn, be related to multiparticle correlation functions:

$$\rho(\vec{r}_1|\{\vec{r}_{i_\alpha}\}) = \rho \frac{g^{(1;\{N_\alpha\})}(\vec{r}_1|\{\vec{r}_{i_\alpha}\})}{g^{(\{N_\alpha\})}(\{\vec{r}_{i_\alpha}\})} \tag{1}$$

where $\rho = N/V$ is the solvent bulk density. Using the PMF expansion for the correlation functions to second order (ie using two- and three-point correlation functions), we obtain

$$\rho(\vec{r}_1|\{\vec{r}_{i_\alpha}\}) = \rho\left[\prod_{\alpha=1}^{M}\prod_{i_\alpha=1}^{N_\alpha} g^{(1;\alpha)}(\vec{r}_1,\vec{r}_{i_\alpha})\right] \times\left[\prod_{\alpha=1}^{M}\prod_{i_\alpha=1}^{N_\alpha}\prod_{\beta=\alpha}^{M}\prod_{\beta=1+\delta_{\alpha\beta}i_\beta}^{N_\beta} \frac{g^{(1;\alpha,\beta)}(\vec{r}_1,\vec{r}_{i_\alpha},\vec{r}_{i_\beta})}{g^{(1;\alpha)}(\vec{r}_1,\vec{r}_{i_\alpha})g^{(\alpha,\beta)}(\vec{r}_{i_\alpha},\vec{r}_{i_\beta})g^{(\beta;1)}(\vec{r}_{i_\beta},\vec{r}_1)}\right] \tag{2}$$

where $\delta_{\alpha\beta}$ is the Kronecker symbol. At this point, Equation 2 represents a general theory and does not rely on any specific model. However, as a simple approximation that retains the most important contributions to the biomolecule's hydration structure (ie hydrogen bonding), we will simplify Equation 2 with regard to the number of different atom types α used. This approximation consists in equating all hydrogen-bond donor or acceptor atoms in the biomolecule to water-oxygen atoms (Hummer & Soumpasis, 1994c). In addition, the nonpolar atoms are represented as excluded volume. This gives the working formula for calculating the water-oxygen density around a biomolecule with n polar atoms and m nonpolar atoms by:

$$\rho_0(\vec{r}|\vec{r}_1,\ldots,\vec{r}_{n+1},\ldots,\vec{r}_{n+m}=\rho_0\prod_{i=1}^{n} g_{OO}^{(2)}(\vec{r},\vec{r}_i)\prod_{j=1}^{n-1}\prod_{k=j+1}^{n}\frac{g^{(3)}(\vec{r},\vec{r}_j,\vec{r}_k)}{g_{OO}^{(2)}(\vec{r},\vec{r}_j)g_{OO}^{(2)}(\vec{r}_k,\vec{r})}$$
$$\times\prod_{\substack{\text{Polar}\\ \text{Hydrogens}}}\chi_{OHO}(\vec{r}_{Donor},\vec{r}_H;\vec{r})\prod_{l=1}^{m}\Theta(||\vec{r}-\vec{r}_{n+l}||-r_c) \tag{3}$$

where ρ_0 is the bulk-water density, $\Theta(x)$ is the Heaviside step function, $|\vec{r}_j-\vec{r}_i|$ is the distance between nonpolar atoms and the point where the density is calculated, and r_c is the distance of closest approach between a water molecule and a nonpolar atom. The set of variables $\{\vec{r}_1,\ldots,\vec{r}_n\}$ and $\{\vec{r}_{n+1},\ldots,\vec{r}_{n+m}\}$ represent the coordinates of the polar and non-polar atoms in the biomolecule, respectively. χ_{OHO} is the three-particles correction to the local density for hydrogen-bond-donor atoms where the bonded pair is represented by OH (at a fixed distance), and the other O represents an oxygen atom in the solvent. The function χ_{OHO} was calculated for a TIP3P bulk water system and is used for all hydrogen-bond-donor and -hydrogen (N-H, O-H) groups in the biomolecule.

In a study of the interface of ice and water, it has been shown that it is both essential and sufficient to retain the triplet term in the density expansion (Hummer & Soumpasis, 1994a). In what follows, we use Equation 3 with $r_c = 3.0$Å, and the oxygen pair and triplet correlation functions $g_{OO}^{(2)}$, $g_{OOO}^{(3)}$, and the $\chi_{OHO}^{(3)}$ functions for the TIP3P model of water (Jorgensen, 1981). These functions have been calculated using extensive (1ns) molecular dynamics simulations of a system of 216 water molecules at 300K and density $\rho_0 = 1\text{g/cm}^3$.

RESULTS

Hydration of A-DNA Oligomers

The atomic coordinates of canonical (Arnott-Hukins) A-DNA structures (Arnott & Hukins, 1972; Soumpasis *et al.,* 1990) with sequence $[d(CG)_5]_2$ and $[d(C_5G_5)]_2$ were used in the computations. These structures were modeled after low-resolution fiber diffraction data and do not contain any sequence-dependent structural variations. Most polymeric DNA molecules under normal conditions exist in a B-family structure. The B to A transitions can be driven by reducing the relative humidity on GC rich DNA fibers, and in solution in Poly(dG).Poly(dC) sequences by adding 1.5M NaCl (Soumpasis *et al.,* 1990). Poly(dA).Poly(dT) and $\text{Poly}[d(AT)]_2$ do not show transitions to A-DNA, but to B'- and D-DNA, respectively. Calculations of the hydration of A-DNA molecules containing AT bases exhibit similar hydration structure as the CG containing sequences presented here and will be described elsewhere.

The water density was calculated at the vertices of a three dimensional grid covering a volume $34 \times 34 \times 44\text{Å}^3$, with a grid spacing of 0.25Å. The triplet correction was applied to all possible triangles with edges ranging from 1.9Å to 6.2Å. The pair contributions were calculated for distances up to 8.0Å. Each calculation takes around 10 minutes CPU time on a Silicon Graphics Indigo^2 workstation. A comparable MD or MC simulation of the same system will take a few weeks on a similar computer.

The neutron diffraction experiments of A-DNA reported by Langan *et al.* (Langan *et al.*, 1992; Forsyth *et al.*, 1992) on *E.coli* DNA fibers show high-probability peaks of finding water molecules at four different regions. These peaks are observed:

- [(I)] between the phosphate-group OA oxygens of neighboring residues,
- [(II)] between the phosphate-group OB oxygens of neighboring residues,
- [(III)] at positions between the deoxyribose sugar and the minor-groove base edges, and
- [(IV)] at the center of the helix, inside the major groove and perpendicular to the helix axis of the molecule.

The refinement of the 3Å resolution data locates the position occupancy of the different peaks. The cylindrical radial coordinates, $\sigma = \sqrt{(x^2 + y^2)}$, for the four peaks listed above are 5.3, 10.5, 12.3 and 1.6Å, respectively. The corresponding occupancies are 1.5, 0.8, 1.3 and 1.6 per base pair of DNA.

To illustrate the main hydration structural features of A-DNA we present all grid points with local water-oxygen density larger than 3 times the bulk-solvent density as small spheres. The radius of the spheres is varied according to the local water density. Densities between 3 and $5\rho_0$, 5 and $7\rho_0$, and larger than $7\rho_0$ correspond to radii of 0.2, 0.4 and 0.6Å, respectively. Figures 1–4 show the hydration structure for the $[d(CG)_5]_2$ and $[d(C_5G_5)]_2$ sequences. Figures 1 and 3 show top views of the hydration structure of $[d(CG)_5]_2$ and $[d(C_5G_5)]_2$ molecules. From these views we can see three main features of the hydration described by Langan *et al.* We observe high-density grid points bridging OA atoms of neighboring phosphate groups (OA is the phosphate-ester oxygen pointing into the helix axis), consistent with the peak labeled I, above. We also observe high-density grid points bridging neighboring phosphate-group OB atoms. However, as it will be explained below, these grid points are making a closer bridge between OB and O3′ of the

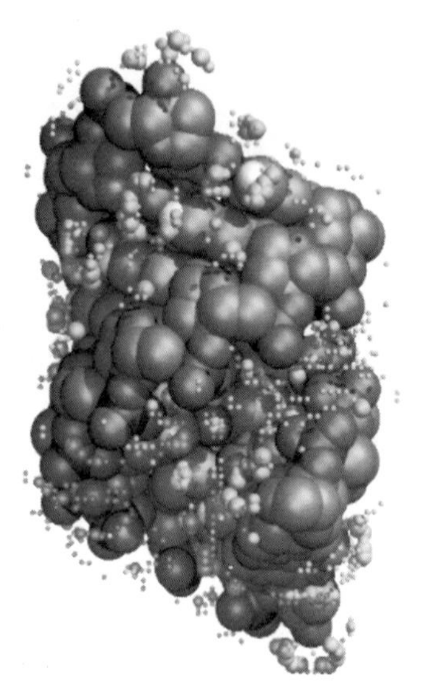

Figure 1. Minor- and major-groove view of $d(GGGGGCCCCC)_2$ in the A-DNA conformation. The high-density points of the local water density calculated in a regular grid are represented by cyan spheres of radii: 0.6Å for ρ larger than $7.0\rho_0$, 0.4Å for ρ between 5 and $7\rho_0$, and 0.2Å for ρ between 3 and $5\rho_0$. DNA atoms are color coded by N=blue, C=grey, H=white, O=red and P=dark blue. (A color version of this figure appears in the color insert following p. 214.)

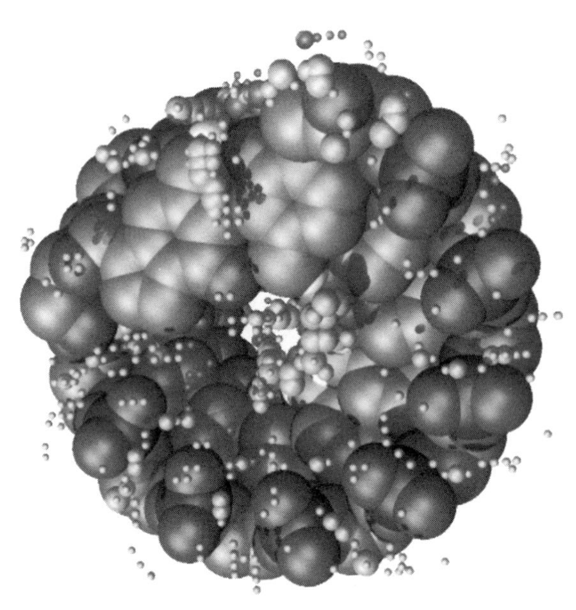

Figure 2. Top view of $d(GGGGGCCCCC)_2$ in the A-DNA conformation. Details as in Figure 1. (A color version of this figure appears in the color insert following p. 214.)

neighboring base in the 5′ to 3′ direction. The position of this high-density region is consistent with the peak labeled II, above. We also observe high-density grid points near the center of the helix, inside the major groove and along the helix axis. This region contains the largest number of grid points with high-density, consistent with the positioning and occupancy numbers found for the peaks labeled IV, above. $[d(AT)_5]_2$ and $[d(A_5T_5)]_2$ molecules show the same hydration structure around the phosphate groups (ie peaks I and II) as the $[d(CG)_5]_2$ and $[d(C_5G_5)]_2$ sequences. The position of the hydration peaks along the helix axis of A-DNA depend on base-pair type (AT vs CG) and sequence.

Figures 2 and 4 show views of the hydration structure in the major and minor grooves (bottom and top of the pictures, respectively). The large cyan spheres arranged vertically along the center of Figure 2, in the major groove, show the position of the high-density grid points associated with peak IV. This pattern is different from the pattern observed, in similar position, for the alternating GpC sequence shown in Figure 4, where the high-density points follow a zig-zag path around the helix axis. On the minor-groove wall, following the sugar-phosphate backbone, we observe high-density grid points consistent with peak II, above. The position and number of high-density grid points obtained depends on the base type (ie purine or pyrimidine) attached to the backbone sugar.

One important feature of the PMF formula presented in Equation 3 is that we can identify the atoms contributing to high water densities at some positions. The pair and triplet correlation functions depend on the segment length and the geometry of the triangles formed among polar atoms. Figure 5 show the pair and selected triplet correction values for a TIP3P water system at 300K and $\rho_0 = 1.0 g/cm^3$. The pair correlation function, $g^{(2)}_{OO}(r)$, gives maximum contributions to the local water density for distances $r \approx 2.8Å$ which corresponds to water-water hydrogen bonding. The triplet correction, $\chi^{(3)}_{OOO}(r,s,t) = g^{(3)}_{OOO}(r,s,t) / g^{(2)}_{OO}(r)\, g^{(2)}_{OO}(s)\, g^{(2)}_{OO}(t)$, is a function of three distances. To illustrate the main properties of the triplet correction, $\chi^{(3)}_{OOO}(r,s,t)$ is plotted for isosceles triangles (ie $\chi^{(3)}_{OOO}(r,r,t)$) as a function of r for different values of t. Notice that $\chi^{(3)}_{OOO}(r,s,t) \approx 1$ for r and t larger than 5.0Å.

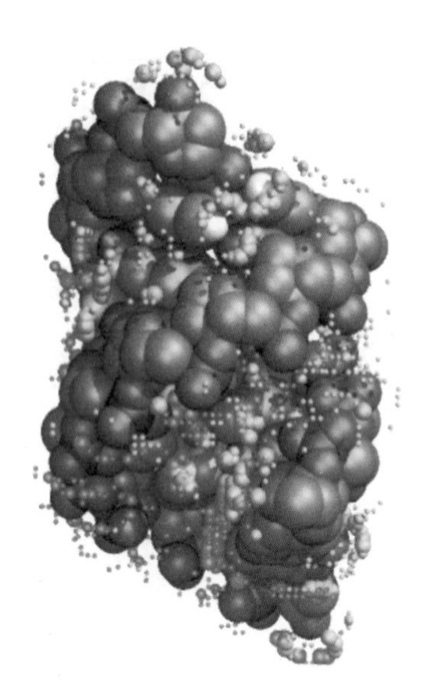

Figure 3. Minor- and major-groove view of $d(GCGCGCGCGC)_2$ in the A-DNA conformation. Details as in Figure 1. (A color version of this figure appears in the color insert following p. 214.)

The first product in Equation 3, known as the Kirkwood Superposition Approximation (KSA), gives a maximum contribution to the local density for multiple occurrences of distances near 2.8Å. However, the triplet correction will reduce the local density to zero if the three sides of any triangle have sides of length around 2.8Å, since $\chi^{(3)}_{OOO}$ (2.8Å, 2.8Å, 2.8Å) ≈ 0. After including the triplet correction terms contained in the second product of Equation 2, large water densities are obtained at grid points participating in many triangles with polar atoms at the other two vertices and sides close to (2.8Å, 2.8Å, 4.5Å) while

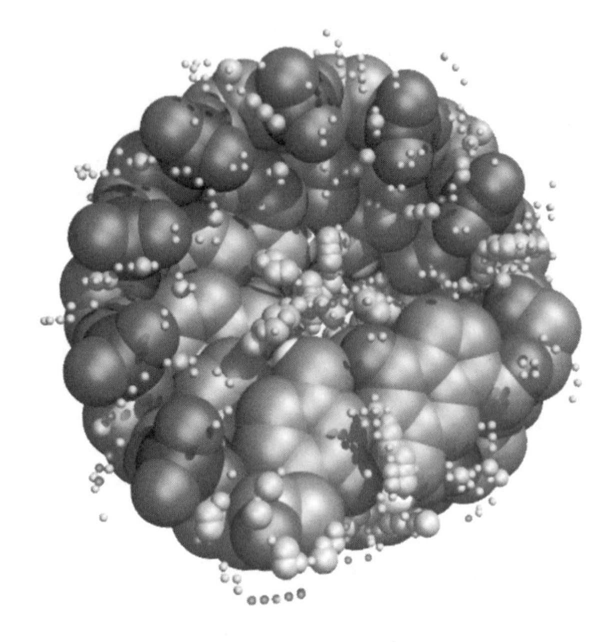

Figure 4. Top view of $d(GCGCGCGCGC)_2$ in the A-DNA conformation. Details as in Figure 1. (A color version of this figure appears in the color insert following p. 214.)

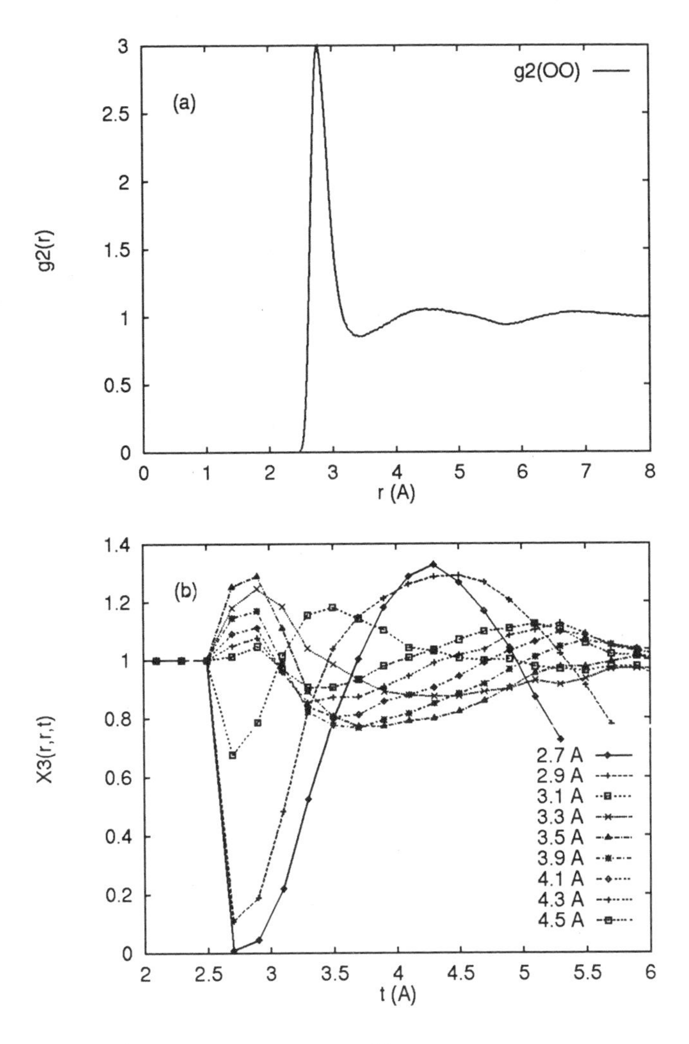

Figure 5. Pair correlation, $g^{(2)}_{OO}$, and triplet correction, $\chi^{(3)}_{OOO}$ (for selected isosceles triangles), functions for TIP3P water at 300K and ρ_0 = 1.0g/cm^3. These two functions are used in Equation 3 to calculate the local density of water on a rectangular grid around a biomolecule.

avoiding equilateral triangles with sides 2.8Å and steric overlaps with biomolecule's atoms. The third product in Equation 3 containing $\chi^{(3)}_{OHO}$ contributes significantly to the local density of water when a grid point, at a distance ≈ 2.8Å is aligned with the donor and hydrogen atoms such that a water molecule at the grid position will form a linear O-H ... X hydrogen bond with a donor atom in the biomolecule.

The hydration structure and the terms contributing to the high water density in regions I-IV listed above will be described next.

Region I: Phosphate-Backbone Atoms

We find high-density at grid points with cylindrical radial coordinates $\sigma \approx 5.7$Å. These grid points show water densities, $\rho \approx 7\rho_0$. The high-density peaks are sequence independent. For a sequence labeled 5′ - $p_iX_ip_{i+1}X_{i+1}$ -3′ there are six polar atoms within 5.0Å of these high-density grid points. These atoms and their distances to the grid point are: OA_i(2.8Å); OA_{i+1}(2.8Å); OB_i(4.1Å); $O3'_i$(4.3Å); $O5'_i$(3.8Å); and $O5'_{i+1}$(4.8Å). A water molecule at this grid position can make a hydrogen bond to OA atoms in neighboring phosphates and is further attracted by other polar atoms in the phosphate groups.

Region II: phosphate-backbone atoms

We find high-density at grid points with cylindrical radial coordinates, $\sigma \approx 10.6$Å. These grid points show water densities, $\rho \approx 10\rho_0$. These high-density grid points are sequence independent. There are six polar atoms within 5.0Å of these high-density grid points. These atoms and their distances to the grid points are: OB_i(2.9Å); $O3'_i$(2.7Å); OA_i(4.76Å); $O5'_i$(3.7Å); OA_{i+1}(3.9Å); and OB_{i+1}(3.8Å). Contrary to the results obtained from the refinement of the fiber diffraction data, we do not obtain OB to OB bridges. Instead we find high-density grid points at hydrogen-bonding distances to OB_i and $O3'_i$ atoms. Grid points within close proximity to OB atoms in consecutive phosphate groups show lower (3–5ρ_0) densities.

Region III: Minor-Groove Walls Near Sugar

We find high-density at grid points with cylindrical radial coordinates, $\sigma \approx 10.0$Å. These grid points show water densities, $\rho \approx 10\rho_0$. These high-density grid points are sequence dependent. For a 5′-G_ipG_{i+1}-3′ sequence there are seven polar atoms within 5.0Å of these high-density peaks. These atoms and their distances to the grid points are: $N2_i$(2.9Å); $N3_i$(4.7Å); $O1'_{i+1}$(4.9Å); $N2_{i+1}$(4.9Å); $N9_{i+1}$(4.4Å); $N3_{i+1}$(3.9Å); and one atom in a base in the complementary strand, $O2_i$(4.1Å). On the other side of the groove, the complementary sequence 5′-C_jpG_{i+1}-3′ also shows high-density grid points, with seven atoms at distances within 5.0Å. These atoms are: $O1'_i$(4.6Å); $O2_i$(3.1Å); $N1_i$(4.4Å); $O5'_{i+1}$(4.5Å); $O1'_{i+1}$(2.9Å); and two atoms in the bases in the complementary strand, $N1_{i+1}$(4.6Å); and $N2_i$(4.8Å).

Region IV: Major Groove, along the Helix Axis

We find high-density at grid points with cylindrical coordinates $\sigma \approx$ 1.0–3.2Å. The position of the high-density grid points is sequence dependent. We describe the position and list the polar atoms in their neighborhood for two sites in $[d(CG)_5]_2$, and three sites in $[d(C_5G_5)]_2$.

5′-G_{i-1} pC_ipC_{i+1}-3′ sequence in $[d(CG)_5]_2$. We find high density grid points with $\sigma \approx$ 1.8Å and densities $\rho \approx 9\rho_0$. We find seven polar atoms within 5.0Å of these high-density grid points. These atoms and their distances to the grid points are: $N7_{i-1}$(4.2Å); $O6_{i-1}$(4.3 Å); $N4_i$(2.7Å); $N3_i$(4.8Å); $N4_{i+1}$(4.1Å); and two atoms in the bases in the complementary strand, $O6_{i+1}$(4.9Å) and $O6_i$(4.3Å). Here $N4_i$ is acting as a hydrogen-bond donor to water, and N4(C_i) is also hydrogen bonded to O6(G_i) in the Watson-Crick base pair.

5′-$G_{i+1}pCipC_{i+1}pC_{i+2}$-3′ sequence in $[d(CG)_5]_2$. We find high-density grid points with $\sigma \approx 1.0$Å and densities $\rho \approx 8\rho_0$. There are ten polar atoms within 5.0Å of these high-density grid points. These atoms and their distances to the grid points are: $O6_{i-1}$(4.8Å); $N4_i$(3.6Å); $N4_{i+1}$(3.8Å); $O6_{i+2}$(4.9Å); $O6_{i+1}$(3.0Å); $N7_{i+1}$(4.4Å); $N1_i$(5.0Å); $O6_i$(2.9Å); $N4_{i-1}$(4.8Å); and $N7_i$ (4.7Å). This pattern is also found in CCC sequences.

$5'$-$G_{i-1}pC_ipG_{i+1}$-$3'$ sequence in $[d(C_5G_5)]_2$. We find high density grid points with $\sigma \approx$ 1.6Å and densities $\rho \approx 18\rho_0$. There are five polar atoms within 5.0Å of these high-density grid points. These atoms and their distances to the grid points are: $O6_{i-1}$ (4.3Å);$N7_{i-1}$(3.9Å); $N4_i$(2.9Å); $O6_{i+1}$(4.8); and one in the complementary strand, $O6_{i+1}$(4.8Å).

$5'$-$C_{i-1}pG_ipC_{i+1}$-$3'$ sequence in $[d(C_5G_5)]_2$. We find high density grid points with $\sigma \approx$ 1.8Å and densities $\rho \approx 16\rho_0$. There are ten polar atoms within 5.0Å of these high-density grid points. These atoms and their distances to the grid points are: $N4_{i+j}$(4.4Å); $N3_i$(4.9Å); $O6_{i+1}$(2.8Å); $N7_i$(2.9Å); $N9_i$(4.8Å); $N1_j$(4.5Å); $N4_{i+1}$(2.6Å); $N3_{i+1}$(4.8Å); with two atoms in the complementary strand, $O6_{i+1}$(4.3Å); and $N4_{i+2}$(4.7Å).

$5'$-G_ipC_{i+1}-$3'$ sequence in $[d(C_5G_5)]_2$. We find high-density grid points with $\sigma \approx$ 3.2Å and densities $\rho \approx 13\rho_0$. There are five polar atoms within 5.0Å of these high-density grid points. These atoms and their distances to the grid points are: $O1_i$(4.8Å); $O6_i$(4.0Å); $N7_{i+1}$(2.9Å); $N9_i$(4.5Å); and $N4_{i+1}$(3.1Å).

CONCLUSIONS

We have shown an efficient and accurate method for calculating the hydrophilic biomolecular hydration. By using the simplest possible implementation of a general formalism based on the PMF expansion of the local density around a biomolecule, we obtain qualitative and quantitative agreement with 3Å neutron fiber diffraction on A-DNA. This simple approximation consists in identifying the effect of various polar groups on the biomolecule's surface by the effect that a similar arrangement of water molecules have on the local structure of water. The contribution of nonpolar atoms, although important in determining the stability of the biomolecule in solution, does not play an important role in increasing the local density of water above the bulk value. This approximation could easily be relaxed from the implementation presented here. Another approximation employed here is to truncate the PMF expansion up to triplet correlation functions. Monte Carlo simulation of a water-ice interface have shown that triplet correlation functions are both necessary and sufficient in order to correctly describe the variations in the local density of water due to inhomogeneities. When compared with simulation studies of systems of equal size in solution, the PMF approach is orders of magnitude faster. By using a closed mathematical formula (Equation 3) to calculate the local density of water, we can determine which atoms are crucial in changing the local density of water at a point nearby. This allows to test the effect of changing sequence or of ligand binding on the hydration structure of biomolecules and biomolecular complexes.

ACKNOWLEDGMENTS

This work was supported by the US Department of Energy.

REFERENCES

Arnott, S., & Hukins, D.W.L., (1972). Optimized parameters for A-DNA and B-DNA. *Biochem. Biophys. Res. Commun.*, 47:1504–1510.

Austin, R.H., Robertson, M.W., & Mansky, P., (1989). Far-infrared perturbation of reaction rates in myoglobin at low temperatures. *Phys. Rev. Lett.*, 62:1912–1915.

Berman, H., (1991). Hydration of DNA. *Curr. Opin. Struct. Biol.*, 1:423–427.

Brooks, C.L., Karplus, M., & Montgomery-Pettitt, B., (1988). Proteins: A theoretical perspective of dynamics, structure and thermodynamics. *Adv. Chem. Phys.*, 71:259pp.

Clementi, E., (1976). Lecture Notes in Chemistry, Vol. 2. Determination of Liquid Water Structure. Coordination Numbers for Ions and Solvation of Biological Molecules. Springer Verlag, Berlin.

Doster, W., Cusack, S., & Petry, W., (1989). Dynamical transition of myoglobin revealed by inelastic neutron scattering. *Nature,* 337:754–756.

Forsyth, V.T., Langan, P., Mahendrasingam, A., Fuller, W., & Mason, S.A., (1992). High angle neutron fiber diffraction studies of DNA. *Neutron News,* 3(4):21–24.

Grimm, H., Stiller, H., Majkrzak, C.F., Rupprecht, A., & Dahlborg, U., (1987). Observation of acoustic umklapp phonons in water-stabilized DNA by neutron scattering. *Phys. Rev. Lett.*, 59:1780–1783.

Hummer, G., & Soumpasis, D.M., (1994a). Computation of the water density distribution at the ice-water interface using the potentials-of-mean-force expansion. *Phys Rev. E.*, 49:591–596.

Hummer, G., & Soumpasis, D.M., (1994b). In *Structural Biology: The State of the Art*, R.H. Sarma and M.H. Sarma, editors. Adenine Press, Schenectady, NY, Vol 2, p273.

Hummer, G., & Soumpasis, D.M., (1994c). Statistical mechanical treatment of the structural hydration of biological macromolecules: results for B-DNA. *Phys. Rev. E.*, 50:5085–5095.

Hummer, G., Garcia, A.E., & Soumpasis, D.M., (1995a). Hydration of nucleic acid fragments: Comparison of theory and experiment for high resolution crystal structures of RNA, DNA and DNA-drug complexes. *Biophys. J.*, 68(5):1639–1652.

Hummer, G., Soumpasis, D.M., & Garcia, A.E., (1995b). Potential-of-mean-force description of ionic interactions and structural hydration in biomolecular systems. In *Nonlinear Excitations in Biomolecules*, M. Peyrard, editor. Springer, Paris. pp83–99.

Jorgensen, W.L., (1981). Transferable intermolecular potential functions for water, alcohols and ethers. Application to liquid water. *J. Am. Chem. Soc.*, 103:335–340. See also Jorgensen, W.L., Chandrasekhar, J., & Madura, J.D., (1983). Comparison of simple potential functions for simulating liquid water. *J. Chem. Phys.*, 79:926–935.

Langan, P., Forsyth, V.T., Mahendrasingam, A., Pigram, W.J., Mason, S.A., & Fuller, W., (1992). A high angle neutron fibre diffraction study of the hydration of the A conformation of the DNA double helix. *J. Biomol. Struct. Dyn.*, 10:489–503.

Levitt, M., (1989). Molecular dynamics of macromolecules in water. *Chemica Scripta*, 29A:197–203.

Lindsay, S.M., Powell, J.W., & Rupprecht, A., (1984). Observation of low-lying Raman bands in DNA by tandem interferometry. *Phys. Rev. Lett.*, 53:1853–1855. For a review see, S.M. Lindsay, (1987). In *Structure and Dynamics of Nucleic Acids, Proteins and Membranes*, E. Clementi and S. Chin, editors. Plenum, New York.

Marky, L.A., & Kupke, D.W., (1989). Probing the hydration of the minor groove of AT synthetic DNA polymers by volume and heat changers. *Biochem.*, 28:9982–9988.

Nemethy, G., & Scheraga, H.A., (1962). The structure of water and hydrophobic bonding in proteins III: The thermodynamic properties of hydrophobic bonds in proteins. *J. Phys. Chem.*, 66:1773–1789.

Otting, G., & Wüthrich, K., (1989). Studies of protein hydration in aqueous solution by direct NMR observation of individual protein-bound water molecules. *J. Am. Chem. Soc.*, 111:1871–1875.

Otting, G., Liepinsh, E., & Wüthrich, K., (1991). Protein hydration in aqueous solution. *Science,* 254:974–980.

Schoenborn, B.P., (1988). The solvent effects in protein crystals. A neutron diffraction analysis of solvent and ion density. *J. Mol. Biol.*, 201:741–749.

Soumpasis, D.M., (1993). In *Computation of Biomolecular Structures. Achievements, Problems and Perspectives*. D.M. Soumpasis and T.M. Jovin, editors. p223. Springer, Berlin.

Soumpasis, D.M., Garcia, A.E., Klement, R., & Jovin, T.M., (1990). The potential-of-mean-force (PMF) approach for treating ionic effects on biomolecular structures in solution. In *Theoretical Biochemistry and Molecular Biophysics*, D.L. Beveridge and R. Lavery, editors. Adenine Press, Schenectady, NY, Vol. 1, pp343–360.

Tao, N.J., (1988) Ph.D. Thesis, Arizona State University, Tempe, Arizona (USA).

Tao, N.J., Lindsay, S.M., & Rupprecht, A., (1987). The dynamics of the DNA hydration shell at gigahetz frequencies. *Biopolymers*, 26:171–188.

Teeter, M.M., (1991). Water-protein interactions: theory and experiment. *Ann. Rev. Biophys. Biophys. Chem.*, 20:577–600.

Tominaga, Y., Shida, M., Kubota, K., Urabe, H., Nishimura, Y., & Tsuboi, M., (1985). Coupled dynamics between DNA double helix and hydrated water by low frequency Raman spectroscopy. *J. Chem. Phys.*, 83:5972–5975.

26

HIGH-LEVEL EXPRESSION AND DEUTERATION OF SPERM WHALE MYOGLOBIN

A Study of Its Solvent Structure by X-Ray and Neutron Diffraction Methods

F. Shu,[1,2,3] V. Ramakrishnan,[2] and B. P. Schoenborn[3]

[1] Physics Department
State University of New York at Stony Brook
Stony Brook, New York 11794
[2] Biology Department
Brookhaven National Laboratory
Upton, New York 11973
[3] Life Science Division
Los Alamos National Laboratory
Los Alamos, New Mexico 87545

ABSTRACT

Neutron diffraction has become one of the best ways to study light atoms, such as hydrogens. Hydrogen however has a negative coherent scattering factor, and a large incoherent scattering factor, while deuterium has virtually no incoherent scattering, but a large positive coherent scattering factor. Beside causing high background due to its incoherent scattering, the negative coherent scattering of hydrogen tends to cancel out the positive contribution from other atoms in a neutron density map. Therefore a fully deuterated sample will yield better diffraction data with stronger density in the hydrogen position. On this basis, a sperm whale myoglobin gene modified to include part of the A cII protein gene has been cloned into the T7 expression system. Milligram amounts of fully deuterated holo-myoglobin have been obtained and used for crystallization. The synthetic sperm whale myoglobin crystallized in $P2_1$ space group isomorphous with the native protein crystal. A complete X-ray diffraction dataset at 1.5Å has been collected. This X-ray dataset, and a neutron data set collected previously on a protonated carbon-monoxymyoglobin crystal have been used for solvent structure studies. Both X-ray and neutron data have shown that there are ordered hydration layers around the protein surface. Solvent shell analysis on the neutron data further has shown that the first hydration layer behaves

Neutrons in Biology, edited by Schoenborn and Knott
Plenum Press, New York, 1996

differently around polar and apolar regions of the protein surface. Finally, the structure of per-deuterated myoglobin has been refined using all reflections to a R factor of 17%.

INTRODUCTION

In the late 80s, Schoenborn and Cheng pointed out that low-order structure factors (small Miller index, hkl) in protein crystal diffraction, which are omitted in conventional procedures for protein structure refinement, are important in two ways. Firstly, they contain information about solvent structure which can be used to derive the distribution of solvent on the protein surface. Secondly, the low-order structure factors can be included in protocols for protein refinement after solvent contributions are subtracted (Schoenborn, 1988; Cheng & Schoenborn, 1990). They demonstrated that two distinguishable hydration shells could be located on the surface of myoglobin, and when low-order reflections were included in the protein refinement, then, the protein's surface features were improved. Since hydrogen diffracts neutrons at the same order of magnitude as do other atoms, but diffracts X-rays much less than others, neutron diffraction data were used in their study. In this work, we address three questions: (i) will the hydration shell positions differ around polar and non-polar protein surface areas; (ii) can the solvent-shell method be applied to X-ray diffraction data, which is commonly used for determining and refining protein structures; and, (iii) how can we obtain enough fully deuterated protein for a neutron diffraction study.

To answer the first question, we aim to contribute to current understanding of solvent - protein interactions, which are known to be very important to protein structure and function, but are still not fully understood (Karplus & Faerman, 1994; Levitt & Park, 1993). For the second question, we try to provide a solvent model that can be used in contemporary refinement procedures of protein structures determined by X-ray crystallography, since a solvent model is needed to obtain a more realistic refinement. Concerning the third question, we want to optimize the power that neutron diffraction has over other methods in locating hydrogen atom positions directly. It is known that hydrogen has a large incoherent neutron scattering factor, and deuterium has a rather small incoherent but a large positive coherent scattering factor (Bacon, 1975), hence, a fully deuterated protein crystal is the only solution to minimize the background from the sample, and therefore, to increase the signal-to-noise ratio in a neutron diffraction study. Considering that more than 50% of all protein atoms are hydrogens, this technique will greatly enhance the neutron technique in studying the positions of hydrogens, the hydration of proteins, and hydrogen-deuterium exchange. However, it is difficult to obtain large amounts of synthetic proteins, particularly in fully deuterated forms (Marston, 1986).

Here, we present a method to obtain a large quantities of per-deuterated sperm-whale myoglobin suitable for neutron diffraction studies. We also discuss our results on differentiating solvent distributions around polar and non-polar surface areas of myoglobin derived from low-order reflections of neutron diffraction data taken on a carbonmonoxymyoglobin crystal, and results on determining solvent structure around per-deuterated myoglobin using X-ray diffraction data and the solvent-shell model.

MATERIALS AND METHODS

Construction of a Fusion Sperm-Whale Myoglobin Gene

The gene for the first 31 N-terminal residues of phage λ cII protein was obtained by digesting plasmid pLcII (Nagai & Thogersen, 1984) with restriction enzymes NdeI and

BamHI. A purified fragment, which had an NdeI site at the 3′ end and a BamHI site at the 5′ end, was cloned into the T7 expression vector pet-13a (Gerchman *et al.*, 1994). This vector subsequently was used in a ligation reaction to make the myoglobin expression plasmid pet13cIIFxMb. The insert of the ligation was constructed such that a blood coagulation factor X_a cleavage site was added to the 3′ end of a synthetic sperm-whale myoglobin gene by the polymerase chain reaction (PCR) (Springer & Sligar, 1987). At the same time, BamHI cloning sites were engineered at the 3′ and 5′ ends of the insert. The correct orientation of myoglobin gene was verified by digesting the ligation products with EcoRI and BamHI according to differences in fragment size. The *E.coli* strain DH5α was used for plasmid manipulations, and standard cloning protocols were followed for the T7 expression system (Studier *et al.*, 1990; Ramakrishnan & Gerchman, 1991).

Expression and Deuteration of Sperm Whale Myoglobin

E.coli BL26(DE3) cells containing the fusion myoglobin plasmid were adapted, stepwise, from rich media to minimal media (2g/L NH_4Cl, 3g/L KH_2PO_4, 6g/L Na_2HPO_4, 0.4% (w/v) D_6-succinate, 2mM $MgSO_4$, 25mg/L kanamycin, 25mg/L $FeSO_4{\cdot}7H_2O$ in 99.6% D_2O). When the cells were in 99.6% D_2O media, they were grown in a closed system where the air supply first was filtered through drierite, and then saturated with 99.6% D_2O. When the optical density of the culture at 600nm reached 1.0, we added to the culture 0.1mM iso-propyl-thio-galactoside (IPTG) dissolved in D_2O to induce the production of fusion myoglobin. The temperature of the incubator was 37°C before induction, and was reduced to 25°C immediately after adding IPTG. The cells were harvested 12~15 hours after induction.

Purification and Digestion of Sperm Whale Myoglobin

Between 6~10g frozen BL26(DE3) cells were thawed at room temperature for about 10 minutes, and resuspended in 50ml lysis buffer (50mM Tris-HCL pH 8.0, 25% (w/v) sucrose, 5mM EDTA.2Na, 1mM dithiothreital(DTT), and 0.05mM phenyl-methane-sulfonyl-fluoride(PMSF)) by homogenizing at 4°C. The cells were lysed for 45 minutes after addition of 50mg lysozyme and 0.08% sodium-deoxycholate. The lysed cells then were treated with 1mg DNAase I in the presence of 10mM $MgCl_2$ and 1mM $MnCl_2$. Fifteen minutes afterward, polyethylenamine was added drop by drop to a final concentration of 0.3%. Finally, the portion containing soluble apo-fusion myoglobin was separated from the extract by centrifuging at 12,000g for 30 minutes. Usually, a second extraction was performed.

Most of the over-expressed myoglobin was in the apo form. For its reconstitution with heme, the pH of the protein solution was increased to 9.0 by adding concentrated CH_3NH_2. Solid heme from Sigma was dissolved in a minimal volume of 0.2M NaOH, and added to the protein solution until the ratio of absorption of the protein solution at 410nm and at 280nm stopped increasing. The supernatant of the reconstituted myoglobin was adjusted to pH 6.0 using concentrated acetate, and then the sample was loaded on a 2.5cm × 30.0cm S-Sepharose Fast-Flow (Pharmacia) column equilibrated with 50mM Na/K PO_4 pH6.0/.05mM PMSF/1mM DTT. The fusion myoglobin was eluted by a linear 0–0.4M NaCl gradient. Fractions containing myoglobin were collected, concentrated to a volume of about 10ml, and dialyzed in digestion buffer (50mM Tris-HCl pH 9.0/.1M NaCl/1mM $CaCl_2$). Before digestion, the protein sample was further concentrated to more than 8mg/ml. Trypsin from Sigma, instead of factor X_a, was used to liberate genuine myoglobin (Varadarajan *et al.*, 1985); 90mg of trypsin were used for 1g of myoglobin. The reaction mixture was incubated at room temperature for 12 to 16 hours; then, the trypsin was inhibited by 1mM PMSF.

Myoglobin was separated from the partially digested products was on a 2.5cm × 35cm S-Sepharose Fast-Flow (Pharmacia) column equilibrated with 50mM (Na/K) PO_4 pH6 / 0.05mM PMSF / 0.05mM benzamidine. Myoglobin eluted at ~0.24M NaCl, and the partially digested proteins came off at ~0.16M NaCl when a 0–0.4M NaCl linear gradient was applied. The sample was applied to a 2.5cm × 120cm S-100 filtration column equilibrated with 50mM Na/K PO_4 pH6 / 0.05mM PMSF / 0.5M NaCl for final purification.

All above procedures were carried out on ice or in 4°C cold room, unless otherwise indicated.

Crystallization

Crystals for diffraction studies were grown using the sitting-drop vapor diffusion method in 9 well-depression plates sealed in plastic boxes. Each well contained 100~200μl of 30mg/ml protein in 45% saturated $(NH_4)_2SO_4$ /50mM (Na/K)PO_4, pH 7.4. Surrounding the plates and inside the boxes were 25ml 70% $(NH_4)_2SO_4$ that served as the precipitant. The protein solution was equilibrated overnight against the precipitant, then was seeded with a 10^{-4} dilution of a small crystal that was finely crushed in 100μl 65% saturated $(NH_4)_2SO_4$ /50mM (Na/K)PO_4, pH 7.4.

Diffraction Data Measurement

A complete set of 1.5Å X-ray data was taken on a single per-deuterated myoglobin crystal. A 300mm Mar image plate at the MarResearch Laboratory (Poway, CA) was used as the detector, and Cu K_α radiation, generated from a Rigaku rotating anode X-ray generator, was used as the source for X-rays. The data was processed with Mosflm, and scaled with the Rotavata and Aagrovata programs from the CCP4 package. This data was used to determine and refine the structure of per-deuterated myoglobin. It also was used to test the feasibility of using X-ray data to observe solvent shell structures described earlier from neutron diffraction data (Cheng & Schoenborn, 1990). The neutron data had been collected by Schoenborn and coworkers on a myoglobin crystal soaked in 90% D_2O (Norvell *et al.*, 1975), and was used in this work for polar and non-polar solvent shell analysis.

Analysis Method for Solvent Structure

About 27% to 77% of the space inside a protein crystal is filled with solvent, mainly water molecules (Matthews, 1968). Except for protein-protein contacts, the protein surface is exposed to an aqueous environment. Solvent-exposed protein surfaces in crystals provide an appropriate data base for examining protein-water interactions.

In X-ray diffraction, the positions of water molecules are located as positive peaks in difference Fourier electron-density maps. Similarly, the positions of water can be located in difference Fourier neutron-density maps. Of these two approaches, neutron diffraction has some advantages for studying water structures. Unlike X-ray diffraction where hydrogen (or deuterium) has a much smaller scattering length than other elements, in neutron diffraction it has an equal, if not larger, scattering length than other atoms. Although water sites determined in this way by different authors using different refinement procedures do not always match even for the same protein, atomic pair distribution functions can be calculated from these positions. It gives a statistical view on how protein and water molecules interact.

The atomic pair distribution function $g_{WA}(r)$ of solvent atoms, W, around protein atoms of type A is defined as:

$$g_{WA}(r) = 1/\left(4\pi r^2\right) dN_{WA}(r)/dr$$

where r is the distance between solvent atoms and protein atoms, and $N_{WA}(r)$ is the number of W atoms within a sphere of radius r around atom A. In this study, W stands for the oxygen atom in a water molecule; A can be polar atoms (P) or apolar atoms (A) in the protein.

Water sites determined during refinement procedures represent only a fraction of the water molecules on the protein surface, and normally, they are tightly bound to the protein. An alternative way to study solvent-protein interactions is to extract solvent structural information from low-order diffraction data.

In diffraction experiments, structural information on the bulk solvent mainly is contained in reflections of resolutions below 4Å (Bragg & Perutz, 1952; Raghavan & Schoenborn, 1984). Therefore, the total structure factor $\mathbf{F}_t$ for each Miller index hkl, can be calculated as

$$\mathbf{F}_t(hkl) = \mathbf{F}_p(hkl) + \mathbf{F}_s(hkl) \qquad (d > 4\text{Å})$$

$$\mathbf{F}_t(hkl) = \mathbf{F}_p(hkl) \qquad (d < 4\text{Å})$$

where $\mathbf{F}_s(hkl)$ and $\mathbf{F}_p(hkl)$ are the calculated structure factors for the solvent, and the protein parts, respectively. $\mathbf{F}_p(hkl)$ is calculated in the conventional way when a protein model is available.

Plain Solvent-Shell Model. $\mathbf{F}_s(hkl)$ is evaluated as a series of spatial shells of equal thickness, projecting outward from the protein's surface (Cheng & Schoenborn, 1990). The protein's surface is defined by van der Waals radii of surface protein atoms. The values of these radii used to determined the protein boundary are optimized to give the best fit for amino acid volumes. Since the positions of hydrogen are missing in protein models determined by X-ray diffraction, the values for non-hydrogen atoms are larger than their canonical values. Table I gives the values used in this work for neutron and X-ray diffraction studies.

The individual structure factors for each of the solvent shells are calculated on a three-dimensional grid. Each grid point in a shell is assigned to a scattering length estimated from an averaged scattering density, ρ, and associated with a B-factor, termed liquidity factor. These factors are the same for all grid points in a particular solvent shell. The scattering density defines the average scattering power of the particular shell. The liquidity factor describes the extent of solvent disorder in terms of the root mean-square displacement of a solvent grid in the shell; it is equivalent to a temperature factor. The structure factor for shell n, thus, is

$$\mathbf{F}_{s_n}(hkl) = \rho_n \exp\left(-B_n \sin^2\theta/\lambda^2\right) \sum_i \exp\left[-2\pi i\left(hx_i + ky_i + lz_i\right)\right]$$

Table I. The pseudo van der Waals radii used to define the protein surface for solvent-shell analysis. The values are scaled to give the best fit amino acid volumes

	Plain solvent-shell model		Polar/non-polar solvent-shell model
Atom Type	Neutron Study (Å)	X-ray Study (Å)	Neutron Study (Å)
C	1.7	2.4	1.7
N	1.5	2.4	1.5
O	1.4	2.4	1.4
S	1.5	1.5	1.8
H	1.6	—	1.2
D	1.6	—	1.2

where x_i, y_i, and z_i are the solvent grid coordinates within the shell n, ρ_n and B_n are the average scattering density, and the liquidity factor of the shell, respectively.

The best values of ρ_n and B_n are determined by conducting a two dimensional search in ρ_n and B_n space to minimize the R factor or χ^2 while keeping the ρ and B values for other shells fixed. Reflections that have Bragg d-spacing bigger than 4Å are used. R and χ^2 are defined as

$$R = \sum_{hkl} |\Delta \mathbf{F}(hkl)| \Big/ \sum_{hkl} |\mathbf{F}_o(hkl)|$$

$$\chi^2 = \sum_{hkl} |\Delta F(hkl)|^2 / \sigma^2$$

$$\Delta \mathbf{F}(hkl) = \left(F_o - |\mathbf{F}_p|\right) - \left| \sum_{n=1,m} \mathbf{F}_{s_n}(\rho_n, B_n) \right|$$

where n is the shell number, σ is the estimated standard deviation for each reflection, and ρ_n, B_n are, respectively, the scattering density and liquidity factor of the nth solvent shell.

Once the solvent structure factors for each shell have been calculated, the bulk solvent structure factor is determined as

$$\mathbf{F}_s = \sum_{n=1,m} \mathbf{F}_{s_n}\left(\rho_n^{\text{best}}, B_n^{\text{best}}\right)$$

where n is the shell number, and m is the total number of shells that are required to fill the space between protein molecules.

Polar/Non-Polar Shell Model. In principle, the idea is similar to that described above; differences lie in the way the solvent shells are defined. Here, the solvent volume is defined as polar or non-polar, depending on the type of atom at the protein surface. Solvent areas that are within 4.0Å from centers of O or N atoms are defined as polar solvent, regardless of presence of C atoms; areas that are beyond 4.0Å of any polar atoms are defined as either polar or non-polar, depending on the type of atom from whose center the solvent point has the shortest distance. Solvent regions are partitioned further into polar or non-polar spatial shells of 0.3Å thickness extending outward from the centers of surface atoms.

We used the second method for neutron diffraction data analysis to study the difference in solvent distribution on the surface of myoglobin. The first method was used in the analysis of X-ray diffraction data to test its feasibility for such X-ray data; then, the resulting solvent model was used to refine the structure of per-deuterated myoglobin.

Refinement of Protein Structure

Standard protocols for slowcool, B refinement, and occupancy refinement in X-PLOR (Brünger, 1993) were used. When the R-factor was reduced below 20%, solvent analysis was performed using diffraction data with a resolution below 4.0Å. The resulting solvent structure factors were subtracted from the observed reflections to give the solvent-free structure factor $\mathbf{F}_n$(hkl); they were added to the calculated protein structure factor to give the total structure factor $\mathbf{F}_t$(hkl). $\mathbf{F}_n$(hkl), defined below, was used in the subsequent steps of protein refinement; $\mathbf{F}_t$(hkl), also defined below, was used to calculate electron density maps in which the protein surface and solvent features was enhanced due to including low-order reflections and solvent phases (Cheng & Schoenborn, 1991).

$$\mathbf{F}_t(\mathrm{hkl}) = \mathbf{F}_p(\mathrm{hkl}) + \mathbf{F}_s(\mathrm{hkl})$$

$$\mathbf{F}_n(\mathrm{hkl}) = \mathbf{F}_o(\mathrm{hkl}) - \mathbf{F}_s(\mathrm{hkl})$$

where $\mathbf{F}_p$(hkl), $\mathbf{F}_s$(hkl), and $\mathbf{F}_o$(hkl) are the structure factors calculated for protein, calculated for solvent, and the observed ones, respectively.

RESULTS

High-Level over-Expression of Sperm-Whale Myoglobin

Fusion sperm-whale myoglobin was over-expressed in *E.coli* to a very high level, in both protonated and deuterated media; comprised 40–50% of the total cellular protein. Figure 1 shows the level of expression for different plasmids both in the pUC19 (9) and the T7 expression systems. About 80% of the over-expressed protein was soluble. We obtained about 40 to 50mg pure per-deuterated myoglobin from a 1*l* culture.

Solvent Distribution around Polar and Non-polar Protein Surfaces

Analysis of the polar and non-polar solvent shell was carried out on neutron data. The liquidity factor, B, of each shell is plotted in Figure 2 which shows marked variations in their liquidity factors on different surface areas. Places with a low B factor indicate the presence of an ordered solvent structure; they lie at 2.7Å and 3.4Å from the center of surface polar and non-polar protein atoms.

Refinement of the X-ray Structure

The starting model used for determining and refining the per-dueterated myoglobin was a carbon-monoxymyoglobin structure refined using neutron diffraction data to R = 11.7% (Cheng & Schoenborn, 1991). Water molecules from the structure were removed

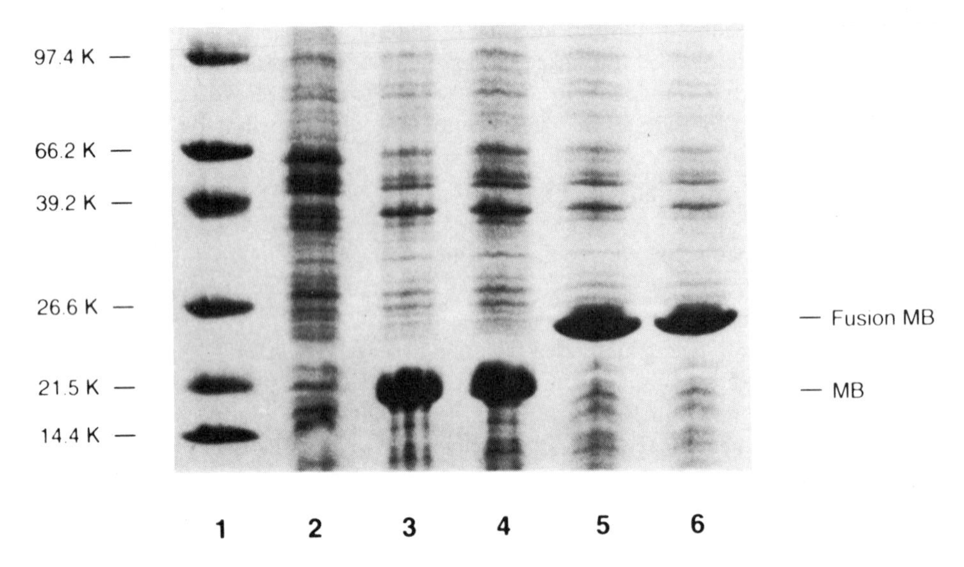

Figure 1. Levels of over-expression of the synthetic myoglobin in different expression vectors as visualized on a 5% - 25% PAGE gel, stained with Coomassie blue. Lane 1 contains the molecular weight marker. Lane 2 contains a cell lysate of the culture containing plasmid pUC19Mb in its late saturation phase (kindly donated by Sligar). Lanes 3–4 contain cell lysates of the culture containing plasmid pet13Mb three hours after induction. Lanes 5–6 contain cell lysates of the culture containing plasmid pet13cIIFxMb three hours after induction. The level of expression is substantially higher in the T7 expression system (lanes 3–6). The plasmid pet13cIIFxMb was used in this work (see Discussion).

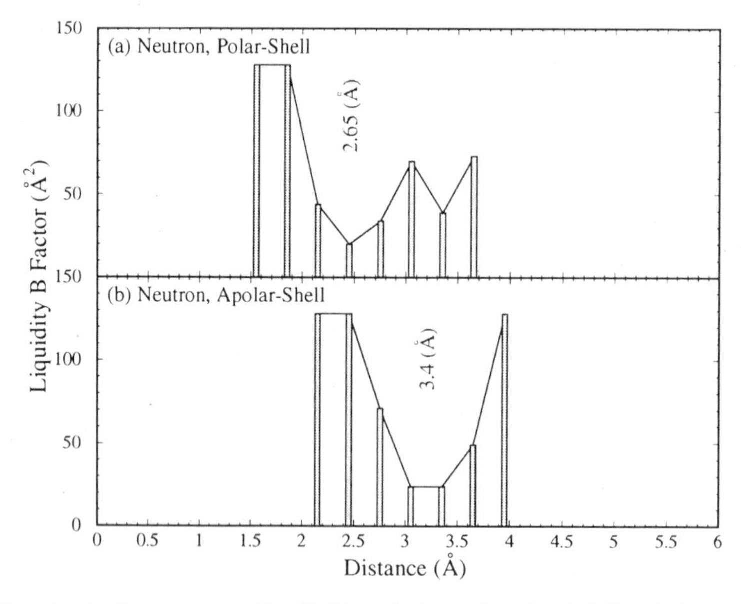

Figure 2. The ordered solvent structure, identified by polar/non-polar solvent-shell analysis on neutron data of myoglobin. (a) Liquidity factor B of solvent shells around polar protein surface areas; (b) Liquidity factor B of solvent shells around non-polar protein surface areas. The distance is measured from the center of non-hydrogen atoms at the protein's surface: C for non-polar shells, N or O for polar shells.

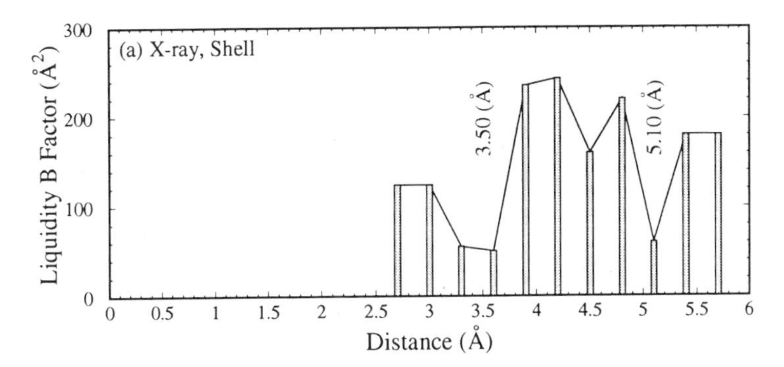

Figure 3. The ordered solvent structure, identified by plain solvent-shell analysis on X-ray data of per-deuterated myoglobin. The distance is measured from the center of non-hydrogen atoms at the protein surface.

before the refinement. The met-per-deuterated myoglobin was refined first using reflections between 1.8Å - 6.0Å resolutions. After 3 cycles of X-PLOR refinement including slowcool, 51 water molecules were added to the model, and the crystallographic R-factor dropped from 30.1% to 18.8%. The plain solvent-shell then was analyzed.

Water molecules in the model were excluded when protein's surface was defined. Fourteen 0.3Å thick shells were defined. The best values for the scattering density and B factor of each shell were determined during 5 cycles of least-square fitting using reflections between 4.0Å - 18.8Å resolutions. Figure 3 shows the B factors. Relatively low B factors are observed at distances 3.5Å and 5.1Å from the center of non-hydrogen protein atoms. In the previous neutron study, low B factors occurred at distances of 2.43Å and 4.1Å from the center of the protein's surface atoms, including hydrogens. Considering the fact that protein surfaces are covered by hydrogens, there should be about a 1Å difference in the location of comparable shells between the two studies. After such adjustment, the results are in good agreement. This suggests that this analysis is applicable to X-ray data, and can be used routinely for X-ray structure refinement.

The above solvent-shell analysis was applied to the model that included the 51 water molecules. The resulting $\mathbf{F}_s$(hkl) was used in further refinement. Figure 4 shows the crystallographic R-factor before and after using solvent-corrected $\mathbf{F}_n$(hkl)'s. It is clear that the R-factor was reduced dramatically at low resolution when solvent-shell refinement was applied. When examining differences between the ($2F_o$-F_t, ϕ_t) and ($2F_n$-F_p, ϕ_p) maps, we found that the former revealed more solvent features than the latter (Figure 5). Therefore, peaks in the (F_o-F_t, ϕ_t) map with values higher than 3.5σ (σ is the rms deviation from the mean difference-intensity), were accepted as water molecules in the following protein refinement. After an additional 20 cycles of X-PLOR standard procedure in which slowcool was replaced by positional refinement using solvent-free reflections between 1.5Å - 18.8 Å, the R factor became 17.1%. In this final model, there were 88 water molecules and 2 sulfate ions.

Atomic Pair Distribution Function

The 88 water molecules determined from the per-deuterated myoglobin structure refinement were used for calculating two different atomic pair distribution functions, g_{WP} and g_{WA}. Water molecules in contact with protein N or O atoms were used to assess g_{WP}

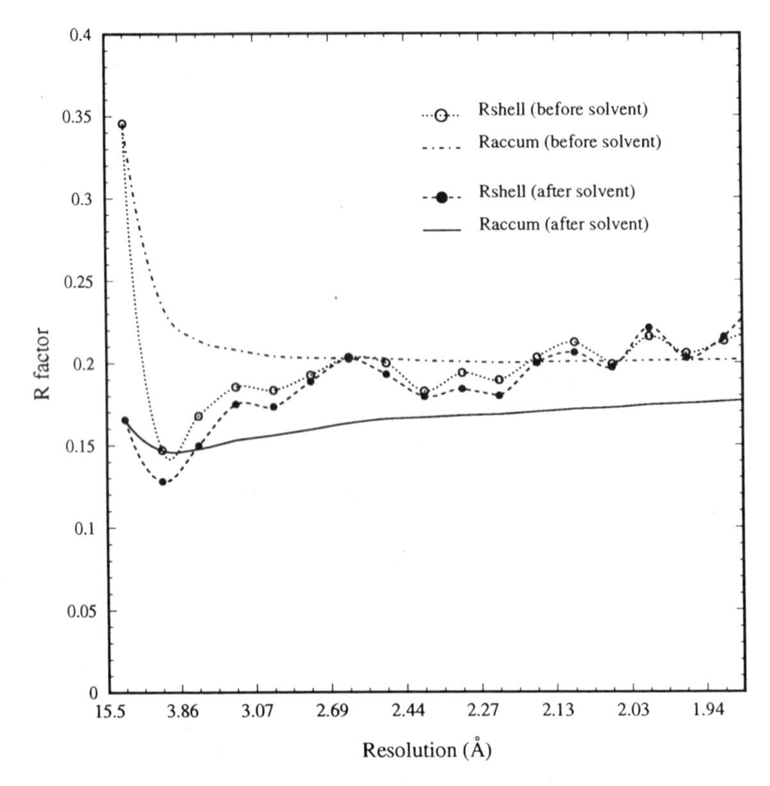

Figure 4. Crystallographic R factor before and after using solvent-corrected reflections.

between water O and protein polar atom; those in contact with protein C atoms were used to evaluate g_{WA} between water O and protein non-polar atoms (Figure 6). A few peaks were distinguishable from the background, especially the first one at 2.8Å in g_{WP}, which represents the interaction between water molecules and protein atoms by hydrogen bonding, and also the first one at 3.6Å in g_{WA}, which represents the interaction between water molecules and protein atoms by van der Waals contacts. These values agree well with a molecular dynamics study on a dipeptide in water, in which 2.85Å and 3.7Å were reported, respectively (Rossky & Karplus, 1978).

DISCUSSION

Importance of Fusion Myoglobin

During the search for the best system for preparing the neutron samples, we tried expressing the synthetic myoglobin gene in the pUC19 and in the T7 expression system, in both its intact and fusion forms. The T7 system clearly expressed the gene at levels at least 10 times higher than the pUC19 system. By using the fusion myoglobin, we were able to cleave off the initial methionine attached to the first myoglobin residue in the expressed protein. This helped us to overcome problems in crystallization, and also yielded a crystal form ($P2_1$), which was identical to that of the wild-type myoglobin. It was reported that

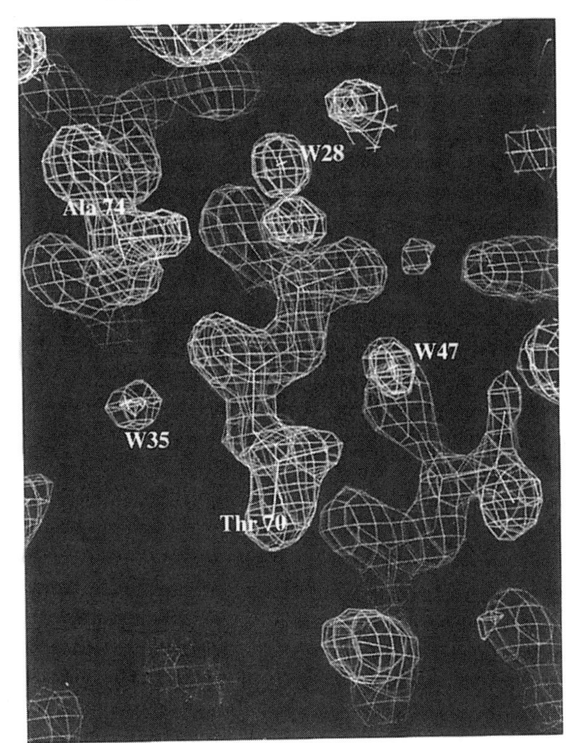

Figure 5. Comparison of $(2F_o\text{-}F_t,\phi_t)$ and $(2F_n\text{-}F_p,\phi_p)$ maps. Both maps are drawn at 1.5σ ($0.72e^-/Å^3$), where σ is the rms deviation from the mean electron-density in each map calculation. The $(2F_o\text{-}F_t,\phi_t)$ map is in blue; the $(2F_n\text{-}F_p,\phi_p)$ map is in magenta. Stronger electron densities at water sites W28, W35, and W47 in the $(2F_o\text{-}F_t,\phi_t)$ map, but similar at protein region, indicate that solvent features are revealed better after solvent phases are included in the map calculation. (A color version of this figure appears in the color insert following p. 214.)

the over-expressed myoglobin with methionine crystallized in P_6 form (Phillips *et al.*, 1990), and in small sizes.

Importance of Induction at Low Temperature

The common problem that accompanies high-level expression of a eukaryotic gene in *E. coli* is the solubility of the protein (Marston, 1986). We had the same problem when we expressed myoglobin in its intact or fusion forms. This issue was greatly relieved when we induced the cell at 25°C instead of 37°C and with less IPTG, 0.1mM instead of 0.4mM. When we followed standard inducing procedures in the T7 system about 30% of the expressed myoglobin was soluble; the ratio increased to about 85% when the modified procedure was used, and when multiple extractions were made.

Ordered Hydration Layers Surrounding Proteins

Solvent structure has been studied by X-rays, neutron diffraction, NMR and molecular dynamics (Karplus & Faerman, 1994; Levitt & Park, 1993). Opinions on why and how water molecules interact with the protein surface atoms differ. From our study, ordered features in the solvent region outside myoglobin were evident (Figure 2 and Figure 3), which supported the idea that water sites near to the protein surface's were occupied constantly by solvent molecules. The results from polar/non-polar solvent shell analysis on neutron data further shows that the distances at which the first minimal B factor of polar and non-polar shells appeared were different, 2.65Å and 3.4Å respectively. These distances are very close to those where water molecules have the largest probability of being

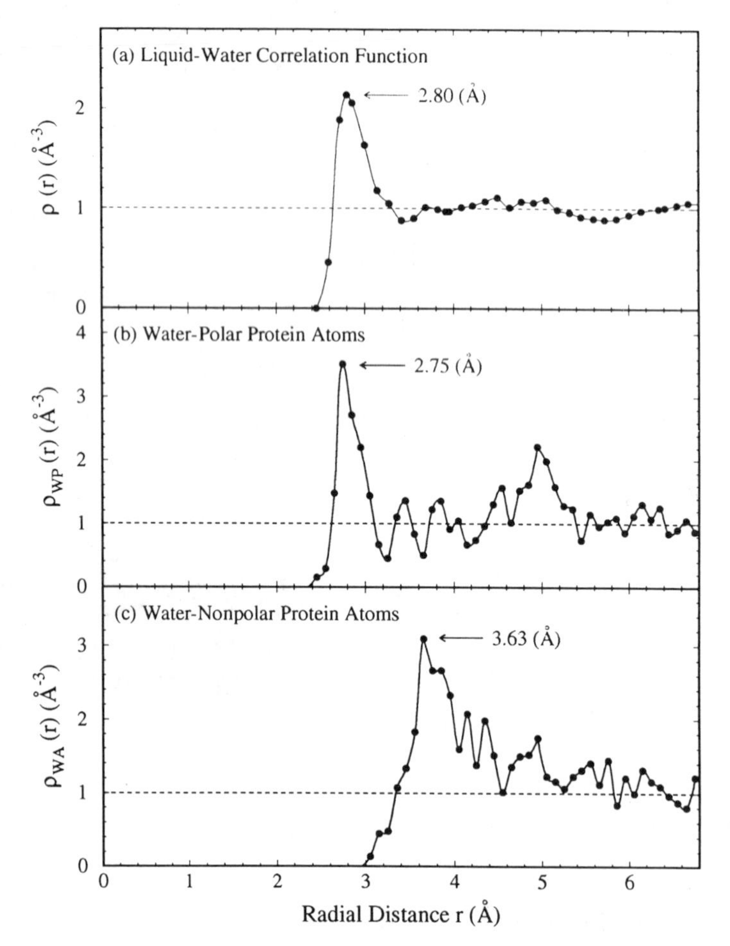

Figure 6. Atomic pair distribution functions. (a) Atomic pair distribution function for liquid water from X-ray study (Narten & Levy, 1972). (b) Atomic pair distribution function $g_{WP}(r)$ of water O and protein polar atoms. (c) Atomic pair distribution function $g_{WA}(r)$ of water O and protein non-polar atoms. The first peak of $g_{WP}(r)$ occurs at 2.75Å, which is the same as the peak position of liquid water distribution function. The peak in liquid water distribution function corresponds to the strong hydrogen-bonding interactions between O atoms in liquid water. The $g_{WA}(r)$ has its first peak at 3.6Å, which is close to the van der Waals contact distance between O atoms and nonhydrogen atoms.

near to polar and non-polar protein atoms according to our studies of atomic pair distribution functions (Figure 6). When the interactions between those water molecules and protein atoms were examined, based on the refined per-deuterated myoglobin structure, we found that majority of water molecules near protein polar residues formed hydrogen bonds with either the main chain N's and O's or with side atoms; those near non-polar residues did not form hydrogen bonds directly, but were hydrogen-bonded to nearby water molecules which, in turn, formed hydrogen bonds with protein atoms or other water molecules. We also found that water molecules that were directly in contact with protein atoms were at van der Waals distances. All these results suggested that layers of hydration existed around myoglobin—the hydration layer formed at the polar protein surface region was due to hydrogen bonding interaction, and that formed at the non-polar surface region was due to van der Waals interactions, which was further strengthened by hydrogen-bond interactions between nearby water molecules. Our study did not seem to support the geometric mechanism (Kuhn *et al.*, 1992) in governing water molecule distribution around proteins not did it agree with the crystal-packing mechanism (Levitt & Park, 1993). Agreement of water sites (within 0.5Å) of our refined structure and other myoglobin structures in the Protein Data Bank was about 50% - 70%, similar to what has been reported for other proteins (Finer-Moore *et al.*, 1992; Zhang & Matthews, 1994). We believe that the difference was due to a complex of reasons. First, the appearance of water density in an electron density map is affected by temperature factors. These factors, in turn, can be affected by different refinement packages. The criteria that researches use for acceptance of a water molecule also differs. We suggest that the water sites from different structures each may only represent a portion of the water molecules surrounding the protein's surface. A similar conclusion was reached from a molecular dynamic study of myoglobin hydration (Lounnas & Pettitt, 1994).

Inclusion of the Procedure for Solvent Shell Analysis in Structure Refinement

The inclusion of the solvent contribution in the refinement of per-deuterated myoglobin structure from X-ray diffraction data has made improvement in the protein structure refinement, especially at the small $\sin\theta/\lambda$ end. Compared to the previous neutron study (Cheng & Schoenborn, 1991), the ($2F_o$-F_t, ϕ_t) map did not reveal new protein features but only resolved new features of the solvent. This is probably due to the fact that the X-ray data used here had a resolution of 1.5Å at which protein structure already was well determined, or X-ray data, in general, were not as sensitive as neutron data to this solvent method. However, improvement of electron density at water sites was distinguishable (Figure 5), and might be more substantial when the data resolution is worse. The procedure for plain solvent shell analysis was reported to have been incorporated into the X-PLOR protein refinement package (Jiang & Brünger, 1994).

SUMMARY

In this work, a fusion sperm-whale myoglobin gene with part of the λ cII protein gene was cloned into the T7 expression system, and highly over-expressed in *E.coli*, both in protonated and deuterated media. Soluble apo-fusion myoglobin was obtained when the gene was induced at 25°C. After its reconstitution with heme, and cleavage with trypsin, milligram amounts of holo-myoglobin were obtained, and used for crystallization. The synthetic sperm-whale myoglobin crystallized in the $P2_1$ space group isomorphously with the native protein crystal. The crystals were large enough for both X-ray and neutron studies.

X-ray diffraction data were collected on a per-deuterated met-myoglobin crystal. The crystal structure was refined to a final R factor of 17.1% with a combination of X-PLOR and solvent-shell procedures using all the reflections in the data, including low-order reflections. The refined structure has 88 water molecules, and 2 sulfate ions.

Two ordered hydration shells were identified from the X-ray diffraction data. Distribution of solvent around protein polar/non-polar regions within the first hydration shell was further differentiated from previous neutron diffraction data using polar/non-polar solvent shell procedures. The results suggest that solvent molecules form strong hydrogen bonds with polar protein atoms, and hydrogen-bond networks at van der Waals contact distances from non-polar protein atoms. Consistent results also were obtained from analysis of atom-pair distribution functions of water oxygen atoms and protein polar and non-polar atoms from the refined structure of per-deuterated myoglobin, which have their first peaks at 2.75Å and 3.63Å, respectively.

In conclusion, this study demonstrated that (i) the first hydration layer distributes differently at the polar and non-polar surfaces - 2.7Å from the centers of polar protein atoms and 3.4Å from the centers of non-polar protein atoms; (ii) X-ray data is sensitive to the solvent structure and can be used to give a sensible solvent model that can be used for protein structure refinement; (iii) sufficient amount of fully deuterated protein can be obtained for neutron crystallography by expressing the protein in *E.coli* grown in fully deuterated minimal medium.

ACKNOWLEDGMENTS

We thank Sue Ellen Gerchman, Helen Kycia, and Vito Graziano for their able technical assistance, and Dr. J. Dunn for valuable suggestions and discussions. We are also indebted to Dr. R. Hamlin in the MarResearch (Poway, CA) for access to his mar detector for data collection.

REFERENCES

Bacon, G.E., (1975). *Neutron Diffraction*, Clarendon Press, Oxford, pp38–39.

Bragg, W.L., & Perutz, M.F., (1952). The external form of the haemoglobin molecule. I. *Acta Cryst.*, 5:277–283.

Brünger, A.T., (1993). In *X-PLOR Version 3.1. A System for X-ray Crystallography and NMR*, Yale University, New Haven, CT.

Cheng, X.D., & Schoenborn, B.P., (1990). Hydration in protein crystals: A neutron diffraction analysis of carbon-monoxymyoglobin. *Acta Cryst.*, B46:195–208.

Cheng, X.D., & Schoenborn, B.P., (1991). Neutron diffraction study of carbon-monoxymyoglobin. *J. Mol. Biol.*, 220:381–399.

Finer-Moore, J.S., Kossiakoff, A.A., Hurley, J.H., Earnest, T., & Stroud, R.M., (1992). Solvent structure in crystals of trypsin determined by X-ray and neutron diffraction. *Proteins: Struct. Funct., Genet.*, 12:203–222.

Gerchman, S.E., Graziano, V., & Ramakrishnan, V., (1994). Expression of histon genes in *E.coli*. *Protein Expression and Purification*, 5:242–251.

Jiang, J.S., & Brünger, A.T., (1994). Protein hydration observed by X-ray diffraction. Solvation properties of penicillopepsin and neutraminidase crystal structures. *J. Mol. Biol.*, 243(1):100–115.

Karplus, P.A., & Faerman, C., (1994). Ordered water in macromolecular structure. *Curr. Opin. Struct. Biol.*, 4(5):770–776.

Kuhn, L.A., Siani, M.A., Pique, M.E., Fisher, C.L., Getzoff, E.D., & Tainer, J.A., (1992). The interdependence of protein surface topography and bound water molecules revealed by surface accessibility and fractional density measures. *J. Mol. Biol.*, 228:13–22.

Levitt, M., & Park, B.H., (1993). Water: Now you see it, now you don't. *Structure*, 1:223–226.

Lounnas, V., & Pettitt, B.M., (1994). A connected-cluster of hydration around myoglobin: correlation between molecular dynamics simulations and experiment. *Proteins: Struct. Funct. Genet.*, 18:133–147.

Marston, F.A.O., (1986). The purification of eukaryotic polypeptides synthesized in *Escherichia coli*. *Biochem. J.*, 240:1–12.

Matthews, B.W., (1968). Solvent content of protein crystals. *J. Mol. Biol.*, 33:491–497.

Nagai, K., & Thogersen, H.C., (1984). Generation off β-globin by sequence-specific proteolysis of a hybrid protein produced in *Escherichia coli*. *Nature*, 309:810–812.

Narten, A.H., & Levy, H.A., (1972). In: *Water-A Comprehensive Treatise*, (editor F. Franks) 1:311–332.

Norvell, J.C., Nunes, A.C., & Schoenborn, B.P., (1975). Neutron diffraction analysis of myoglobin: structure of the carbon monoxide derivative. *Science*, 190:568–570.

Phillips, G.N.Jr., Arduini, R., Springer, B.A., & Sligar, G., (1990). Crystal structure of myoglobin from a synthetic gene. *Proteins: Struct. Funct. Genet.*, 7:358–365.

Raghavan, N.V., & Schoenborn, B.P., (1984). The structure of bound water and refinement of acid metmyoglobin. In: *Neutrons in Biology, Basic Life Sciences* (editor B.P. Schoenborn) Plenum Press, New York 27:247–259.

Ramakrishnan, V., & Gerchman, S.E., (1991). Cloning, sequencing, and overexpression of genes for ribosomal proteins from *Bacillus stearothermophilus*. *J. Biol. Chem.*, 266:880–885.

Rossky, P., & Karplus, M., (1978). Solvation. A molecular dynamics study of a dipeptide in water. *J. Amer. Chem. Soc.*, 10:1913–1937.

Schoenborn, B.P., (1988). Solvent effect in protein crystals. A neutron diffraction analysis of solvent and ion density. *J. Mol. Biol.*, 201:741–749.

Springer, B.A., & Sligar, S.G., (1987). High-level expression of sperm whale myoglobin in *Escherichia coli*. *Proc. Natl. Acad. Sci. USA*, 84:8961–8965.

Studier, F.W., Rosenberg, A.H., Dunn, J.J., & Dubendorff, J.W., (1990). Use of T7 RNA polymerase to direct expression of cloned gene. *Methods Enzymol.*, 185:60–89.

Varadarajan, R., Szabo, A., & Boxer, S.B., (1985). Cloning, expression in *Escherichia coli*, and reconstitution of human myoglobin. *Proc. Natl. Acad. Sci. USA*, 82:5681–5684.

Zhang, X.J., & Matthews, B.W., (1994). Conservation of solvent-binding sites in 10 crystal forms of T4 lysozyme. *Protein Science*, 3:1031–1039.

MYOGLOBIN SOLVENT STRUCTURE AT DIFFERENT TEMPERATURES

B. V. Daniels,[1] B. P. Schoenborn,[2] and Z. R. Korszun[1]

[1] Biology Department
Brookhaven National Laboratory
Upton, New York 11973
[2] Life Sciences Division
Los Alamos National Laboratory
Los Alamos, New Mexico 87545

ABSTRACT

The structure of the solvent surrounding myoglobin crystals has been analyzed using neutron diffraction data, and the results indicate that the water around the protein is not disordered, but rather lies in well-defined hydration shells. We have analyzed the structure of the solvent surrounding the protein by collecting neutron diffraction data at four different temperatures, namely, 80, 130, 180, and 240K. Relative Wilson Statistics applied to low resolution data showed evidence of a phase transition in the region of 180K. A plot of the liquidity factor, B_{sn}, versus distance from the protein surface begins with a high plateau near the surface of the protein and drops to two minima at distances from the protein surface of about 2.35Å and 3.85Å. Two distinct hydration shells are observed. Both hydration shells are observed to expand as the temperature is increased.

INTRODUCTION

Water plays an essential role in determining protein structure, dynamics and function. Free-energy and volume changes (Kuntz & Kauzmann, 1973) are manifest in myoglobin when its charged groups interact with water. In hemoglobin, allosteric regulation is dependent upon the binding of water molecules when making the transition between T and R conformations (Perutz, 1970; Colombo *et al.,* 1992). Poorly hydrated lysozyme remains inactive without a minimum amount of water present (Yang & Rupley, 1979). Its specific heat capacity is solvent dependent, responding nonlinearly at low water content. Once present, water surrounds the protein with a hydration layer that differs from the bulk dynamically and thermodynamically (Rupley *et al.,* 1983). Water that hydrogen bonds to the

Neutrons in Biology, edited by Schoenborn and Knott
Plenum Press, New York, 1996

protein backbone has a specific heat that differs from the bulk solvent (Careri *et al.*, 1980). Water heat capacity at high water-to-protein content approximates that of ice below 0°C and that of pure water above 0°C (see Kuntz & Kauzmann, 1974 for references). Early studies (Takashima & Schwan, 1965; Rosen, 1963) of the dielectric constant of a protein-water system revealed a slow increase until the sample had at least 10% water, increasing rapidly thereafter until about 25% water was in the sample.

As a result of the evidence of the importance of water in the solvation of globular proteins, the question arises as to how the water is interacting with the protein. A study done on the radius of gyration of myoglobin determined that it is larger than what would be expected for the protein itself (Ibel & Stuhrmann, 1975), suggesting that there are water molecules that are tightly bound to the protein and, hence, move along with it. These are easily localized in X-ray and neutron crystallography. Electron microscope studies (Kellenberger, 1978) also provide evidence for a tightly bound hydration layer. Certainly, the first step in the hydration process (Rupley *et al.,* 1983) which involves water interactions with charged groups of the protein is consistent with this picture. The second step involves a change in the heat capacity of water that binds to backbone carbonyl and amide groups, while in the third stage clusters of water molecules form which eventually cover the surface of the protein. What ultimately results from these steps is an ordered hydration layer that eventually diffuses into the bulk. One model of hydration (see Edsall & McKenzie, 1983, for references) classifies three types of water. Type I is bulk water. Type II water is found around nonpolar residues. It is partially bound, and has a short residence time. Water molecules that form hydrogen bonds with the polypeptide, generally at the surface, are of the Type III category, strongly bound, and have long residence times.

Macromolecular X-ray and neutron crystallography have been successfully used to determine the location of strongly bound water molecules and water sites that are, on average, occupied by water molecules that may be exchanging. Unlike the time- and space-averaged crystallographic results, NMR studies present a picture of hydration on a fast time-scale, and provide information on the dynamics of exchange of water. NMR, however, is not sensitive to the changes in the position of a water molecule that occurs over a longer time scale. Thus, although NMR agrees well with crystallography in that there is a small number of well-defined water molecules whose positions in the crystal structure and in solution are the same, there is disagreement about protein surface hydration (Otting *et al.,* 1991). However, molecular dynamics simulations of myoglobin (Lounnas & Pettitt, 1994) also reveal multiple solvation layers extending from the protein surface. In addition, Monte Carlo simulations (Parak *et al.,* 1992), give excellent agreement with crystallographic analysis of hydration structure around the protein.

Based on a neutron diffraction study of carbonmonoxy myoglobin, Cheng & Schoenborn (1990), showed two distinct hydration layers at positions of approximately 2.35Å and 3.85Å from the protein surface. In this model, water surrounding the protein can be described as a 'liquid-like' hydration sphere with properties that gradually approach those of the bulk liquid as the distance from the surface of the protein increases. In Schoenborn's model (1988), the solvent region is divided on a fine grid and the grid points are grouped into regions that form radially expanding shells around the protein. An average scattering density for the solvent is assigned at each grid point and the contribution of each shell to the total coherent intensity is calculated. The observed structure factor modulus for the system, F_o, is the sum of the structure factors for the protein, $\mathbf{F}_p$, and the solvent, $\mathbf{F}_s$. In this model

$$F_o = |\mathbf{F_p} + \mathbf{F_s}|$$

where the structure factor for the solvent is summed over n shells

$$F_s = \sum_n F_{sn}$$

and

$$F_{sn} = \rho_{sn} \exp(-B_{sn} \sin^2\theta / \lambda^2) \times \sum_{xyz} \exp[-2\pi i(hx + ky + lz)_n]$$

$\exp(-B_{sn} \sin^2\theta/\lambda^2)$ is the Debye-Waller factor, θ is the Bragg angle and λ is the wavelength. ρ_{sn} is the average scattering density of the solvent and xyz are the grid solvent coordinates of the nth shell surrounding the protein. B_{sn} has been interpreted as a 'liquidity factor' and is a measure of the coherent scattering from the solvent. Although low resolution data are typically ignored in data analysis, the hydration spheres contribute significantly to coherent scattering at low resolution and an analysis of these data provides information on the general distribution of solvent within the crystal.

Several X-ray crystallographic studies as a function of temperature have shown that various aspects of protein structure, dynamics, and function are affected. For example, X-ray analysis of ribonuclease-A at different temperatures (Tilton *et al.*, 1992) gives evidence of a biphasic behavior of the protein, with a phase change occurring in the vicinity of 180K. Egg-white lysozyme also exhibits structural changes with varying temperature (Young *et al.*, 1994). X-ray analysis on myoglobin at different temperatures (Parak *et al.*, 1987) indicates that the lattice vectors a and b vary linearly with temperature, and c behaves non-linearly above a temperature of 180K. In order to further characterize the physical and chemical nature of hydration of globular proteins, we have initiated a neutron diffraction study on myoglobin as a function of temperature.

METHODS

5Å resolution neutron diffraction data on a single myoglobin crystal (space group $P2_1$, a = 64.7Å, b = 31.1Å, c = 34.8Å, β = 105.7°) were collected on beam line H3A at Brookhaven National Laboratory's High Flux Beam Reactor. The crystal was frozen via direct insertion into liquid nitrogen. Since myoglobin has a rather high tolerance to temperature shock, the only cryo-protectant used was to first dip the crystal in mineral oil. This proved sufficient, since the Bragg reflections showed no evidence of crystal damage. The wavelength of radiation was 1.61Å. Experimental temperatures were 80, 130, 180, and 240K. These data were reduced using MADnes (Pflugrath, 1987), and scaled using FBSCALE (Kabsch, 1988) and subsequently merged. Reduced data were analyzed using software developed by Schoenborn (1988). The conventional crystallographic R-factor for these data sets is in the range of 37% prior to the inclusion of the solvent terms and drops to about 22% after the solvent contribution to the coherent Bragg scattering is included.

RESULTS AND DISCUSSION

Wilson statistics were calculated for each data set. A plot of B versus temperature, shows a sharp increase near 180K (Figure 1), indicating a phase transition in this region. A similar transition has been observed in other globular protein-water systems, and is thought to be a glass transition. Figure 2 is a plot of the liquidity factor, B_{sn}, versus shell number, n, obtained from our solvent analysis for the 80K data. The general features in this figure are typical of the results obtained at all temperatures. The high relative liquidity factor of 512 near the protein surface is due to the van der Waals' separation between the protein surface and the first hydration layer. As the distance from the protein surface increases, the liquidity factory reaches a minimum, rises again, drops to a second minimum, and finally reaches a plateau. Each minimum represents a hydration layer with a high probability of finding a water molecule, which contributes to the coherent scattered intensity. The final plateau corresponds to the bulk which shows no contribution to the coherent scattered intensity of the low resolution reflections. The two distinct hydration layers are evident at each of the temperatures analyzed. The existence of two such layers is in good agreement with the room temperature results of Cheng and Schoenborn (1990) on carbonmonoxymyoglobin, where hydration shells at approximately 2.35Å and 3.85Å from the protein surface are observed, as well as with Monte Carlo simulations (Parak *et al.*, 1992).

At each temperature, the barrier between these hydration layers was fit to a Gaussian curve. This yielded information on the width of the barrier. The position of the Full Width at Half Maximum (FWHM) versus temperature (Figure 3), were plotted. Figures 3 shows that the hydration barrier increases in width as a function of temperature. Two linear fits to the data were calculated, one using all temperatures and a second omitting the 180K data point. The residual for the latter fit was significantly better and this least-squares line is shown as a dotted line in the figure. Based on this straight line fit to the data, the plot of FWHM versus temperature yielded a slope of 1.38×10^{-4}Å/°K which is close to that observed for the increase in the O—O separation of 1.54×10^{-4}Å/°K (Eisenberg & Kauzmann, 1969) for pure ice at atmospheric pressure in this temperature range. Figure 3

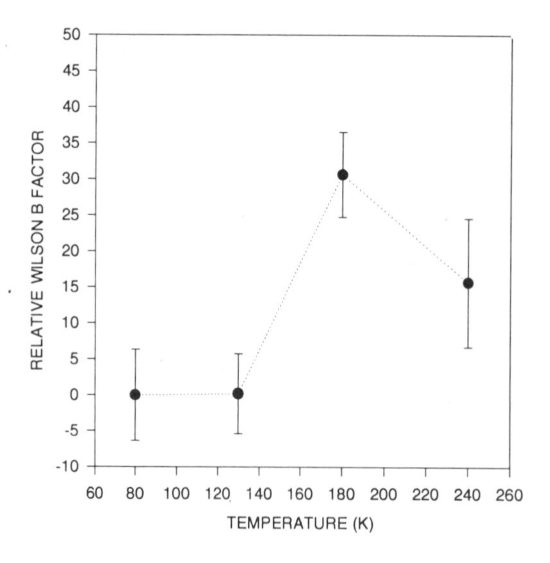

Figure 1. A plot of the relative Wilson B factor versus temperature. B is obtained from the expression $\ln [\langle I \rangle / \Sigma f^2] = \ln K - 2 B [\sin^2 \theta / \lambda]$. The plot indicates a phase transition at 180K. Error bars shown are the standard errors calculated from a least squares fit to the data at each temperature.

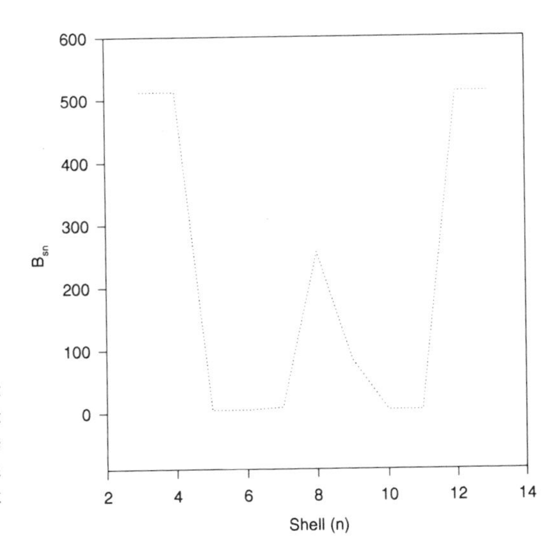

Figure 2. Liquidity factor, B_{sn} vs. distance at 80K. Two hydration layers are observed. The first plateau represents the hard sphere contact and the peak characterizes the barrier between hydration layers, while the second plateau results from bulk water.

also provides evidence that the observed phase transition manifests itself in the solvent as well as in the protein.

The linear thermal expansion coefficient, α, has been estimated from this model. In going from 80K to 240K the protein is calculated to expand by approximately 0.3Å based on the shift in the first plateau, which corresponds to the van der Waals surface of the protein, to longer distance (Figure 1). Myoglobin can be described as an ellipsoid of approximately 10Å × 10Å × 19Å and assuming an average radius of 13Å, α is estimated to be 1.4×10^{-4}Å/°K. This value is higher than, but comparable to, the value of 1.15×10^{-4}Å/°K obtained by Fraunfelder *et al.* (1987). The major reason for this higher value is that the solvent grid used in this study (0.3Å) was probably a little too coarse to obtain a more accurate value for α. Nevertheless, the agreement is good and may be considered as an independent validation of the shell solvent model. Future work will include data collection

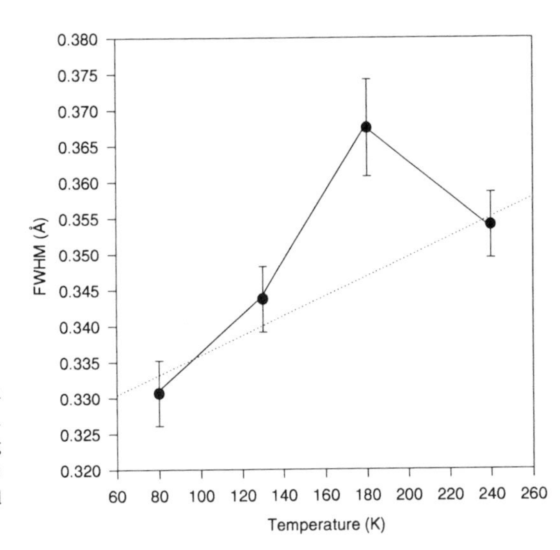

Figure 3. FWHM vs. temperature. All data points are joined by solid line segments. A straight line fit to the data (dotted line), excluding the 180K data point, yields a slope of 1.38×10^{-4}Å/°K. Errors are estimated from the standard error of the slope of the least squares line.

and analysis at more closely spaced temperatures in the 180K region and the determination of high resolution structures at the interesting temperatures.

CONCLUSION

In conclusion, our results indicate that the shell solvent model provides a good physical approximation to the coherent scattering contribution of the solvent to low resolution Bragg reflections, and can be used as a correction to the structure factor amplitudes of these reflections, allowing their incorporation in structure refinement.

ACKNOWLEDGMENTS

This work is supported by the Office of Health and Environmental Research of the United States Department of Energy, and by an NSF neutron user grant MCB-93 18839

REFERENCES

Careri, G., Gratton, E., Yang, P.H., & Rupley, J.A., (1980). Correlation of IR spectroscopic, heat capacity, diamagnetic susceptibility and enzymatic measurements on lysozyme powder. *Nature,* 284:572–573.

Cheng, X., & Schoenborn, B.P., (1990). Hydration in protein crystals. A neutron diffraction analysis of carbonmonoxymyoglobin. *Acta Cryst.,* B46:195–208.

Colombo, M.F., Rau, D.C., & Parsegian, V.A., (1992). Protein solvation in allosteric regulation: A water effect on hemoglobin. *Science,* 256:655–659.

Edsall, J.T., & McKenzie, H.A., (1983). Water and protein. II. The location and dynamics of water in protein systems and its relation to their stability and properties. *Adv. Biophys.,* 16:53–183.

Eisenberg, D., & Kauzmann, W., (1969). In *The Structure and Properties of Water.* Oxford University Press, p74 *et seq.*

Fraunfelder, H., Hartmann, H., Karplus, M., Kuntz, I.D., Kuriyan, J., Parak, F., Petsko, G.A., Ringe, D., Tilton, R.F., Connolly, M.I., & Max, N., (1987). Thermal expansion of a protein. *Biochemistry,* 26:254–261.

Ibel, K., & Stuhrmann, H., (1975).Comparison of neutron and X-ray scattering of dilute myoglobin solutions. *J. Mol. Biol.,* 93:255–265.

Kabsch, W., (1988). Evaluation of single-crystal X-ray diffraction data from a position-sensitive detector. *J. Appl. Cryst.,* 21:916–924.

Kellenberger, E., (1978). High resolution electron microscopy. *Trends Biochem. Sci.,* 3(6):135–137.

Kuntz, I.D., & Kauzmann, W., (1974). Hydration of proteins and polypeptides. *Adv. Protein Chem.,* 28:239–345.

Lounnas, V., & Pettitt, B.M., (1994). Distribution function implied dynamics versus residence times and correlation: solvation shells of myoglobin. *Proteins: Structure, Function and Genetics,* 18:148–160.

Otting, G., Liepinsh, E., & Wuthlich, K., (1991). Protein hydration in aqueous solution. *Science,* 254:974–980.

Parak, F., Hartmann, H., Aumann, K.D., Reuscher, H., Rennekamp, G., Bartunik, H., & Steigemann, W., (1987). Low temperature X-ray investigation of structural distributions in myoglobin. *Eur. Biophys. J.,* 15:237–249.

Parak, F., Hartmann, H., Schmidt, M., Corongiu, G., & Clementi, E., (1992). The hydration shell of myoglobin. *Euro. Biophys. J.,* 21:313–320.

Perutz, M.F., (1970). Stereochemistry of cooperative effects in haemoglobin. *Nature,* 228:726–734.

Pflugrath, J.W., & Messerschmidt, A., (1987). Munich Area Detector NE Systems, V. 27.

Rosen, D., (1963). Dielectric properties of protein powders with adsorbed water. *Trans. Faraday Soc.,* 59:2178–2191.

Rupley, J.A., Gratton, E., & Careri, G., (1983). Water and globular proteins. *Trends Biochem. Sci.,* 8:18–22.

Schoenborn, B.P., (1988). The solvent effect in protein crystals. A neutron diffraction analysis of solvent and ion density. *J. Mol. Biol.,* 201:741–749.

Takashima, S., & Schwan, H.P., (1965). Dielectric dispersion of crystalline powders of amino acids, pepides, and proteins. *J. Phys. Chem.,* 69:4176–4182.

Tilton, R.F., Dewan, J.C., & Petsko, G.A., (1992). Effects of temperature on protein structure and dynamics: X-ray crystallographic studies of the protein ribonuclease-A at nine different temperatures from 98 to 320K. *Biochemistry,* 31:2469–2481.

Wang, C.X., Bizawrri, A.R., Xu, Y.W., & Cannistraro, S., (1994). Molecular dynamics of copper plastocyanin: simulations of structure and dynamics as a function of hydration. *Chem. Phys.,* 813:155–166.

Yang, P., & Rupley, J.A., (1979). Protein-water interactions. Heat capacity of the lysozyme-water system. *Biochemistry,* 18:2654–2661.

Young, A.C., Tilton, R.F., & Dewan, J.C., (1994). Thermal expansion of hen egg-white lysozyme. Comparison of the 1.9Å resolution structures of the tetragonal form of the enzyme at 100K and 298K. *J. Mol. Biol.,* 235:302–317.

DETERMINATION OF PROTEIN AND SOLVENT VOLUMES IN PROTEIN CRYSTALS FROM CONTRAST VARIATION DATA

John Badger*

Rosenstiel Basic Medical Sciences Research Center
Brandeis University
Waltham, Massachusetts 02254

ABSTRACT

By varying the relative values of protein and solvent scattering densities in a crystal, it is possible to obtain information on the shape and dimensions of protein molecular envelopes. Neutron diffraction methods are ideally suited to these contrast variation experiments because H/D exchange leads to large differential changes in the protein and solvent scattering densities and is structurally non-perturbing. Low resolution structure factors have been measured from cubic insulin crystals with differing H/D contents. Structure factors calculated from a simple binary density model, in which uniform scattering densities represent the protein and solvent volumes in the crystals, were compared with these data. The contrast variation differences in the sets of measured structure factors were found to be accurately fitted by this simple model. Trial applications to two problems in crystal structure determination illustrate how this fact may be exploited. (i) A translation function that employs contrast variation data gave a sharp minimum within 1–9Å of the correctly positioned insulin molecule and is relatively insensitive to errors in the atomic model. (ii) An *ab initio* phasing method for the contrast variation data, based on analyzing histograms of the density distributions in trial maps, was found to recover the correct molecular envelope.

INTRODUCTION

One of the most important and unique aspects of neutron diffraction methods for the study of biological structures is the opportunity, through controled deuteration, to vary the

---

* Present address: Biosym Technologies Inc., 9685 Scranton Road, San Diego CA 92121.

Neutrons in Biology, edited by Schoenborn and Knott
Plenum Press, New York, 1996

average scattering densities of the protein and solvent parts of the sample. With X-rays the average scattering density for protein is ~0.43e/Å^3 and the average scattering density for pure water is 0.33e/Å^3 (*ie* the protein/solvent *contrast* is 0.1e/Å^3). Since X-ray scattering factors are proportional to atomic numbers, crystallographic contrast variation experiments with X-rays require the introduction of high concentrations of strongly scattering ions or other solutes into the crystal to significantly increase the solvent scattering density. For example, the electron density of a 4M solution of ammonium sulphate is 0.41e/Å^3. Recently, the use of tunable synchrotron X-radiation sources with anomalously scattering ions has been investigated as an alternative method for changing the solvent scattering (Fourme *et al.*, 1994) but high concentrations of ions are still required for reliable measurements. With neutron diffraction, the great sensitivity of the solvent scattering density to the extent of deuteration may be exploited to provide very large contrast variation scattering differences. Furthermore, potential problems with crystal stability, specific solute binding and non-isomorphism that may occur when ions or other small molecules are introduced into the crystal solvent are avoided.

Neutron contrast variation measurements have been used to locate disordered components in known protein crystal structures (for example, Roth *et al.*, 1989; Timmins *et al.*, 1992) and for low resolution structure determinations (Bentley *et al.*, 1984). In recent years there has been a revival in interest in developing *ab initio* methods for analyzing very low resolution diffraction data in order to determine protein molecular envelopes (Lunin *et al.*, 1990; Subbiah, 1991). Several reports at the recent American Crystallographic Association meeting in Atlanta were devoted to this problem (Podjarny *et al.*, 1994; Roth & Pebay-Peyroula, 1994; David & Subbiah, 1994; Schluzen *et al.*, 1994; Filman *et al.*, 1994; Lunin *et al.*, 1994). Carter *et al.* (1990) have provided cogent theoretical arguments for the advantages of contrast variation measurements over single sets of structure factor data for determining molecular envelopes in macromolecular crystals. By analyzing data collected at several protein/solvent scattering contrasts it is possible to derive a structure factor amplitude which corresponds to the protein/solvent molecular envelope, without any contribution from density fluctuations within the protein or solvent volumes. Since the structure factors corresponding to the molecular envelope are small except at very low resolution, maps computed from these structure factors do not contain significant artifacts due to series termination errors and the analysis can be justifiably restricted to a small number of low order terms. A host of direct phasing approaches including the tangent formula (Carter *et al.*, 1990), the Sayre equation (Sayre, 1972), density histogram analysis (Lunin *et al.*, 1990) and 'dot condensation' (Subbiah, 1991) may be more appropriately applied to structure factors that represent the Fourier transforms of binary density maps of the protein and solvent volumes than to conventional data. In this report I describe the application of methods for low resolution structure determination which make use of contrast variation data.

ACCURACY OF THE MOLECULAR ENVELOPES DETERMINED FROM CONTRAST VARIATION DATA

We have measured neutron diffraction data from cubic insulin crystals (space group $I2_13$, a = 78.9Å) in which the solvent is ~55% and ~66% deuterated. An inspection of structure factor data from these two crystals shows that the contrast variation differences are only very significant below 20Å resolution and are small beyond 15Å resolution (see Badger *et al.*, 1995 for a tabulation of data and Figure 1). For this particular space group,

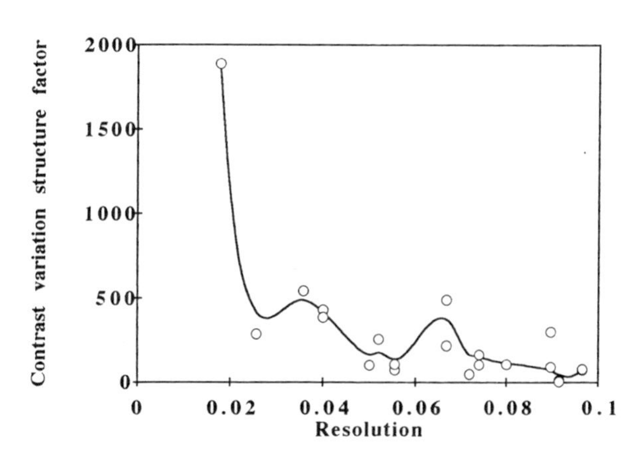

Figure 1. Differences in centric structure factor amplitudes, ΔF_{obs}, for cubic insulin crystals with different H/D contents plotted as a function of resolution. The resolution scale is given in reciprocal-Ångstrom units. The open circles mark the difference between individual structure factors and a smooth curve is fitted to these data points. Beyond the node at ~15Å resolution the contrast variation effect is small.

in which continuous solvent channels and rows of protein molecules run parallel to the crystal cell axes, the contrast variation effects are dominated by the lowest order centric data. The atomic structure of this crystal form has been refined against 1.7Å X-ray diffraction data (Badger *et al.*, 1991) and is, therefore, accurately known relative to the low resolution of the contrast variation differences. For comparison with the observed data, contrast variation differences were calculated from atomic models in which density inside the protein molecular envelope was set to zero and density outside the envelope was set to a constant level. When a protein envelope defined by the atomic van der Waals radii was expanded by 0.75Å, there was very good agreement of the contrast variation differences, ΔF_{obs}, with the calculated structure factors for the molecular envelope, ΔF_{calc}. Using a reliability index similar to the crystallographic Cullis R value to compare observed and calculated centric structure factors viz:

$$R = \frac{2\left(\sum |\Delta F_{obs} - \Delta F_{calc}|\right)}{\left(\sum \Delta F_{obs} + \Delta F_{calc}\right)}$$

R was ~0.18 for the 11 centric reflections to 15Å resolution. With only two sets of data there is a possible ambiguity in the value of ΔF_{obs} depending on whether the structure factor vectors are parallel or antiparallel. (When the solvent scattering density is substantially above or below the protein scattering density in both data sets the vectors will probably be parallel). In the calculation of the Cullis R value this ambiguity is easily resolved by choosing the value of ΔF_{obs} that is closest to ΔF_{calc}.

These comparisons show that the Fourier transform of the simple binary density model accurately fits the measured contrast variation difference amplitudes. A difference density map computed from these contrast variation difference data accurately represents the shape of the protein molecular envelope in the crystal (Figure 2) and a histogram of the density values in this map is strongly bimodal, with maxima corresponding to the low density protein region and the high density solvent region (Figure 3). This example of contrast variation difference data with a known structure illustrates the main arguments in favor of contrast variation data for the determination of molecular envelopes given by Carter *et al.* (1990); (i) the important contrast variation differences are confined to a small number of low order structure factors, (ii) the contrast variation differences agree well with the Fourier transform of a binary density map which accurately represents the protein

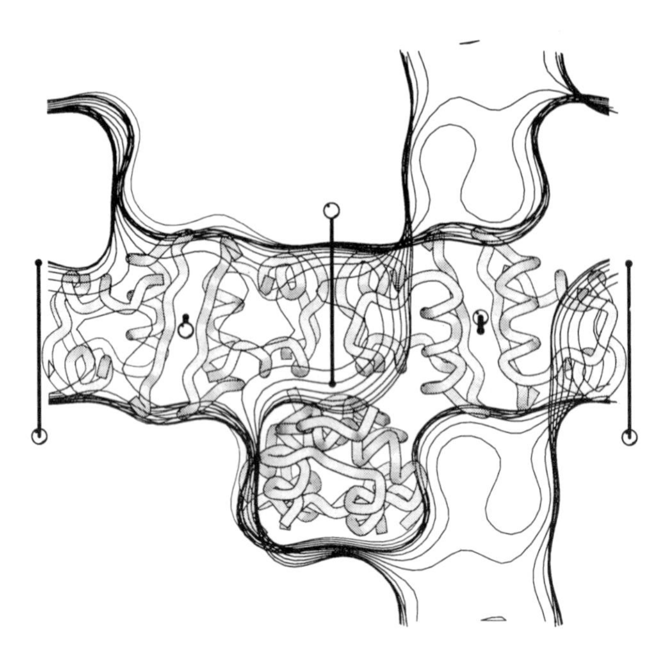

Figure 2. Density map from contrast variation data. This figure shows the density map computed from contrast variation structure factor differences, ΔF_{obs}, for the 11 centric terms to 15Å resolution. Phases were obtained from the Fourier transform of a model in which density inside the protein molecular envelope was set to zero and density outside the envelope was set to a constant. The atomic model (Badger *et al.*, 1991 and entry 9INS in the Brookhaven Protein Data Bank) is represented by a C_{α} trace and was drawn with the MOLSCRIPT program (Kraulis, 1991). One horizontal row of four molecules (the width of the unit cell) and parts of one row of molecules perpendicular to the page are shown.

molecular envelope, (iii) the magnitudes of the contrast variation differences are not necessarily be proportional to the magnitudes of the original structure factors - some structure factors are much more sensitive than others to variations in the protein/solvent scattering contrast.

APPLICATION TO TRANSLATION SEARCHES

Crystallographic techniques for determining a new structure using a known homologous structures as a trial model are becoming increasingly useful as the data base of solved structures grows in size. Methods for correctly orienting the trial model in the crys-

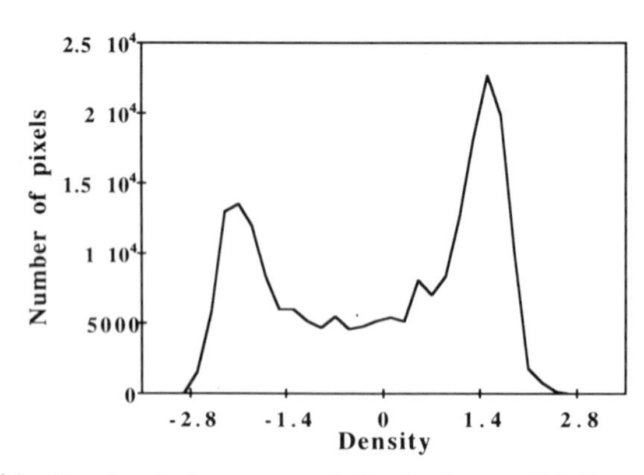

Figure 3. Histogram of density values in the contrast variation density map. This histogram shows that the distribution of density values in the map shown in Figure 2 is strongly bimodal, with maxima corresponding to the average densities of the solvent and protein volumes. The density values are on an arbitrary scale.

tal cell have become quite powerful (Fujinaga & Read, 1987; Yeates & Rini, 1990; Brunger, 1990) but algorithms for positioning this model in the crystal cell ('translation functions') frequently fail to give clear-cut solutions (Lattman, 1985). The main problem with determining the solution of a translation function is that relatively small differences between the co-ordinates of the true and trial structures may lead to poor agreement between observed and calculated structure factors. When the agreement between observed and calculated structure factors is poor, the solution for the correctly positioned trial model is not easily distinguished from false solutions for models which are positioned completely incorrectly. In principle, the sensitivity of the structure factor data to co-ordinate errors may be reduced by analyzing relatively low resolution data but, because of difficulties in accounting for bulk solvent scattering, these data are usually excluded from translation function calculations.

In the preceding section it was noted that structure factors derived from a binary density representation of the insulin atomic model gave good agreement with the contrast variation difference data. This result suggests that it might be possible to use the low resolution contrast variation difference data in a translation search, avoiding some of the problems with traditional translation functions. To investigate this possibility, I have carried out translation searches with the neutron diffraction contrast variation data. The cubic insulin crystal structure was originally solved assuming that the two fold axis in the cubic space group would generate an insulin dimer similar to the dimer found in the asymmetric unit of the 2Zn insulin crystal (Dodson *et al.*, 1978). Thus, the crystallographic search problem for a correctly oriented but unpositioned trial model was reduced to a 1-dimensional positional search along the X-axis. I have repeated this 1-dimensional search using the neutron contrast variation data with the correctly oriented cubic insulin structure (Figure 4a) and with molecules 1 and 2 from the 2Zn insulin crystal asymmetric unit (Figure 4b). The test with the cubic insulin structure should give a perfect solution within experimental errors and investigates the possibility of false minima. The tests with the two insulin monomers from the 2Zn insulin crystal are reasonably challenging examples of translation functions with inaccurate trial models since these two structures (Baker *et al.*, 1988) differ significantly from the cubic insulin monomer. For example, when monomer 1 of the 2Zn insulin crystal is superimposed on the cubic insulin monomer, the positions of 11 of the 51 C_α atoms differ by more than 1Å and 8 side chains are oriented with different χ_1 rotomer angles. In the comparison with monomer 2 the positions of 19 C_α atoms were found to differ by more than 1Å and 11 χ_1 rotomer angles are altered (Badger, 1992).

Figure 4 shows that the translation functions carried out with the low resolution contrast variation data were able to place all three trial models within 1–2Å of the correct position (defined here to be at X = 0). The very high crystal symmetry in this space group results in one additional R value minima when the trial models are placed at $X = \pm 1/2$ (Figure 4a). At this position the structure factor amplitudes for the centric terms are not distinguishable from their values at the correct position and only the phase angles of the purely imaginary terms change by 180 degrees. This ambiguity in the calculation of structure factor amplitudes for the low resolution centric terms is an inherent problem in this space group; however, this false solution may be easily distinguished from the correct solution by conventional calculations of the acentric structure factors at high resolution. With the exception of this ambiguity, the translation functions are characterized by a complete absence of significant false solutions. Other tests (not shown) indicate that, because of the low resolution of the structure factor data, the results of the translation search are insensitive to errors in the orientation of the trial model. In contrast to these calculations with low resolution contrast variation data, Dodson *et al.* (1978) found that the low resolu-

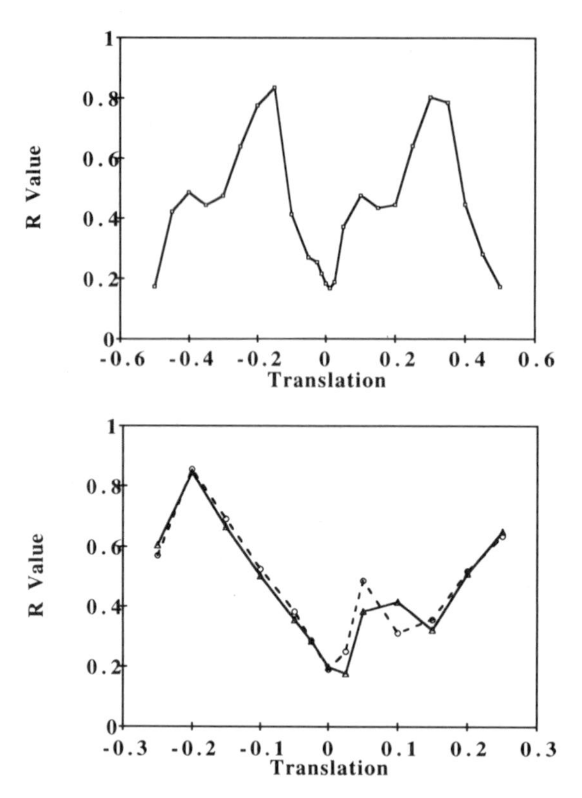

Figure 4. Translation searches with contrast variation data. Binary density models of the cubic insulin crystal were constructed in which the insulin model was placed at various positions along the X-axis. In these graphs, the positions along the X-axis are given as fractions of the unit cell (a = 78.9Å) with the correct solution defined to be at X = 0. The R value compares calculated structure factors , ΔF_{calc}, to 15Å resolution with the centric contrast variation structure factor differences, ΔF_{obs}, as described in the text. (a) shows a complete translation search with the refined cubic insulin structure (b) shows the portion of the translation search between X = -1/4 and X = +1/4 performed with molecules 1 (heavy line) and 2 (dashed line) from the 2Zn insulin crystal asymmetric unit.

tion structure factors in their X-ray data could not be reliably included in the R factor search. The test calculations shown in Figure 4 show that translation searches with contrast variation data could be used to select the correct solution from a set of minima found by a conventional translation search or to replace the conventional translation function. The method would be most worthwhile where the trial model was too inaccurate for a conventional translation function to succeed. The accuracy of the solution should be within the range of convergence of model refinement algorithms that make use of higher resolution data and lead to a successful structure determination.

DIRECT DETERMINATIONS OF MOLECULAR ENVELOPES

An exciting possibility for structure determination from contrast variation data, applicable without any trial model, is to use direct phasing procedures to determine the protein molecular envelope in the crystal. The identification of the protein molecular envelope provides a restraint for phase refinement calculations at atomic resolution (Schevitz *et al.*, 1981; Wang, 1985) and an independent experimental approach to envelope determination should make the application of this restraint more reliable. A reliable method for the direct determination of the molecular envelope could also provide information on protein domain organization and surface topology at an early stage in structure determination and would be useful with poorly diffracting crystals.

The direct phasing method for use with contrast variation data that I have investigated uses density histograms for evaluating phase sets. The numerical index for determining the likely correctness of a trial map was:

$$M(L,H)^2 = \frac{\sum(\rho(L)-\langle\rho(L)\rangle)^2}{N(L)} + \frac{\sum(\rho(H)-\langle\rho(H)\rangle)^2}{N(H)}$$

which is the sum of the mean square density variations over separate sets of low (L) and high (H) density pixels. This function is a minimum for an ideal binary density histogram with spikes at density values $\langle\rho(L)\rangle$ and $\langle\rho(H)\rangle$ containing N(L) and N(H) pixels. Thus, a map which has uniform but different density values in the solvent and protein volumes (for example the correct map shown in Figure 3) should give a low value of M(L,H). In the case of the cubic insulin crystal the protein volume, N(L), is set to 40% of the crystal volume and the solvent volume, N(H), is set to 60% of the crystal volume.

In each of four trials, eight randomly phased maps were computed from the centric contrast variation structure factors, ΔF_{obs}, to 15Å resolution (*ie* phases were set to 0/180 degrees for reflections with purely real components and -90/+90 degrees for reflections with purely imaginary components). Since the contrast variation differences are small for most of the acentric terms, these were not included in the map calculations. The strong [011] reflection was set to the correct phase of +90 degrees to avoid the alternative branch of solutions discussed in the section on translation functions. The map from each of the four trials with the lowest value of M(L,H) was selected. To refine each of these solutions, the contrast variation structure factor amplitudes were arranged in order of magnitude and, in turn, each phase was reversed to see if changing its value would reduce M(L,H). This process was repeated until no further improvement was obtained.

From the four sets of maps, two of the final phase sets were identical to the set of phases obtained from the atomic model (M(L,H) = 0.866). The two slightly poorer solutions (M(L,H) = 0.868) differed only in that the weak [020] and [042] reflections have the wrong phases and the density map is very similar to the correct map. Further trials with 96 randomly phased maps did not produce any map with a lower value of M(L,H) than the correct map and all maps with significantly lower values of M(L,H) than the average random map ($\langle M(L,H)\rangle$ = 1.515) had correct phases for the strongest reflections. Thus, this algorithm was able to produce the same set of phases as was obtained from the known atomic model from a small number of trials.

Trials of this method with simulated data from several other crystal structures suggest that the success in determining the protein/solvent envelope for the cubic insulin crystal structure is due to the strong demarcation of the low resolution map into protein-containing tubes and wide solvent channels. Density histograms from low resolution maps of small protein crystals that contain only small solvent cavities do not show the same bimodal appearance as the cubic insulin crystal. Thus, it is probably only appropriate to try this type of method for crystal structures with cell dimensions greater than ~75Å. The direct determination of the protein/solvent envelope for the cubic insulin crystal structure involved only 11 centric structure factor amplitudes. Although in this application the correct map was identified by refining a very small number of randomly phased maps, it would have been quite feasible to compute the low resolution maps from all possible phase combinations (2048 maps) for such a small problem (Mariani *et al.*, 1988). On the type of computers now available in most crystallographic laboratories it should be possible to evaluate 10^3–10^4 low resolution maps in a few days of CPU time. Nevertheless, for

Figure 5. Direct determination of protein envelope for citrate synthase at 25Å resolution. (a) Contoured sections of the unit cell ~12Å thick viewed down the c-axis showing the protein/solvent envelope obtained from atomic coordinates at 25Å resolution. The protein is represented by the C_{α} trace for all molecules in this portion of the crystal. (b) Preliminary results of a direct determination of the protein/solvent envelope with no initial phase information. The main body of the protein is contained within the envelope but there are some errors in the precise determination of the protein/solvent boundary.

more general applications it is necessary to consider if similar methods are applicable with larger amounts of data without incurring a prohibitive increase in computing time.

Model calculations with structure factors generated from the atomic co-ordinates for citrate synthase (Brookhaven Protein Data Bank entry 1CTS, space group $P4_12_12$ with a = b = 77.4Å, c = 196.4Å, solvent volume ~55%) have been carried out to test this histogram based phasing method with a larger crystal structure. A calculation of the molecular envelope from this model resulted in 137 structure factors to 15Å resolution and showed that several acentric reflections had relatively large structure factor amplitudes. Thus, these calculation of randomly phased maps included the acentric terms with arbitrary phases. The value of M(L,H) obtained from the correct 15Å map was 2.314. This value is much lower than the best map (M(L,H)=2.795) generated from a calculation of 508 random maps and is also significantly outside the range of values that were obtained (the average value of M(L,H) was ~3.2 and only 5 maps had values of M(L,H) smaller than 2.9). Visual inspections of the maps with lowest values of M(L,H) show significant correlations with the correct protein/solvent envelope. Therefore although not exhaustive, these calculations suggest that the histogram analysis with contrast variation data may provide a unique solution to this trial determination of the protein/solvent envelope. When there are a very large number of phase sets to explore it is possible to apply sophisticated algorithms for determining more accurate solutions from analysis of a large collection of randomly phased maps (Lunin *et al.*, 1990; Sjolin *et al.*, 1991). Following the suggestion of Lunin *et al.* (1990) I averaged the five best solutions together (~1% of the total number of maps generated). When Fourier filtered to 25Å resolution, the resulting map, derived with no structural data except the estimated solvent volume, shows significant correlation with the expected protein/solvent envelope (Figure 5). This is a first attempt at this phasing problem and does not fully exploit the power of M(L,H) to distinguish the correct map. Improvements in this result are to be expected with more substantial investigation into the phasing method.

CONCLUDING SUMMARY

Because the dialysis of crystals with deuterated solvent is usually non-perturbing it should be possible to collect contrast variation data from most protein crystals. Furthermore, since these experiments require the measurement of only a relatively small number of low resolution reflections the requirements for very large crystals and long experimental runs that handicap high resolution neutron crystallography are substantially reduced (Timmins *et al.*, 1992). Comparisons of the cubic insulin crystal structure with the contrast variation data show that these measurements may provide quite accurate information on the protein molecular envelope, albeit at low resolution. The potential for contrast variation data to contribute to the structure determination process has been demonstrated both previously (Carter *et al.*, 1988) and in this report. These methods could well provide practical solutions to a wide range of structural problems.

ACKNOWLEDGMENTS

This work was supported by grant CA 47439 from the National Cancer Institute to Prof. D.L.D. Caspar. Neutron diffraction measurements were carried out at beamline H3A at the High Flux Beam Reactor at Brookhaven National Laboratory in collaboration with

Dr Richard Korszun (*cf* Badger *et al.,* 1995). The work at Brookhaven National Laboratory was supported by a grant from the Office of Health and US Department of Energy (R.K.) and an NSF user grant.

REFERENCES

Badger, J., (1992). Flexibility in crystalline insulins. *Biophys J.,* 61:816–819.

Badger, J., Harris, M.R., Reynolds, C.D., Evans, A.C., Dodson, E.J., Dodson, G.G., & North, A.C.T., (1991). Structure of the pig insulin dimer in the cubic crystal. *Acta Cryst.,* B47:127–136.

Badger, J., Kapulsky, A., Caspar, D.L.D., & Korszun, R., (1995). Neutron diffraction analysis of the solvent accessible volume in cubic insulin crystals. *Nature Structural Biology,* 2:77–88.

Baker, E.N., Blundell, T.L., Cutfield, J.F., Cutfield, S.M., Dodson, E.J., Dodson, G.G., Hodgkin, D.M.C., Hubbard, R.E., Isaacs, N.W., Reynolds, C.D., Sakabe, K., Sakabe, N., & Vijayan, N.M., (1988). The structure of 2Zn pig insulin at 1.5Å resolution. *Phil. Trans. Roy. Soc.Lond.,* B319:369–456.

Bentley, G.A., Lewit-Bentley, A., Finch, J.T., Podjarny, A.D., & Roth, M., (1984). Crystal structure of the nucleosome core particle at 16Å resolution. *J. Mol. Biol.,* 176:55–75.

Brünger, A.T., (1990). Extension of molecular replacement: a new search strategy based on patterson correlation refinement. *Acta Cryst.,* A46:46–57.

Carter, C.W., Crumley, K.V., Coleman, D.E., Hage, F., & Bricogne, G., (1990). Direct phase determination for the molecular envelope of tryptophanyl-tRNA synthetase from *bacillus stearothermophilus* by X-ray contrast variation. *Acta Cryst.,* A46:57–68.

David, P., & Subbiah, S., (1994). Low-resolution real-space envelopes: The application of the condensing protocol to the *ab initio* macromolecular phase problem of a variety of examples. *ACA Annual Meeting* abstract TRN12.

Dodson, E.J., Dodson, G.G., Lewitova, A., & Sabesan, M., (1978). Zinc-free cubic pig insulin: crystallization and structure determination. *J. Mol. Biol.,* 125:387–396.

Filman, D.J., Miller, S.T, & Hogle, J.M., (1994). A genetic algorithm for the *ab initio* phasing of icosahedral viruses. *ACA Annual Meeting* abstract M09.

Fourme, R., Shepard, W., L'Hermite, G., & Kahn, R., (1994). The multiwavelength anomalous solvent contrast method (MASC) in macromolecular crystallography. *ACA Annual Meeting* abstract TRN08.

Fujinaga, M., & Read, R.J., (1987). Experiences with a new translation-function program. *J. Appl. Cryst.,* 20:517–521.

Kraulis, P.J., (1991). MOLSCRIPT: A program to produce both detailed and schematic plots of protein structures. *J. Appl. Cryst.,* 24:946–950.

Lattman, E.E., (1985). Use of the rotation and translation function. *Methods Enzymol.,* 115:55–77.

Lunin, V.Y., Urzhumtsev, A.G., & Skovoroda, T.P., (1990). Direct low-resolution phasing from electron-density histograms in protein crystallography. *Acta Cryst.,* A46:540–544.

Lunin, V.Y., Lunina, N.L., Petrova, T.E., Vernoslova, E.A., Urzhumtsev, A.G., & Podjarny, A.D., (1994). A Monte Carlo approach to low resolution phasing in protein crystallography. *ACA Annual Meeting* abstract PTR03.

Mariani, P., Luzzati, V., & Delacroix, H., (1988). Cubic phases of lipid-containing systems. Structure analysis and biological implications. *J. Mol. Biol.,* 204:165–189.

Podjarny, A.D., Urzhumtsev, A., & Navaza, J., (1994). On the solution of the molecular replacement problem at very low resolution: application to a ribosome model. *ACA Annual Meeting* abstract TRN09.

Roth, M., Lewit-Bentley, A., Michel, H., Deisenhofer, J., Huber, R., & Oesterhelt, D., (1989). Detergent structure in crystals of a bacterial photosynthetic reaction center. *Nature,* 340:659–662.

Roth, M., & Pebay-Peyroula, E., (1994). Phasing at low resolution using direct methods in protein crystallographic studies. *ACA Annual Meeting* abstract TRN10.

Sayre, D., (1972). On least-squares refinement of the phases of crystallographic structure factors. *Acta Cryst.,* A28:210–227.

Schevitz, R.W., Podjarny, A.D., Zwick, M., Hughes, J.J., & Sigler, P.B., (1981). Improving and extending the phases of medium- and low-resolution macromolecular structure factors by density modification. *Acta Cryst.,* A37:669–677.

Schluzen, F., Volkmann, N., Thygesan, J., Hansen, H.A.S., Harms, J., Bennett, W.S., & Yonath, A., (1994). Attempts at low resolution phasing of ribosomal crystals by MASC. *ACA Annual Meeting* abstract TRN15.

Sjolin, L., Prince, E., Svensson, L.A., & Gilliland, G.L., (1991). *Ab initio* phase determination for X-ray diffraction data from crystals of a native protein. *Acta Cryst.,* A47:216–223.

Subbiah, S., (1991). Low-resolution real-space envelopes: An approach to the *ab initio* macromolecular phase problem. *Science,* 252:128–133.

Timmins, P.A., Poliks, B., & Banaszak, L., (1992). The location of bound lipid in the lipovitellin complex. *Science,* 257:652–659.

Wang, B.-C., (1985). Resolution of phase ambiguity in macromolecular crystallography. *Methods Enzymol.,* 115:90–112.

Yeates, T.O., & Rini, J.M., (1990). Intensity-based domain refinement of oriented but unpositioned molecular replacement models. *Acta Cryst.,* A46:352–359.

DNA HYDRATION STUDIED BY NEUTRON FIBER DIFFRACTION

W. Fuller,[1] V. T. Forsyth,[1] A. Mahendrasingam,[1] P. Langan,[1] W. J. Pigram,[1] S. A. Mason,[2] and C. C. Wilson[3]

[1] Department of Physics
Keele University
Keele, Staffordshire ST5 5BG, United Kingdom
[2] Institut Laue Langevin
Avenue des Martyrs, 156X 38042, Grenoble Cedex, France.
[3] Rutherford Appleton Laboratory
Chilton, Didcot, Oxon OX11 OQX, United Kingdom

ABSTRACT

The development of neutron high angle fiber diffraction to investigate the location of water around the deoxyribonucleic acid (DNA) double-helix is described. The power of the technique is illustrated by its application to the D and A conformations of DNA using the single crystal diffractometer, D19, at the Institut Laue-Langevin, Grenoble and the time of flight diffractometer, SXD, at the Rutherford Appleton ISIS Spallation Neutron Source. These studies show the existence of bound water closely associated with the DNA. The patterns of hydration in these two DNA conformations are quite distinct and are compared to those observed in X-ray single crystal studies of two-stranded oligodeoxynucleotides. Information on the location of water around the DNA double-helix from the neutron fiber diffraction studies is combined with that on the location of alkali metal cations from complementary X-ray high angle fiber diffraction studies at the Daresbury Laboratory SRS using synchrotron radiation. These analyses emphasize the importance of viewing DNA, water and ions as a single system with specific interactions between the three components and provide a basis for understanding the effect of changes in the concentration of water and ions in inducing conformational transitions in the DNA double-helix.

INTRODUCTION

Fibrous polymers exhibit great variety in the degree of order with which the molecules comprising them are arranged. This ranges from so-called amorphous polymers in which the various molecules shown no preferred orientation and little regularity in their side-by-side packing to highly oriented crystalline polymers which consist of a large number of crystallites each with the order of a single crystal oriented with their c-axes closely parallel to the fiber axis but in random orientation about this direction. The work described here is concerned with such highly ordered fibers in which the arrangement of the DNA double-helices within each crystallite can be described using standard crystallographic nomenclature. It is natural therefore for the structural analysis of these fibers to be based on the methods typically used in the analysis of single crystals of biological macromolecules. Of even greater significance for our work has been the opportunity to exploit the experimental arrangements developed for neutron single crystal diffraction analysis.

Some perspective on the problems associated with developing neutron high angle fiber diffraction is provided by first considering what can be achieved using available facilities for X-ray high angle fiber diffraction. Figure 1 shows an X-ray fiber diffraction pattern recorded on beamline 7.2 at the Daresbury Synchrotron Radiation Source (SRS) from a fiber of the synthetic two-stranded polydeoxynucleotide poly [d(A-T)].poly [d(A-T)] in the D conformation. This synthetic DNA contains an alternating sequence of adenine (A) and thymine (T) residues along each polynucleotide strand. The fiber had a diameter of ~150μm and the diffraction peaks can be seen to extend to d spacings of <2.5Å. Using data from patterns such as this, the structures of the A, B, C, D and Z (or S) conformation of the DNA double-helix have been refined, with the position of the phosphorus atoms (ie the atoms in the DNA which scatter X-rays most strongly) located to within a few tenths of an angstrom (Fuller *et al.*, 1965; Langridge *et al.*, 1960; Marvin *et al.*, 1961; Davies & Baldwin, 1963; Arnott *et al.*, 1980). More recently isomorphous replacement and superposition techniques have been used to locate the position of alkali metal cations which neutralize the negative charges carried by DNA phosphate groups (Forsyth *et al.*, 1993; Langan *et al.*, 1993). These studies did not, however, provide convincing evidence for ordered water within any of the structures. This gap in our knowledge is of particular significance since a number of conformational transitions within the DNA double-helix can be readily induced by changing the relative humidity of the fiber environment and hence the degree of hydration of the DNA double helix. Indeed it has been

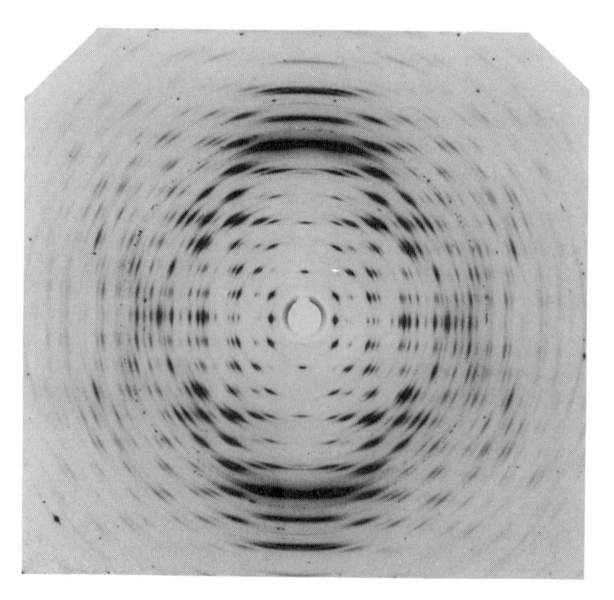

Figure 1. X-ray high angle fiber diffraction pattern from the D conformation of the potassium salt of poly [d(A-T)].poly [d(A-T)] at 57% relative humidity, recorded on photographic film at the Daresbury Synchrotron Radiation Source with an exposure time of 5 minutes.

possible to monitor the stereochemical pathways followed in such transitions using time-resolved X-ray fiber diffraction (Mahendrasingam *et al.*, 1986). Humidity driven structural transitions within the DNA double-helix are generally reversible and highly reproducible with regard to the relative humidity of the fiber environment. It is, therefore, reasonable to expect that there are highly specific interactions between water and the DNA double-helix which are characteristic of the various DNA conformations. The techniques whose development and application are reviewed here were developed to investigate such possibilities by exploiting the marked difference in scattering power for neutrons of hydrogen and deuterium.

EXPERIMENTAL TECHNIQUES

The most formidable experimental difficulties faced in developing neutron high angle fiber diffraction for investigating DNA structures stem from the relatively low flux of available neutron beams and the consequent requirement to use a large beam (which fortunately can be obtained) as compared with those available from synchrotron or laboratory based rotating anode X-ray sources. This is reflected in the dimensions of the specimens used in our neutron diffraction studies. These typically consist of a parallel array of fibers, each ~150μm in diameter and ~5mm in length, with an area of ~5 × 5mm^2 perpendicular to the neutron beam and up to 3mm thick. While it is possible to draw fibers with diameters greater than this, the degree of crystallinity and orientation obtained within them tends to decrease significantly with increasing diameter. Therefore increasing the dimensions of a specimen to make them a better match to those of the neutron beam was better achieved by increasing the number of fibers from which it was composed rather than by using fibers of larger diameter. In our most recent experiments the specimen contained about 450 fibers. The quality was monitored by recording the X-ray high angle fiber diffraction pattern from selected fibers.

Neutron high angle fiber diffraction patterns were recorded on the single crystal diffractometer, D19, at the Institut Laue-Langevin, Grenoble and the time-of-flight diffractometer, SXD, at the Rutherford Appleton ISIS Pulsed Spallation Neutron Source. The specimen was mounted on a standard goniometer head inside a purpose designed gas-tight can with windows for the incident and exit neutron beam of 10μm thick melinex on D19 and aluminium foil on SXD. Scattering of the incident beam by gas within the can was minimized by replacing air by helium. The relative humidity of the specimen environment was controled by passing the helium through an appropriate saturated salt solution. Exchange of H_2O in the specimen by D_2O was achieved by replacing the salt solution made up in water with one made up in D_2O. The changes in the diffraction pattern following H_2O/D_2O exchange were highly significant, reproducible and reversible. The time for these changes to be complete was typically 4 or 5 hours and progress was monitored by continuously observing the change in intensity of a strong feature in the diffraction pattern. These changes are of course due not only to the replacement of H_2O by D_2O but also to replacement by deuterium of hydrogen atoms attached to nitrogen atoms in the DNA bases.

The wavelength of the monochromatic neutron beam used on D19 at the ILL reactor source was either 1.54Å or 2.42Å depending on the strategy of data collection chosen for the experiment. The neutron high angle fiber diffraction pattern was recorded using a 64° (vertical with 256 pixels) by 4° (horizontal with 16 pixels) gas-filled multi-wire detector. Diffraction data was collected as a series of strips which after correction for detector response and placing on a common scale were merged to produce a pattern like those in Figures 2 and 3. The time taken to record a single strip was typically 4 hours.

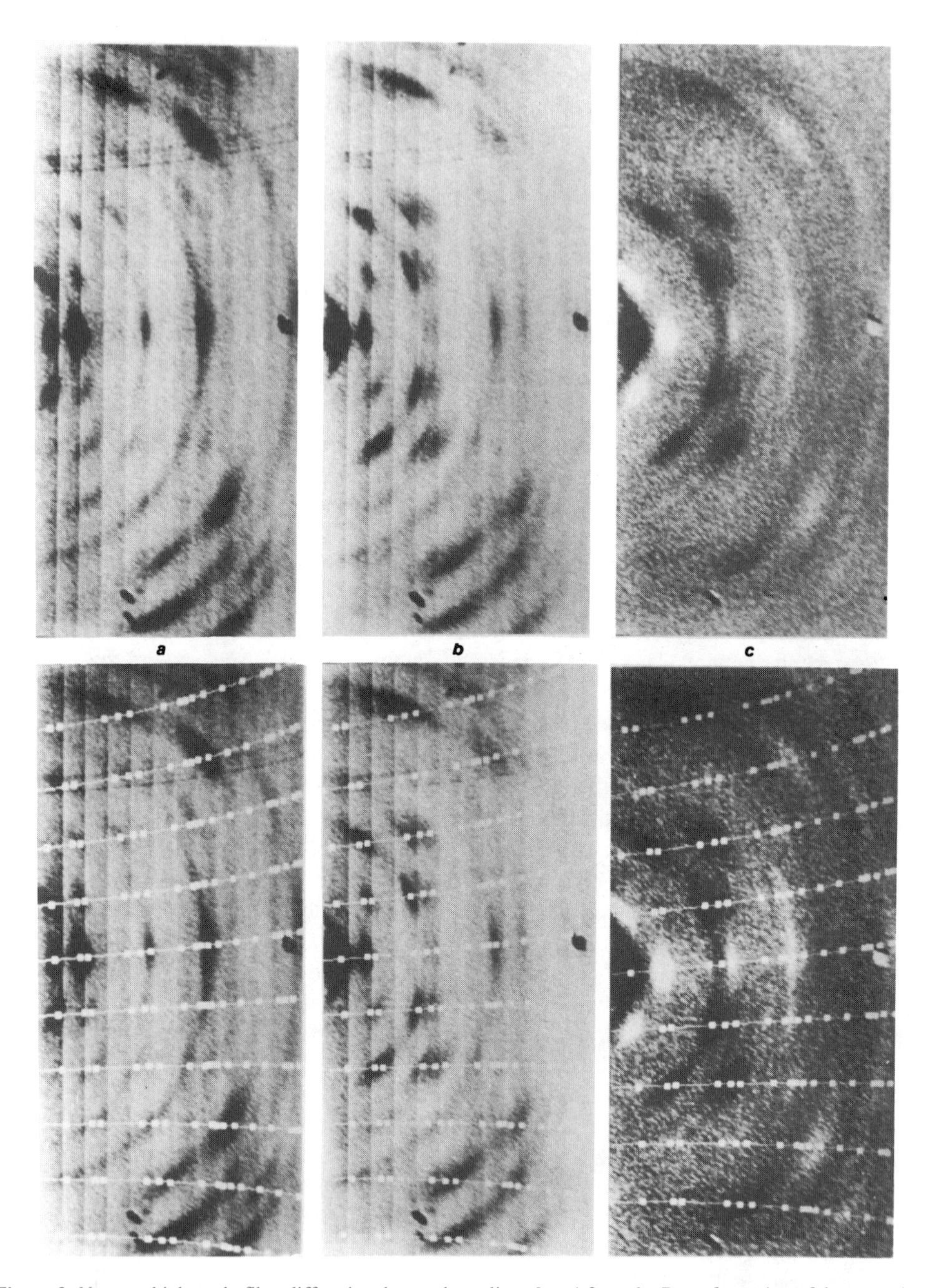

Figure 2. Neutron high angle fiber diffraction data on layer lines 0 to 4 from the D conformation of the potassium salt of poly [d(A-T)].poly [d(A-T)], recorded on instrument D19 at the Institut Laue-Langevin (a) with D_2O surrounding the DNA, (b) with H_2O surrounding the DNA and (c) the difference between (a) and (b). In the lower half of the figure the positions of Bragg reflections in these patterns are indicated by white squares.

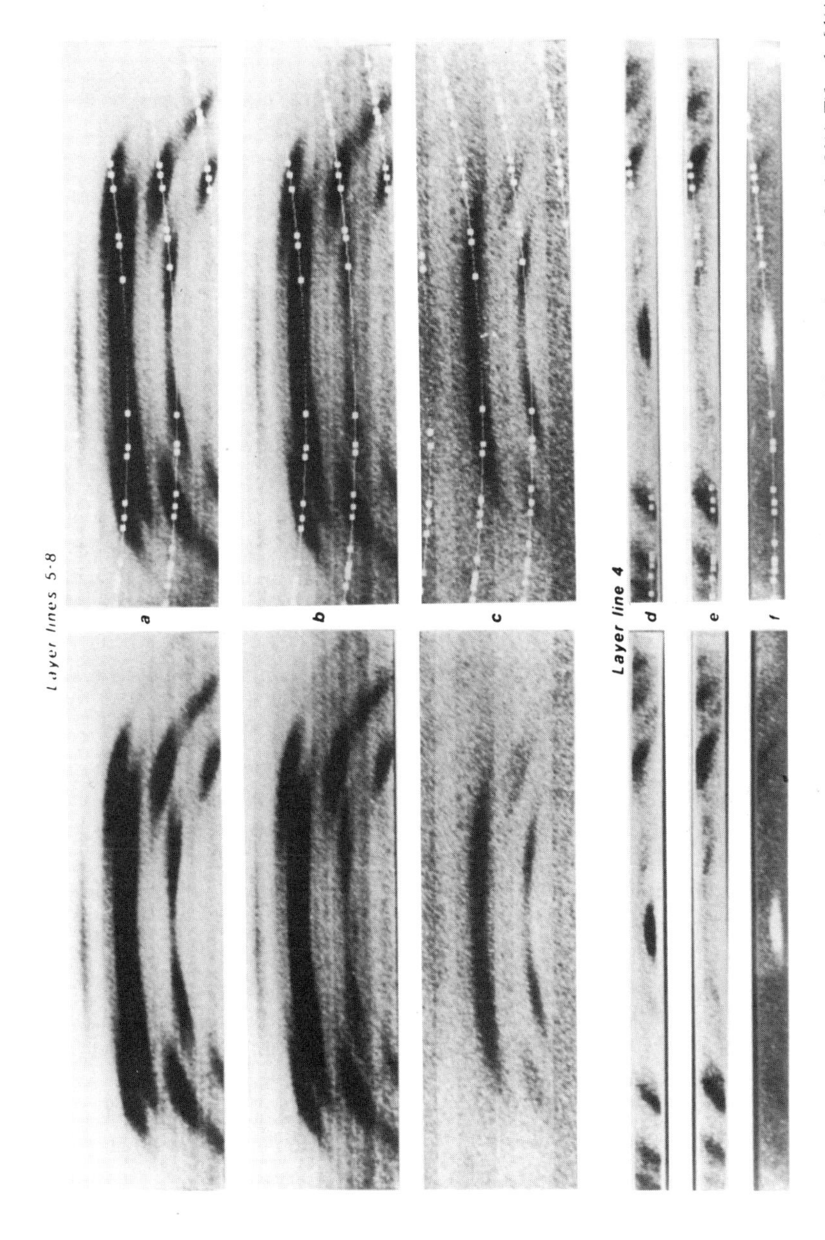

Figure 3. Neutron high angle fiber diffraction data on layer line 4, and layer lines 5 to 8 from the D conformation of the potassium salt of poly [d(A-T)].poly [d(A-T)], recorded on instrument D19 at the Institut Laue-Langevin, (a) and (d) with D_2O surrounding the DNA, (b) and (e) with H_2O surrounding the DNA and (c) and (f) the difference between patterns (a) and (b) and between (d) and (e) respectively. In the right hand side of the figure the positions of Bragg reflections in these patterns are indicated by white squares. The data on layer line 4 is of interest because in addition to illustrating reflections visible on layer line 4 in Figure 2 it shows the marked change in the meridional diffraction following D_2O/H_2O substitution.

The wavelength of the neutrons used in the analyses performed using the time of flight spectrometer SXD at the ISIS pulsed source typically ranged from ~0.5Å to ~9Å. The variation of intensity as a function of wavelength within the incident neutron beam was determined by recording the scattering from a standard specimen of vanadium. Pulses occurred with a frequency of 50Hz. The radiation scattered during each pulse of incident radiation was recorded on a flat detector 192mm (64 pixels) × 192mm (64 pixels) in area as a series of 645 time slices. The data in each time slice corresponded to a spread of neutron wavelengths with the wavelength recorded in each pixel depending on the distance of the pixel from the specimen. Purpose designed software was used to 'collapse' the data in the complete set of time slices into a single fiber diffraction pattern. This process also involved correction for the variation of intensity with wavelength in the incident beam and variation in reflectivity and absorption of the sample as a function of wavelength.

ANALYTICAL TECHNIQUES

The distribution of localised water around the DNA double helix corresponding to changes in diffraction following H_2O/D_2O replacement was imaged by calculating classical Fourier difference syntheses. In these syntheses the amplitudes were either $|F_D|-|F_H|$ or $|F_D|-|F_C|$ and the phases were α_c where $|F_D|$ and $|F_H|$ were respectively the observed structure factor amplitudes for a particular reflection when D_2O and H_2O were surrounding the DNA, $|F_C|$ and α_c were respectively the neutron structure factor amplitude and phase calculated for the model of the DNA double helix itself (ie not including ions and water in the structure) as refined against the X-ray high angle fiber diffraction data.

At the resolution of the data typically recorded during these studies, the hydrogen and oxygen atoms within a water molecule can, to a first approximation, be treated as a single scattering center with a scattering length (in units of 10^{-12}cm) of -0.17 for H_2O and of 1.91 for D_2O. Therefore the Fourier difference synthesis calculated with $|F_D|-|F_C|$ as amplitudes can be expected to show similar features to that calculated with $|F_D|-|F_H|$. Comparison of these two syntheses provides a check on the quality of the data and the validity of assumptions made in these investigations. However, using $|F_D|-|F_C|$ rather than $|F_D|-|F_H|$ has an important advantage because the precision of the difference amplitude is not lowered by the incoherent hydrogen scattering from water surrounding the DNA and from hydrogens bound to nitrogens in the base pairs which are exchanged when H_2O is replaced by D_2O.

The image of water localised around the DNA double helix derived using Fourier difference syntheses can be refined using standard crystallographic techniques. This involves identifying peaks in the difference maps and associating with each of them (x,y,z) coordinates, thermal parameters and an occupancy. These peaks can then be added to the model of DNA to provide an 'improved heavy atom' and $|F_C|$ and α_c recalculated. A new Fourier difference synthesis can then be calculated and the whole process of 'improving the heavy atom' repeated until ideally the calculated Fourier difference map is featureless.

In parallel with such a real space refinement, the structure of water localised around the DNA can be refined by the reciprocal space technique of least squares in which the positional, thermal and occupancy parameters of the various water peaks are refined by minimising the residual which is a measure of the difference in $|F_D|-|F_C|$ summed over all structure factors. An important aspect of such refinement is an investigation of the extent to which water associated with the DNA double helix is assumed to retain the symmetry of the DNA. For a DNA structure with N nucleotides per turn, assuming that the dyad symmetry between the two polydeoxynucleotide strands is preserved, relaxing the constraint of helical symmetry in the water structure will result in an increase in the number of water parameters by a factor of N over a helically constrained model. Such a large decrease in the ratio of observations to parameters will inevitably result in an improved crystallographic residual. However, the significance of such an improvement must be evaluated by applying the Hamilton significance test (Hamilton, 1965).

THE D CONFORMATION OF DNA

The D conformation of DNA is observed for two-stranded synthetic polydeoxynucleotides with alternating purine and pyrimidine nucleotides along each chain - typically poly [d(A-T)].poly [d(A-T)] but also poly [d(I-C)].poly [d(I-C)h] in which the alternating residues are inosine and cytosine. The D form is favoured by low humidity in the fiber environment (eg 44% to 57%) and an excess of ~0.5 ion pairs per phosphate of alkali halide (eg NaCl but also other salts in which the cation is Li^+, K^+ and Rb^+ and the anion is F^-, Cl^-, Br^- and I^-). From density measurements the number of water molecules per nucleotide-pair is ~10. The effect of D_2O/H_2O exchange on the neutron high angle fiber diffraction data recorded on D19 at the ILL from a specimen containing ~40 fibers, each ~150μm in diameter and 3mm in length, is illustrated in Figures 2 and 3. The Fourier difference map calculated using $|F_D|$-$|F_H|$ as difference structure factor amplitudes shows water localised predominantly in the minor groove of the D conformation (Figure 4).

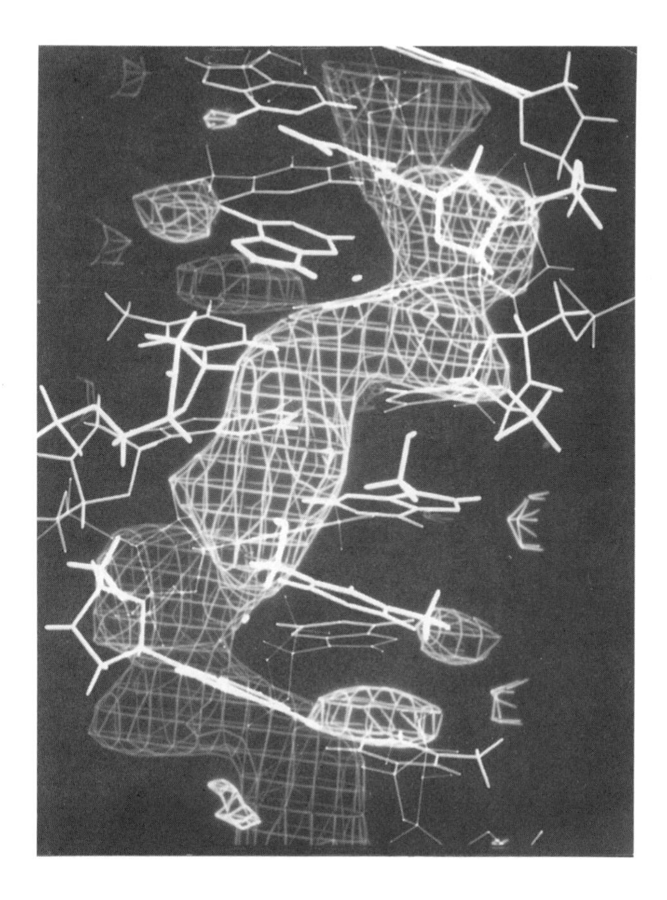

Figure 4. Difference Fourier map generated from the D_2O and H_2O datasets from D-DNA illustrated in Figures 2 and 3. Particularly noticeable in this image is the network of hydration which runs down the narrow minor groove of the DNA double helix. (A color version of this figure appears in the color insert following p. 214.)

This investigation of the hydration of the D form of DNA was the first application of neutron high angle fiber diffraction to investigate the location of water around a conformation of the DNA double helix and there was a concern to scrutinise the results cautiously (Fuller *et al.*, 1989; Forsyth *et al.*, 1989; Forsyth *et al.*, 1990; Forsyth *et al.*, 1992). However, confidence in the analysis was given by the fact that the peaks in the difference Fourier were in regions not occupied by DNA atoms nor in conflict with the positions proposed for alkali metal ions on the basis of isomorphous X-ray fiber diffraction analyses (Forsyth *et al.*, 1990; Forsyth *et al.*, 1993). The minor groove in the D conformation is the narrowest observed in any of the principal DNA conformations and might therefore be seen as a region where localised water and ions could be expected to play a crucial role in stabilising the conformation by minimising repulsion between negatively charged phosphate groups in opposite polydeoxynucleotide strands.

It would be useful to be able to compare these observations on the location of water and ions around the D conformation of a two stranded synthetic polydeoxynucleotide with structures observed in X-ray single crystal determinations of two-stranded oligodeoxynucleotides. This is not possible because to date there has been no report of a D-type structure in an oligodeoxynucleotide crystal structure. However, in this context it is interesting to note that although the minor groove is narrower in the D than in the B form of the DNA double helix, there are general similarities between the conformations of the two structures. These are paralleled by similarities between localised water observed in the minor groove of the D-form and the spine of hydration in the minor groove proposed from oligonucleotide single crystal studies as a characteristic feature of the B conformation (Drew & Dickerson, 1981).

THE A CONFORMATION OF DNA

The A conformation is observed as a low humidity form of both naturally occurring and synthetic polydeoxynucleotides and is favoured by little or no excess salt (typically an alkali halide) within the fiber. The dramatic and readily reversible structural transition from the fully crystalline A form to the semi-crystalline B form when the relative humidity of the fiber environment is raised from 75% to 92% was the first well-defined conformational transition within the DNA double helix to be reported (Franklin & Gosling, 1953). The characterisation of the factors responsible for inducing this transition in terms of interatomic interactions remains one of the most challenging problems facing the application of molecular mechanics and molecular dynamics in the investigation of the conformational stability and flexibility of DNA. The role of water associated with the DNA double helix is clearly central to the development of this understanding and, therefore, the neutron high angle fiber diffraction techniques developed for studying the location of water around the D-conformation have been applied in an analysis of the A to B conformational transition.

In these studies a specimen consisting of ~100 fibers of calf thymus DNA each ~200μm in diameter and 5mm long with a salt content which favoured the assumption of the A-conformation at 75% relative humidity was assembled. Exchange of H_2O for D_2O in the specimen and the control of the relative humidity of the fiber environment was achieved using the same technique as that used in the study of the D-form. The neutron high angle fiber diffraction data recorded with D_2O surrounding the DNA is compared with that for H_2O in Figures 5 and 6. As observed for the D-form, the changes in structure-factor amplitudes following replacement of H_2O and D_2O extend over the whole diffrac-

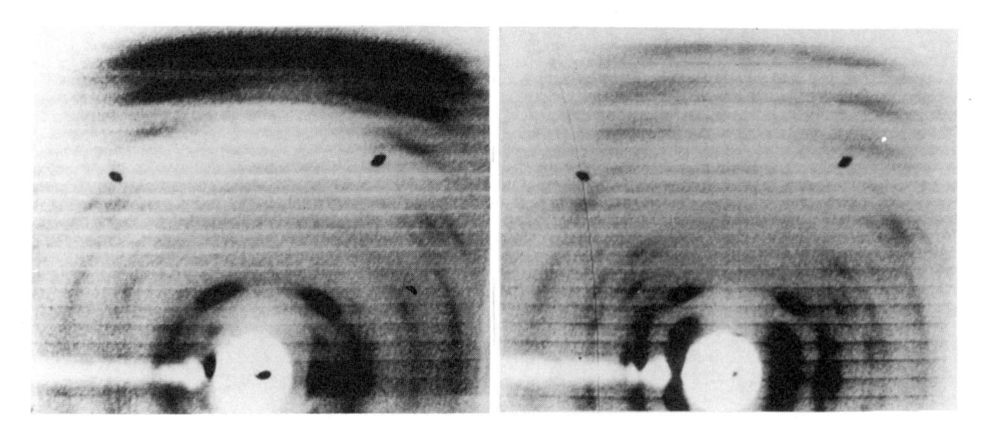

Figure 5. Comparison of neutron fiber diffraction data recorded from the A conformation of the sodium salt of DNA at 75% relative humidity on instrument D19 at the Institute Laue-Langevin (a) with D_2O surrounding the DNA and (b) with H_2O surrounding the DNA.

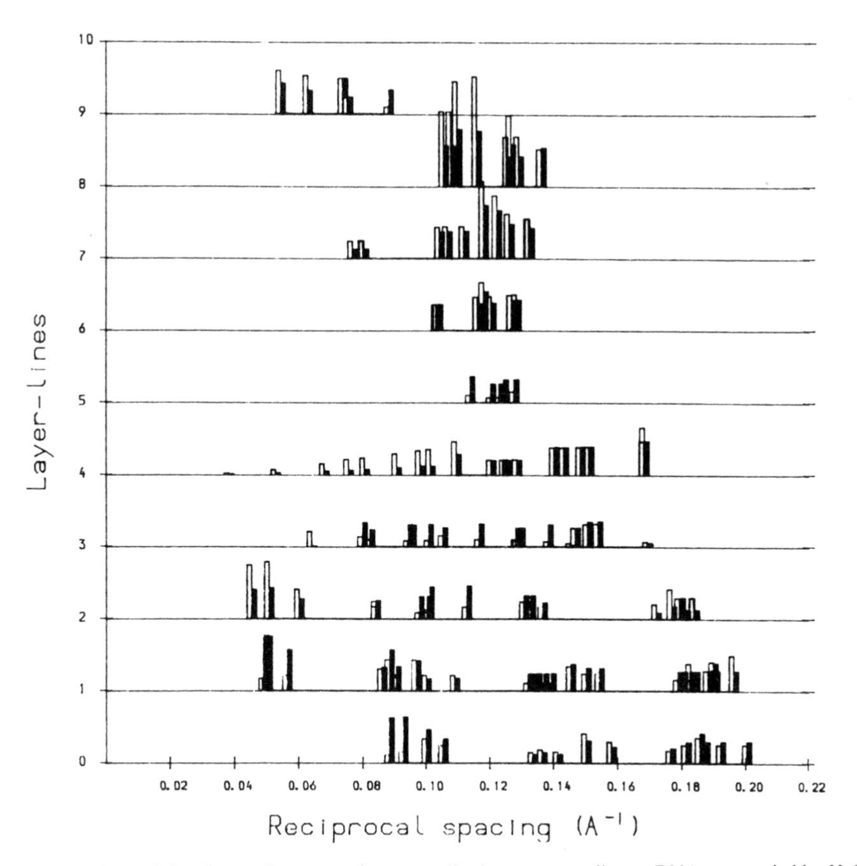

Figure 6. Comparison of the observed structure factor amplitudes corresponding to DNA surrounded by H_2O (F_H) (full bars) and DNA surrounded by D_2O (F_D) (open bars) from the A form of the sodium salt of DNA based on the diffraction data in Figure 5. Overlaid bars correspond to accidentally overlapping reflections.

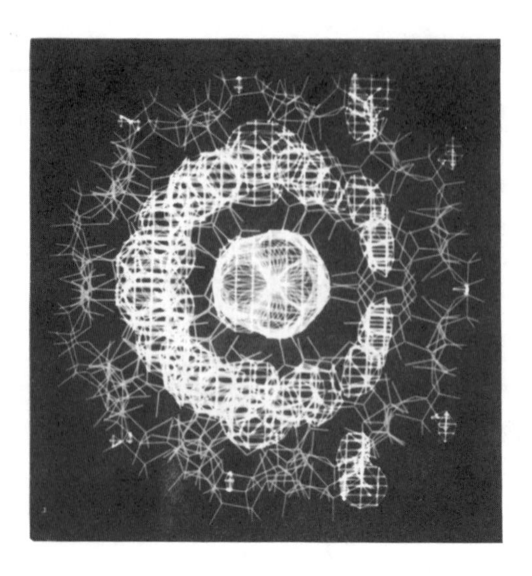

Figure 7. A projection down the helix axis of the Fourier difference map generated using the D_2O and H_2O datasets from A-DNA illustrated in Figure 5. (A color version of this figure appears in the color insert following p. 214.)

tion pattern and are clearly significant, highly reproducible and fully reversible. The amplitude differences $|F_D|-|F_C|$ were combined with phases α_c and an image of the water surrounding the DNA calculated (Forsyth *et al.*, 1990; Langan *et al.*, 1992a; 1992b). $|F_D|$ was the observed structure factor amplitude recorded when D_2O surrounded the DNA, and $|F_C|$ and α_c were the structure factor amplitude and phase calculated for the model refined using high angle X-ray diffraction data.

Figure 7 shows a projection down the c-axis of the A-DNA unit cell of the water associated with a DNA molecule. The two most obvious features are a central disc of density, approximately 5Å in diameter, and a circular ring of density with a diameter of approximately 11Å.

The central disc can be readily related to the fact that in the A conformation the base-pairs are displaced by ~5Å from the helix axis towards the major groove as compared with their position in the B conformation where the helix axis passes close to the center of the base pairs. As a consequence water associated with the edges of the bases which are exposed in the major groove of the DNA double-helix has the overall character of a helical ribbon with the same pitch as the DNA but at such a small radial distance from the helix axis that it approximates to a cylindrical column which appears in projection in Figure 7 as a solid central core.

Figure 8 shows a view at right angles to the helix axis of water peaks associated with the DNA. Peaks of density in Figure 8 lying in the wide major groove of the A conformation are closely associated with the charged oxygens of the phosphate groups which, in this conformation, point into the major groove. Figure 9 illustrates an off-helix axis view of the distribution of water in the wide groove of the DNA.

As with the D form, the water structure in the Fourier difference syntheses of the A conformation is in stereochemically credible positions with respect to atoms in the DNA and to cations neutralising the DNA phosphates, as determined by X-ray high angle fiber diffraction (Langan *et al.*, 1993) and illustrated in Figure 10. The pattern of water structure observed in the large groove of the A conformation shows marked similarities to the pattern observed in X-ray single crystal structure determinations of oligodeoxynucleotides in A-type conformations (Conner *et al.*, 1982).

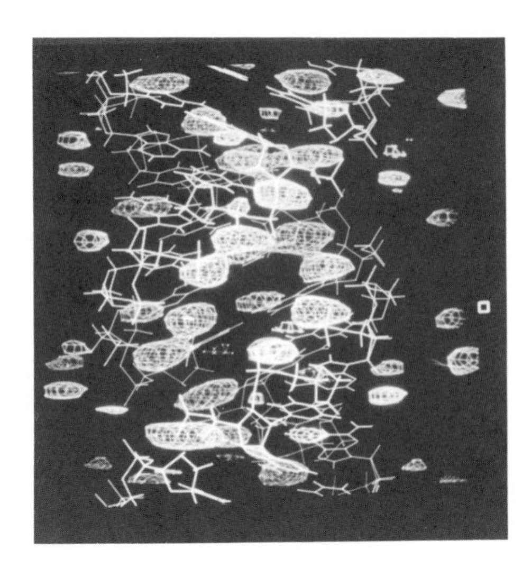

Figure 8. Water peaks associated with the A-form of DNA viewed perpendicular to the helix axis in the Fourier difference map generated using the D_2O and H_2O datasets illustrated in Figure 5. (A color version of this figure appears in the color insert following p. 214.)

The distribution of water as revealed in the Fourier difference syntheses of the A conformation exhibits a marked helical periodicity and we have used a least squares analysis to investigate the extent to which deviations from helical symmetry are significant. In this analysis four families of water peaks were identified, ie (i) water associated with the edge of the bases facing the major groove and forming the central core of density observed in Figure 7 (family 4), (ii) peaks associated with the phosphates in the major groove of the DNA observed in projection as a ring in Figure 7 and discrete peaks in Figure 8 (family 1), (iii) and (iv) peaks associated with the outside of the DNA double helix in positions where the water can form bridges between neighbouring DNA molecules (families 2 and 3) (Figure 8). A least squares refinement without helical constraints of the 4×11 peaks associated with these 4 families resulted in a lower residual than when the parameters were constrained with the peaks comprising each family assumed to lie on a helix with the same pitch, helix sense and number of residues per turn as the DNA A-form. However, on the basis of the Hamilton significance test (Hamilton, 1965) this reduc-

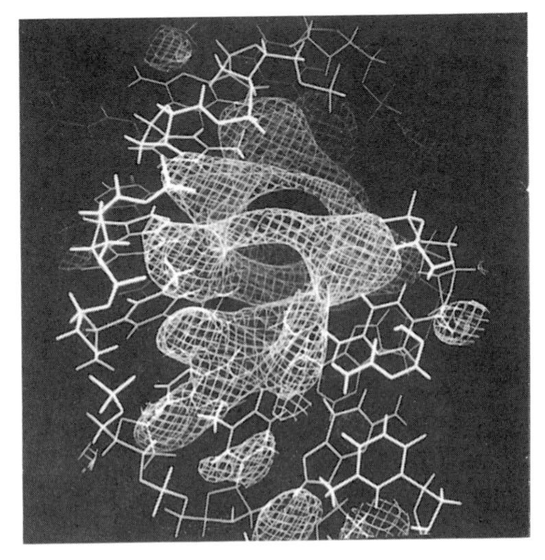

Figure 9. An off-helix axis view of the location of water in the wide groove of the A-form of the DNA double helix. (A color version of this figure appears in the color insert following p. 214.)

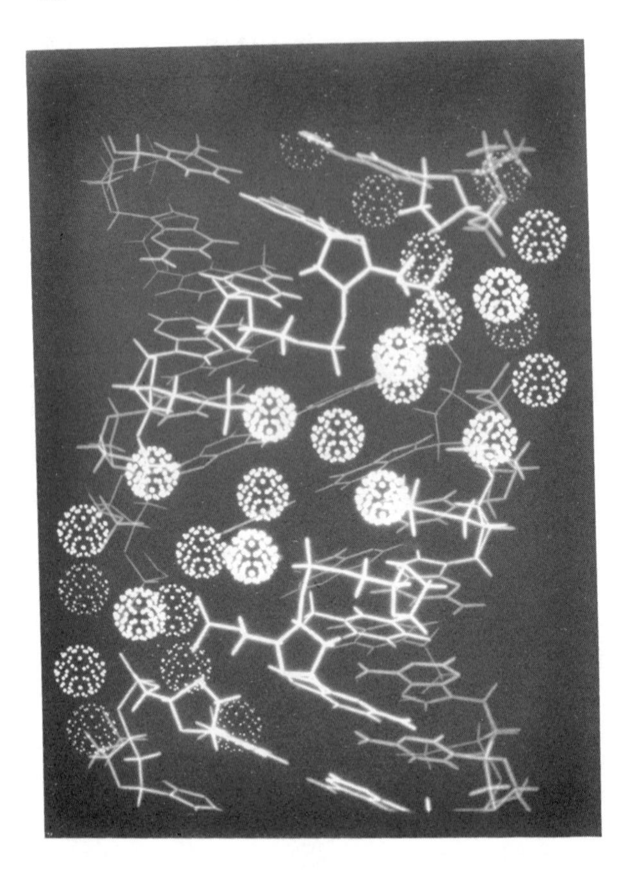

Figure 10. The distribution of ions (shown as stippled spheres) with respect to the major groove of the A form of the DNA double helix. (A color version of this figure appears in the color insert following p. 214.)

tion was not significant when related to the 11-fold increase in the number of parameters. This least squares refinement was remarkably well behaved with similar final parameters being obtained whether all parameters were refined together or whether the parameters describing each family of water peaks were refined separately. The final parameters were also independent of the order in which the parameters for the four families were refined.

The four families of water peaks with the helically constrained parameters were added to the original A form model of DNA and a new set of neutron structure factors with amplitude $|F_C|$ and phase α_c calculated. The Fourier difference map calculated with $|F_D|-|F_C|$ as amplitudes and α_c as phases showed no significant peaks.

STUDIES OF THE A FORM PERDEUTERATED E.COLI DNA

The unit cell of the A form of DNA contains ~20 water molecules per nucleotide pair. While these molecules are readily replaced by D_2O and similarly the 4 hydrogens per average base-pair attached to ring nitrogens and amino groups of the bases are replaced by deuterium, the remaining 16.5 hydrogens per average base-pair attached to carbon atoms in the sugar and base groups are not exchanged in the presence of D_2O. Incoherent scattering by the 40% of the hydrogens which are not exchanged results in increased background and in absorption effects which place an upper limit on specimen volume, and hence limits the statistical accuracy of the diffraction data. To avoid these effects we have extracted

DNA from *E.Coli* cells grown on deuterated media. In this DNA essentially all the hydrogens attached to carbons were replaced by deuterium It has not been possible to make a direct comparison between data recorded with this deuterated DNA and that recorded from normally hydrogenated DNA because of the three year shut-down of the ILL reactor during the period this material has been available. However, fiber diffraction data has been recorded using the time-of-flight diffractometer, SXD, at the ISIS spallation neutron source. The specimen consisted of ~450 fibers each ~250μm in diameter and ~5mm long and data was collected over a period of 6 days. During this experiment SXD was still undergoing development particularly with regard to the detector system. As a consequence the data was of a lower quality than is expected to be obtainable in future experiments. This was particularly marked for larger d spacing because of the relatively small number of longer wavelength neutrons. This situation, and work on large scale structures generally can be expected to benefit from improved access to an effective cold source at ISIS. Despite these limitations in data quality, Fourier difference syntheses using data recorded on SXD showed similar features to those observed in Fourier difference syntheses calculated using data recorded on D19 (Langan *et al.*, 1995; Forsyth *et al.*, this volume).

CONCLUSIONS

The neutron high angle fiber diffraction studies described here provide the most direct evidence so far reported that water occupies specific sites on the DNA double-helix and further that these sites differ markedly for the two DNA conformations studied to date. This information together with that from complementary X-ray studies on the location of ions provides the basis for treatment of ions, water and DNA as a highly coordinated interacting system. The observation that the two families of water peaks most closely associated with the A conformation occupy regular helical positions with the 11-fold screw symmetry of the DNA rather than the spacegroup C2 symmetry of the crystal unit cell is of major significance. In particular, this observation is relevant to the long-standing debate on the extent to which the structure of the A form is determined predominantly by crystal packing forces or by interactions within an extended helical repeating unit which embraces associated ions and water in addition to the nucleotide residue.

The applications of the neutron high angle fiber diffraction technique described in this review have not only made important contributions to understanding the role of water in stabilising the DNA double helix but they have stimulated the application of the technique in investigations of the stereochemistry of other fibrous polymers. Studies are already underway at the ISIS spallation source on the carbohydrates cellulose (Kroon-Batenburg *et al.*, 1994) and hyaluronic acid (A.Derieu, personal communication) and the ability of this technique to locate at atomic resolution hydrogens substituted by deuterium in an organic synthetic polymer of technological interest is illustrated by work at the Institut Laue-Langevin, Grenoble on poly (aryl-ether-ether-ketone) (PEEK) (Mahendrasingam *et al.*, 1992).

ACKNOWLEDGMENTS

This work was supported by the Science and Engineering Research Council through grants GR/E/19954, GR/J/86643 and GR/H/67959, by an EMBO grant (to VTF) to support the extraction of deuterated DNA at the EMBL, Grenoble and by generous allocations of beamtime at the Institut Laue-Langevin reactor source, Grenoble and the Rutherford Appleton ISIS Spallation Source. We are grateful to M.G. Davies, G. Dudley, E.G.T. Greasley, G. March, M. Wallace, M. Daniels and H. Moors for technical support and help with preparation of the manuscript.

REFERENCES

Arnott, S., Chandrasekaran, R., Birdsall, D.L., Leslie, A.G.W., & Ratliff, R.L., (1980). Left-handed DNA helices. *Nature*, 283:743–745.

Conner, B.N., Takano, T., Tanaka, S., Itakura, K., & Dickerson, R.E., (1982). The molecular structure of d(ⁱCpCpGpG), a fragment of right-handed double helical A-DNA. *Nature*, 295:294–299.

Davies, D.R., & Baldwin, R.L., (1963). X-ray studies of two synthetic DNA copolymers. *J. Mol. Biol.*, 6:251–255.

Drew, H.R., & Dickerson, R.E., (1981). Structure of a B-DNA dodecamer III. Geometry of hydration. *J. Mol. Biol.*, 151:535–556.

Forsyth, V.T., Mahendrasingam, A., Pigram, W.J., Greenall, R.J., Bellamy, K.A., Fuller, W., & Mason, S.A., (1989). Neutron fibre diffraction study of DNA hydration. *Int. J. Biol. Macro.*, 11:236–240.

Forsyth, V.T., Mahendrasingam, A., Langan, P., Pigram, W.J., Stevens, E.D., Al-Hayalee, Y., Bellamy, K.A., Greenall, R.J., Mason, S.A., & Fuller, W., (1990). Neutron and X-ray scattering: Complementary techniques. *Inst. Phys. Conf. Ser.*, 101:237–248.

Forsyth, V.T., Mahendrasingam, A., Langan, P., Al-Hayalee, Y., Alexeev, D., Pigram, W.J., Fuller, W., & Mason, S.A., (1992). High angle neutron fibre diffraction studies of the distribution of water around the D form of DNA. *Physica B*. 180 & 181:737–739.

Forsyth, V.T., Langan, P., Mahendrasingam, A., Mason, S.A., & Fuller, W., (1993). Conference Proceedings of the Italian Physical Society: Water-Biomolecular Interactions, 43:231–234.

Forsyth, V.T., Langan, P., Whalley, M.A., Mahendrasingam, A., Wilson, C.C., Giesen, U., Dauvergne, M.R., Mason, S.A., & Fuller, W., (1995). Time of flight Laue fiber diffraction studies of perdeuterated DNA. (this volume).

Franklin, R.E., & Gosling, R.G., (1953). The structure of sodium thymonucleate fibres. I. The influence of water content. *Acta Cryst.*, 6:672–677.

Fuller, W., Wilkins, M.H.F., Wilson, H.R., Hamilton, L.D., & Arnott, S., (1980). The molecular configuration of deoxyribonucleic acid IV. X-ray diffraction study of the A Form. *J. Mol. Biol.*, 12:60–80.

Fuller, W., Forsyth, V.T., Mahendrasingam, A., Pigram, W.J., Greenall, R.J., Langan, P., Bellamy, K., Al-Hayalee, Y., & Mason, S.A., (1989). The location of water around the DNA double-helix. *Physica B*, 156 & 157:468–470.

Hamilton, W.A., (1965). Significance tests on the crystallographic R factor. *Acta Cryst.*, 18:502–510.

Kroon-Batenburg, L.M.J., Scheurs, A.M.M., & Wilson, C.C., (1994). ISIS Annual Report A434.

Langan, P., Forsyth, V.T., Mahendrasingam, A., Alexeev, D., Fuller, W., & Mason, S.A., (1992a). A neutron diffraction study of the distribution of water in the A form of the DNA double helix. *Physica B*, 180 & 181:759–761.

Langan, P., Forsyth, V.T., Mahendrasingam, A., Pigram, W.J., Mason, S.A., & Fuller, W., (1992b). A high angle neutron fibre diffraction study of the hydration of the A conformation of the DNA double helix. *J. Biomol. Struct. Dyn.*, 10:489–503.

Langan, P., Forsyth, V.T., Mahendrasingam, A., Alexeev, D., Mason, S.A., & Fuller, W., (1993). Conference Proceedings of the Italian Physical Society: Water-Biomolecular Interactions, 43:235–238.

Langan, P., Forsyth, V.T., Mahendrasingam, A., Giesen, U., Dauvegne, M-Th., Mason, S.A., Wilson, C.C., & Fuller, W., (1995). Neutron fibre diffraction of DNA hydration. *Physica B* (in press).

Langridge, R., Marvin, D.A., Seeds, W.E., Wilson, H.R., Hooper, C.W., Wilkins, M.H.F., & Hamilton, L.D., (1960). The molecular configuration of deoxyribonucleic acid II. Molecular models and their Fourier transform. *J. Mol. Biol.*, 2:38–64.

Mahendrasingam, A., Forsyth, V.T., Hussain, R., Greenall, R.J., Pigram, W.H., & Fuller, W., (1986). Time-Resolved X-ray Diffraction Studies of the B⇔D Structural Transition in the DNA Double Helix. *Science*, 233:195–197.

Mahendrasingam, A., Al-Hayalee, Y., Forsyth, V.T., Langan, P., Fuller, W., Oldman, R.J., Blundell, D.J., & Mason, S.A., (1992). Neutron diffraction studies of the structure of PEEK. *Physica B*, 180 & 181:528–530.

Marvin, D.A., Spencer, M., Wilkins, M.H.F., & Hamilton, L.D., (1961). The molecular configuration of deoxyribonucleis acid III. X-ray diffraction study of the C form of the lithium salt. *J. Mol. Biol.*, 3:547–565.

30

TIME-OF-FLIGHT LAUE FIBER DIFFRACTION STUDIES OF PERDEUTERATED DNA

V. T. Forsyth,[1] P. Langan,[3*] M. A. Whalley,[1] A. Mahendrasingam,[1]
C. C. Wilson,[4] U. Giesen,[2†] M. T. Dauvergne,[3*] S. A. Mason,[2] and W. Fuller[1]

[1] Department of Physics
Keele University
Keele, Staffordshire ST5 5BG, United Kingdom
[2] Institut Laue-Langevin
BP156 38042, Grenoble Cedex, France
[3] EMBL Outstation
c/o Institut Laue-Langevin
BP156 38042, Grenoble Cedex, France
[4] ISIS
Rutherford Appleton Laboratory
Chilton, Didcot, Oxon OX11 0QX, United Kingdom

ABSTRACT

The diffractometer SXD at the Rutherford Appleton Laboratory ISIS pulsed neutron source has been used to record high resolution time-of-flight Laue fiber diffraction data from DNA. These experiments, which are the first of their kind, were undertaken using fibers of DNA in the A conformation and prepared using deuterated DNA in order to minimise incoherent background scattering. These studies complement previous experiments on instrument D19 at the Institut Laue Langevin using monochromatic neutrons. Sample preparation involved drawing large numbers of these deuterated DNA fibers and mounting them in a parallel array. The strategy of data collection is discussed in terms of camera design, sample environment and data collection. The methods used to correct the recorded time-of-flight data and map it into the final reciprocal space fiber diffraction dataset are also discussed. Difference Fourier maps showing the distribution of water around A-DNA calculated on the basis of these data are compared with results obtained using data recorded from hydrogenated A-DNA on D19. Since the methods used for sample preparation, data collection and data processing are fundamentally different for the

* Present address: Institut Laue-Langevin.
† Present address: Boehringer-Mannheim, Tutzing, Bavaria, Germany.

monochromatic and Laue techniques, the results of these experiments also afford a valuable opportunity to independently test the data reduction and analysis techniques used in the two methods.

INTRODUCTION

The use of neutron scattering in high resolution fiber diffraction studies of DNA has generated new opportunities that are of particular importance for the location of solvent around the various conformations that the DNA can adopt. The first high-angle fiber diffraction studies of DNA were undertaken on the instrument D19 at the ILL (Fuller *et al.*, 1989; Forsyth *et al.*, 1989; Langan *et al.*, 1992a; 1992b) and are described in detail by Fuller *et al.*, (this volume). These studies exploited the ability to isotopically replace H_2O surrounding the DNA by D_2O and to use the large difference in the scattering power of these isotopes in imaging the distribution of ordered water around the DNA.

The development of time-of-flight Laue fiber diffraction for the study of DNA has required special attention to sample preparation and data collection techniques. It is intrinsically difficult to make accurate comparisons between the flux available at a reactor neutron source on an instrument such as D19 with that on an instrument such as SXD using the ISIS pulsed source. However, it is clear that the current arrangement in which SXD operates with a water moderator at a temperature of 316K provides significantly less useable flux than that available on D19. To a certain extent this limitation is offset by the availability of sensitive detector which capture a large angular range of the diffraction data and which can be placed close to the sample. However, a significant increase in sample size is also required to take advantage of the beam size at SXD.

One of the biggest problems encountered in performing these experiments and indeed neutron experiments involving biological samples generally, occurs as a result of incoherent scattering arising from the presence of hydrogen atoms in the sample. This results in elevated levels of background scattering in the recorded diffraction patterns and places limitations both on the accuracy with which data can be measured and also on the maximum useable sample size. The effect is particularly severe for data recorded from DNA fibers which are hydrated with H_2O but is also very significant even when DNA is hydrated with D_2O since there are still substantial numbers of non-exchangeable hydrogen atoms that are covalently bound to the DNA sugars and bases. In this respect, it is clear that for both monochromatic and Laue neutron fiber diffraction studies, large improvements in the accuracy of the recorded data can be obtained by the use of perdeuterated DNA obtained from specially cultured cells grown in deuterated media. This is clearly significant for the Laue fiber diffraction experiments described here where there was a special need to optimise the signal-to-noise ratio for the recorded data.

EXPERIMENTAL METHODS

Production of Perdeuterated A-DNA

The DNA used in this study was obtained from MRE600 *E.Coli* cells grown on deuterated media which were cultured during the course of an EMBO Fellowship held at the EMBL Outstation in Grenoble. A number of separate batches of cells were used for the DNA preparation. These were all cultured using deuterated succinate carbon sources in media containing either 98% or 95% D_2O. Deuterated double-stranded DNA was purified from these cells using standard methods to remove protein, RNA and single-stranded DNA. The purified DNA was tested using high-angle X-ray fiber diffraction to identify the conditions in which highly crystalline A-DNA diffraction patterns could be reproducibly obtained from this material. This was a particularly important step in characterizing the perdeuterated material since it was found that the conformational polymorphism of deuterated DNA as a function of salt strength and hydration was rather different from that observed for natural hydrogenated DNA.

Sample Preparation

The comparatively low flux available on instrument SXD meant that the dimensions of samples used needed to be of the order of four times larger than those which have been used previously on D19 at the ILL. The sample used in these studies was prepared by arranging approximately 450 fibers into a parallel array. The fibers were individually mounted within a 8mm × 8mm window cut out of a cadmium sheet. The construction of the sample allowed for the interdigitation of fibers mounted on opposite sides of the aperture in the cadmium sheet and had the advantage of being able to accommodate the lengthening of the fibers as the sample was humidified. The glue which was used to attach fibers to the solid cadmium sheet was masked using cadmium foil. The whole arrangement was designed therefore to maximise the flux on the sample whilst also ensuring that any part of the beam which did not strike the sample was absorbed by the cadmium mount assembly.

Sample Environment and Data Collection

The application of the SXD diffractometer to the study of single crystal systems has been described (Wilson, 1994). Since DNA conformation is highly sensitive to hydration, a purpose designed sample enclosure was used to allow the relative humidity of the sample environment to be controled whilst at the same time reducing the effect of air scatter on the data. Figure 1 shows a schematic outline of the experimental system used in this work. The sample enclosure itself was made from aluminium, and had thin aluminium foil windows at the front and rear of the chamber to allow the passage of the incident and diffracted neutrons. The incident neutron beam was approximately 8mm in diameter and closely matched the size of the DNA sample on the cadmium mount. The relative humidity of the sample environment was regulated by passing dry argon gas through a saturated salt solution of sodium chlorate in D_2O. The humidified gas was passed through the sample chamber via the inlet and outlet pipes illustrated in Figure 1. This arrangement allowed the relative humidity of the fiber environment to be closely regulated at 75% which produced optimal crystallinity in the A-DNA sample. The relative humidity of the fiber environment was measured using a Vaisala HMP 4UT probe located within the sample enclosure.

The detector used in this study was a flat zinc sulphide scintillator position-sensitive detector (Wilson, 1992) which was positioned with its center at 30° to the incident beam and located in the equatorial plane 32cm distant from the sample position. The detector was 19.2cm × 19.2cm in size and had 64 × 64 pixels. The spread of neutron wavelengths available on SXD covers a wavelength range from ~0.5Å to ~9.0Å with the peak of the intensity distribution occurring at ~1.14Å. The pulsed nature of the ISIS source allows the wavelength of diffracted neutrons arriving at the detector to be determined by measuring

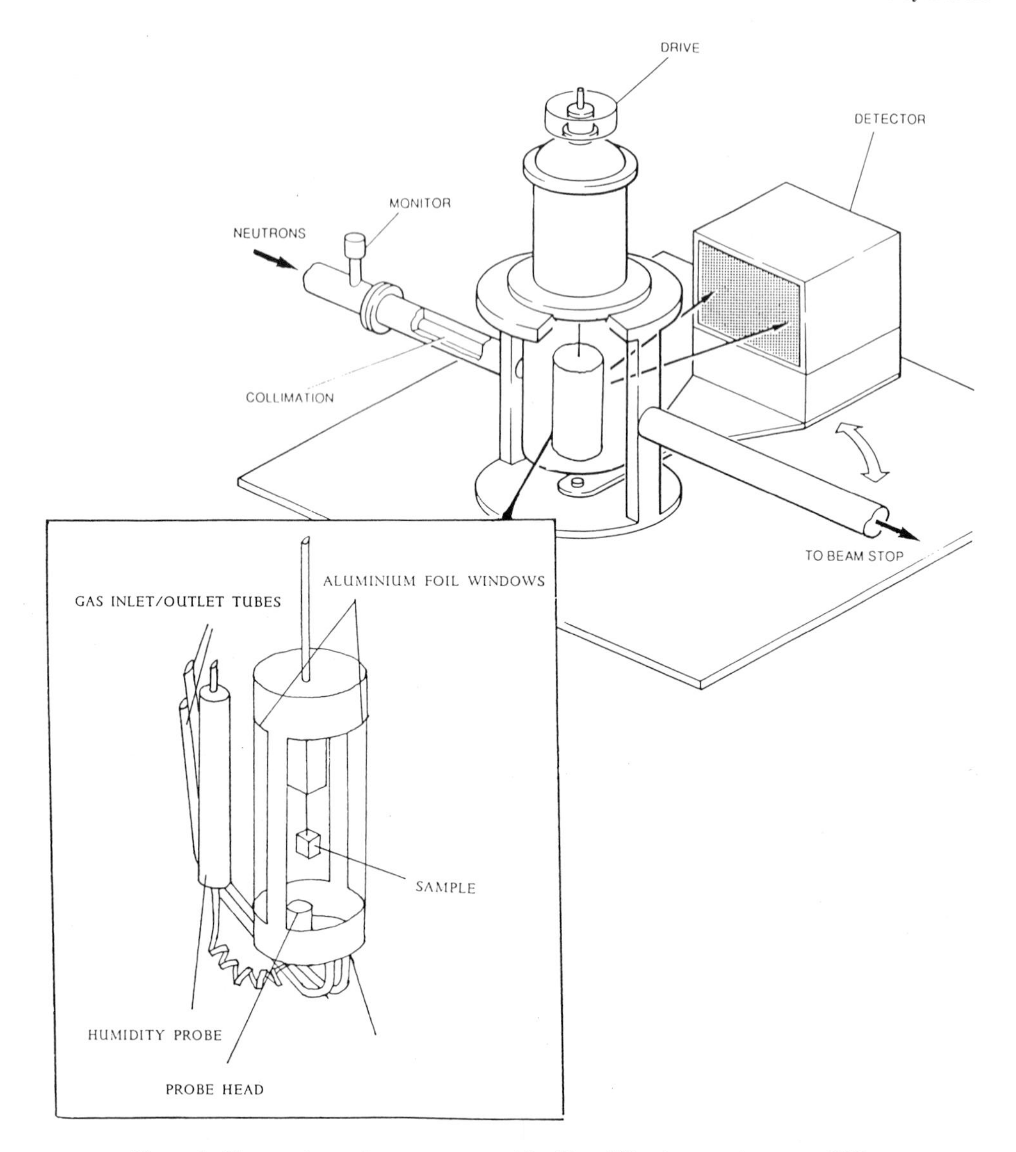

Figure 1. The experimental arrangement used for fiber diffraction experiments on SXD.

their 'time-of-flight' at the detector relative to the proton pulse hitting the target. In the experiments described here, 645 'time channels' were used to cover this range of wavelengths. Every pixel on the detector therefore produced a corresponding time-of-flight spectrum containing 645 points.

Data were collected from perdeuterated A-DNA, hydrated with D_2O, over a period of 6 days. During this time, datasets were recorded with the fiber axis in two different orientations such that diffraction data for the entire unique region of reciprocal space could be recorded. The sample was positioned first with the fiber axis vertical and second with the fiber axis horizontal. In each case data were recorded over a period of three days.

Data Processing

The processing of the time-of-flight data recorded from the SXD data was undertaken using purpose-designed software and involved two separate stages. During the first stage the individual time-of-flight spectra were processed to remove background and to account for detector response, the variation with wavelength of the incident neutron beam intensity, and the neutron reflectivity. Of these corrections by far the most important and difficult is the algorithm used to model and remove the background. A number of methods to account for the effect of background variation in the time-of-flight spectra are currently being tested. These include methods by which (i) diffraction features are located by detecting statistically significant deviations in the time-of-flight spectra and data fitted by a polynomial. This method has so far produced the cleanest time-of-flight spectra but is highly susceptible to the generation of artefacts, particularly at low resolution where the counting statistics are poor; (ii) peaks are identified in the data and Gaussian profiles fitted to these peaks. This approach eliminates some of the problems associated with the use of polynomials in methods (i) above; (iii) data from a vanadium standard are used to correct for background. The main advantage of using such a standard is that it mirrors the neutron distribution and more genuinely accounts for detector variation effects. However good vanadium counting statistics are required for this method to work well and these are typically obtained at the expense of counting time for the main sample; and (iv) DNA diffraction is extracted from the spectra using Fourier transform techniques. This method is still being evaluated but may provide a novel and effective approach to the problem. Figure 2 illustrates the application of the first of these methods to the correction of a single time-of-flight spectrum.

In the second stage, each of the corrected time-of-flight spectra are binned into reciprocal space to form an image which can then be processed using standard software for

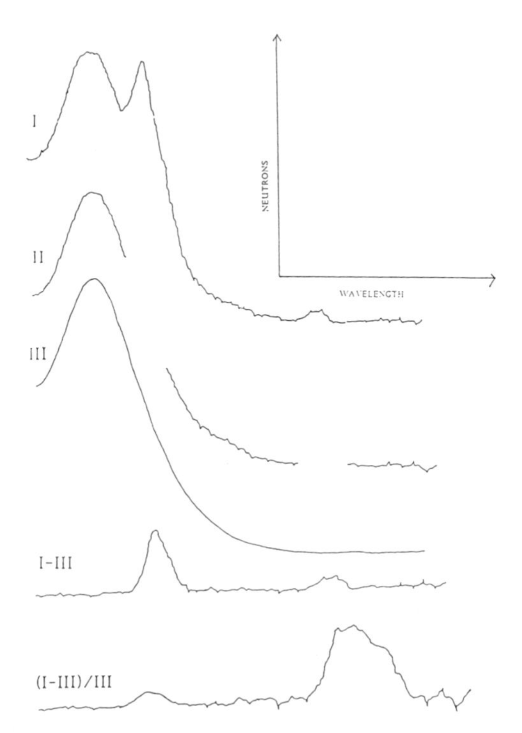

Figure 2. The correction of individual time-of-flight spectra. Plot I shows the raw data, exhibiting two diffraction peaks. Plot II shows the remainder of this spectrum after removing the diffraction features. Plot III shows a polynomial fit to Plot II. The last two plots illustrate how the polynomial fit to the background is used to correct the raw data.

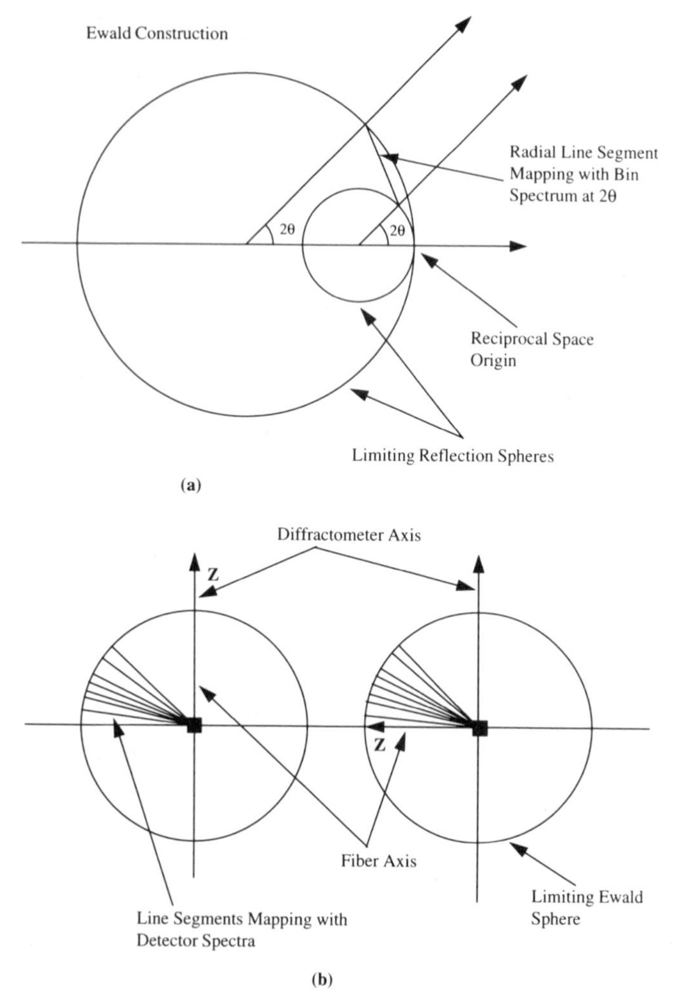

Figure 3. (a) The Laue diffraction geometry maps individual time-of-flight spectra into radial line segments in reciprocal space. (b) The recording of unique region of reciprocal space using two orientations of the fiber axis (the picture on the left shows the arrangement with the fiber axis vertical and that on the right with the fiber axis horizontal).

determining intensities. The diffraction geometry for these experiments is illustrated in Figure 3a and shows how individual time-of-flight spectra mapping over a range from λ_{min} to λ_{max} result in radial line segments in reciprocal space. The diagrams shown in Figure 3b illustrate how the radial line segments recorded for the two orientations of the fiber axis can be used to record data covering most of the unique region of reciprocal space.

RESULTS

The final reciprocal space image of the data recorded from the A form of deuterated A-DNA is shown in Figure 4. This dataset contains data processed for both sample orientations and was obtained by binning all time-of-flight data after correction using the poly-

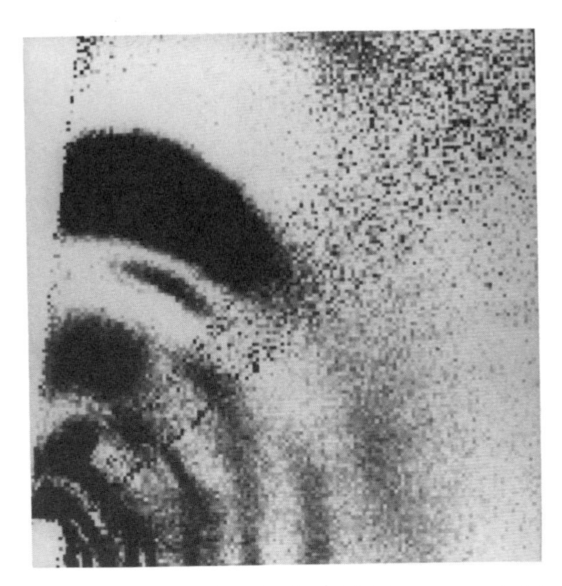

Figure 4. Reciprocal space image computed from the time-of-flight data recorded for the A form of deuterated DNA on instrument SXD.

nomial fitting methods described in the previous section. The image shows diffraction data extending from 3Å to a low resolution limit of approximately 25Å. Sharp Bragg reflections are clear throughout the image. Since the intensity distribution of the incident neutron results in very few long wavelength neutrons contributing to the diffraction pattern, the counting statistics are not uniform throughout the image and there are particular difficulties associated with accurately determining low resolution data in these experiments.

The diffracted intensities determined in these experiments have been measured and used to compute difference Fourier maps aimed at locating ordered water around the DNA. A view of one of these maps is shown in Figure 5. This image shows good correspondence with previous data recorded from hydrogenated DNA using D19 at the ILL as can be seen from a comparison of Figure 5 in this paper with Figure 8 in the paper by Fuller *et al.* (this volume). In particular, the water peaks located between successive O1 phosphate oxygen atoms along the DNA backbone are clearly apparent.

CONCLUSION

The experiments described here illustrate the scope of high resolution time-of-flight Laue fiber diffraction in the study of DNA hydration. The results which have been obtained using SXD are consistent with earlier experiments undertaken at the ILL, although the SXD experiments require significantly larger sample volumes. The current configuration on instrument SXD allows data to be recorded down to a low resolution limit of approximately 25Å. However, despite the advantages which are available at very high resolution, the lower flux available at the long wavelength end of the neutron spectrum places limitations on the quality of the diffraction dataset at low resolution. Poor counting statistics therefore make it more difficult to record the weaker reflections in the innermost

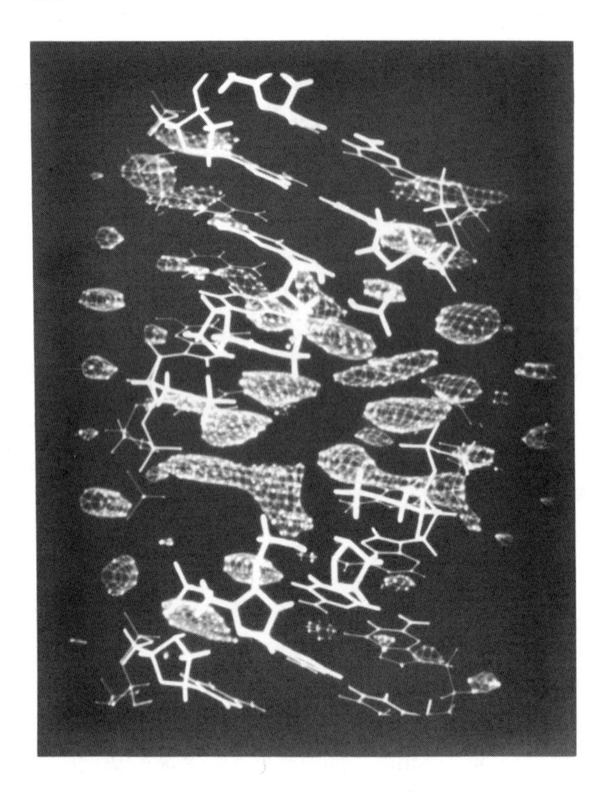

Figure 5. Difference Fourier map calculated using data recorded at SXD. Peaks associated with water located between successive O1 phosphate oxygen atoms are clearly visible. (A color version of this figure appears in the color insert following p. 214.)

region of the pattern although it is evident from the SXD pattern shown in Figure 4 that the stronger of these reflections are adequately represented in the data.

Since the methods used in the processing of time-of-flight data are fundamentally different from those used in the processing of monochromatic data, it is particularly important to take advantage of the restart of the ILL reactor so that a direct comparison of the two methods of data collection for identical DNA material can be made, as well as a critical evaluation of the data processing methods described above for the treatment of SXD data. It is clear that further developments in instrumentation which are planned at ISIS will have important consequences for experiments of this type. These include enhanced detector facilities for use on SXD and also the use of neutrons from a cold moderator which will shift the peak of the neutron intensity distribution to longer wavelengths.

ACKNOWLEDGMENTS

This work was supported under EPSRC grant number GR/H67959. Support from EMBO in the form of a short term Fellowship to VTF is gratefully acknowledged. We also thank Dr R. Leberman and other staff at the EMBL Outstation in Grenoble for invaluable assistance during the period of this Fellowship. We thank technical staff in the Keele University Physics Department workshop for constructing equipment that was essential to this work. We also thank Mrs H.E. Moors for secretarial assistance and Mr M. Daniels for photographic work.

REFERENCES

Forsyth, V.T., Mahendrasingam, A., Pigram, W.J., Greenall, R.J., Bellamy, K.A., Fuller, W., & Mason, S.A., (1989). Neutron fibre diffraction study of DNA hydration. *Int. J. Biol. Macr.,* 11:236–240.

Fuller, W., Forsyth, V.T., Mahendrasingam, A., Pigram, W.J., Greenall, R.J., Langan, P., Bellamy, K.A., Al-Hayalee, Y.A., & Mason, S.A., (1989). The location of water around the DNA double helix. *Physica B,* 156 & 157:468–470.

Langan, P., Forsyth, V.T, Mahendrasingam, A., Pigram, W.J., Mason, S.A., & Fuller, W., (1992a). A high angle neutron fibre diffraction study of the hydration of the A conformation of the DNA double helix. *J. Biomolec. Struct. Dyn.,* 10:489–503.

Langan, P., Forsyth, V.T., Mahendrasingam, A., Alexeev, D., Fuller, W., & Mason, S.A., (1992b). A neutron diffraction study of the distribution of water in the A form of the DNA double helix. *Physica B,* 180 & 181:759–761.

Wilson, C.C., (1992). In *Neutrons, X-rays and Gamma Rays*, SPIE Proceedings, J.M. Carpenter and D.F.R. Mildner editors 1737:226–234.

Wilson, C.C., (1994). In *Collaborations in Crystallography*, Rutherford-Appleton Report RAL-94-031, C.C. Wilson and V.I. Simonov editors, pp141–158.

31

THE CHEMICAL REACTIVITY AND STRUCTURE OF COLLAGEN STUDIED BY NEUTRON DIFFRACTION

T. J. Wess, L. Wess, and A. Miller

Department of Biological and Molecular Sciences
University of Stirling
Stirling FK9 4LA
Scotland
United Kingdom

ABSTRACT

The chemical reactivity of collagen can be studied using neutron diffraction (a non-destructive technique), for certain reaction types. Collagen contains a number of lysine and hydroxylysine side chains that can react with aldehydes and ketones, or these side chains can themselves be converted to aldehydes by lysyl oxidase. The reactivity of these groups not only has an important role in the maintenance of mechanical strength in collagen fibrils, but can also manifest pathologically in the cases of aging, diabetes (reactivity with a variety of sugars) and alcoholism (reactivity with acetaldehyde). The reactivity of reducing groups with collagen can be studied by neutron diffraction, since the crosslink formed in the adduction process is initially of a Schiff base or keto-imine nature. The nature of this crosslink allows it to be deuterated, and the position of this relatively heavy scattering atom can be used in a process of phase determination by multiple isomorphous replacement. This process was used to study the following: the position of natural crosslinks in collagen; the position of adducts in tendon from diabetic rats *in vivo* and the *in vitro* position of acetaldehyde adducts in tendon.

INTRODUCTION

Neutron diffraction has the advantage over many techniques, including electron microscopy, in that collagen is in a hydrated state. The deuteration of a regularly packed structure such as tendon collagen at specific locations in each molecule, allows the samples to be treated as multiple isomorphous derivatives. This process requires addition to

Neutrons in Biology, edited by Schoenborn and Knott
Plenum Press, New York, 1996

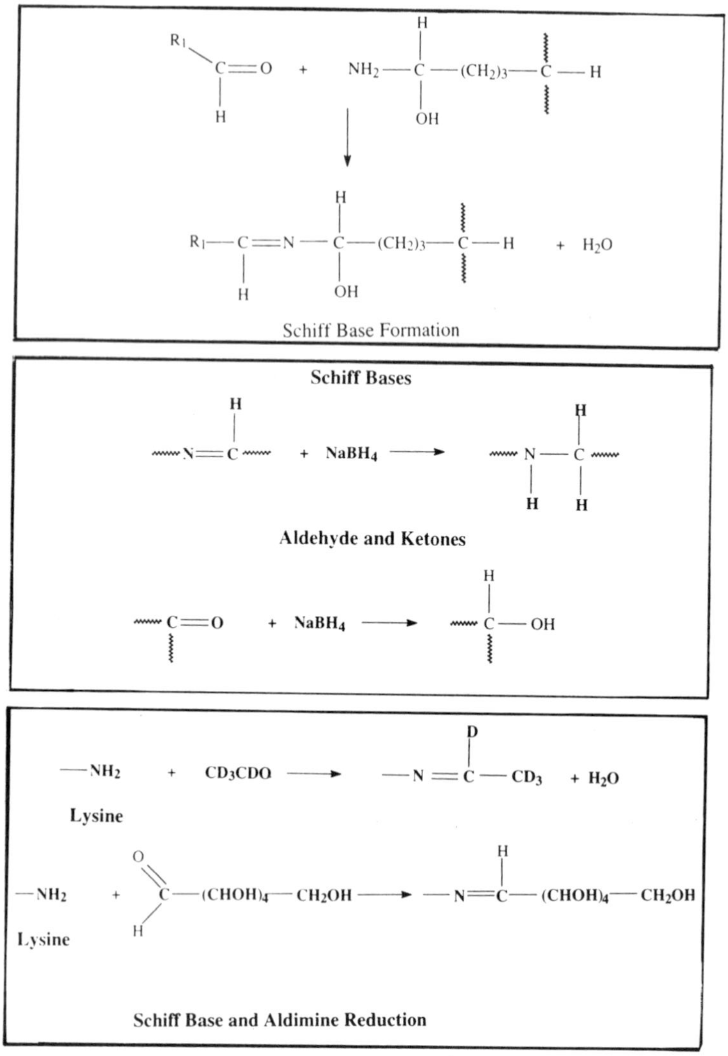

Schiff Base Formation

Schiff Bases

Aldehyde and Ketones

Schiff Base and Aldimine Reduction

Figure 1. This figure is composed of three parts: (a) The formation of a Schiff base; (b) The reaction of Schiff bases and aldehydes and ketones with sodium borohydride; (c) Reaction of lysine with deuterated acetaldehyde and reaction of lysine with glucose.

the sample of strongly scattering atoms in a specific manner. The neutron scattering cross-section of deuterium is large and positive (Bacon, 1975), compared to constituent atoms in collagen, and therefore deuteration fulfils this role. The neutron diffraction intensities of derivative samples can be used with control sample intensities in a uniaxial crystallographic structure determination.

Collagen has a number of side chains which are known to react with aldehydes and ketones and can also be themselves converted to aldehydes and be reactive. This can lead to a number of situations, where the collagen forms inter- and intra- molecular crosslinks, where there is reactivity between collagen and various sugars (leading to diabetes) and where there is reactivity between collagen and acetaldehyde as in the case of alcoholism (see Figure 1 for a detailed view of typical collagen adduct formation).

The study reported here is two fold; namely an investigation into the position of collagen glycation adducts, and the location of acetaldehyde binding sites in the axial unit cell of type I collagen. The tissue used was rat tail tendon which has a well defined axial structure as judged by both X-ray (Fraser *et al.*, 1983; Bradshaw *et al.*, 1989), neutron diffraction (Wess *et al.*, 1990) and electron microscopy (Meek *et al.*, 1979). Diabetic rat tail tendon was used as the *in vivo* glycation labeling system, and acetaldehyde reacted tendon was used as the *in vitro* acetaldehyde labeling system.

The collagen under study was type I collagen from rat tail tendon because it is the most abundant collagen in animals and is well characterized. Knowledge of the sequence and structure of type I collagen provides a firm basis for investigation.

General Reactivity of Collagen with Other Molecules

The reaction between protein and sugars (non enzymatic glycation) has been known since the turn of the century (Maillard, 1912). The biological significance of reactions between reducing sugars and protein were initially overlooked, however, more recently the relevance of these reactions with respect to conditions such as aging and diabetes has indicated that glycation of protein is important. A review of these effects with respect to collagen is by Reiser (1991).

Acetaldehyde is the highly reactive oxidation product of ethanol. It is therefore of importance in the understanding of changes that occur due to alcoholism. The potential of this molecule to alter the structure, development and maintenance of cells and organs is apparent. The impairment of control of collagen production in the liver, and the arrangement of collagen fibres are features common to fibrosis in alcohol induced liver disease. Most of the research conducted on collagen acetaldehyde interactions has been based on the increased deposition of collagen in clinically assessed alcoholics.

It has been reported that the hydroxylysine residues are more readily glycated than the lysine residues in a protein. The postulated reason for this being that the presence of the hydroxyl group on hydroxylysine may facilitate the attachment of sugar through enhancement of the ε-NH2 nucleophilicity (Perejda *et al.*, 1984). Le Pape also remarked on the likelihood of hydroxylysine as a preferential glycation site (Le Pape *et al.*, 1984). The glycation process continues with maturation of the Schiff base linked group. Amadori rearrangements occur in the case of aldose sugars, whilst a Heyn's rearrangement may occur with ketose sugar (Reynolds, 1969).

To date investigation of glycation especially with respect to collagen has been limited to the estimation of glycation using a variety of techniques. These usually involve degradation of the protein and calorimetric chemical procedures specific for the detection of ketoimine linked sugars (Brennan, 1989), radiolabeling of sugar or the lysine residues (Perejda *et al.*, 1984) or fluorimetric analysis of the mature glycated crosslink (Richard *et al.*, 1991). The specific tritiation of keto and ketoimine linkages formed in proteins upon glycation has also revealed the nature and level of glycation in samples. These techniques have been used to investigate the *in vivo* and *in vitro* glycation of protein including collagen (Robins & Bailey, 1972). The use of *in vitro* glycation allows the complex process of glycation to be controled and the number of sugars presented to a tissue to be limited.

Glycation of Collagen in Aging and Diabetes

In general, the results from studies such as these indicate that the amount of glycated sugar rises in conditions such as diabetes and in aging. The biomechanical effects of glycation have also been investigated (Kent *et al.*, 1985), and alterations have been made to the micro-crystalline structure of tendon upon *in vitro* glycation (Tanaka *et al.*, 1988). Changes in the lateral packing of skin with diabetes have been reported (James *et al.*, 1991). The position of collagen glycation within the axial unit cell has not been extensively investigated. Radiochemical labeling of glycation sites has been used to locate glycation sites in CNBr fragments (Le Pape *et al.*, 1984). CNBr fragments are well characterised peptide fragments of the collagen molecule where cyanogen bromide has caused peptide bond cleavage at methionine residues.

Previous research by our group has indicated possible sites of glycation in control (non diabetic) tendon samples (Wess *et al.*, 1990). The technique used was the specific deuteration of borohydride reducible groups in collagen with sodium borodeuteride ($NaBD_4$). The products of glycation are included in this, together with the immature reducible crosslinks characteristic of collagen.

Acetaldehyde Reactivity with Collagen in Alcoholism

Investigation of acetaldehyde reaction with collagen has been limited to the estimation of adduct formation using a variety of techniques, usually involving degradation of the protein and calorimetric chemical or radioactive procedures (Donohue *et al.*, 1983). Electron microscopy has been used to study the reaction of collagen with acetaldehyde (Cox *et al.*, 1973).

MATERIALS AND METHODS

Preparation of Glycated and Acetaldehyde Reacted Samples

Sample Preparation for Glycation. The rats used in this study were from the Edinburgh colony of spontaneously diabetic, insulin dependent BB rats (BB/E). Both control (non-diabetic), and diabetic tendon samples were obtained from rats as described above. Tail tendon samples were dissected from sufficient rat tails to obtain 0.75g of tendon (wet weight).

Sample Preparation for Acetaldehyde Labeling

Native tendon samples were obtained from 3–4 month old Wistar rats. Tail tendon samples were dissected from sufficient rat tails to obtain 0.75g of tendon (wet weight) for each labeling condition. The tendons from all the rats used were combined in order to reduce any possible changes in crosslinking between animals although this has not been found to be significant in previous experiments.

Deuterated acetaldehyde (CD_3CDO) was used to enhance the scattering length of acetaldehyde in neutron diffraction. Tendon samples were incubated with 150μl of deuterated acetaldehyde to a final concentration of 1gm/l. The samples were incubated for one hour in sealed tubes at 2–40°C as the boiling point of acetaldehyde is 200°C, after which time the samples were reduced with sodium borodeuteride.

Periodate Reactions

In order to investigate in more detail the nature and position of glycation in tendon, periodate degradation was performed on both control (non diabetic) and diabetic samples.

In a reaction volume of 20ml, 0.75g of tendon sample (diabetic or control (non diabetic)) were equilibrated in 15mM citrate/phosphate buffer pH 5 at room temperature for at least 10 minutes and reacted with l0mM sodium periodate for 30 minutes. Samples were subsequently borodeuterated for 25 minutes as described below. This is a modified Smith degradation on intact tendon without subsequent hydrolysis.

The reaction of periodate is thought to occur at both crosslinkage sites and glycation sites (Robins, 1983). The purpose of the reaction was to increase the number of reducible sites at any one crosslinking or glycation locus. The basis for this is the action of periodate to oxidise and cleave between adjacent diol groups on sugars. The resultant aldehydo-groups produced, will be reducible by $NaBD_4$. Robins found that although the hexitol lysine groups are expected to be susceptible to periodate cleavage, the yield of lysine is low. This is probably due to the reductive addition of some of the products of the oxidation reaction (Robins, 1983). It can also be expected that periodate will react with the proteoglycan found in rat tail tendon, and will also react with hydroxylysine glycosides (sugars reacted enzymatically with collagen). Both of these groups are found at specific locations in the axial projection of a fibril (Spiro, 1969; Scott & Orford, 1981). The oxidation of a sugar polymer will produce a polyaldehyde which will be readily reducible by borohydride. It is possible that the released formates will react with other groups in the sample (Feeney *et al.,* 1975).

Borodeuteration

Borodeuteride was used as the agent to specifically deuterate at crosslinks and glycation sites. In a volume of 5ml, 70mg of sodium borodeuteride was reacted with 0.75g wet weight tendon for either 6 or 25 minutes at pH 7.4 in 0.15M phosphate buffered saline (PBS). The reaction was stopped by copious washing with buffer to remove unreacted borodeuteride.

Sodium borohydride reduction of collagen has been used to isolate the immature components of collagen crosslinks. Both Schiff base (aldimine) and ketoimine crosslinks are reducible by this reagent. The aldehyde precursors of crosslinks, allysine and hydroxyallysine are also reducible.

Cyanoborodeuteration

Sodium cyanoborodeuteride can be used in a similar way to sodium borodeuteride, however the selectivity of this reagent at pH 4 makes it a useful tool to distinguish between linkage types in glycated material. In this case 70mg of cyanoborodeuteride was added to 0.75g of tendon as in the case above. Cyanoborodeuteride was reacted with the tendon for 25 minutes. This is a more gentle reducing agent than borodeuteride; however we maintained the reaction for this length of time to see if differences can be produced between the two techniques. According to Rucklidge, (Rucklidge *et al.,* 1983) the action of cyanoborodeuteride at pH 4 will facilitate the reduction of Amadori rearranged products preferentially to the reduction of Schiff bases since many of these will be broken at this pH.

Borch reported the ability of this reagent to crosslink at low pH (Borch *et al.,* 1971). The advantage of conducting reductions at this pH was that the specific labeling of aldehyde and keto groups would occur. The Schiff base crosslinks between collagen molecules, and between lysine residues and glycated sugar residues will begin to be cleaved at this pH. Cyanoborodeuteride has been used to specifically reduce acid stable, reducible crosslinks in collagen (Robins, 1983).

NEUTRON DIFFRACTION

Tendon samples were aligned in a suitable sample cell that allowed the samples to be tensioned in order to remove the inherent crimp, whilst maintaining an environment that did not allow the samples to dry. All diffraction data were collected at the Institut Laue Langevin, Grenoble. The small angle scattering instrument D11 was used to collect data on a 64 × 64 element detector (Ibel, 1976). Each sample was placed in the path of the collimated neutron beam of wavelength 10Å. Two camera lengths were used in order to obtain an optimum number of diffraction peaks with sufficient peak to peak resolution. These were 5.6 meters to obtain orders 1–3, and 2.8 meters for orders 2–8. This limits the resolution of data to 670/8Å.

Data Analysis

The data were collected for control (non diabetic) samples, diabetic samples and their derivatives respectively, and also for acetaldehyde reacted and control samples and their derivatives. These were analysed using a suite of programmes written especially for the purpose in VMS Fortran.

The programs allowed the intensity and therefore amplitude of each meridional Bragg peak to be determined by integration of each order and then making a suitable subtraction of local background. This was done for all orders of diffraction at all camera lengths. Related datasets were then scaled in order to obtain the first eight intensities of each sample. A standard correction with respect to order number was then applied as conducted by Hulmes (Hulmes *et al.*, 1980). This is significant since the wavelength of neutrons used means that the intersection of a reflection with the Ewald sphere decreased as order number increased.

Data scaling between control and derivative samples has often caused problems with structure determination by diffraction. In the case here, all intensity sets were scaled to a first order of 10000. We previously used the intensity of the transmitted beam in native and derivative samples to determine a scaling factor (Wess *et al.*, 1990). Since the samples were in an aqueous environment, changes in the amount of water present would alter the transmission values to a greater extent than the addition of a relatively small amount of deuterons. We therefore attempted to combine scaling by transmission measurement with scale factors for derivatives obtained by using difference Patterson maps.

The nature of a one dimensional Patterson map means that the origin should always be the maximum. This corresponds to a vector length of 0.0 ie all self-self vector density is superimposed at this point. If the maximum density of a difference Patterson map is not at this point, then an error in scaling is indicated. Using a range of scale factors from 1.0 to 1.5 at intervals of 0.01 an iterative process was used to calculate a difference Patterson function for each value until the 0.0 length vector had maximal density. This gave an initial minimum scaling value for each derivative dataset. Typically, this was in the range of 1.02–1.09. This value was compared to the scaling factor obtained by transmission measurements, typically 1.05–1.15. In the following rounds of refinement and structure determination, the minimum scaling factor was used as a lower limit and the scaling factor was allowed to float up to 0.06 more than this minimum value for each derivative until maximum agreement of individual phases for each structure factor between derivatives was obtained.

Analysis of difference Patterson maps allowed the vector distances between deuterated sites to be determined. A process of model building was then conducted and the proposed position of deuteration within a 670Å axial unit cell was tested by comparison of an autocorrelation of the proposed structure to the respective difference Patterson map.

The agreement of peak positions and size between the difference Patterson profile and autocorrelation profile was assessed. Any necessary alterations made to the proposed deuteration model structure were made in order to obtain an optimal agreement. When this was obtained for a derivative, the cyclic process of phase refinement could be used to improve the proposed deuterium labeling maps and to obtain real space projections of control and derivative samples. This procedure was conducted as described previously (Wess *et al.*, 1990) until optimal fits were made between the position of peaks in the autocorrelation and difference Patterson maps.

The reliability of a phase determined by this method was made by assigning a relative figure of merit to each phase depending on the closure error of phase determination at each round of refinement. This allowed the more reliable phases to be favorably weighted. Since the process of difference map generation uses information from the control and derivative intensities only, an R factor estimation for the structures given would be meaningless.

RESULTS FROM THE ANALYSIS OF THE POSITION OF GLYCATION OF COLLAGEN

In any structural investigation, the correctness of a structure is best judged by the correlation of structural data of a meaningful physicochemical model. Although the resolution of the study here is relatively low, the presence and reactivity of sugar adducts in normal and diabetic tissues can be distinguished. Results are shown here as the difference maps between the control neutron scattering profile and each derivative. These are shown in Figure 2. The use of a large number of derivatives has allowed the reliability of the first 8 meridional phases of collagen to be improved.

DISCUSSION OF FIGURE 2

In the case of each discussion presented here, the collagen molecule is regarded as the molecule projected onto a D-repeat of 234 amino acids. Therefore when a labeling event is described as being in the N-terminal region etc. the labeling has occurred at a point on a segment of the collagen molecule at, or in register with the collagen molecule.

Each map on Figure 2 is labeled with a number and short title, the numbers correspond to the bracketed numbers given for the interpretation of each map given below. In each case, the difference map represents the difference in neutron scattering density between the control (non diabetic) sample with no borodeuteride labeling, and the control (non diabetic) sample that has been labeled with deuterons according to the methods given.

- (Map 1) Diabetic 25 minutes $NaBD_4$: Diabetic tendon exhibits the majority of labeling in the region around the N-terminal. The c3 band also appears to be heavily labeled. This corresponds to regions where known crosslinks and hydroxylysine residues are found. The C-terminal region seems to be less labeled than the N-terminal. This could correlate to the lower amount of hydroxylysine residues in this region. There seems to be little deuterium labeling in the overlap

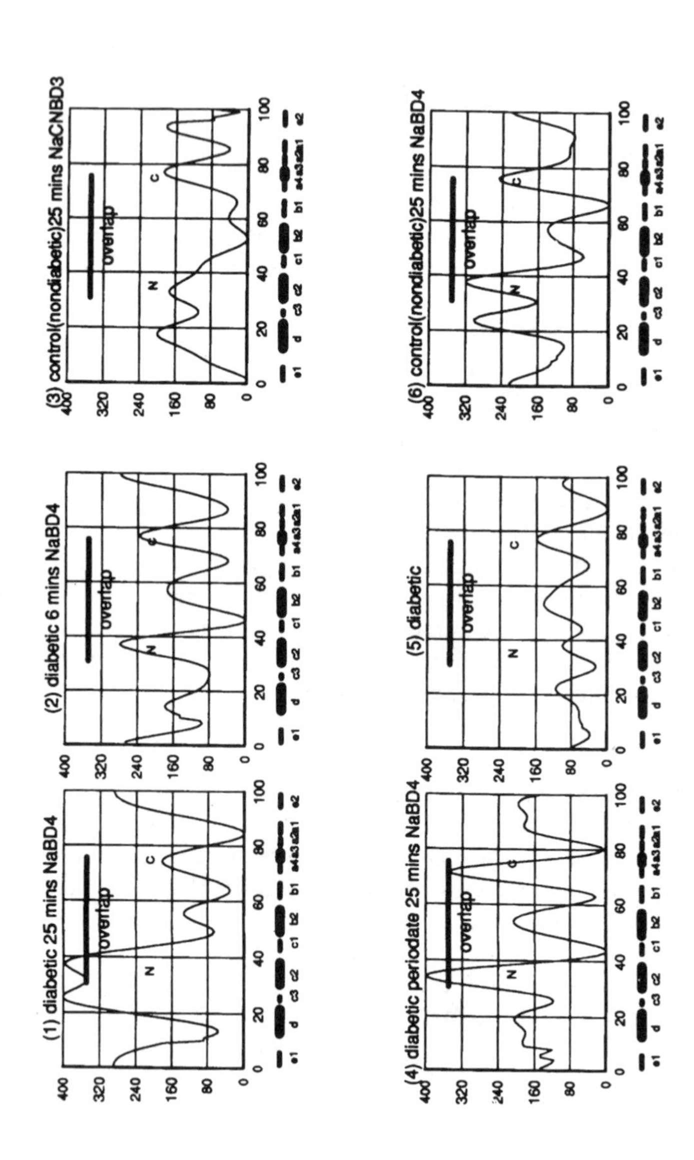
(1) diabetic 25 mins NaBD4
(2) diabetic 6 mins NaBD4
(3) control(nondiabetic)25 mins NaCNBD3
(4) diabetic periodate 25 mins NaBD4
(5) diabetic
(6) control(nondiabetic)25 mins NaBD4
overlap
N
C
400
320
240
160
80
0
20
40
60
80
100
e1 d c3 c2 c1 b2 b1 a4a3a2a1 e2

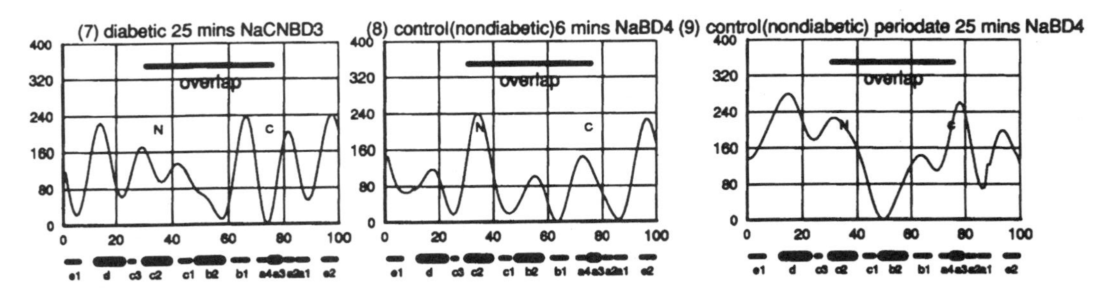

Figure 2. All nine of the difference maps obtained in this study are shown here, a full description of reaction conditions is given in the text for each sample. Each difference map represents a projection in real space of the neutron scattering density of collagen in one 670Å D-repeat. The D-repeat consists of gap and overlap regions, the overlap region is marked in each map by the black line above the density profile. The capital letters in the middle of each map indicate the position of the N- and C-termini. The highest difference density point of each plot depends on the scaling factor used to determine the structure. Maps such as diabetic no label had a small heavy atom scale factor, and therefore the plot here appears relatively flat compared to diabetic periodate 25 minutes $NaBD_4$ which had the largest scaling factor. The sites of deuteration are regarded as being most likely at lysine or hydroxylysine amino acids and their derivatives. The bands below each map correspond to the electron density bands found after positive staining with heavy atoms such as phosphotungstic acid. These are the regions which contain most lysine residues, and are therefore of interest in this study. Bands which are of double thickness represent regions that not only contain lysine, but also contain hydroxylysine residues.

region between the N and C regions. The gap region has a large peak centred between the el and e2 bands, indicating that non-enzymatic glycation occurs preferentially in the gap region. No hydroxylysine is found in this region, the evidence from this experiment infers that lysine is involved in glycation in this part of the gap region.

- (Map 2) Diabetic 6 minutes $NaBD_4$: The labeling of collagen at the region preceding the N-terminal is less than that in Map 1, indicating that the chemical nature or accessibility of this region is different to the region around the N-terminal ie band c2. After 6 minutes, the difference between the density of labeling at the N- and C-terminal regions is not great. This can be contrasted with the case of Map 1 where after 25 minutes the N-terminal region predominates, indicating that the initial reactivity at the N- and C- termini is similar with respect to reduction by borodeuteride.
- (Map 3) Control (non diabetic) 25 minutes $NaCNBD_3$: Tendon collagen labeled for 25 minutes with cyanoborodeuteride shows a large wide peak of density at the D band region, cyanoborodeuteride would not be expected to reduce the proteoglycan in this region, but indicates that more keto or aldehyde groups may be present in this region than previously thought.
- (Map 4) Diabetic periodate 25 minutes $NaBD_4$: This gave a complicated pattern indicating labeling of crosslinks, allysine and hydroxyallysine, glycated groups, proteoglycan and enzymatically linked carbohydrate. A stronger peak is observable at the C-terminal region when compared to Map 1. Labeling at the N- and C-terminal is greater than the peaks observable in the gap region. In rat tail tendon the position of enzyme directed glycosylation is regarded as principally being at position HYL 87. This residue lies in the band a3, and also corresponds to the C terminal peak on projection.
- (Map 5) Diabetic (no deuteration label): The comparison of this profile to Map 1 and Map 2 is of use since it may indicate that although sites of glycation may be present, they may have become non-reducible, or are less accessible to labeling than other residues. The map indicates labeling at N- and C-termini as well as those regions known to contain hydroxylysine residues. The relatively large peak in the overlap region indicates that some glycation may occur but is not accessible to the borodeuteride in the relatively short labeling times.
- (Map 6) Control (non diabetic) 25 minutes $NaBD_4$: The data produced from this labeling has been discussed by Wess (Wess *et al.*, 1990). Compared with that of the 1990 map this shows consistency. The only alteration is a movement of the peak in the gap region to the el, e2 region. In comparison to Map 1 the reduced peak of the map presented here corresponding to band c3 is reduced. Map 1 indicates a relatively larger peak in the gap region.
- (Map 7) Diabetic 25 minutes $NaCNBD_3$: This indicates a lower level of labeling generally. This indicates that cyanoborodeuteride is a more gentle reducing agent. In comparison to Map 1 tissue treated in this way shows a reduced labeling at the N-terminal. It is therefore possible to speculate that fewer aldehydo or keto groups exist in this region. If all the diabetic maps can be regarded as being dominated by the sugar attachment, then less labeling at the N-terminal region may indicate that sugar attachment in this region is predominantly in a Schiff base form, which is cleaved at pH 4. Cleavage of lysyl oxidase mediated crosslinks would not theoretically alter the deuteration profile in this region, since the result of the

cleavage would be allysine or hydroxyallysine both of which contain a reducible aldehyde group.

- (Map 8) Control (non diabetic) 6 minutes $NaBD_4$: In the context of comparison to the other maps here, labeling after 6 minutes indicates little difference to the diabetic tendon.
- (Map 9) Control (non diabetic) periodate 25 minutes $NaBD_4$: Under the reaction conditions used to produce this map, we expect the lysyl oxidase mediated crosslinks, any low level of glycation, proteoglycan and enzyme directed glycosylation adducts to be reduced. The presence of proteoglycan in discrete bands in tendon, located principally at the D band may lead to the high background in this map.

DISCUSSION OF GLYCATION OF COLLAGEN

The use of diabetic and control (non diabetic) tendon deuterated in a number of novel ways has provided a more reliable phase determination for the collagen molecule in tendon, as well as indicating the differences in reactivity of different chemical groups on the collagen molecule in health and disease. The use of sodium borodeuteride/cyanoborodeuteride allows specific sites of interest to be investigated in a non-destructive manner. In the case of conventional degradative biochemical analysis of collagen crosslinks and glycation adducts, the possibility always exists that what is being studied is also being degraded. Using this technique it is possible to investigate these structures and their position *in situ*. The possibility that glucose is not the only glycating agent has gained favor. This technique highlights regions of tendon which are deuterated when the tissue is obtained from diabetic animals compared to controls. The nature of the glycating agent is not distinguishable in this case, and therefore all glycation events are detectable if at sufficient levels. The use of periodate prior to borodeuteration in a modified Smith degradation also allows the position of proteoglycan molecules in well ordered tissues such as tendon to be studied in a fully hydrated state. A combination of this study with essential conventional chemical analysis of collagen glycation provides a powerful investigation into the molecular basis of diabetes related conditions and aging. The neutron diffraction data alone can only indicate the position and density of deuteration in the axial unit cell of collagen, this has to be analysed in the light of chemical evidence. This study opposes evidence proposed by Brennan (1989), however it must be stressed that the diabetic state induced in the two systems was significantly different with respect to the general health of the animals and the metabolic stress to which they were subjected. We believe that the study shown here resembles a more typical insulin controled hyperglycemic state. The techniques used here do not indicate the presence of mature non-reducible crosslinks or the advanced Maillard reaction products. It may however be possible to detect these in a non-destructive manner using X-ray diffraction.

RESULTS FOR THE ANALYSIS OF ACETALDEHYDE LABELING OF COLLAGEN

Although the resolution of the study here is relatively low, the presence and reactivity of acetaldehyde and inherent crosslinks in collagen can be distinguished. As we have found in previous studies, (Bradshaw *et al.*, 1989; Wess *et al.*, 1990) the differences be-

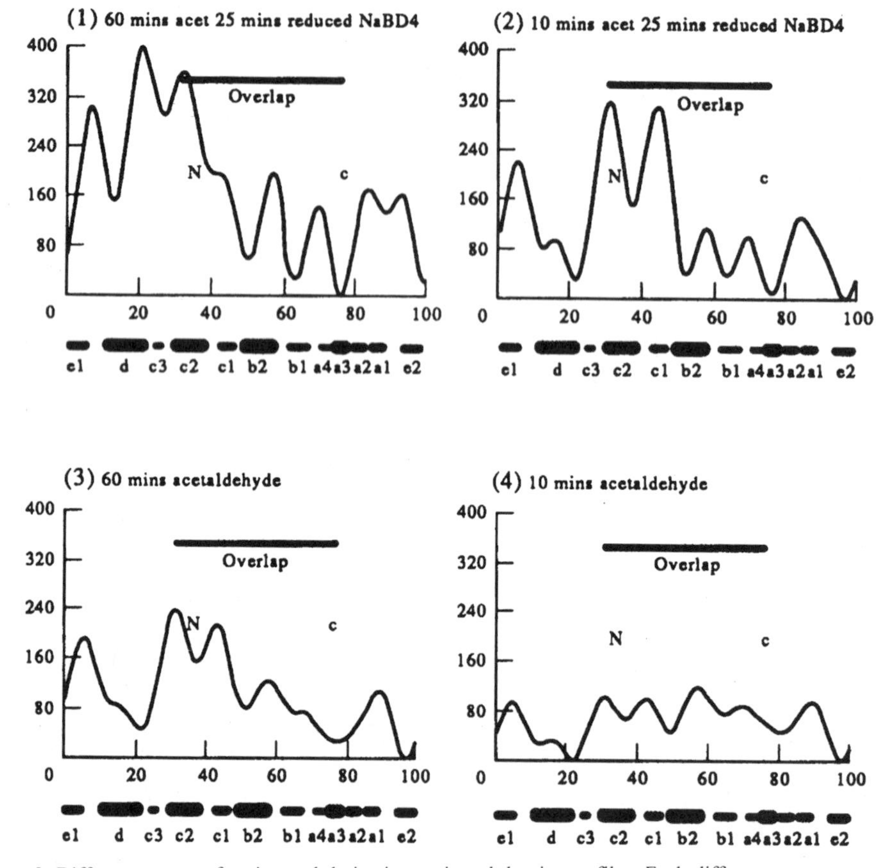

Figure 3. Difference maps of native and derivative projected density profiles. Each difference map represents a projection in real space of the neutron scattering density of collagen in one 670Å repeat. Map 1: 60 minutes acetaldehyde and 25 minutes reduction; Map 2: 10 minutes acetaldehyde and 25 minutes reduction; Map 3: 60 minutes acetaldehyde and no reduction; Map 4: 10 minutes acetaldehyde and no reduction. The axially projected regions of positive charge are shown as bands under each map.

tween the native and derivative (in this case acetaldehyde treated tendon samples) projected density profiles is by showing the difference profile of the two. These are shown in Figure 3. Each profile is discussed below.

ANALYSIS OF RESULTS

The relative scaling of the maps was determined relative to Map 1 (60 minutes acetaldehyde and 25 minutes reduction with sodium borodeuteride pH7.4). This difference map had the largest scaling factor for the contribution of the deuterated part of the map in phase refinement. The subsequent relative scaling of other maps to this employed the scaling factors used for the deuteron contribution in the final round of phase refinement. Under each map, the axially projected regions of positive charge are shown as bands, they are notated in the classical electron microscopy manner. Regions that contain hydroxylysine residues are represented as thicker bands since they are thought to be more reactive with keto containing groups, and are known to be important in crosslink formation.

- (Map 1) 60 minutes acetaldehyde and 25 minutes reduction with $NaBD_4$: The labeling of rat tail tendon with deuterated acetaldehyde and subsequent reduction of all Schiff base and ketoimine crosslinks formed, reveals a map with one dominant area of density. This is around the N-terminal telopeptide region. At this resolution, it is difficult to resolve the area of acetaldehyde labeling from the crosslink containing region of the N-terminus. However, the increase in the deuterated labeling intensity in this region indicates that a significant proportion of labeling occurs here. The potential of acetaldehyde to alter N-terminal crosslinking properties in collagen fibrillogenesis is therefore drawn to our attention. Less labeling seems to occur at the C-terminal region, indicating that although the deuteration of crosslinks and inherent aldehydes is occurring in this region, fewer acetaldehyde labeling events are occurring.
- (Map 2) 10 minutes acetaldehyde with 25 minutes reduction with $NaBD_4$: This map indicates that after a relatively short labeling period with acetaldehyde but with subsequent reduction of the tendon sample, the reduction of the crosslinks predominates. When compared to Map 4 where no reduction took place, it can be seen that the region close to the N-terminal is more heavily deuterated. The reduction and stabilisation of acetaldehyde crosslinks can be seen in this map when compared to Map 4. This would be expected since a number of extra deuterons are being introduced into the acetaldehyde adducts.
- (Map 3) 60 minutes acetaldehyde and no reduction: After 60 minutes of acetaldehyde labeling, without reduction using sodium borodeuteride, the small peaks seen in Map 4 are enhanced compared to the background. These peaks mainly occur in the gap region of collagen which will contain more accessible residues in the collagen molecule. This may account for the lack of electron density in the negatively stained gap region as observed by electron microscopy. An appreciable amount of neutron scattering density can be seen at the region of the N-terminal. This coincides with the hydroxylysine-rich c2 band which was observed to be labeled heavily in diabetic tendon (Wess *et al.*, 1993).
- (Map 4) 10 minutes acetaldehyde labeling and no reduction: The labeling of tendon collagen with acetaldehyde for a short time such as this indicates that few deuteron containing molecules are adducted to the tendon. The small humps observed in the map do, however, indicate that the gap region in this timescale is the major region to be labeled. Labeling around the N-terminal is more appreciable after 60 minutes. This effect may be due to steric hindrance. The time allowed in this experiment may not allow the acetaldehyde to penetrate the more dense telopeptide regions.

DISCUSSION OF ACETALDEHYDE LABELING OF COLLAGEN

The use of neutron diffraction to analyse the adduction of acetaldehyde to collagen has revealed the following points.

- Neutron diffraction allows the acetaldehyde residue position to be viewed *in situ* since it is a non-destructive method of investigating the position of adduction events in collagen. Previous structural studies conducted using electron microscopy are only able to distinguish changes in the electron density of a heavy atom stained samples as modified by acetaldehyde labeling. An inherent problem of electron microscopy is also the change in the axial unit cell parameters, the reduction of the axial period from 670Å to 640Å on sample preparation, indicates that a number of changes have occurred and the native molecule is no longer being studied. The position of acetaldehyde in the axial unit cell would also only be revealed by X-ray diffraction if a suitable heavy atom was used that would label sites where acetaldehyde has not adducted, and would be unable to label the positions where acetaldehyde was located. Neutron diffraction is a unique technique for the *in situ* analysis of collagen labeling.

- The non-reduced and reduced sample comparison allows the correlation between the deuterated acetaldehyde binding and the formation of a Schiff base or ketoimine link base at this location to be made. The potential reactivity of acetaldehyde with collagen may serve to alter the biomechanical properties of preformed collagen fibrils by the introduction of extra crosslinks into the fibril in a relatively uncontroled manner. With respect to fibril formation the presence of acetaldehyde may block the naturally occurring lysine crosslinking sites of the collagen molecule. The increased observed labeling at and around the N and C termini of the collagen molecule by acetaldehyde may be significant in altering crosslinking patterns in the collagen fibril since these parts of the molecule are known to contain the major natural crosslinking sites in collagen.
- The analysis of acetaldehyde labeling of a wet collagen sample *in situ* indicates that no change in the 670Å D-spacing of collagen was observed. Therefore although acetaldehyde was seen to label collagen at a number of different locations, no gross changes have altered the axial periodicity of the collagen unit cell (Wess *et al.,* 1994).

There is evidence that acetaldehyde may react with collagen at intracellular as well as at extracellular locations (Baraona *et al.,* 1993). Thus the acetaldehyde may alter the formation of collagen triple helices and intramolecular crosslinks before fibril formation. These effects will be paramount in the changes occurring to collagen in alcohol-induced liver disease.

ACKNOWLEDGMENTS

We are indebted to Dr Joyce Baird and Dr Mark Lindsay of the Metabolic Unit, Department of Medicine, Western General Hospital, Edinburgh for the diabetic rats and advice. We gratefully acknowledge the help of Dr P.D. Adams of the University of Edinburgh for advice with computing and Dr P. Timmins our local contact at the ILL Grenoble.

REFERENCES

Bacon, G.E., (1975). *Neutron Diffraction*, Oxford, Clarendon Press.

Baraona, E., Liu, W., Ma, X.-L., Svegliati-Baroni, G., & Lieber, C.S., (1993). Acetaldehyde-collagen adducts in N-nitrosodimethylamine induced liver cirrhosis in rats. *Life Sciences,* 52:1249–1255.

Borch, R.F., Bernstein, M.D., & Durst, H.D., (1971). The cyanohydridoborate anion as a selective reducing agent. *J. Am. Chem. Soc.,* 93:2897–2904.

Bradshaw, J.P., Miller, A., & Wess, T.J., (1989). Phasing the meridional diffraction pattern of type I collagen using isomorphous derivatives. *J. Mol. Biol.*, 205:685–694.

Brennan, M., (1989). Changes in the cross-linking of collagen from rat tail tendon due to diabetes. *J. Biol. Chem.*, 264:20953–20960.

Cox, R.W., Grant, R.A, & Kent, C.M., (1973). An electron microscope study of the reaction of collagen with some monoaldehydes and bifunctional aldehydes. *J. Cell Sci.*, 12:933–949.

Donohue, T.M., Tuma, D.J., & Sorrell, M.F., (1983). Binding of metabolically derived acetaldehyde to hepatic proteins *in vitro*. *Lab. Investigation*, 49:226–229.

Fraser, R.D.B., MacRae, T.P., & Miller, A., (1983). Molecular conformation and packing in collagen fibrils. *J. Mol. Biol.*, 167:497–521.

Hulmes, D.J.S., Miller, A., White, S.W., Timmins, P.A., & Berthet-Colominas, C., (1980). Interpretation of the low angle meridional neutron diffraction patterns from collagen-fibres in terms of the amino acid sequence. *Int. J. Biol. Macromol.*, 2:338–345.

Ibel, K., (1976). Neutron small angle camera D11 at high flux reactor, Grenoble. *J. Appl. Cryst.*, 9:630–643.

James, V.J., Delbridge, L., McLennan, S.V., & Yue, D.K., (1991). Use of X-ray diffraction in the study of human diabetic and aging collagen. *Diabetes*, 40:391–394.

Kent, M.J.C., Light, N.D., & Bailey, A.J., (1985). Evidence for glucose-mediated covalent crosslinking of collagen after glycosylation *in vitro*. *Biochem. J.*, 225:745–752.

Le Pape, A., Guitton, J.-D., & Muh, J.-P., (1984). Distribution of non-enzymatically bound glucose in *in vivo* and *in vitro* glycosylated type I collagen molecules. *FEBS Letts.*, 170:23–27.

Maillard, L.C., (1912). Action des acides amines sur les sucres : formation de melanoidines par voie methodique. *C. R. Acad. Sci.*, 154:66–68.

Meek, K.M., Chapman, J.C., & Hardcastle, R.A., (1979). Staining pattern of collagen fibrils - Improved correlation with sequence data. *J. Biol. Chem.*, 254:10710–10714.

Perejda, A.J., Zaragoza, E.J., Eriksen, E., & Uitto, J., (1984). Non-enzymatic glucosylation of lysyl and hydroxylysyl residues in type I and type II collagens. *Collagen Related Res.*, 4:427–439.

Reiser, K.M., (1991). Non-enzymatic glycation of collagen in aging and diabetes. *Proc. Soc. Exp. Biol.*, 196:17–29.

Reynolds, T.M., (1969). In *Symposium on foods carbohydrates and their roles*. H.W. Schultz, R.F. Cain and R.W. Wrolstad, editors. Avi Westport Conn. pp219–252.

Richard, S., Tamas, C., Sell, D.R., & Monnier, V.M., (1991). Tissue specific effects of aldose reductase inhibition on fluorescence and crosslinking of extracellular matrix in chronic galactosemia. Relationship to pentosidine crosslinks. *Diabetes*, 40:1049–1056.

Robins, S.P., (1983). Analysis of the crosslinking components of collagen and elastin. *Methods of Biochemical Analysis*, 28:329–379.

Robins, S.P., & Bailey, A.J., (1972). Age related changes in collagen : The identification of reducible lysine carbohydrate condensation products. *Biochem. Biophys. Res. Comm.*, 48:76–84.

Rosenberg, H., Modrak, J.B., Hassing, J.M., Al-Turk, W.A., & Stohs, S.J., (1979). Glycosylated collagen. *Biochem. Biophys. Res. Comm.*, 91:498–501.

Rucklidge, G.J., Bates, G.P., & Robins, S.P., (1983). Preparation and analysis of the products of non-enzymatic protein glycosylation and their relationship to crosslinking of proteins. *Biochim. Biophys. Acta.*, 747:165–70.

Scott, J.E., & Orford, C.R., (1981). Dermatan sulphate-rich proteoglycan associates with rat tail tendon collagen at the D band in the gap region. *Biochem. J.*, 197:213–216.

Spiro, R.G., (1969). Characterisation and quantitative determination of hydroxylysine linked carbohydrate units of several collagens. *J. Biol. Chem.*, 244:602–612.

Tanaka, S., Avigad, G., Brodsky, B., & Eikenberry, E.F., (1988). Glycation induces expansion of the molecular packing of collagen. *J. Mol. Biol.*, 203:495–505.

Wess, T.J., Bradshaw, J.P., & Miller, A., (1990). Cross-linkage sites in type I collagen fibrils studied by neutron diffraction. *J. Mol. Biol.*, 203:1–5.

Wess, T.J., Wess, L., Miller, A., Lindsay, R.M., & Baird, J.D., (1993). The *in vivo* glycation of diabetic tendon collagen studied by neutron diffraction. *J. Mol. Biol.*, 230:1297–1303.

Wess, T.J., Wess, L., & Miller, A., (1994). The *in vitro* binding of acetaldehyde to collagen studied by neutron diffraction. *Alcohol and Alcoholism*, 29:403–409.

32

IN SITU SHAPE AND DISTANCE MEASUREMENTS IN NEUTRON SCATTERING AND DIFFRACTION

Satoru Fujiwara and Robert A. Mendelson

Cardiovascular Research Institute and
Department of Biochemistry and Biophysics
University of California
San Francisco, California 94143

ABSTRACT

Neutron scattering combined with selective isotopic labeling and contrast matching is useful for obtaining *in situ* structural information about a selected particle, or particles, in a macromolecular complex. The observed intensities, however, may be distorted by inter-complex interference and by scattering-length-density fluctuations of the (otherwise) contrast-matched portions. Methods have been proposed to cancel out such distortions (Hoppe's method, the Statistical Labeling Method, and the Triple Isotopic Substitution Method). With these methods as well as related unmixed-sample methods, structural information about the selected particle(s) can be obtained without these distortions. We have generalized these methods so that, in addition to globular particles in solution, they can be applied to *in situ* structures of systems having underlying symmetry and/or net orientation as well. The information obtainable from such experiments is discussed.

INTRODUCTION

Neutron scattering and diffraction are powerful techniques for obtaining *in situ* structural information about members of a macromolecular complex. The combination of selective isotopic labeling and contrast matching allow one to obtain *in situ* shape and distance information about selected particles in a macromolecular complex. However, interference effects from scattering-length density fluctuations in the density-matched portions may not be negligible for small labeled particles with low contrast. Also, for solution-scattering measurements at high concentrations, the resulting patterns may be distorted by inter-complex interference.

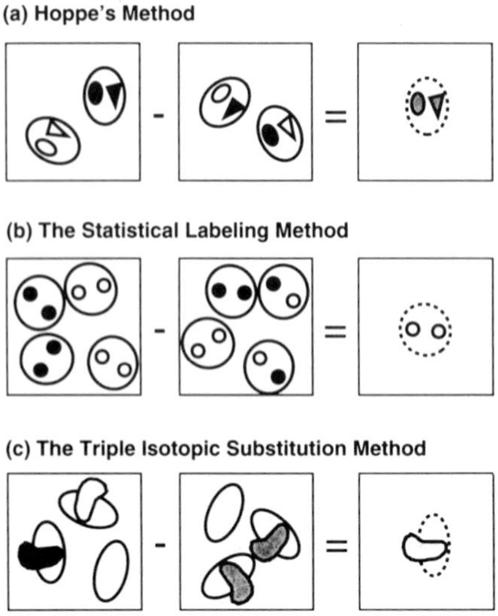

(a) Hoppe's Method

(b) The Statistical Labeling Method

(c) The Triple Isotopic Substitution Method

Figure 1. Schematic diagrams of Hoppe's method (a), the Statistical Labeling Method (b), and the Triple Isotopic Substitution Method (c).

Special methods to cancel out such distorting interference effects have been proposed. In one such method, Hoppe (1972; 1973) proposed that the interference function between two nonidentical labeled particles in a reconstituted complex could be obtained without inter-complex interference. One of the two samples needed (Figure 1a) is prepared with a one-to-one mixture of doubly labeled complexes and unlabeled complexes. The other sample contains a one-to-one mixture of two singly labeled complexes. The difference intensity between these samples yields the inter-particle interference function. A variation of Hoppe's method, which we term the Statistical Labeling Method (SLM), was proposed for obtaining the interference function from two identical labeled particles in a reconstituted complex in solution (Kneale *et al.,* 1977). Here the one-to-one mixture of two singly labeled particles in Hoppe's method is replaced by a mixture of complexes in which the isotopically labeled particles are statistically distributed (Figure 1b). Pavlov and Serdyuk (Pavlov & Serdyuk, 1987; Serdyuk & Pavlov, 1988) have introduced a powerful solution-scattering method for analysis of the *in situ* shape of particles. In their Triple Isotopic Substitution Method (TISM), three levels of isotopic labeling are employed to generate the two samples used for solution scattering (Figure 1c). One sample consists of a mixture of particles having the two extremes of degree of labeling. The other sample contains particles isotopically labeled to an intermediate extent. The difference intensity yields the *in vacuo* solution-scattering intensity pattern of the selected particle (in the multicomponent complex).

In addition to these methods, which employ mixtures of isotopically labeled samples, unmixed-sample methods have also been proposed (Engelman & Moore, 1972; Curmi & Mendelson, 1991). Here complexes containing particles having different degrees of labeling or substitution are each separately exposed to the neutron beam. Proper subtraction of the scattering intensities from these samples cancels out all contributions from the components other than the particles of interest, but inter-particle interference is not eliminated.

To date, experimental applications of Hoppe's method and the TISM have been reported only for globular complexes in solution. Hoppe's method has been applied to the determination of the protein arrangement in the 30S ribosomal subunit (eg Capel *et al.*, 1987), RNA polymerase (Stöckel *et al.*, 1979; Stöckel *et al.*, 1980a; 1980b), and distance measurements in the 50S ribosomal subunit (Hoppe *et al.*, 1975). The TISM has been applied to solution-scattering studies of the bacterial polypeptide elongation factor Tu (Pavlov *et al.*, 1991; Serdyuk *et al.*, 1994) and 50S ribosomal subunits (Harrison *et al.*, 1993).

Many important biological systems, such as filamentous structures in muscles, viruses and membranes, form macromolecular complexes having underlying order. Oriented samples are sometimes available from such ordered systems. It is therefore important to know if and how the above methods are applicable to ordered and oriented systems. Hoppe claimed that his method is applicable to any ordered and oriented system. However, Pavlov and Serdyuk derived the TISM only for systems having no strict underlying symmetry, such as globular particles in solution. The SLM was also aimed at globular particles in solution. To understand possibilities and limitations of these methods, we have developed a reciprocal-space expression describing the scattering intensity of particles embedded in a macromolecular matrix (Fujiwara & Mendelson, 1994). We have shown that all of these methods can be generalized to systems having underlying order and to oriented systems. Here we describe the theory and discuss experimental possibilities.

THEORY

The Intensity from Isotopically Labeled Particles Embedded in a Matrix

Consider a 'large' particle consisting of a matrix portion and n (≥ 2) 'small' particles. The scattering amplitude of this large particle, $F_{LP}(\mathbf{S})$, can be described as:

$$F_{LP}(\mathbf{S}) = F_M(\mathbf{S}) + \sum_{j=1}^{n} F_{SP}^{j}(\mathbf{S})\exp(2\pi i\mathbf{S}\mathbf{r}_j) \tag{1}$$

where $F_M(\mathbf{S})$ and $F_{SP}^{j}(\mathbf{S})$ are the scattering amplitudes of the matrix portion and the j-th small particle, respectively, $\mathbf{S}$ is the reciprocal-space vector, and $\mathbf{r}_j$ is the vector from the center-of-mass of the large particle to the center-of-mass of the small particle at the j-th site. The small particles can have identical structures, nonidentical structures, or both. These scattering amplitudes are the Fourier transforms of the scattering-length density contrast, ie the difference between the scattering-length density of particles and that of the solvent (assumed to be constant).

Assume a system consisting of N_T structurally identical large particles. If the small particles are either isotopically labeled or unlabeled, the scattering intensity from this system is averaged over the distribution of the labeled small particles as well as orientations of the large particles. If the distribution of labeled small particles in one large particle does not affect those in other large particles, and if the large particles having various distributions of labeled small particles are randomly distributed, the scattering intensity is expressed as:

$$I(\mathbf{S}) = N_T\{<<|F_{LP}(\mathbf{S})|^2>_{labeling}>_{orientation} + <(1/N)\sum_{l\neq m}^{N}\sum_{l\neq m}^{N}<F^*_{LP}(\mathbf{B}_l^{-1}\mathbf{S})>_{labeling}<F_{LP}(\mathbf{B}_m^{-1}\mathbf{S})>_{labeling}\exp(2\pi i\mathbf{S}\mathbf{R}_{lm})>_{orientation}\} \quad (2)$$

Here N represents the number of large particles in one region of coherence (see Fujiwara & Mendelson, 1994), $\mathbf{B}_l$ denotes the rotation matrix specifying the orientation of the l-th large particle relative to some standard orientation, $\mathbf{R}_{lm}$ is the distance between the centers-of-mass of the l-th large particle and the m-th large particle, $\langle\ \rangle_{labeling}$ denotes averaging over the distribution of labeled small particles, and $\langle\ \rangle_{orientation}$ denotes averaging over orientations of large particles. Equation 2 requires no assumptions concerning the positions and orientations of the large particles, so it can be applied to systems having any degree of order, from globular particles in solution to single crystals. This equation is the basic equation for the derivation of the SLM, Hoppe's method, and the TISM.

The Statistical Labeling Method

Figure 2 shows the scheme for the preparation of the two samples used in the SLM. The 'mixture' sample is prepared by mixing the large particles bearing only labeled small particles and those bearing only unlabeled small particles, in a ratio of δ : 1 - δ. The other 'randomly labeled' sample is prepared by reconstituting the large particles from the matrix portion and a mixture of labeled and unlabeled small particles. Here the ratio of the labeled small particles to the unlabeled small particles is δ : 1 - δ. Assuming that the behavior of the small particles is not changed by labeling, labeled small particles are statistically

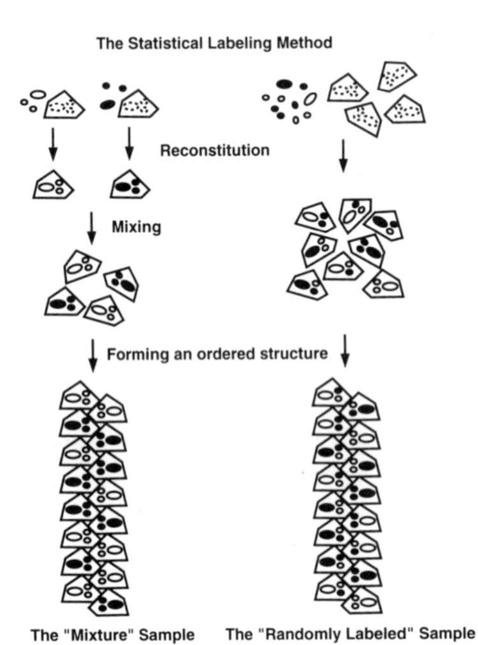

Figure 2. Scheme of the sample preparation for the Statistical Labeling Method. Here the large particle consists of the matrix portion represented by a polygon and three small particles, two of which are identical. After the mixing or reconstitution, the large particles form an ordered structure consisting of 16 large particles. Adapted in part from Fujiwara and Mendelson (1994).

distributed in the large particles. It is also assumed that no exchange of the small particles occurs.

Scattering intensity from the mixture sample, $I_{L+U}(\mathbf{S})$, and that from the randomly labeled sample, $I_{RL}(\mathbf{S})$, can be derived by calculating $\langle |F_{LP}(\mathbf{S})|^2\rangle_{labeling}$ and $\langle F_{LP}(\mathbf{S})\rangle_{labeling}$ for each sample and by substituting them into Equation 2. The final expressions are:

$$\begin{aligned} I_{L+U}(\mathbf{S}) = N_T[&< \delta|F_{LP,L}(\mathbf{S})|^2 + (1-\delta)|F_{LP,U}(\mathbf{S})|^2 >_{orientation} + <(1/N)\sum_{l\neq m}^{N}\sum_{l\neq m}^{N}\{\delta F^*_{LP,L}(\mathbf{B}_l^{-1}\mathbf{S}) \\ &+ (1-\delta)F^*_{LP,U}(\mathbf{B}_l^{-1}\mathbf{S})\} \times \{\delta F_{LP,L}(\mathbf{B}_m^{-1}\mathbf{S}) \\ &+ (1-\delta)F_{LP,U}(\mathbf{B}_m^{-1}\mathbf{S})\}\exp(2\pi i\mathbf{S}\mathbf{R}_{lm}) >_{orientation}] \end{aligned} \tag{3}$$

and:

$$\begin{aligned} I_{RL}(\mathbf{S}) = N_T[&< \delta|F_{LP,L}(\mathbf{S})|^2 + (1-\delta)|F_{LP,U}(\mathbf{S})|^2 >_{orientation} \\ &- \delta(1-\delta) < \sum_{j\neq k}^{n}\sum_{j\neq k}^{n}\Delta F^{j*}_{SP}(\mathbf{S})\Delta F^{k}_{SP}(\mathbf{S})\exp(2\pi i\mathbf{S}\mathbf{r}_{jk}) >_{orientation} \\ &+ <(1/N)\sum_{l\neq m}^{N}\sum_{l\neq m}^{N}\{\delta F^*_{LP,L}(\mathbf{B}_l^{-1}\mathbf{S}) + (1-\delta)F^*_{LP,U}(\mathbf{B}_l^{-1}\mathbf{S})\} \\ &\times \{\delta F_{LP,L}(\mathbf{B}_m^{-1}\mathbf{S}) + (1-\delta)F_{LP,U}(\mathbf{B}_m^{-1}\mathbf{S})\}\exp(2\pi i\mathbf{S}\mathbf{R}_{lm}) >_{orientation}] \end{aligned} \tag{4}$$

where $F_{LP,L}(\mathbf{S})$ and $F_{LP,U}(\mathbf{S})$ denote the scattering amplitudes of the large particle in which all small particles are labeled and of the large particle in which all small particles are unlabeled, respectively, and $\mathbf{r}_{jk}$ denotes the vector between j-th and k-th small particles in the large particle. Assuming that δ, the scattering-length density of the solvent and the concentration of the large particles are the same in both samples, the intensity difference yields the interference function:

$$\begin{aligned} \Delta I(\mathbf{S}) &= I_{L+U}(\mathbf{S}) - I_{RL}(\mathbf{S}) \\ &= N_T\delta(1-\delta) < \sum_{j\neq k}^{n}\sum_{j\neq k}^{n}\Delta F^{j*}_{SP}(\mathbf{S})\Delta F^{k}_{SP}(\mathbf{S})\exp 2\pi i\mathbf{S}\mathbf{r}_{jk}) >_{orientation} \end{aligned} \tag{5}$$

where $\Delta F_{SP}{}^{k}(\mathbf{S})$ is the difference between the scattering amplitude from the labeled k-th small particle and that from the unlabeled k-th small particle.

The derivation described here has the following implications. This generalized SLM is applicable to any system having any degree of underlying order, from solution to crystals, at any concentration. The difference intensity does not suffer sampling effects due to inter-large-particle interference, but rather has a diffuse, continuous nature, even in oriented systems such as fibers or single crystals. Also, the number of the small particles can be more than two, independent of whether they are identical, nonidentical or a mixture of both. The SLM can be applied to systems that contain 'foreign' molecules as well as large particles. In addition, the SLM can be employed with a solvent having any D_2O content. However, the optimal condition for maximum signal intensity is that the scattering-length density of the solvent is matched to that of the unlabeled particles and $\delta = 1/2$.

Hoppe's Method

Hoppe's method can also be derived by a similar reciprocal-space approach. Here the number of labeled particles is two, and δ is 1/2. The difference intensity between the two samples can be shown to be:

$$\Delta I(\mathbf{S}) = N_T < Re\{\Delta F_{SP}^{1*}(\mathbf{S})\Delta F_{SP}^{2}(\mathbf{S})\exp(2\pi i\mathbf{S}\mathbf{r}_{12})\} >_{\text{orientation}} \tag{6}$$

Thus, as Hoppe showed by a real-space approach using Patterson functions, his method can be applied to any ordered and oriented system. Note that the intensity in Equation 6 is twice that in Equation 5 for n = 2 and $\delta = 1/2$. Thus, for the measurements with nonidentical small particles, Hoppe's method is preferable to the SLM.

The Triple Isotopic Substitution Method

A general form of the TISM can be derived from the equations describing the generalized SLM. Here the particle of interest can be regarded as one large particle. Each hydrogen atom, which is replaceable by a deuterium atom when the particle is deuterated, can be regarded as one small particle. The randomly labeled sample contains particles having a degree of deuteration δ. This corresponds to the intermediately deuterated particle. The fully labeled large particles and the unlabeled large particles in the mixture sample denote the fully deuterated particles and the protonated particles, respectively. $\Delta F_{SP}{}^{k}(\mathbf{S})$ in Equation 5 corresponds to the difference in the scattering length between deuterium and hydrogen. The difference intensity is therefore:

$$\Delta I(\mathbf{S}) = N_T\delta(1-\delta)(\Delta b)^2 < \sum_{j\neq k}^{n}\sum_{j\neq k}^{n}\exp(2\pi i\mathbf{S}\mathbf{r}_{jk}) >_{\text{orientation}} \tag{7}$$

where Δb is the difference in scattering length between deuterium and hydrogen. The term inside the angle bracket represents the Fourier transform of the sum of the vectors connecting the positions of replaceable hydrogens within a single particle. Thus, this equation is sensitive to the shape of the particle of interest free from any interparticle interference, matrix fluctuations, solvent composition or solvent effects. Again, no assumptions were made concerning the degree of underlying order and orientation in the system, so this generalized TISM is exact and applicable to systems having *any* degree of underlying order or orientation.

Unmixed-Sample Methods

The SLM, Hoppe's method, and the TISM are applicable to any ordered and oriented system. The key to canceling out inter-complex interference and density-fluctuation effects is that large particles having a different distribution of labeled small particles are physically mixed (Hoppe, 1973; Pavlov *et al.*, 1991; Fujiwara & Mendelson, 1994). In the SLM and Hoppe's method, it is assumed that no exchange of the small particles occurs. If exchange does occur, the randomization of the mixture sample precludes the use of these methods. In such cases, unmixed-sample methods should be employed. The unmixed-sample method related to Hoppe's method (Engelman & Moore, 1972) requires the measurements of four samples (one doubly labeled, two singly labeled, and one unlabeled). The unmixed-sample method related to the SLM (Fujiwara & Mendelson, 1994) requires the measurements of three samples (fully labeled, unlabeled, randomly labeled). After completion of the measurements, the sum of the scattering intensities from the two singly labeled samples (or the scattering intensity from the randomly labeled sample) is subtracted from the sum of the scattering intensities from the doubly labeled (or fully labeled) sample and the unlabeled sample with proper normalization factors. This yields the scattering intensity from the labeled particles alone; all contributions from the other components are canceled out. However, interference between the small particles in different complexes is not eliminated, so measurements should be done in dilute solutions in which inter-complex interference is negligible. Alternatively, such unmixed-sample methods may be useful for highly ordered and oriented systems. A unmixed-sample method related to the TISM was proposed in order to extract the diffraction intensities from small particles embedded in a matrix (Curmi & Mendelson, 1991). Here proper subtraction of the scattering intensities from three samples containing particles with different degrees of deuteration yields the diffraction pattern arising only from the particles of interest. Since the difference intensity contains only the terms from the small particles, the diffraction from the small particles in a truly 'invisible' matrix can be obtained. If the lattice parameters of the matrix are known, structural information about the particles of interest can be extracted.

EXPERIMENTAL POSSIBILITIES

We have shown that the SLM and the TISM, as well as Hoppe's method, can be generalized to systems having any degree of order or orientation, from solution to crystals. This increases the number of systems that can be studied and the amount of information that can be obtained. We describe below how these methods might be practically applied to ordered model systems.

Figure 3 shows an example of applications to distance measurements in an ordered system. In this model example, the large particles, reconstituted from the matrix portion and the two small particles, bind to the foreign matrix to form the higher-order structure. The SLM (for identical small particles) and Hoppe's method (for nonidentical small particles) can be employed to measure the interference function between the small particles. The samples required (in the SLM) are shown in Figure 3b. If the samples have no net orientation, a spherically averaged interference function (solid curve in Figure 3c) between the small particles within one large particle can be obtained without contributions from vectors between small particles in different large particles and the foreign matrix. Experiments employing the SLM for an ordered system have been reported (Fujiwara & Mendelson, 1994; Fujiwara *et al.*, 1994). The interference function between a pair of regulatory light chains within one myosin molecule were measured under ionic conditions where myosin forms an ordered filamentous structure. Here a meridional reflection arising from inter-myosin interference and seen in the individual intensities was canceled out in the resulting interference functions.

The TISM can also be employed if small particles having three different levels of deuteration are prepared. In this case, two small particles within one large particle are treated as one small particle. The scattering intensity from a pair of the small particles in one large particle (dashed curve in Figure 3c) is obtained. Here the scattering intensity obtained from the TISM at $s = 0$ is twice that of the interference function generated by the SLM at $s = 0$ (assuming the same deuteration level of the highly deuterated particles in

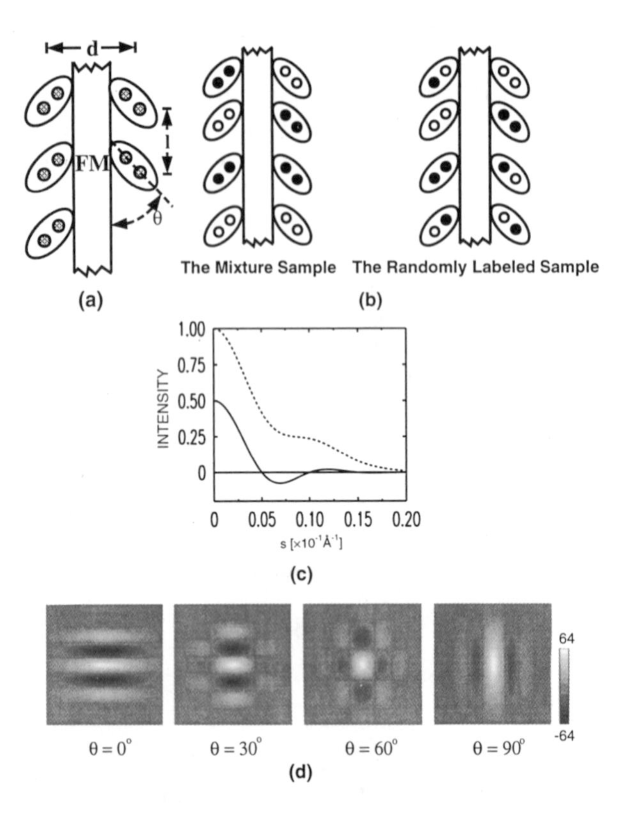

Figure 3. (a) A model polymer consisting of many 'large' (ellipsoidal) particles each containing isotopically labeled 'small' particles (here n = 2). The large particles bind to a 'foreign' matrix (FM). The patterns in (c) and (d) were calculated for a pair of identical spherical particles having radii of 30Å and a separation of 100Å. (b) The two samples required in the SLM. (c) The interference function (———) as would be obtained by applying the SLM to unoriented samples. The scattering intensity (– – –) from a pair of small particles in the ellipsoid, as would be obtained by treating the pair of small particles as one and applying the TISM. (d) Two-dimensional interference functions as would be obtained by applying the SLM to oriented samples. R_{max} and Z_{max} in these patterns are $0.02Å^{-1}$. The patterns are sensitive to the angle between the filament axis and the vector between the centers of the small particles (θ) as well as to the interparticle separation. All computed patterns in (c) and (d) are independent of the lattice parameters *l* and *d*. Adapted in part from Fujiwara and Mendelson (1994).

both cases). From this viewpoint the TISM would seem the preferred method. However, the TISM requires a considerably greater biochemical and experimental effort than does the SLM (or Hoppe's method). One must prepare labeled particles with two different degrees of deuteration in addition to protonated samples and ancillary experiments to determine the degree of deuteration of the two deuterated small particles for correct mixing of protonated and highly deuterated particles. Thus for such distance measurements, the SLM or Hoppe's method may be preferred.

If the system is oriented as in a fiber diffraction experiment, two dimensional interference patterns are obtained as shown in Figure 3d. The two-dimensional pattern is quite sensitive to the angle between the filament axis and the vector between the small particles in one ellipsoid. If ancillary information about the relative positions of the small particles in the ellipsoid is available, it may be possible to monitor the orientation of the ellipsoids relative to the filament axis.

Another useful application of the SLM may be to complexes containing many identical small particles. An example of such a system is shown in Figure 4a. Here many identical small particles bind to a matrix, which is assumed to have helical symmetry. The matrix and all small particles are now to be considered one large particle. The mixture sample in this case is a mixture of matrices (fully) decorated with labeled small particles and matrices (fully) decorated with unlabeled small particles. In the randomly labeled sample labeled small particles and unlabeled small particles bind randomly to the matrices (Figure 4b). For unoriented samples the one-dimensional interference function (Figure 4c) contains information about the radial position of the small particles, their separation along

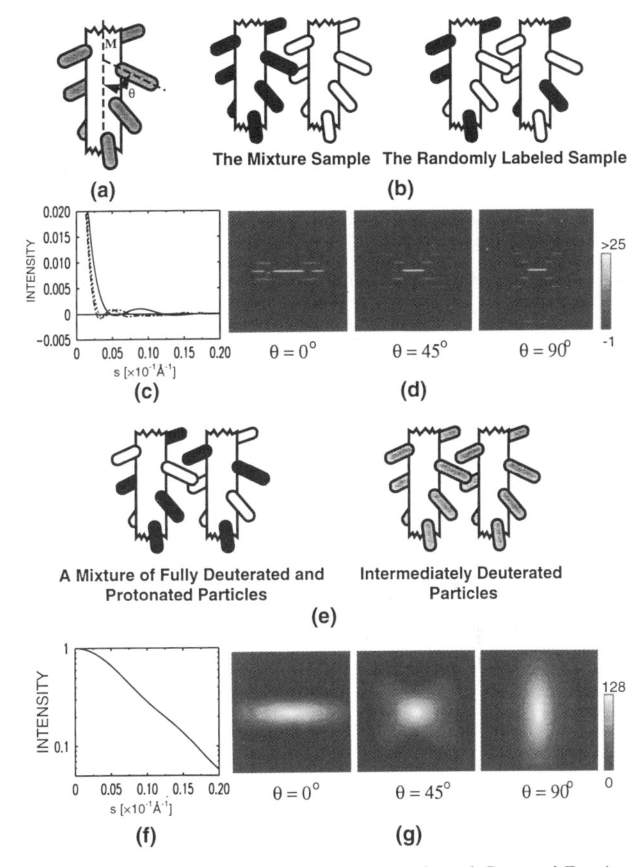

Figure 4. (a) A model polymer consisting of many identical small particles (shaded) bound to a (helical) matrix M. Calculations in (c), (d), (f), and (g) below were done with the following structure. The small particles are on a 13/6 two-start helix having a repeat of 357Å (like F-actin). Each small particle is comprised of three co-axial spheres (radius = 20Å) bound to M at a helical radius of 40Å. (b) The two samples required in the SLM. (c) Spherically averaged interference functions obtained using the SLM on unoriented samples. Here θ is defined as the angle between the helix axis and the long axis of the small particles. Interference functions for θ = 0° (———), 45° (- - - -) and 90° (—-—) are shown. (d) Two-dimensional interference functions as would be obtained by application of the SLM to the oriented samples for θ = 0°, 45°, 90°. The patterns depend on the lattice parameters as well as the shape of the small particles. (e) The two samples required in the TISM. (f) The spherically averaged scattering intensity from the small particles as would be obtained by applying the TISM. (g) Two-dimensional intensity patterns as would be obtained from the TISM when the samples are oriented. R_{max} and Z_{max} in the patterns in (d) and (g) are 0.02Å^{-1}. Adapted in part from Fujiwara and Mendelson (1994).

the filament axis, some shape information and, for asymmetric small particles as modeled here, some orientational information. In the low-s region the cross-sectional radius of gyration may be obtained by Guinier analysis. The higher-s region contains the remainder of the information which, depending on the availability of ancillary information, may be extractable. If the system can be oriented, such information is more readily obtained from the shape of the two-dimensional interference function. Orientational and some shape information about the small particles, as well as lattice parameters, are contained in layer-line difference intensities (Figure 4d).

In an application of the TISM to this model system, protonated and fully deuterated small particles bind to matrices randomly in one sample, while the other sample contains matrices decorated with intermediately deuterated small particles, as shown in Figure 4e. If the samples have no net orientation, the *in vacuo* spherically averaged scattering intensity from the particle of interest, free from *all* inter-particle interference, is obtained (Figure 4f). If the system is oriented and the particle of interest is asymmetric, the orientational information as well as shape information is obtained from a two-dimensional pattern (see Figure 4g). The orientational information here is cylindrically averaged around an axis parallel to the helical axis and through the particle's center-of-mass. These

Table I. Some useful applications of the Statistical Labeling Method, Hoppe's Method, and the Triple Isotopic Substitution Method

Number of Small Particles in One Complex (n)	Type of Small Particle	Method[a]	Information Obtainable (Type of Samples)
1	—	TISM	Shape (solutions)
2[b]	Nonidentical	Hoppe's Method[c]	Distance (solutions and oriented samples)
	Identical	SLM	Orientation (oriented samples)
>2	Identical[d]	SLM	Positional information (solutions and oriented samples)
			Shape and orientational information (oriented samples)
		TISM	Shape of small particle (solutions and oriented samples)
			Orientation (oriented samples)

[a]The table lists the preferred method. Other possible methods are discussed in the footnotes.

[b]The TISM can also be employed here. However, as discussed in the text, the biochemical effort is greater than for Hoppe's method or the SLM.

[c]The SLM can also be employed here. However, the signal intensity from the SLM is a half of that of Hoppe's method.

[d]The SLM and the TISM can also be applied to n > 2 systems where the small particles are nonidentical or a mixture of identical and nonidentical. For investigating the quaternary structures of the large particles having nonidentical small particles, label triangulation based on Hoppe's method (Hoppe, 1973; Hoppe *et al.*, 1975; Stöckel *et al.*, 1979; Stöckel *et al.*, 1980a; 1980b; Capel *et al.*, 1987) is a powerful technique. For a mixture of nonidentical and identical small particles, label triangulation using combination of Hoppe's method and the SLM (Stöckel *et al.*, 1980b) should be useful.

patterns are sensitive only to intra-small-particle vectors. On the other hand, the patterns in Figure 4d are sensitive only to inter-small-particle vectors. Thus, the patterns in Figure 4g are sensitive to the shape and orientation of the small particle while the patterns in Figure 4d are sensitive to the position and orientation of the small particles.

Possible applications of the SLM, Hoppe's method, and the TISM are summarized in Table I. Here n = 1 corresponds to the solution experiments employing the TISM shown in Figure 1; n = 2 corresponds to the application of Hoppe's method or the SLM shown in Figure 1 or to the system shown in Figure 3; and n > 2 corresponds to a system shown in Figure 4.

Although various experimental applications of these methods are possible, derivations of these methods rely on assumptions which must be fulfilled. In all of these methods, it is assumed that the behavior of the small particles is not altered by isotopic labeling. In addition, for the SLM and Hoppe's method, it is assumed that no significant exchange of the small particles occurs. The effects of violations of these assumptions are discussed elsewhere (Fujiwara & Mendelson, 1994). An important requirement of these methods is that the concentrations of the large particles, scattering-length density of the solvent ρ_{sol} and the mixing ratio δ are the same in the two requisite samples. The errors introduced by deviation from this requirement are discussed elsewhere (Pavlov *et al.,* 1991; Harrison *et al.,* 1993; Ramakrishnan & Moore, 1981; Moore & Engelman, 1977). When applying the methods described here it is important to assess, for each experimental system, the effects of any deviations from the ideal system.

Studies employing the methods described here provide information which might not be readily obtainable using other methods. Although some of these methods can be readily implemented with current techniques, others, such as studies of oriented samples, may prove to be more technically demanding. However, with the prospect of more intense neutron sources and the rapid progress in instrumentation and methodology of preparing deuterated samples, these methods should be increasingly useful.

ACKNOWLEDGMENTS

This work was supported in part by a Postdoctoral Fellowship to S.F. from the American Heart Association, California Affiliate, and by NIH (R01-AR39710, P01-AR42895) and NSF (DMB-876091) grants to R.A.M.

REFERENCES

Capel, M.S., Engelman, D.M., Freeborn, B.R., Kjeldgaard, M., Langer, J.A., Ramakrishnan, V., Schindler, D.G., Schneider, D.K., Schoenborn, B.P., Sillers, I.-Y., Yabuki, S., & Moore, P.B., (1987). A complete mapping of the proteins in the small ribosomal subunit of *Escherichia coli*. *Science*, 238:1403–1406.

Curmi, P.M.G., & Mendelson, R.A., (1991). Neutron diffraction intensities from arrays of isotopically substituted particles in an invisible matrix. *J. Appl. Cryst.*, 24:312–315.

Engelman, D.M., & Moore, P.B., (1972). A new method for the determination of biological quaternary structure by neutron scattering. *Proc. Natl. Acad. Sci. USA*, 69:1997–1999.

Fujiwara, S., & Mendelson, R.A., (1994). The statistical labeling method and *in situ* neutron scattering and diffraction measurements on ordered systems. *J. Appl. Cryst.*, 27:912–923.

Fujiwara, S., Stone, D.B., & Mendelson, R.A., (1994). Measurement of inter-RLC distance in scallop myosin by neutron scattering. *Biophys. J.*, 66:a76.

Harrison, D.H., May, R.P., & Moore, P.B., (1993). Measurement of the radii of gyration of ribosomal components *in situ* by neutron scattering. *J. Appl. Cryst.*, 26:198–206.

Hoppe, W., (1972). A new X-ray method for the determination of the quaternary structure of protein complexes. *Israel J. Chem.*, 10:321–333.

Hoppe, W., (1973). The label triangulation method and the mixed isomorphous replacement principle. *J. Mol. Biol.*, 78:581–585.

Hoppe, W., May, R., Stöckel, P., Lorenz, S., Erdmann, V.A., Wittmann, H.G., Crespi, H.L., Katz, J.J., & Ibel, K., (1975). Neutron scattering measurements with the label triangulation method on the 50 S subunit of *E. coli* ribosomes. *Brookhaven Symp. Biol.*, 27:IV 38–48.

Kneale, G.G., Baldwin, J.P., & Bradbury, E.M., (1977). Neutron scattering studies of biological macromolecules in solution. *Q. Rev. Biophys.*, 10:485–527.

Moore, P.B., & Engelman, D.M., (1977). Model calculations of protein pair interference functions. *J. Mol. Biol.*, 112:228–234.

Pavlov, M.Yu., & Serdyuk, I.N., (1987). Three-isotopic-substitutions method in small-angle neutron scattering. *J. Appl. Cryst.*, 20:105–110.

Pavlov, M.Yu., Rublevskaya, I.N., Serdyuk, I.N., Zaccaï, G., Leberman, R., & Ostanevich, Yu.M., (1991). Experimental verification of the triple isotopic substitution method in small-angle neutron scattering. *J. Appl. Cryst.*, 24:243–254.

Ramakrishnan, V.R., & Moore, P.B., (1981). Analysis of neutron distance data. *J. Mol. Biol.*, 153:719–738.

Serdyuk, I.N., & Pavlov, M.Yu., (1988). A new approach in small-angle neutron scattering: a method of triple isotopic substitutions. *Makromol. Chem.*, 15:167–184.

Serdyuk, I.N., Pavlov, M.Yu., Rublevskaya, I.N., Zaccaï, G., & Leberman, R., (1994). The triple isotopic substitution method in small angle neutron scattering. Application to the study of the ternary complex EF-Tu·GTP·aminoacyl-tRNA. *Biophys. Chem.*, 53:123–130.

Stöckel, P., May, R., Strell, I., Cejka, Z., Hoppe, W., Heumann, H., Zillig, W., & Crespi, H.L., (1980a). The core subunit structure in RNA polymerase holoenzyme determined by neutron small-angle scattering. *Eur. J. Biochem.*, 112:411–417.

Stöckel, P., May, R., Strell, I., Cejka, Z., Hoppe, W., Heumann, H., Zillig, W., & Crespi, H.L., (1980b). The subunit positions within RNA polymerase holoenzyme determined by triangulation of centre-to-centre distances. *Eur. J. Biochem.*, 112:419–423.

Stöckel, P., May, R., Strell, I., Cejka, Z., Hoppe, W., Heumann, H., Zillig, W., Crespi, H.L., Katz, J.J., & Ibel, K., (1979). Determination of intersubunit distances and subunit shape parameters in DNA-dependent RNA polymerase by neutron small-angle scattering. *J. Appl. Cryst.*, 12:176–185.

33

THE DETERMINATION OF THE *IN SITU* STRUCTURE BY NUCLEAR SPIN CONTRAST VARIATION

Heinrich B. Stuhrmann[1] and Knud H. Nierhaus[2]

[1] GKSS Forschungszentrum
D-21502 Geesthacht, Germany
[2] Max-Planck-Institut für Molekulare Genetik
AG Ribosomen, Ihnestraße 73, D-14195 Berlin, Germany

ABSTRACT

Polarized neutron scattering from polarized nuclear spins in hydrogenous substances opens a new way of contrast variation. The enhanced contrast due to proton spin polarization was used for the *in situ* structure determination of tRNA of the functional complex of the *E.coli* ribosome.

INTRODUCTION

Although large biological particles are still not easily amenable to a structure determination at atomic resolution, considerable progress has been made on the way towards this goal. Both microscopic and diffraction techniques have pushed the limits of structural resolution to lower values. This is well illustrated by the example of the ribosome. A resolution of 30Å has been achieved by electron microscopy. From X-ray diffraction a model of the small subunit of ribosomes from *thermus thermophilus* at 20Å resolution has been obtained by Yonath (Berkovitch-Yelin *et al.*, 1992), and its resolution could well become extended to 10Å resolution in the near future.

Neutron scattering has very much contributed to this development. The exchange of hydrogen (mainly ^{1}H) by its heavier isotope ^{2}H (= deuterium, D) is known to introduce a strong change of the scattering amplitude. The high abundance of hydrogen in biomolecules and its relatively easy isotopic exchange made it a strongly favored staining technique in macromolecular structure research by neutron scattering. Regions in a unit cell of a crystal or in a particle marked by H-D exchange are called *labels*. The site and the size of labels can be tailored to the requirements of a scientific project, in many cases.

Neutrons in Biology, edited by Schoenborn and Knott
Plenum Press, New York, 1996

The easiest and cheapest way of isotopic substitution is offered by the exchange of H_2O by heavy water, D_2O. This method has been used both in neutron small-angle scattering from solutions (for a review see Perkins, 1988) and in neutron diffraction (Bentley & Mason, 1981). For particles of fairly uniform scattering density the change of the solvent mimics a change of the scattering density of the solute (Babinet's principle). The size of the label is equivalent to the volume of the dissolved particle. For particles of non-uniform scattering density, neutron scattering varies with the contrast of the particle in a more complicated way, which can be associated with long range fluctuations of the scattering density inside the dissolved particle. Although this method is likely to produce models the resolution of which is hardly better than half the radius of the label, it has been widely used both in small-angle scattering from solutions, and in diffraction from single crystals. The structural studies on ribosomes are an example (Berkovitch-Yelin *et al.*, 1992; Stuhrmann *et al.*, 1976).

The more laborious but also more rewarding way of labeling particles for neutron scattering studies, is site-directed (or specific) deuteration of parts of a large structure (Moore *et al.*, 1977). These labeled regions are usually small compared to the total volume of the particle. A protonated ribosomal protein, eg L3 (M = 23kD), incorporated in a deuterated ribosomal particle (M = 2,300kD) comprises only 1% of the total particle volume. Most of the existing specific deuterated particles which have been prepared for neutron scattering experiments are ribosomes.

Why ribosomes? Ribosomes are outstanding for at least two reasons. Their function is the production of proteins in each living cell ie they fulfil the last step of gene expression. Secondly, the dissociation of the ribosome into its components (54 ribosomal proteins of the *E.coli* ribosome, three ribosomal RNAs) and their reconstitution to a functional complex including two tRNAs and mRNA is the essential prerequisite for specific deuteration. Each of these components may be supplied in deuterated or protonated form before reconstitution (Nierhaus *et al.*, 1983; Vanatalu *et al.*, 1993).

Why still more neutron scattering studies on ribosomes? This question will be raised even more as results from X-ray diffraction from single crystals become more promising. An answer is, specific deuteration creates contrast, where X-ray diffraction cannot. Neutron scattering studies from specifically deuterated particles reveal the structure of the label in its native environment ie its *in situ* structure. These studies may provide an answer to structural parameters like the shape of the label and its site inside the host particle. If the task of the structural study is defined in this way, then an *a priori* knowledge of the structure of the host particle becomes crucial. So far, models from electron microscopy studies have been used. More recently, well resolved ribosome models from X-ray synchrotron radiation studies are being produced. This paper will show that the results of electron microscopy and single crystal diffraction are an essential ingredient in the analysis of neutron small-angle scattering data and that they become even more useful with advanced techniques of contrast variation (Stuhrmann *et al.*, 1995).

Polarized neutron scattering from dynamic polarized targets is an advanced method of contrast variation. It has first been used by White and coworkers at the Institute Laue-Langevin (ILL) at Grenoble (Leslie *et al.*, 1980). The target material was a single crystal of lanthanum magnesium nitrate ($La_2Mg_3(NO_3)_{12}.24H_2O$) (LMN) doped with $^{142}Nd^{3+}$. The proton spins of the crystal water were dynamically polarized and studied by polarized neutron diffraction. The intensity of the diffracted neutron not unexpectedly turned out to be most sensitive to the structure and polarization of the hydrogen atoms of the crystal water. It was concluded that the use of nuclear polarization as a method of labeling protons in a complicated unit cell might have advantages for determining the proton structure eg the proton configuration around the active site in an enzyme (Leslie *et al.*, 1980). Although this method did not find a continuation at the ILL, it must be regarded as a highlight in neutron physics of the seventies which initiated further experiments of polarized neutron scattering by polarized nuclear spin targets elsewhere in the mid-eighties.

The renaissance of this technique was due to a new class of polarized target materials in high energy scattering experiments. Using a dilution refrigerator ($T < 0.5K$) and a 2.5T magnetic field, proton spin polarizations up to 98% were obtained in Cr(V) doped alcohols, like butanol or 1,2 propandiol (de Boer & Niinikoski, 1973). Contrary to the crystalline LMN, these target materials were frozen liquids and, moreover, well-known solvents in cryobiochemistry (Douzou, 1977). It is therefore not surprising that the proton spins in frozen solutions of macromolecules were polarized and studied by polarized neutron small-angle scattering. In 1986, first results from contrast variation of protonated particles in deuterated solvents by proton spin polarization were obtained at the GKSS Research Centre, Geesthacht, in collaboration with CERN, Geneva (Knop *et al.*, 1986). In 1987, frozen solutions of protonated crown ethers were studied at KENS, Tsukuba (Koghi *et al.*, 1987). Proton spin dependent neutron small-angle scattering studies from block copolymers have been started at the Orphee reactor, Saclay (Glättli *et al.*, 1989).

Modern polarized target stations allow selective depolarization of dynamic polarized targets by NMR saturation (Stuhrmann *et al.*, 1995). Starting from a dynamic polarized target with all non-spinless nuclei polarized (though to different extent) the proton spins may be depolarized and leaving all other nuclear spin systems (eg those of ^{2}H, ^{14}N if any) polarized. Hence, the same sample will exhibit different scattering density distributions depending on the choice of the nuclei polarized.

The impetus for the construction of a polarized target facility at GKSS came from ribosomal structure research - more precisely from the concept of the 'transparent' ribosome developed by Nierhaus (Nierhaus *et al.*, 1983). A ribosome particle showing no contrast and hence no small-angle neutron scattering would be the ideal background for the determination of the *in situ* structure of a labelled component, like that of a ribosomal protein. The problem with this method is that it requires a very precise matching of the scattering density of the unlabelled rRNA and of the unlabeled ribosomal proteins to that of the solvent. An eventually unmatched contrast between the total protein of the ribosome and the rRNA would give rise to a scattering intensity which cannot be separated from the scattering intensity of the label, unless the contrast of the label is varied (see DATA ANALYSIS). Variation of the degree of isotopic substitution of the label would be a way to solve this problem. The more powerful alternative is protein spin polarization. The expected gain in intensity of polarized neutron scattering from protonated labels in a deuterated ribosome (and deuterated solvent) is more than one order of magnitude with respect to the corresponding effect due to mere isotopic substitution.

A series of polarized neutron scattering experiments were carried out in order to verify the expected gain in intensity. To start with, solutions of serum albumin (Knop *et al.*, 1989) and of apoferritin (Knop *et al.*, 1992), were studied. Specifically deuterated ribosomal particles followed (Knop *et al.*, 1991). In the first stage, the volume of the label amounted to about one half of the total particle, like that of protonated rRNA in the deuterated large ribosomal subunit. In a further step, the size of the label was decreased to that of a single ribosomal protein (Zhao & Stuhrmann, 1993). Small RNAs like tRNA binding to the ribosome intermittently during protein synthesis so far escaped the characterization by neutron small-angle scattering. There are two reasons for this: (i) the contrast of protonated tRNA in a deuterated environment is lower than that of protonated proteins and (ii) the occupation density of tRNA in ribosomes is hardly higher than 0.4 ie in more than 60% of the particles, the translational apparatus of the ribosome is 'empty'. Both factors lower the contribution of tRNA to the total neutron small-angle scattering intensity of the sample and hence jeopardized attempts to study the *in situ* structure of ribosome bound tRNA by mere isotopic substitution. This paper will describe how the increased contrast available in polarized neutron scattering from proton spin polarized targets overcame this problem.

MATERIALS AND METHODS

Protonated tRNA in a Deuterated Ribosome

The 70S functional complex containing two protonated tRNA and a protonated mRNA fragment (46 nucleotides) were prepared by specific binding of the respective tRNA ligands to nearly fully deuterated, native ribosomes. The average deuteration of the ribosome was about 0.94. The length of the mRNA chain was equal to that sequence of the mRNA shown to be covered by the ribosome (Rheinberger *et al.,* 1990). The nucleotide sequence of the mRNA was chosen in such a way that the ribosome could be fixed in the middle of the mRNA chain by binding $tRNA_f^{Met}$ to the ribosomal P-site and N-acetylated Phe-$tRNA^{Phe}$ (peptidyl-tRNA-analog) to the ribosomal A-site. At this point the ribosomes were in the pretranslocational state. The functional state was quantified by the relative puromycin sensitivity of the bound N-acetyl Phe-$tRNA^{Phe}$ (Rheinberger *et al.,* 1990).

Polarized Target Material

The target was a dilute solution of ribosomes. It was prepared as follows: 0.5ml of the aqueous ribosome solution (2 wt.%) was dialyzed against a buffer (100mM imidazol $C_3H_4N_2$, 10mM $MgCl_2$, 100mM KCl in D_2O). 13mg of sodium bis(2-ethyl-2-hydroxybutyrato)-oxochromate(V) dihydrate, $Na[Cr(C_6H_{10}O_3).2H_2O]$ were dissolved in 0.15ml D_2O. This radical was synthesized from dry sodium dichromate and 2-ethyl-2-hydroxybutyric acid (EHBA) in acetone (Krumpole & Rocek, 1985). 60mg of the concentrated solution of KCl (1.6M) and $MgCl_2$ (0.16M) in D_2O were added to 820mg deuterated glycerol, $C_3D_8O_3$ (98.6% D). The average deuteration of the solvent was 0.97.

The components were mixed in the following way. The ribosome solution in the deuterated buffer was added to the deuterated glycerol and stirred for 5min. Then the concentrated solution of EHBA-Cr(V) was rapidly diluted in the ribosome solution and frozen immediately in a liquid nitrogen cooled copper mould, as the EHBA-Cr(V) complex decomposes at pH 7 with a halftime of 10min. Therefore the time between addition of EHBA-Cr(V) to the ribosome solution and freezing the sample had to be kept short, typically to less than 1min. The sample was a dark red coloured, transparent glass.

Polarization of the Target Material

The frozen sample ($3 \times 17 \times 17mm^3$) was introduced into the cavity of the dilution refrigerator (Figure 1) (Niinikoski & Udo, 1976). The nuclear spins of the sample material were aligned with respect to an external magnetic field by dynamic nuclear spin polarization (DNP). Applying microwave power into the multimode cavity at B = 2.5T, the microwave frequencies of 69.0GHz and 69.3GHz yield positive and negative polariza-

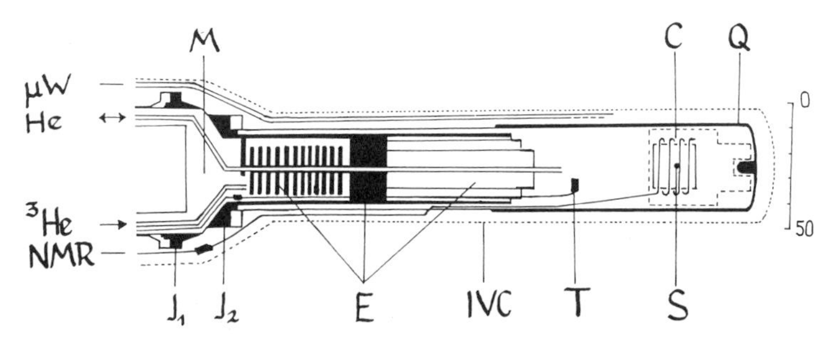

Figure 1. The sample cell for nuclear spin polarization. The scale is given in mm. The polarized sample (S) with dimensions of $17 \times 17 \times 2.8mm^3$ is cooled by superfluid helium (1K > T > 0.1K). It is thermally connected to the mixing chamber (M) by a heat exchanger (E) which has been built at CERN. Cooling is achieved by dilution of liquid ^3He in the immobile phase of superfluid ^4He. The maximum circulation rate is 20mM ^3He/sec. The sample cell is a steel cylinder of 30mm diameter which is connected to a quartz tube (Q) of the same diameter. The proton and deuteron spin resonance (NMR) are measured with the same coil (C). The microwave guide (μW) is in close contact with the cylindrical inner vacuum chamber (IVC) made of steel. The inner vacuum chamber and the sample cell are disconnected from the refrigerator when the sample is loaded. The frozen sample is put into the open cell, kept at liquid nitrogen temperature, and then it is closed by the indium joint J2. The inner vacuum chamber is closed by another indium joint outside the frame of the figure. J1 is opened if the mixing chamber has to be accessed (eg if one of the connectors to the various thermometers (T) is defective; only one is shown).

tions, respectively. The proton spin polarization was determined by comparing the enhanced nuclear magnetic resonance (NMR) signal with the thermal equilibrium signal. The deuteron spin polarization was calculated from the asymmetry of the resonance profile. The final values of proton spin polarization vary between 70% and 90%. Those of the deuteron polarization are lower, typically between 15% and 25% as it would be expected at equal nuclear spin temperature. The polarization of ^{14}N nuclei is about half that of the deuterons.

All non-spinless nuclei are affected by DNP. In the target material described above only the nuclear spins of the hydrogen isotopes ^1H and ^2H needed to be considered. Pure proton spin targets were obtained by selective depolarisation of the deuteron spins (Stuhrmann *et al.*, 1995). This was achieved by a radio frequency (rf) sweep of 200kHz bandwidth across the NMR profile at 16.4MHz. The polarized deuteron spin target was obtained in a similar way. An rf-sweep of 200kHz bandwidth across the proton spin resonance profile at 106.4MHz depolarized a major fraction of the polarized proton spin system. The efficiency of proton spin depolarization in deuterated material will be discussed elsewhere.

At temperatures below 150mK the relaxation time of the polarized targets was of the order of several weeks. Thus, the nuclear spin polarization was almost constant during a two day neutron scattering experiment. In a run of one week, the sample was studied as a proton target, as a deuteron target and as a dynamic polarized target without selective depolarization.

Polarized Neutron Small-Angle Scattering

The experiments were done at the small-angle instrument SANS-1 at the research reactor FRG1 of GKSS (Figure 2) (Knop *et al.*, 1992). Thermal neutrons were monochromatized by a mechanical velocity selector and polarized by total reflection from magnetized surfaces (Stuhrmann, 1993). The wavelength distribution centered at 8.5Å had a fwhm of

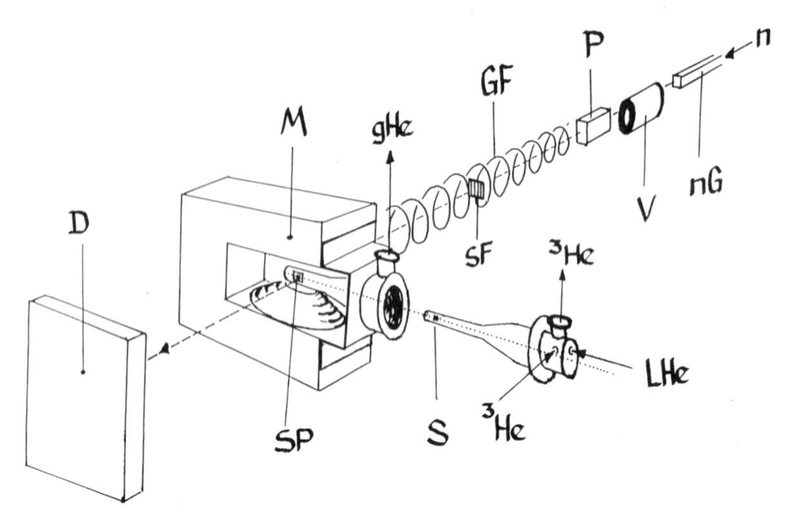

Figure 2. The instrument SANS-1 (GKSS) for polarized neutron scattering by polarized nuclear spin targets (a schematic view). The cold neutrons (n) from the reactor FRG1 leaving the neutron guide (nG) are monochromatized by the velocity selector (V) and polarized by total reflection from the magnetized surfaces of a set of super mirrors (P). The polarization of the neutron beam is maintained by a magnetic guide field (GF). A flat coil spin flipper (SF) may invert the polarization direction of the neutron beam. The neutron guides of the 9m collimator (not shown) are in the solenoid. The sample (S) in the nose of the horizontal dilution refrigerator is not yet in the final position (SP) between the poles of the 5 ton C-shaped electromagnet (M). For further details of the sample environment see Figure 1. The inlets and outlets of helium and of the closed ^{3}He circuit are marked by LHe = liquid helium supply, hHe = gaseous helium, ^{3}He = gaseous ^{3}He, respectively. The microwave guide and the NMR cable (not shown) are connected to dilution refrigerator once it is inserted in the heat shield fixed to the electromagnet. The neutron scattering intensity is recorded by a position sensitive area detector (D).

0.8. The Q range from 0.009 to 0.28$\mathring{A}^{-1}$ was large enough to observe both the characteristic dimensions of the ribosome with its average diameter of 250Å, and those of its label which are almost an order of magnitude smaller. The polarization direction of the neutron beam was changed after each measurement of 1000sec by a flat coil spin flipper. The scattering intensity was recorded by an area detector.

DATA ANALYSIS THEN AND NOW

The determination of the *in situ* structure of labels from the very beginning in the early seventies concentrated on the label itself and deliberately disregarded its connection to the host particle. The strategy of data taking was defined in such a way that the scattering intensity of the label would come out cleanly. The determination of distances between labels is an example which has been proposed independently by Hoppe (Hoppe, 1972) and by Engelman and Moore (Engelman & Moore, 1972). The elimination of the influence of the host particle on the scattering intensity was achieved in the following way. Let M be the scattering amplitude of the host particle (eg ribosome) and L1 and L2 the amplitudes of the labels (eg ribosomal proteins). Then the amplitudes M, M+L1, M+L2, and M+L1+L2 give rise to the scattering intensities I_1, I_2, I_3 and I_4, respectively.

$$
\begin{aligned}
I_1 &= |M|^2 \\
I_2 &= |M|^2 + 2Re[ML_1^*] + |L_1|^2 \\
I_3 &= |M|^2 + 2Re[ML_2^*] + |L_2|^2 \\
I_4 &= |M|^2 + 2Re[M(L_1 + L_2)^*] + |L_1 + L_2|^2
\end{aligned} \tag{1}
$$

The sum scattering intensities I_1+I_4 diminished by the sum I_2+I_3 is twice the product of the amplitudes L_1, and L_2, more precisely $2Re[L_1L_2^*]$. All terms containing the amplitude of M cancel.

The product $2Re[L_1L_2^*]$ is the Fourier transform of the distinct correlation function of the labels 1 and 2, which the set of vectors connecting the volume elements of label 1 with those of label 2. The average length of the connecting vectors is close to the distance between the centers of mass of the labels. The determination of interprotein distances is the basic step in label triangulation (Moore *et al.*, 1975).

The method of triple isomorphous substitution (TIS) works in a similar way (Serdyuk & Pavlov, 1988). The label is assumed to be present in three different contrasts marked by 0, x/2 and x. The scattering intensity of three samples differing in the contrast of their labels is measured.

$$
\begin{aligned}
I_1 &= |M|^2 \\
I_2 &= |M|^2 + 2\frac{x}{2}Re[ML^*] + \frac{x^2}{4}|L|^2 \\
I_3 &= |M|^2 + 2xRe[ML^*] + x^2|L|^2
\end{aligned} \tag{2}
$$

The sum of the scattering intensities I_1+I_2 diminished by $2I_2$ is half the scattering function $|L|^2$ of the label. Again, all terms containing the amplitude M cancel. Both methods are also applicable to dense systems (Serdyuk & Pavlov, 1988).

Triple isotopic substitution is a special case of label contrast variation. Variation of the contrast of the label is achieved even more efficiently in polarized neutron scattering by nuclear spin polarized samples (see below). The scattering function of the label can be obtained in a similar way and only one sample is necessary. The latter fact is the reason for a difference between these methods. Interparticle scattering will cancel in TIS (like in double labeled particles mentioned above) whereas it will not in nuclear spin contrast variation. However, both methods yield similar results at low concentrations of the solute, where interparticle scattering becomes negligible.

Including the cross term $Re[ML^*]$ in the analysis of small-angle scattering is an extension of the methods described above. The meaning of the cross term becomes more transparent when its Fourier transform is considered, which is the ensemble of vectors connecting volume elements of the label with those of its host particle. Depending on what is known about M, the cross-term $Re[ML^*]$ can be exploited more or less successfully in terms of the site and orientation of the label with respect to the host particle. If structural details of the host particle are missing then we are left with the fact that it is there, and that it will have a center of mass which is a reference and possible origin of the coordinate system.

From a more rigorous treatment of the basic scattering functions using a multipole expansion, it appears that the site and orientation of the label can be determined in more detail when the structure of the host particle, and hence its amplitude M, is known to some extent (Svergun & Stuhrmann, 1991). This method was chosen to analyse data from nuclear spin contrast variation.

Basic Scattering Functions from Nuclear Spin Contrast Variation

For a particle consisting of M nuclei (N out of them are non-spinless) one defines:

$$U(\mathbf{Q}) = \sum_{j=1}^{M} b_j e^{i\mathbf{Q}.\mathbf{r}_j}$$

$$V(\mathbf{Q}) = \sum_{j=1}^{N} P_j I_j B_j e^{i\mathbf{Q}.\mathbf{r}_j} = P_H V_H(\mathbf{Q}) + P_D V_D(\mathbf{Q}) \tag{3}$$

where U(**Q**) is the amplitude of the unpolarized sample and V(**Q**) the increase of the amplitude due to nuclear polarization P. **Q** is the momentum transfer. $\mathbf{r}_j$ is the position of the nucleus j with the spin I_j and the scattering lengths b_j and B_j. The indices H and D refer to the proton spin target and deuteron spin target respectively. According to Abragam and Goldman (1982) the intensity of coherent polarized neutron scattering is:

$$S(\mathbf{Q}) = |U(\mathbf{Q})|^2 + 2p\mathit{Re}\left[U(\mathbf{Q})V^*(\mathbf{Q})\right] + |V(\mathbf{Q})|^2 \tag{4}$$

where p denotes the neutron spin polarization. A dilute solution of identical particles gives rise to small-angle scattering which can be described in a way analogous to Equation 2. Averaging Equation 4 with respect to the solid angle yields the intensity I(**Q**):

$$I(\mathbf{Q}) = 2\pi^2 \sum_{l=0}^{\infty} \sum_{m=-l}^{+l} |U_{lm}(\mathbf{Q})|^2 + 2p\mathit{Re}\left[U_{lm}(\mathbf{Q})V^*_{lm}(\mathbf{Q})\right] + |V_{lm}(\mathbf{Q})|^2 \tag{5}$$

where a multipole expansion of the amplitude has been used. For U_{lm}(**Q**) for instance, we have:

$$U_{lm}(Q) = \sqrt{\frac{2}{\pi}} i^l \sum_{j=1}^{M} b_j j_l(Qr_j) Y^*_{lm}(\theta_j, \phi_j) \tag{6}$$

where Q = |**Q**| = $4\pi/\lambda \sin\theta$, j_l is the spherical Bessel function of l-th order, and the Y_{lm} are the spherical harmonics. The site of the j-th nucleus is given by its radial distance r_j and the polar angles θ_j and ϕ_j. For dilute solutions of identical particles, I(Q) is proportional to the scattering from one particle. This equation is quite handy for the calculation of scattering profiles from complex particles.

Coherent neutron scattering depends on the polarization of neutron spins and nuclear spins as described by Equation 4 and Equation 5. There are three basic scattering functions differing in their dependence on nuclear and neutron spin polarization. When there is no nuclear polarization, only $|U(\mathbf{Q})|^2$ is observed. Nuclear spin polarization adds $|V(\mathbf{Q})|^2$, even if unpolarized neutrons are used (Koghi *et al.*, 1987; Glättli *et al.*, 1989). With polarized neutrons the cross term $2\mathit{Re}[U(\mathbf{Q})V^*(\mathbf{Q})]$ is observed from polarized samples. The basic scattering functions mentioned above are obtained from measurements of neutron scattering using different polarizations of both the neutron beam and the target nuclei. The intensity difference $\Delta\sigma = [I(\uparrow\uparrow) - I(\downarrow\uparrow)]/P$ is twice the cross term, $4\,Re[U(\mathbf{Q})V^*(\mathbf{Q})]$.

The analysis of the data started from a model which consisted of two parts:

- the low resolution model of the ribosome known from electron microscopy (Frank *et al.,* 1991), and
- the label, the *in situ* structure of which needed to be determined.

As the number of structural elements which can be derived from small-angle scattering of randomly oriented particles is limited to the order of 10, the information obtainable will include hardly more than the location and orientation of the label with respect to the ribosome. The amplitude of the ribosome $M(\mathbf{Q})$ and of the label $L(\mathbf{Q})$ are developed as a series of harmonics. Both are normalized to the volume of their structures at $Q = 0$. After multiplication of M and L with the contrast c (or k for the spin dependent part) the amplitudes U and V are obtained in units of a scattering length:

$$\begin{aligned} U(\mathbf{Q}) &= c_M M(\mathbf{Q}) + c_L L(\mathbf{Q}) \\ V(\mathbf{Q}) &= k_M M(\mathbf{Q}) + k_L L(\mathbf{Q}) \end{aligned} \tag{7}$$

As the volume of the ribosome is two orders of magnitude larger than that of the label the amplitude of the ribosome will dominate. The intensity of $|V|^2$ would be lower to that of $|U|^2$ by four orders of magnitude and hence escape the observation by neutron scattering. The choice of deuterated ribosomes in a deuterated glycerol/water mixture avoided this situation as the contrast c_M of the ribosome with respect to the solvent and the polarization dependent contrast k_M turned out to be very small (Figure 3). The contrasts c_M and k_M were determined from polarized neutron small-angle scattering of the deuterated, unlabeled ribosome (Stuhrmann, 1993).

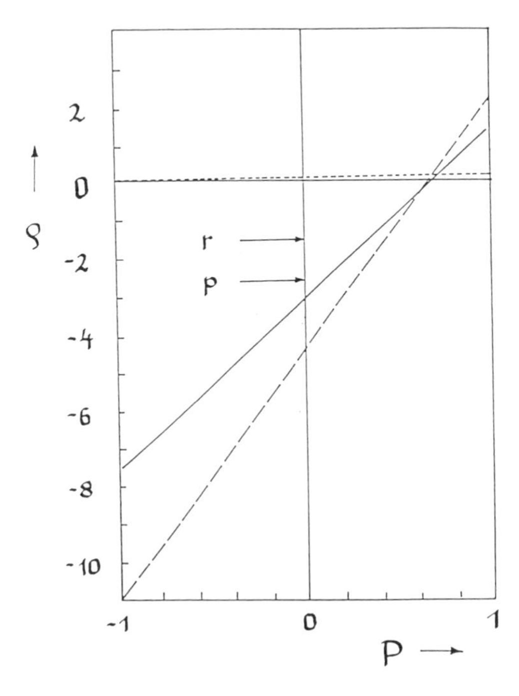

Figure 3. Polarization dependent contrasts, ρ of protonated labels and of the deuterated ribosome in a mixture of deuterated glycerol and heavy water as described in the text. The contrast is given in units of $10^{10}cm^{-2}$. Full line: protonated tRNA, dashed line: protonated proteins, dotted line: deuterated ribosome (mean deuteration 0.94). For comparison the contrast of proteins (p) and tRNA (r) in a H_2O/D_2O mixture containing 92% D_2O is shown. The scattering density of this solvent matched that of appropriately deuterated rRNA and ribosomal proteins (Nierhaus *et al.,* 1983).

The contrast c_L of the protonated label in the unpolarized state and its polarization added contrast k_L were calculated from the chemical composition using the spin dependence of the scattering lengths of the hydrogen isotopes:

$$b_H = (-0.374 + 1.456pP_H)10^{-12}\,\text{cm}$$
$$b_D = (+0.667 + 0.270pP_D)10^{-12}\,\text{cm} \tag{8}$$

These relationships lead to the correct cross-sections in Equations 4 or 5 if p = +1 or p = -1. The polarization dependent contrast of the protonated labels and of the deuterated rRNA is shown in Figure 3. The polarization added contrast exceeded considerably the contrast of the label in the unpolarized state. We also note that the contrast of a protonated label in a mixture of deuterated glycerol and heavy water (1/1) is larger than it is in heavy water.

The basic scattering functions calculated from a model of the ribosome and of its label, were compared with the experimental data. The quality of the fit was described by:

$$\Delta = \sum_{i=1}^{N} \frac{(I_{exp}(Q_i) - I_{calc}(Q_i))^2}{\sigma_i^2} \tag{9}$$

The sum was extended over the scattering angles at which neutron scattering intensities have been measured and over the basic scattering functions. σ_i is the statistical error of the measurement at Q_i.

RESULTS AND DISCUSSION

As the *in situ* structure of the label and its site and orientation with respect to the ribosome will be determined the structure of the latter needs to be known. Moreover, the sample will be studied in the frozen state. It must be made sure to what extent the low resolution structure of the ribosome might change due to freezing the solution. As DNP requires the presence of paramagnetic centers, the homogeneity of the polarization dependent contrast needs to be investigated.

The Unlabeled, Protonated Ribosome

Polarized neutron scattering was measured from the frozen spin target of protonated ribosomes in a mixture of deuterated glycerol and heavy water. The basic scattering functions obtained from the proton spin target were very similar, except for the sign. The similarity is due to the fact that the ratio between the contrast of rRNA and that of the ribosomal proteins in the unpolarized state is rather close to the corresponding value of the polarization added contrasts (Figure 3). The negative sign of the cross term $2Re[UV^*]$ told us that the sign of U was different from that of V_H. As the proton concentration of the solute was higher than that of the solvent, the proton spin polarization added contrast (or proton spin contrast) was positive and so was $V_H(0)$. Hence, U(0) was negative. The profile of neutron scattering from the frozen target shown in Figure 4, is in good agreement with data obtained at room temperature (Stuhrmann *et al.*, 1978). The radius of gyration of this particle is R = 95Å.

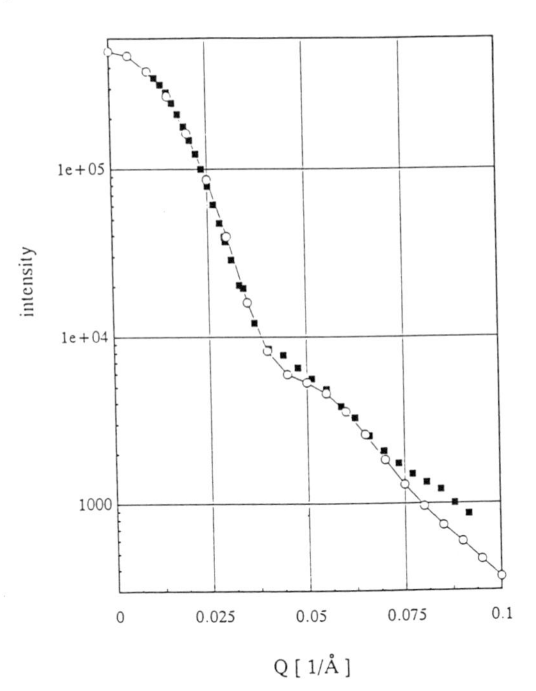

Figure 4. Small-angle neutron scattering from the protonated ribosome in a deuterated solvent. Squares: The frozen solution of ribosomes in a mixture of glycerol and water. The same results are obtained at room temperature using heavy water as a solvent. Open circles: Small-angle scattering from the shape of the ribosome as obtained from electron microscopic studies (Frank *et al.*, 1991). This scattering curve has been convoluted with the divergence of the primary beam and the wavelength distribution of the incident neutrons. The scattering intensity scale in neutrons per 1000 sec per cm^2 relates to a special set up of the neutron small-angle scattering instrument SANS-1 at GKSS: collimation length: 2m, cross section of the neutron guide: 30× $40mm^2$, irradiated area of the sample: 15 × $15mm^2$; distance from sample to detector: 0.72m, pixel size of the detector: 9 × $9mm^2$.

A positive cross term $2Re[U\, V_D^*]$ was observed when the deuteron spin target was studied by polarized neutron scattering. As the amplitude U was negative, the polarization added amplitude V_D must be negative as well. This was expected as the density of deuterons inside the volume of the ribosome was much lower than in the solvent. The deuteron spin contrast of the protonated ribosome in the deuterated solvent was negative. It remained negative even in highly deuterated ribosomes as the density of deuterons in nucleic acids and proteins was lower than that in deuterated solvents.

Both the ribosomal proteins and the rRNA contained only a small number of deuterons. Their deuteron spin contrasts were nearly equal ie the amplitude V_D was close to that of the shape of the ribosome, whereas the contrast of proteins contributing to U was higher than that of rRNA. The radius of gyration deduced from the cross term of neutron scattering from the deuteron spin target is 91Å.

Shape scattering of the ribosome could be obtained from $|V_D|^2$ of the deuteron target. Due to the low values of deuteron spin polarization ($|P_D| < 0.3$) and that of the deuteron spin contrast, the accuracy of this basic scattering function was so poor that it did not allow any detailed analysis.

We note in passing that a relatively uniform contrast of the ribosomal particle is achieved when the deuterated ribosome is dissolved in a protonated solvent. The radius of gyration of the shape of the frozen ribosomal particle is 89Å.

The small-angle scattering profile calculated from the ribosome model obtained by Frank *et al.* (1991) looks rather similar to the small-angle scattering curves of the ribosome solution both in frozen and in liquid state (Figure 4). A good fit was achieved when the radius of the model from electron microscopy matched the corresponding value from small-angle scattering, and when a smooth variation of the density across the surface of the model was introduced. This corresponded to a rugged surface of the ribosomal shape which was not resolved in detail by neutron small-angle scattering.

The Unlabeled, Deuterated Ribosome

The intensity of neutron small-angle scattering from deuterated ribosomes in a deuterated solvent is much weaker than that observed from the protonated ribosome in the same solvent. The contrast of the unpolarized ribosome changed from $-3.4 \times 10^{10}cm^{-2}$ to $+0.2 \times 10^{10}cm^{-2}$. Hence, the intensity of forward scattering dropped by a factor 300 (Stuhrmann, 1993). The same decrease in intensity was observed with the other basic scattering functions.

The Functional Complex

The protonated complex of two tRNAs (M = 2 × 28 000) and mRNA (M = 15000) in the deuterated ribosome gave rise to a contribution to neutron scattering shown in Figure 4, which is small compared to that of protonated proteins in the large subunit of the ribosome. There are two reasons for this:

i. The contrast of RNA in a deuterated solvent is lower by a factor 1.5 due to the lower hydrogen concentration (Figure 3). The same holds for the proton spin polarization added contrast.
ii. The occupation density of the ribosomal binding sites by tRNAs does not exceed 40%.

This results in a decrease of the scattering intensity from the tRNA/mRNA label by an order of magnitude with respect to protonated protein labels in ribosomal subunits. Nevertheless, polarized neutron scattering from the proton spin target of the functional complex labeled in its two tRNAs and the mRNA fragment clearly revealed the basic scattering functions of proton and deuteron spin contrast variation. The change of the neutron scattering intensity with the polarization of the incident neutron beam, [I(↑↑)-I(↓↑)], was shown for both the proton and deuteron spin target (Figure 5). The cross terms differ

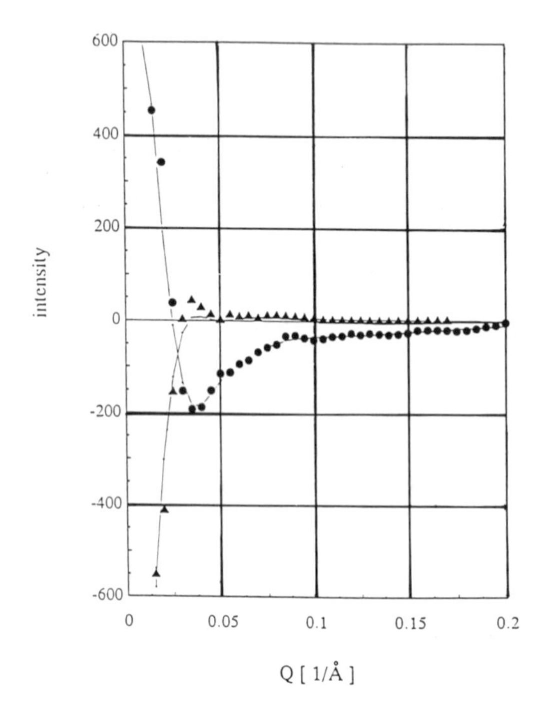

Figure 5. Basic scattering functions of the proton spin target and the deuteron spin target of the functional complex of the ribosome: the cross terms. The intensity is given in the same units as in Figure 4. Circles: Proton spins polarized, triangles: deuteron spins polarized. Actually the differences [I(↑↑)-I(↓↑)] are shown at equal nuclear spin temperature. The polarizations of the proton spins and deuteron spins were 0.7 and 0.15, respectively.

in the sign at very small Q, and so do the contrasts induced by polarization of protons and deuterons, respectively. As the density of deuteron spins in the solvent was higher than that of the solute, the deuteron spin contrast was negative. Hence the proton spin contrast was positive and so was U(0). Also at slightly wider angles, the profiles of the two cross terms looked rather different. The intensity of UV_H decreased less strongly with Q than that of UV_D. This is due to the fact that the nuclear polarization dependent contrast of the tRNA is considerably higher in the proton spin target. At $Q > 0.1Å^{-1}$ the intensity of coherent neutron scattering from the ribosome was mainly due to the amplitude of the tRNAs.

The determination of the *in situ* structure of the two tRNAs was done in two steps. In a first step the center of mass of the $(tRNA)_2$-mRNA complex was determined. The model of the complex was that of a sphere. A radius of gyration of 33Å has been used. The coordinates of the center of mass of the protonated label were varied stepwise all over the volume and the near neighbourhood of the ribosome particle, and the root mean square deviation Δ of the calculated basic scattering functions from those determined by the experiment was calculated. The lowest Δ was achieved at the interface between the ribosomal subunits (Figure 6). There was another not so deep minimum of Δ which was also located at the interface. The distance between the center of the label and the origin of the ribosome is close to 50Å.

In a further step we tried to determine the orientation the tRNAs with respect to the ribosome. To achieve this task, two facts were helpful: (i) the structure of tRNA was known to atomic resolution (Sussman *et al.,* 1978), and (ii) the extremes of the structure of tRNA must be rather close to each other (Figure 7). There was also an estimation of the angle between the planes of the tRNAs, which may be as large as 90° (Lim *et al.,* 1992).

At low resolution the structure of the L-shaped tRNA will be described by four points (Figure 7). There was no significant difference between the scattering curve of the full set of about 1600 atoms of tRNA and that of the 4 spheres at $Q > 0.2Å^{-1}$. The mRNA fragment (20% of the mass of the $(tRNA)_2$-mRNA complex) has been omitted. The orientation of the tRNA-complex was varied over the whole range of the Eulerian angles while the center of the label was kept at the parameters of Frank's model ($r = 60Å$, $\theta = 2.35$, $\phi =$

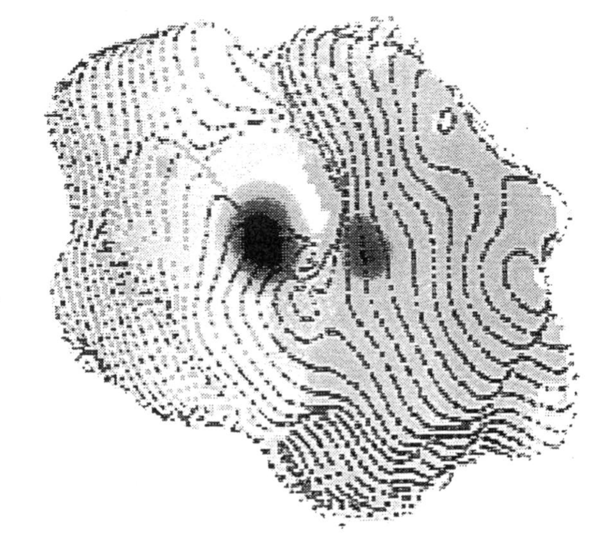

Figure 6. Search for the center of mass of the tRNA-complex bound to the ribosome. The map of deviations, Δ, as defined by Equation 9, shows most probable sites (dark spots) of tRNA near the interface between the large subunit (dark grey) and the small subunit (light grey). The contour lines are those of the ribosome model suggested by Frank (Frank *et al.,* 1991).

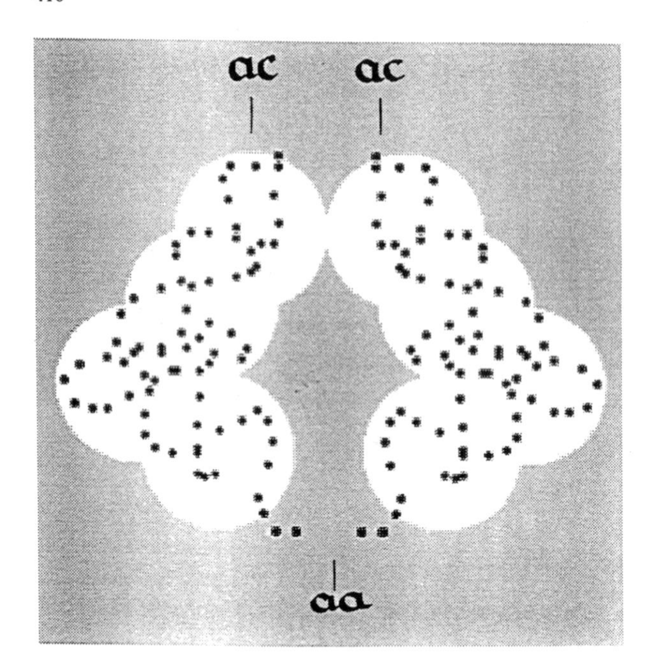

Figure 7. The anticodons (ac) of the two tRNAs and the aminoacyl group (aa) of the tRNA at the extremes of the structure must be rather close as the former bind to adjacent codons on the mRNA and the latter was near the peptidyltransferase center of the ribosome. The planar mutual orientation of the two tRNAs shown in the figure fold to an almost orthogonal orientation. Points denote phosphorus atoms of the tRNA. The large circles present a low resolution description of the tRNA.

0.92). The values of Δ varied between 1.67 and 1.87. Considering Δ values not greater than 1.69, several possible orientations of the axis of the tRNA-complex resulted in one of them being predominant (Figure 8). The anticodon of the tRNAs were close to the neck of the small ribosomal subunit, whereas the aminoacyl group approached the base of the central protuberance of the large ribosomal subunit (Figure 9). This is in agreement with biochemical observations (Lim *et al.*, 1992). A further analysis of the neutron scattering data and of their biochemical interpretation will be published elsewhere (Nierhaus *et al.*, 1995).

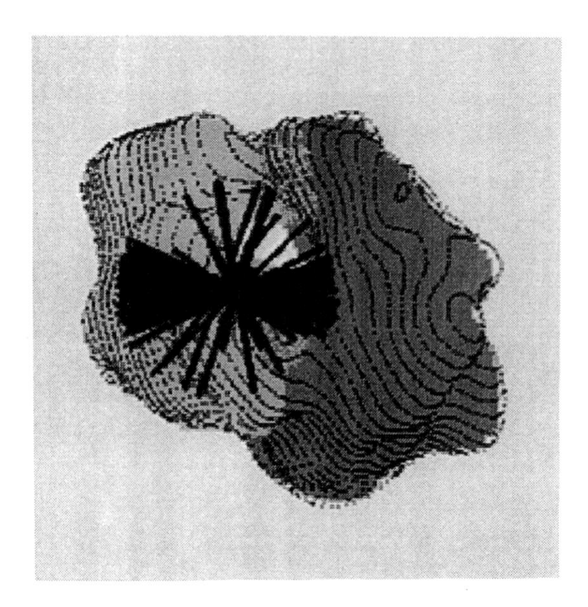

Figure 8. Possible orientations of the (ac-aa) axis of the tRNAs (Figure 7) with respect to the ribosome as obtained from the basic scattering functions of neutron scattering using the ribosome model proposed by Frank (Frank *et al.*, 1991).

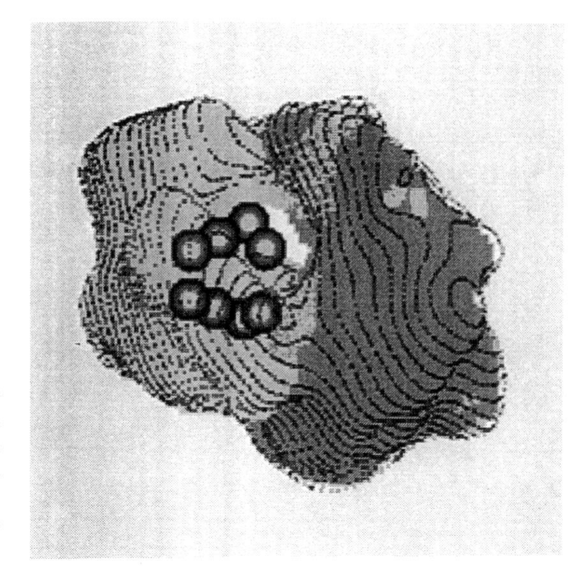

Figure 9. The site and orientation of the two tRNAs in the ribosome. The tRNAs are given in a low resolution description of four spheres (Figure 7). The anticodon is close to the neck of the small subunit whereas the aminoacyl group of tRNA approaches the central protuberance of the large subunit.

CONCLUSION

Polarized neutron scattering from polarized nuclear spin targets of biological origin extend the possibilities of the methods of isotopic substitution widely used in macromolecular structure research. The increase in contrast by proton spin polarization is very helpful in those cases where the relatively low hydrogen content (like in tRNA) and the relative small size (like that of mRNA) exclude the use of conventional methods of neutron scattering.

The method of nuclear spin contrast variation is a nondestructive labeling method. Using good glass formers as a solvent (like mixtures of glycerol and water, with a glycerol content above 0.5) and rapid freezing is a good way to preserve the properties of biomolecules. Performance of neutron scattering at temperatures below 1K and storing the samples in liquid nitrogen minimizes the damage to the sensitive biological material.

Hence, neutron scattering data can be accumulated over a long period of time in order to increase their accuracy. The results from neutron scattering of a small label like protonated mRNA would greatly improve at a more powerful neutron source. The latter is also a necessary prerequisite for polarized neutron diffraction from polarized nuclei in protein crystals.

ACKNOWLEDGMENTS

It is evident that the construction and running of the complicated instrument needed the co-operation of many people. We thank everyone who contributed to this project. We wish to enumerate some of them: T. Niinikoski, A. Rijilart, M. Rieubland and Ch. Pollicella from CERN helped us to install a polarized target station of CERN at the reactor of the GKSS Research Centre. The dilution refrigerator has been modified at CERN to meet the requirements of neutron scattering. The neutron spin polarizer is that of O. Scharpf from the Institut Laue-Langevin, Grenoble. The small-angle instrument with its area detector was constructed at the technical workshop of GKSS. The skill of W. Müller and M.

Marmotti is traceable in more recently built parts of the system. The project has been supported in its initial stage by the Bundesministerium für Forschung und Technologie.

REFERENCES

Abragam, A., & Goldman, M., (1982). In *Nuclear Magnetism: Order and Disorder*, International Series of Monographs in Physics, J.A. Krumhansl, W. Marshal and D.H. Wilkinson, editors. Clarendon Press, Oxford.

Bentley, G.A., & Mason, S.A., (1981). The crystal structure of triclinic lysozyme by neutron diffraction. In *Structural Studies on Molecules of Biological Interest*. G. Dodson, J.P. Glusker and D. Sayre editors. pp246–255. Clarendon Press, Oxford.

Berkovitch-Yelin, Z., Benett, W.S., & Yonath, A., (1992). Aspects in structural studies on ribosomes. *Crit. Rev. Biochem. Mol. Biol.*, 27:403–444.

Beyer, D., Skripkin, E., Wadzack, J., & Nierhaus, K.H., (1995). How the ribosome moves along the mRNA during protein synthesis. *J. Biol. Chem.*, 269:30713–30717.

de Boer, W., & Niinikoski, T.O., (1973). Dynamic proton polarization in propanediol below 0.5K. *Nucl. Instr. Methods*, 114:495–498.

Douzou, P., (1977). *Cryobiochemistry, An Introduction*. Academic Press, London.

Engelman, D.M., & Moore, P.B., (1972). A new method for the determination of quaternary structures by neutron scattering. *Proc. Natl. Acad. Sci. USA*, 69:1997–1999.

Frank, J., Penczek, P., Grasucci, R., & Srivastava, S., (1991). Three dimensional reconstruction of the 70S *Escherichia coli* ribosome in ice: the distribution of ribosomal RNA. *J. Cell Biol.*, 115:597–605.

Glättli, H., Fermon, C., & Eisenkremer, M., (1989). Small-angle neutron scattering with nuclear polarization on polymers. *J. Physique*, 50:2375–2388.

Hoppe, W., (1972). A new method for the determination of quaternary structures. *Israel J. Chem.*, 10:321–333.

Knop, W., Nierhaus, K.H., Nowotny, V., Niinikoski, T.O., Krumpolc, M., Rieubland, M., Rijllart, A., Schärpf, O., Schink, H.-J., Stuhrmann, H.B., & Wagner, R., (1986). Polarized neutron scattering from dynamic polarized targets of biological origin. *Helvetica Physica Acta*, 50:741–746.

Knop, W., Schink, H.-J., Stuhrmann, H.B., Wagner, R., Wenow-Es-Souni, M., Schärpf O., Krumpolc M., Niinikoski, T.O., Rieubland, M., & Rijllart, A., (1989). Polarized neutron scattering by polarized protons of bovine serum albumin in deuterated solvent. *J. Appl. Cryst.*, 22:352–362.

Knop, W., Hirai, M., Olah, G., Meerwinck, W., Schink, H.-J., Stuhrmann, H.B., Wagner, R., Wenkow-Es-Souni, M., Zhao, J., Schärpf, O., Crichton, R.R., Krumpolc, M., Nierhaus, K.H., Niinikoski, T.O., & Riillart, A., (1991). Polarized neutron scattering from dynamic polarized targets in biology. *Physica B*, 173:275–290.

Knop, W., Hirai, M., Schink, H.-J., Stuhrmann, H.B., Wagner, R., Zhao, J., Schärpf, O., Crichton, R.R., Krumpolc, M., Nierhaus, K.H., Niinikoski, T.O., & Rijilart, A., (1992). A new polarized target for neutron scattering studies on biomolecules: First results from apoferritin and the deuterated 50S subunit of *E.coli* ribosomes. *J. Appl. Cryst.*, 25:155–165.

Koghi, M., Ishida, M., Ishikawa, Y., Ishimoto, S., Kanno, Y., Masiake, A., Masuda, Y., & Morimoto, K., (1987). Small-angle neutron scattering from dynamically polarized hydrogeneous materials. *J. Phys. Soc. Japan*, 56:2681–2688.

Krumpolc, M., & Rocek, J., (1985). Chromium (V) oxidations of organic compounds. *Inorganic Chemistry*, 24:617–621.

Leslie, M., Jenkin, G.T., Hayter, J.B., White, J.W., Cox, S., & Warner, G., (1980). Precise location of hydrogen atoms in complicated structures by diffraction of polarized neutrons from dynamically polarized nuclei. *Phil. Trans. R. Soc. Lond.*, B290:497–503.

Lim, V., Venclovas, C., Spirin, A., Brimacombe, R., Mitchell, P., & Müller Florian, (1992). How are TRNAs and mRNA arranged in the ribosome? An attempt to correlate the stereochemistry of the tRNA-mRNA interaction with constraints imposed by the ribosomal topography. *Nucleic Acids Research*, 20:2627–2637.

Moore, P.B., Langer, J.A., Schoenborn, B.P., & Engelman, D.M., (1977). Triangulation of proteins in the 30S ribosomal subunit of Escherichia Coli. *J. Mol. Biol.*, 112:199–239.

Nierhaus, K.H., Lietzke, R., May, R.P., Novotny, V., Schulze, H., Simpson, K., Wurmbach, P., & Stuhrmann, H.B., (1983). Shape determination of ribosomal proteins *in situ*. *Proc. Natl. Acad. Sci. USA*, 80:2889–2893.

Nierhaus, K.H., Wadzack, J., Burkhardt, N., Jünemann, J., Meerwinck, W., & Stuhrmann, H.B., (1995). manuscript in preparation.

Niinikoski, T.O., & Udo, F., (1976). 'Frozen spin' polarized target. *Nucl. Instr. Methods*, 134:219–233.

Perkins S., (1988). Structural studies of proteins by high-flux X-ray and neutron solution scattering. *Biochem. J.*, 254:313–327.

Rheinberger, H.J., Geigenmüller, U., Gnirke, A., Hausner, T.-P., Remme, J., Sauyama, H., & Nierhaus, K.-H., (1990). In *The Ribosome: Structure, Function and Evolution* (W.E. Hill, A. Dahlberg, R.A. Garret, P.B. Moore, D. Schlessinger and J.R. Warner, editors) American Society for Microbiology, Washington D.C. pp318–330.

Serdyuk, I., & Pavlov, M.Yu., (1988). A new approach in small-angle neutron scattering: a method of triple isotopic substitution. *Makromol. Chem. Macromol. Symp.*, 15:167.

Stuhrmann, H.B., (1993). SANS with polarized neutrons. *Physica Scripta* T49:644–649.

Stuhrmann, H.B., Haas, J., Ibel, K., de Wolf, B., Koch, M.H.J., Prafait, R., & Crichton, R.R., (1976). A new low resolution model of the 50S subunit of *E.coli* ribosomes. *Proc. Natl. Acad. Sci. USA*, 73:2379–2383.

Stuhrmann, H.B., Koch, M.H.J., Haas, J., Ibel, K., & Crichton, R.R., (1978). Determination of the distribution of protein and nucleic acid in the 70S ribosome of *Escherichia coli* and their 30S subunits by neutron scattering. *J. Mol. Biol.*, 119:203–212.

Stuhrmann, H.B., Burkhardt, N., Diedrich, G., Junemann, R., Meerwinck, W., Schmitt, M., Wadzack, J., Willumeit, R., Zhao, J., & Nierhaus, K.H., (1995). Proton and deuteron spin targets in biological structure research. *Nucl. Instr. Methods in Phys. Res.*, A356:124–132.

Sussman, J., Holbrook, S.R., Warant, R.W., Church, G.M., & Kim, S.-H., (1978). Crystal structure of yeast phenylalanine transfer RNA. I. Crystallographic refinement. *J. Mol. Biol.*, 123:607–630.

Svergun, D.I., & Stuhrmann, H.B., (1991). New developments in direct shape determination from small-angle scattering. Part 1 Theory and model calculations. *Acta Cryst.*, A47:736–744.

Vanatalu, K., Paalme, T., Vilu, R., Burkhardt, N., Jünemann, R., May, P., Rühl, M., Wadzack, J., & Nierhaus, K.H., (1993). Large scale preparation of fully deuterated cell components. Ribosome from *Escherichia coli* with high biological activity. *Eur. J. Biochem.*, 216:315–321.

Zhao, J., & Stuhrmann, H.B., (1993). The *in situ* structure of the L3 and L4 proteins of the large subunit of *E.coli* ribosomes as determined by nuclear spin contrast variation. *Journal de Physique IV, Colloque*, C8, 3:233–236.

NEUTRON DIFFRACTOMETER FOR BIO-CRYSTALLOGRAPHY (BIX) WITH AN IMAGING PLATE NEUTRON DETECTOR

Nobuo Niimura

Advanced Science Research Center
Japan Atomic Energy Research Institute
Tokai-mura Ibaraki-ken 319–11, Japan.

ABSTRACT

We have constructed a dedicated diffractometer for neutron crystallography in biology (BIX) on the JRR-3M reactor at JAERI (Japan Atomic Energy Research Institute). The diffraction intensity from a protein crystal is weaker than that from most inorganic materials. In order to overcome the intensity problem, an elastically bent silicon monochromator and a large area detector system were specially designed. A preliminary result of diffraction experiment using BIX has been reported. An imaging plate neutron detector has been developed and a feasibility experiment was carried out on BIX. Results are reported.

INTRODUCTION

The X-ray diffraction of single crystals has supplied knowledge about the atomic structure of proteins, viruses, t-RNA and DNA. Since the structure-function relationship of proteins is dominated by the behavior of hydrogen atoms, it is important to know the structural information of hydrogen atoms. However, it is difficult for X-ray crystallography in biology to give the structural information of hydrogen atoms. On the other hand, neutron diffraction provides an experimental method of directly locating hydrogen atoms.

The most serious disadvantage of neutron diffraction is the low flux of neutrons irradiated on the sample specimen. The diffraction intensity is written as:

$$I \propto I_O * V * A / (v_O)^2 \tag{1}$$

Neutrons in Biology, edited by Schoenborn and Knott
Plenum Press, New York, 1996

where I, I_0, V, A and v_0 are diffraction intensity, incident neutron intensity, the volume of the sample specimen, detector area subtended by the specimen and the volume of the unit cell, respectively. Since normally the unit cell dimension of the protein crystal is about ten times larger than that of most of the inorganic materials, diffracted neutron intensity from the protein crystal becomes very weak.

We have a project to construct a dedicated diffractometer for neutron crystallography in biology (BIX) at the JRR-3M reactor at JAERI. In order to overcome the intensity problem, three items have been considered carefully. These are (i) how to get a large single crystal, (ii) how to get intense neutron flux on a sample position, and (iii) how to get a large area detector system.

This paper reports the specific features of BIX and several preliminary results of diffraction experiments by the use of BIX.

RESTRICTION ON BIX CONSTRUCTION

The most serious restriction is to share a beam with a high resolution powder diffractometer (HRPD). The beam port where the BIX is allocated is 1G and downstream the HRPD has already been installed. The least interference must be imposed for the installation of the BIX. The cross section of 1G beam is 8cm in height and 4cm in width, and for the BIX the bottom of the half height is allowed to be used, that is, 4cm × 4cm in area. Since the HRPD uses the full area of the beam (8cm × 4cm in area), the BIX is not allowed to use a pyrolytic graphite (PG) monochromator to avoid the wavelength filtering effect of PG. There are other several restrictions as follows:

- the monochromator angles are fixed as 32° and 44°
- the place where the BIX is installed is narrow
- the distance between the monochromator and the sample is larger than 3m.

MONOCHROMATOR SYSTEM

In order to solve the intensity problem, an intense neutron flux on a sample position becomes essential and a specially designed monochromator is indispensable. The requirements on the monochromator for neutron crystallography in biology are as follows: (i) Since the volume of a bio-macromolecule single crystal is small (at most $5mm^3$ in volume), a neutron beam size at the sample position must be less than 5mm × 5mm, and (ii) since the unit cell dimension is about 100Å, beam divergence is desired to be less than 0.4°. Above this limit, the adjacent Bragg reflections might overlap each other.

The PG is normally used as a thermal neutron monochromator because of the excellent reflectivity. However, the PG can not be used because of the restriction as mentioned above. In order to meet the requirement of the monochromator for this purpose, we have designed an elastically bent silicon perfect crystal (Mikula *et al.,* 1987; Mikula *et al.,* 1990; Mikula *et al.,* 1992). Figure 1 shows the monochromator system of the BIX which consists of 3 pieces of silicon (7mm in height, 250mm in length and 5mm in thickness, at present another piece of 15mm in height is added). They are bent elastically by tensioned piano wires. The obtained beam divergence is 0.4°. The current neutron flux at the sample position is 5×10^6 n/cm^2/sec. The revised monochromator is scheduled. The details are reported elsewhere (Niimura *et al.,* 1995b).

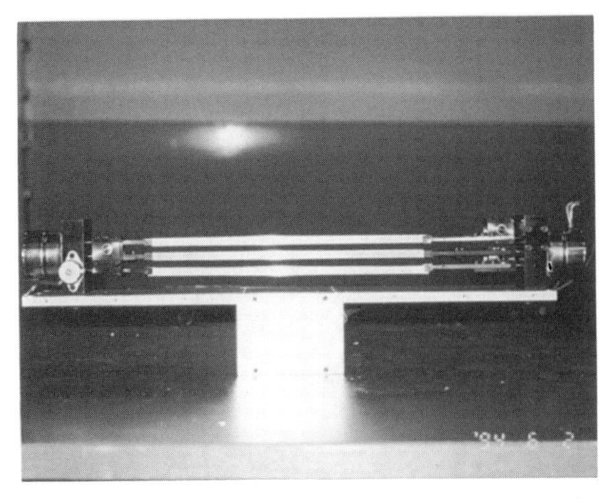

Figure 1. Monochromator system of the BIX. (A color version of this figure appears in the color insert following p. 214.)

CRYSTALLIZATION OF A LARGE SINGLE CRYSTAL

To obtain a large single crystal of bio-macromolecule is not an easy task, and especially for neutron crystallography in biology it is very important to grow a large single crystal.

The success in the structure analysis of biological macromolecules using either X-ray or neutron depends on the availability of proper single crystals. Rational design of the crystal growth based on basic understanding of the growth mechanism is more and more essential.

Small angle neutron scattering (SANS) method was used to study lysozyme solutions with particular interest in an understanding of the crystallization process. We have been able to propose the lysozyme crystallization process, the detail of which will be reported elsewhere (Niimura *et al.,* 1995a).

DETECTOR SYSTEM

In order to enlarge the solid angle subtended by a sample, multiple area detectors are equipped. At the first stage, two conventional gas-filled proportional detectors (ORDELA MODEL 2250N; active area 25cm × 25cm; pixel size 2mm × 2mm) are utilized and they cover 5.5% of the 4π effective solid angle subtended by the specimen.

PERFORMANCE AND PRELIMINARY RESULTS

Figure 2 shows the constructed BIX. Each component is explained in the caption. The preliminary results of Bragg reflections are displayed in Figure 3. The sample is hen egg white lysozyme single crystal, the size of which is $3 \times 2 \times 2mm^3$, and irradiation time is 5min. The neutron wavelength is 1.7Å.

The performance of the BIX is summarized in Table I.

Figure 2. The photograph of the BIX. (a) incident beam flight tube, (b) beam monitor, (c) sample and (d) detectors and their shielding house. (A color version of this figure appears in the color insert following p. 214.)

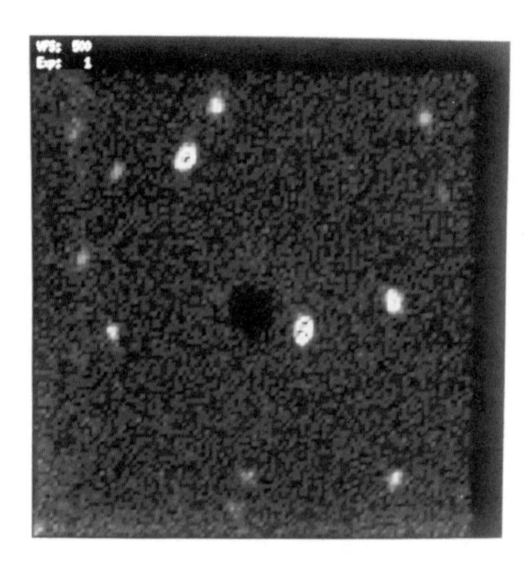

Figure 3. The diffraction patterns of hen egg white lysozyme single crystal recorded on the gas-filled position sensitive detector. (A color version of this figure appears in the color insert following p. 214.)

Table I. The performance parameters of BIX

Monochromator angle	32.0°	44.0°
Monochromator & Wavelength		
elastically horizontal bent Si(111)	1.73 Å	2.35Å
Distance between monochromator and sample	3,000–4,000 mm	
Beam size at sample position	5 mm × 5 mm	
Flux at sample position (currently)	5×10^6 n/cm^2/sec	
Goniometer	HUBER-5042	
Distance between sample and detector	600 mm	
PSD ORDELA-2250N (2 units)		
effective area	250 mm × 250 mm	
positional resolution	2.0 mm	
(Neutron imaging plate: under development)		
Detector shield	70 mm(B_4C) + 50 mm(Resin) (Bg: 0.01 cts/cm^2/sec)	

AN IMAGING PLATE NEUTRON DETECTOR

As part of the project, the development of the more efficient detector such as an imaging plate (IP) for neutrons has been undertaken. Imaging plates are used extensively in X-ray diffraction, and the characteristic features such as 100% detection quantum efficiency for 8–17keV X-rays, a spatial resolution better than 0.2mm (fwhm), a dynamic range of $1{:}10^5$ and no counting rate limitation provide a revolutionary development in the X-ray structure analysis of bio-macromolecule. An X-ray imaging plate could be converted to neutron detector when neutron converter, such as ^{10}B, ^{6}Li or Gd are combined within the imaging plate. We have already published the results obtained with an imaging plate for neutron detector (IP-ND), where neutron converter, ^{6}Li or Gd were mixed with photostimulated luminescence (PSL) material (Niimura *et al.,* 1994). The properties are wide dynamic range $1{:}10^5$ and spatial resolution better than 0.2mm. We found that the quantity of PSL created by one captured neutron depends on the kind of converters.

We have applied our IP-ND for the observation of the Bragg reflections of bio-macromolecules. The IP-ND was placed in front of the BIX detector in the shield box. Figure 4 shows the diffraction pattern obtained from a hen-egg white lysozyme single crystal, the size of which is $2 \times 2 \times 3\text{mm}^3$. The neutron wavelength is 1.7 Å. The distance between the sample and the IP-ND is 500mm, and the exposure time is 1 hour. The size of the IP-ND used is 200mm × 400mm. The specifications for the IP-ND are summarized in Table II.

The preliminary analyses of three reflections (indicated as 1 to 3 in Figure 4) were carried out. The reflections 1, 2 and 3 correspond to strong, medium and weak intensity,

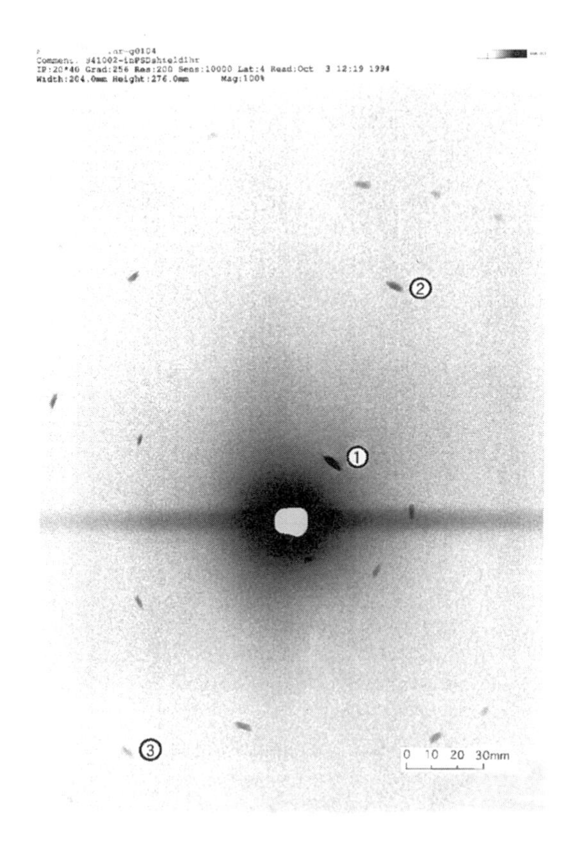

Figure 4. The diffraction patterns of hen egg white lysozyme single crystal recorded on the imaging plate neutron detector. (A color version of this figure appears in the color insert following p. 214.)

Table II. Specifications for the IP-ND developed at JAERI

Converter material	Gd_2O_3
PSL material	BaFX(X:Br,I) (Eu^{2+})
Molar ratio of converter material to PSL material	1:1
Thickness	130 μm
Size	400 mm × 200 mm
Neutron capture efficiency	62% (λ = 1.7Å)

respectively. The enlarged spot and the profiles along the apsis and minor axis of the spot are shown in Figure 5. The (a), (b) and (c) correspond to reflection 1, 2 and 3, respectively. Each profile was obtained by scanning a 1mm wide bar along the axis. The signal-to-noise ratio of the weak reflection is rather good, and a 15 minute exposure may be adequate even for such a weak reflection.

The FWHM of the profiles along the apsis and the minor axis are 4.2 - 4.7mm and 1.6 - 1.9mm, respectively. Since the measurement was carried out with the specimen not scanned, the spot size was the cross-section of the reciprocal point by the Ewald sphere. The apsis length of the spot corresponds to the contribution from the crystal size (1.5 - 2mm) and the mosaic spread (0.3°). The contribution from the IP-ND spatial resolution was negligible.

Our preliminary measurements proved that the IP-ND is feasible for the neutron diffractometer for bio-crystallography. The spatial separation between the Bragg reflections is large (Figure 4) because the camera constant (the distance between the specimen and the IP-ND) was 500mm. In future developments, this distance will be shortened. Since the IP-ND is flexible, a cylindrical shape might be the most appropriate candidate. We are now designing a new diffractometer equipped with the IP-ND as shown in Figure 6. A reader of the IP-ND is also included.

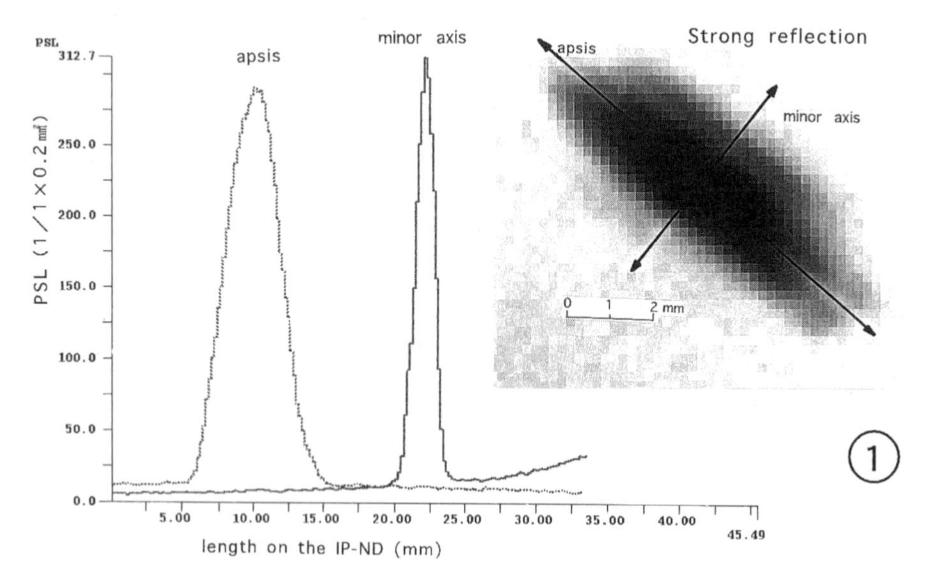

Figure 5. The enlarged spot and the profile along the apsis and the minor axis of the spot. (a), (b) and (c) correspond to the reflection 1, 2 and 3, respectively, in Figure 4. The ordinate value is the PSL per 5 × 1 pixels (1mm × 0.2mm). (A color version of this figure appears in the color insert following p. 214.)

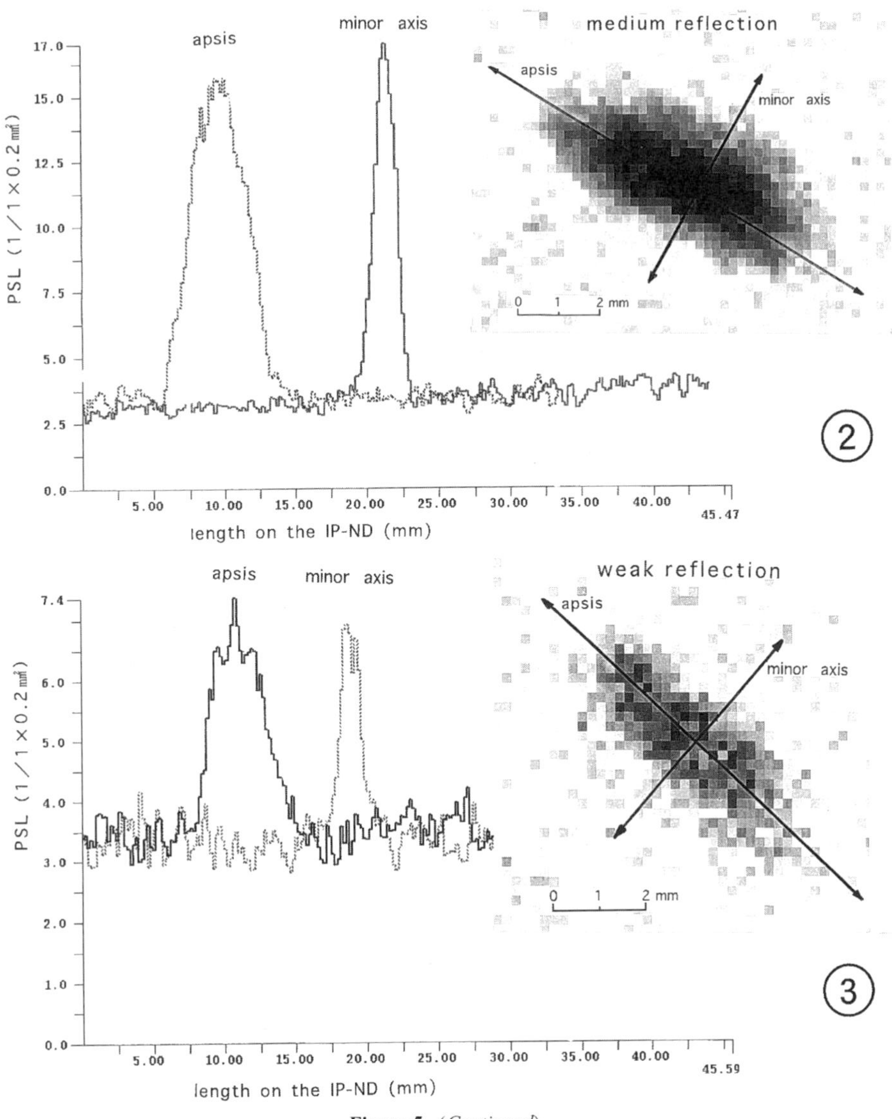

Figure 5. (*Continued*)

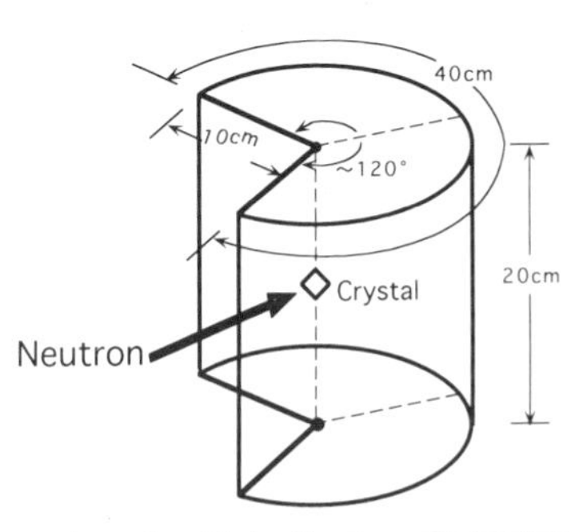

Figure 6. The layout of the new diffractometer equipped with the IP-ND.

REFERENCES

Mikula, P., Lukas, P., & Michalec, R., (1987). An experimental test of an elastically bent silicon crystal as a thermal-neutron monochromator. *J. Appl. Cryst.*, 20:428–430.

Mikula, P., Kruger, E., Scherm, R., & Wagner,V., (1990). An elastically bent silicon crystal as a monochromator for thermal neutrons. *J. Appl. Cryst.*, 23:105–110.

Mikula, P., Lukas, P., Wagner, V., & Scherm, R., (1992). Real and momentum space focussing of neutrons by a cylindrically bent silicon crystal. *Physica B*, 180 & 181:981–983.

Niimura, N., Karasawa, Y., Tanaka, I., Miyahara, J., Takahashi, K., Saito, H., Koizumi, S., & Hidaka, M., (1994). An imaging plate neutron detector. *Nucl. Instr. Methods*, A349:521–525.

Niimura, N., Ataka, M., Minezaki, Y., & Katsura, T., (1995). *Physica B* (in press).

Niimura, N., Tanaka, I., Karasawa, Y., & Minakawa, N., (1995). *Physica B* (in press).

35

A NEUTRON IMAGE PLATE QUASI-LAUE DIFFRACTOMETER FOR PROTEIN CRYSTALLOGRAPHY

F. Cipriani,[1] J. C. Castagna,[1] C. Wilkinson,[1] M. S. Lehmann,[2] and G. Büldt[3]

[1] European Molecular Biology Laboratory
Avenue des Martyrs, 156X, 38042 Grenoble Cedex, France
[2] Institut Laue Langevin
Avenue des Martyrs, 156X, 38042 Grenoble Cedex 9, France
[3] KFA Jülich
52425 Jülich, Germany

ABSTRACT

An instrument which is based on image plate technology has been constructed to perform cold neutron Laue crystallography on protein structures. The crystal is mounted at the center of a cylindrical detector which is 400mm long and has a circumference of 1000mm, with gadolinium oxide-containing image plates mounted on its exterior surface. Laue images registered on the plate are read out by rotating the drum and translating a laser read head parallel to the cylinder axis, giving a pixel size of 200μm × 200μm and a total read time of 5 minutes. Preliminary results indicate that it should be possible to obtain a complete data set from a protein crystal to atomic resolution in about two weeks.

INTRODUCTION

The mean intensity of Bragg reflections in a single crystal experiment is proportional to the product of the incident beam flux and the volume of the sample, but inversely proportional to the volume of the unit cell of the crystal. X-ray techniques have long been used to determine the structures of single crystals containing large molecules of biological significance, which by their nature have large unit cells and are often available only in the form of small crystals. Neutrons come in relatively low flux beams, but have the advantage over X-rays that they can easily detect the positions of the hydrogen atoms in a crystal structure. Since protons and water molecules are vital ingredients of all biological processes, it is clear that neutron diffraction can supply biologically useful information

Neutrons in Biology, edited by Schoenborn and Knott
Plenum Press, New York, 1996

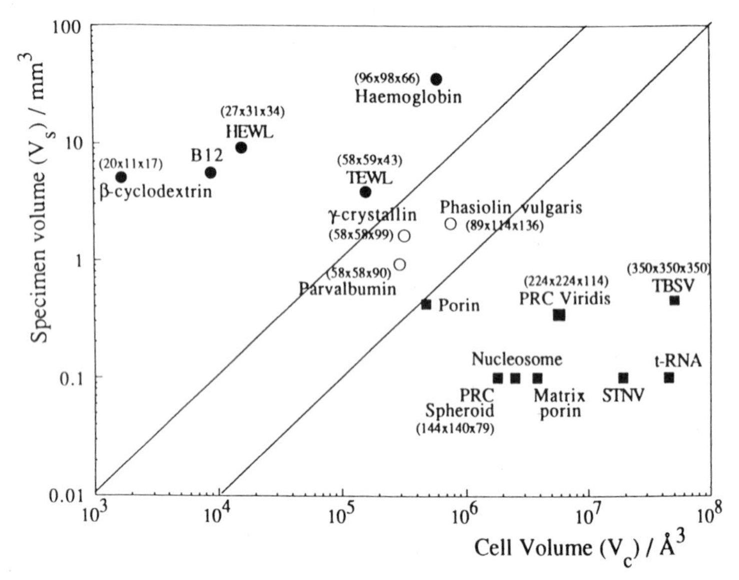

Figure 1. Plot showing the volume of single crystal specimens against unit cell size for experiments carried out on the D19 and DB21 diffractometers at ILL.

which is not available from X-ray diffraction (Schoenborn, 1969; Cheng & Schoenborn, 1991; Wlodawer *et al.*, 1984; Roth *et al.*, 1989; Bouqiere *et al.*, 1993).

In the case of a steady state source, such as the high flux neutron beam reactor at the Institut Laue Langevin (ILL) in Grenoble, in a traditional monochromatic beam experiment ($\Delta\lambda/\lambda\sim$ a few %) on a single crystal of a small protein structure, data collection times are of the order of two to four weeks when the crystal volume is several cubic millimetres. Structures which have been measured are shown in Figure 1, which is an empirical plot of specimen volume V_s against unit cell volume V_c for some protein single crystal experiments which have been performed at ILL.

The filled circles show those which have been successfully carried out to atomic resolution (~2Å) on the monochromatic single-crystal protein diffractometer D19, while the filled squares show experiments of about the same length of time which have been performed at low (~15Å) resolution on the monochromatic diffractometer DB21 using 7.6Å neutrons. The open circles show attempts which have been made to measure to atomic resolution structures with larger unit cells on D19. They could not be completed since the experiments would have taken too long. Neutrons are a scarce resource, and one would like to use a greater fraction of those which are available in the beam from the reactor (using a white beam, Laue method) in order to speed up data collection (Cipriana *et al.*, 1994). This will enable measurements to be extended to smaller volume crystals, or those with larger unit cells, the aim being to reach samples at least up to the second diagonal line of Figure 1.

LAUE DIFFRACTION

Intensity of Laue reflections

The intensity of a Laue spot at wavelength λ is given by:

$$I(hkl) = \phi(\lambda)\frac{V_s}{V_c^{\,2}}F^2(hkl)\left[\frac{\lambda^4}{2\sin^2\theta}\right] \text{ n s}^{-1}$$

and thus:

$$I(hkl) = 2\phi(\lambda)\frac{V_s}{V_c^{\,2}}\left[F(hkl)\lambda d_{hkl}\right]^2$$

where $\phi(\lambda)$ is the differential flux at the specimen, V_s is the specimen volume and d_{hkl} is the Bragg interplanar spacing. The incoherently scattered neutron background from the same crystal is:

$$B_{inc} = \phi(\lambda)\Delta\lambda\frac{V_s}{V_c}\left[\frac{\sum\sigma_{inc}(\lambda)}{4\pi}\right] \text{ n s}^{-1} \text{ sterad}^{-1}$$

where $\Delta\lambda$ is the wavelength range used and $\Sigma\sigma_{inc}(\lambda)$ is the total incoherent background cross-section of the atoms in one unit cell, coming mainly from the hydrogen atoms. It has been shown (Howard *et al.*, 1987) that the incoherent scattering cross section of hydrogen increases approximately linearly with λ and therefore the ratio of the intensity of a reflection to the background on which it sits (signal to noise ratio) is proportional to $\lambda/\Delta\lambda$ $F^2(hkl)/V_c$. This leads to the conclusion that the longest wavelength possible consistent with the desired crystallographic resolution should be used. However, in the case of very large unit cells, the waveband may need to be restricted. This is considered further below.

Comparative Efficiency of Diffractometers as a Function of the Width of Waveband

Let us compare the relative times taken to measure the same data set from a crystal using wavebands of width $\Delta\lambda_1$ and $\Delta\lambda_2$ both at the same mean wavelength. To a first order, the ratio of the number of reflections measured in the two cases is $n = \Delta\lambda_2/\Delta\lambda_1$ and therefore using $\Delta\lambda_1$ to map the **same** data which are obtained using $\Delta\lambda_2$, the mean wavelength must be varied in n steps. Similarly, if the background under a reflection is dominated by the incoherent scattering from the specimen, the background when using $\Delta\lambda_2$ will be n times larger than with $\Delta\lambda_1$. Suppose that the total background under a reflection with $\Delta\lambda_2$ is B counts per second over p pixels and that its mean value is determined from q pixels measured outside the reflection. If the total intensity of a reflection is I counts per second, then the statistical uncertainty $\sigma(It_2)$ in the measurement of I for t_2 seconds is:

$$\sigma(It_2)\left[I + \left(1 + \frac{p}{q}\right)B\right]^{1/2}(t_2)^{1/2}$$

and the fractional uncertainty f is:

$$f = \sigma(It_2)\Big/\left[I + \left(1 + \frac{p}{q}\right)B\right]^{1/2}(t_2)^{1/2}$$

The time taken to achieve this precision is therefore:

$$t_2 = \left[1+\left(\frac{p}{q}\right)\left(\frac{B}{I}\right)\right]\Big/ f^2 I$$

By a similar argument, the time t_1 taken to measure a reflection using bandwidth $\Delta\lambda_1$ to the fractional precision f is:

$$t_1 = \left[1+\left(\frac{p}{q}\right)\left(\frac{B}{nI}\right)\right]\Big/ f^2 I$$

and the ratio r of the times taken to measure the **same** data to the **same** fractional precision is given by:

$$r = (nt_1)\Big/ t_2 = \left[n+\left(1+\frac{p}{q}\right)\left(\frac{B}{I}\right)\right]\Big/\left[1+\left(1+\frac{p}{q}\right)\left(\frac{B}{I}\right)\right]$$

This may be written:

$$r = (nt_1)\Big/ t_2 = 1+\left\{(n-1)\Big/\left[1+\left(1+\frac{p}{q}\right)\left(\frac{B}{I}\right)\right]\right\}$$

and is always greater than 1 when $n > 1$, and in principal there is always a gain in using a wider bandwidth to measure data. In the case when the peak to background ratio $({}^{I}/_{B}) \gg n$, the time gained by using the wider band is n, the ratio of the bandwidths, as one would expect. In the other extreme, when $({}^{B}/_{I}) \gg n$, $r \Rightarrow 1$ and the gain is small.

To use the Laue method to measure Bragg reflections requires a large area detector. The resolution needed is ~1mm and a dynamic range of $\sim 10^5$ is desirable. To gain full advantage from the technique, the detector must have a high detective quantum efficiency and preferably an 'on-line' digital readout. One such detector is the storage phosphor or 'image' plate (Amemiya *et al.,* 1988; Rausch *et al.,* 1992; Wilkinson *et al.,* 1992; von Seggern *et al.,* 1994). Although image-plates have the drawback of being integrating devices with a time consuming reading procedure, the detection surface can be made very large and there is a wide range of freedom in the choice of detector shape. A prototype image-plate diffractometer for neutron measurements has therefore been constructed and is reported here.

NEUTRON IMAGE PLATES

X-ray image plates are flexible sheets about 500μm thick, which have been coated with an approximately 150μm layer of finely powdered photostimuable phosphor combined with an organic binder. The photostimuable material is most commonly BaFBr doped with Eu^{2+} ions and when irradiated with X-rays or γ-rays electrons liberated by ionising Eu^{2+} to Eu^{3+} are trapped in Br vacancy states just below the conduction band to form colour centers. In order to use the plates to detect neutrons, it is necessary to have a neutron converter and the most obvious candidate is Gd, which has a high cross section for thermal neutrons and produces conversion electrons and prompt γ-rays. The Gd scintillator may be used externally with a standard X-ray image plate, or mixed in the phosphor itself in the form of Gd_2O_3 powder. The main advantage at present of a separate converter/image plate system is that X-ray sensitive plates are commercially available. There are, however, a number of drawbacks in using such an arrangement. One is that the converter plate has to be removed before reading, which might be cumbersome for an on-line reading system. Another is that the resolution is reduced compared to the plates containing gadolinium. Gadolinium containing plates are not yet generally available. However, some specimens have been fabricated by Siemens, Erlangen (Rausch *et al.*, 1992) and have been made available to us for testing. Similar plates have been also made by Fuji of Japan (Niimura *et al.*, 1994, *ibid*) The plates provided for us by Siemens contain 20% by weight of Gd_2O_3, and are reported to have a detective quantum efficiency of 18% (von Seggern *et al.*, 1994).

Resolution of External and Intrinsic Converter Plates

Figure 2 shows a comparative test of the two techniques. Both kinds evidently give a resolution which would be sufficient for the study of diffraction from millimeter large crystals, but the plate containing gadolinium has the better resolution and a lower background.

These differences can be explained by the differences in the photo stimulation process which is produced either by γ-rays which are long range or by electrons which are short range. For the case where the converter is separated from the image plate the activation of the phosphor is entirely by γ-rays. The electron cascade produced in the (n,γ) resonance process ends up as additional γ-rays of all energies emitted from the Gd converter. These will all have different path lengths through the phosphor thus giving less sharp response and higher background. When gadolinium is embedded in the image plate the cascade of electrons can directly stimulate the phosphor at short range and the total γ-ray production will be reduced, leading to higher resolution and lower background.

It is possible to get an estimate of the spatial resolution by fitting the data for the Gd containing image plate to the edge response function (Harms & Wyman, 1986):

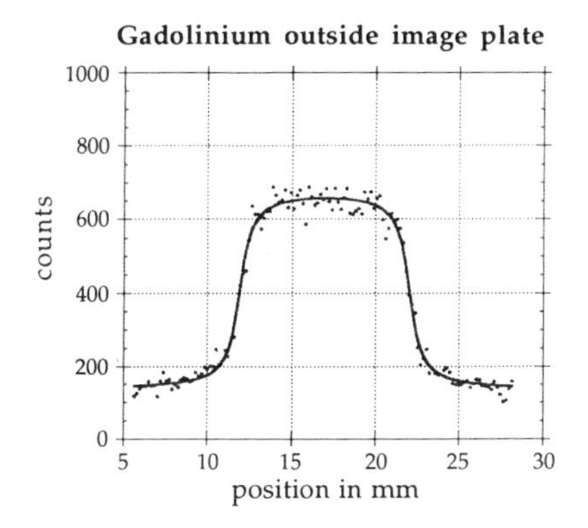

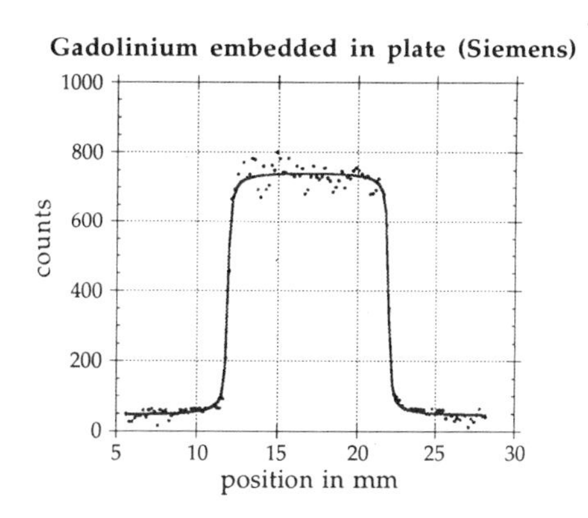

Figure 2. Sections through a 10mm × 10mm uniform neutron beam recorded on (a) Kodak X-ray image plate with a Gd_2O_3 external converter and (b) special plates (Siemens) containing 20% by weight Gd_2O_3.

$$\Gamma(x) = k_1 + k_2 \arctan(\lambda(x - x_0))$$

where $\Gamma(x)$ is the number of counts as a function of the position x on the plate, k_1 and k_2 are related to the scale factor, the absorption coefficient and the thickness of the plate that defines the edge, x_0 is the edge position and λ the resolution parameter. By a least squares fit, a value of λ of $18(1)cm^{-1}$ was found for the case with converter outside the image plate, while the value was $90(5)cm^{-1}$ with Gd embedded in the plate. The full curves in Figure 2 show the $\Gamma(x)$ function in each case. Similar differences have been observed by von Seggern *et al.* (1994).

Activation of Plates

Despite the presence of only a small quantity of Eu in the image plates, it is possible to activate the plates by thermal neutron bombardment as some of the decay processes for the excited states of the Eu^{151} and Eu^{153} isotopes which are present in natural europium have long half-lives. This has the effect that if for example a main neutron beam is allowed to fall upon the image plate, the recorded image of the beam may be read and erased, but will gradually return over a period of time, due to Eu nuclear decay. Experiments in erasing and re-reading the plates over extended periods of time show the half life of this process to be ~6(1)hours. The four possible excited states of the Eu nuclei after activation are Eu^{152m_2} ($\sigma = 3.8$ barns, $\tau = 96$ minutes), Eu^{152m_1} (3000 barns, 9.2hours), Eu^{152} (5000 barns, 13 years) and Eu^{154} (450 barns, 16 years), indicating that it is probably the Eu^{152m_1} channel which is responsible for this 'ghosting' of the original image. This could be avoided by using only Eu^{153} in preparing the image plates, but this would prove expensive. In practice, the ghosting is unlikely to be a serious problem for image recording, as plates are normally erased immediately before exposure and even after 7 hours, the level reached is found to be only ~1% of the original intensity. It is, however, unwise to let the main beam fall on the plates, as it can give an image which lasts for a considerable length of time.

NEUTRON DIFFRACTOMETER

A diagram of the diffractometer is shown in Figure 3.

Eight 200mm × 200mm square image plates (Siemens) are mounted on the outside of a 4mm thick aluminium drum with outside diameter 318.3mm. A horizontal geometry was chosen in order to simplify the bearing mechanics for the cylinder readout. A He-Ne laser is placed horizontally under the reading head and the laser light is transmitted to the drum via mirrors placed to the right and on the moveable head. The blue light which is stimulated by the evacuation of the colour center traps is detected by a photomultiplier tube, which is scanned parallel to the drum axis while the drum rotates at 350rpm, with a readout time of 5 minutes. The two motions are coupled and the integration time of the signal from the photomultiplier tube is arranged to give square pixels 200μm on edge. The drum is 400mm in length, giving some 8 Mpixels over the whole detector and is erased after readout by gas discharge lamps situated on the opposite side of the drum to the read head. The data collection by the instrument and the display and manipulation of images is controled by a Macintosh *Quadra* computer.

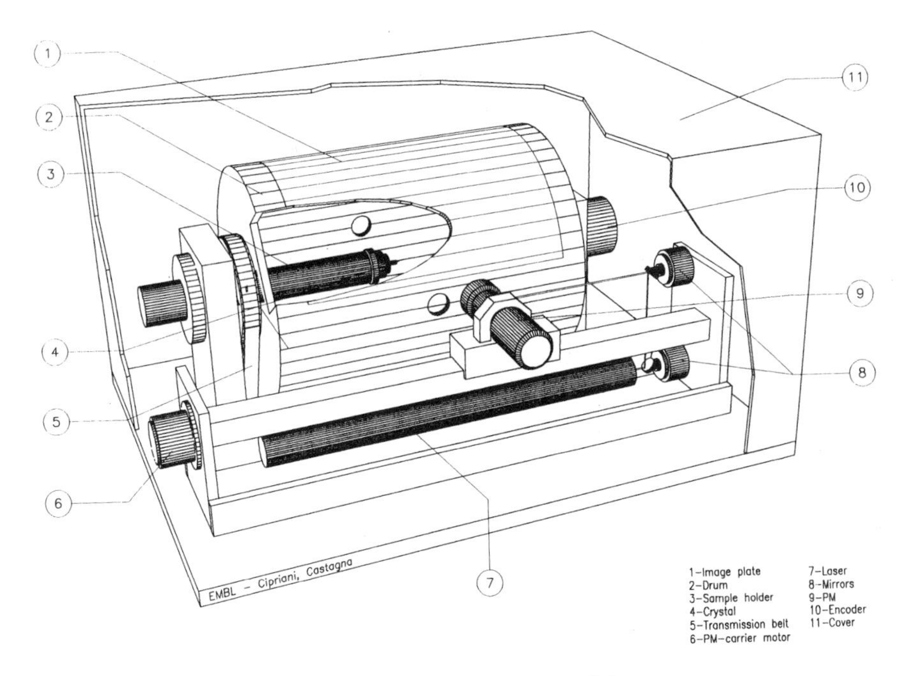

Figure 3. The on-line image plate neutron diffractometer.

First Laue Images with the Diffractometer

Diffraction patterns from several protein crystals have been recorded using the cold white beam H142 at the ILL, Grenoble. One example is shown in Figure 4, which is a pattern from a triclinic crystal (V_c ~ 26400Å^3) of hen egg-white lysozyme of volume ~1.4mm^3 taken with 150 minutes exposure.

The data has been corrected for inhomogeneities in the storage phosphor and the image opened out to show the small angle section at the center. The borders of the image plates are easily visible and reflections are visible right up to scattering angles Γ= 144°, the upper limit of the plates. Using the unit cell dimensions, the pattern has been indexed and the positions of reflections have been fitted with a root mean square positional accuracy of 200μm over the whole pattern, showing that the image is almost free from distortion. Reflections are visible to a d spacing of 3Å, and integration and analysis of the pattern is now under way. The 3Å limitation is mainly due to the level of the neutron incoherent background scattering, which is high since the full white beam with neutron wavelengths from 2.5 to 8.5Å was used. The aim is now to reduce the spread of the wavelength distribution, reduce the background, and measure data down to 2Å resolution.

CONCLUSIONS

Preliminary experiments with an on-line image plate neutron Laue diffractometer show that it functions correctly, considerably increasing the data collection rate for large scale crystal structures. It is clear that neutron image plate detectors will be useful in situations where an integrating diffractometer is situated in a low γ background environment

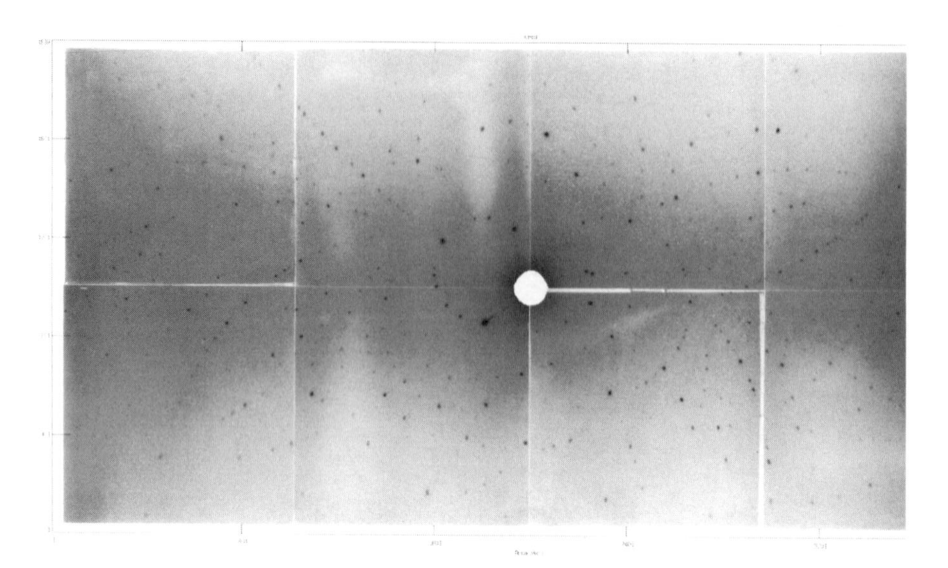

Figure 4. Laue diagram from a crystal of triclinic hen egg-white lysozyme. (A color version of this figure appears in the color insert following p. 214.)

and requires infrequent read-out (eg Laue diffraction, small-angle neutron scattering, powder diffraction) at a steady state source.

ACKNOWLEDGMENTS

We are grateful to Dr H. von Seggern of Siemens Research Laboratories, Erlangen for the supply of neutron sensitive image plates, and Didier Richard of ILL for his continuing help in the manipulation and visualisation of images. The development of this detector was financially supported by the German Budesminsterium für Bildung Wissenschaftforschung und Technology (BMBF) under grant number 03-BU3FUB awarded to one of us (G. Büldt, KFA Jülich).

REFERENCES

Amemiya, Y., Matsushita, T., Nakagawa, A., Satow, Y., Miyahara, J., & Chikawa, J., (1988). Design and performance of an image plate system for X-ray diffraction study. *Nucl. Instr. Meth.*, A266:645–653.

Bouqiere, J.P., Finney, J.L., Lehmann, M.S., Lindley, P.F., & Savage, H.J.F., (1993). High resolution neutron study of vitamin B_{12} CoEnzyme at 15 Kelvin: Structure analysis and comparison with the structure at 279 Kelvin. *Acta Cryst.*, B49:79–89.

Cheng X., & Schoenborn, B.P., (1990). Hydration in protein crystals. A neutron diffraction analysis of Carbonmonoxymyoglobin. *Acta Cryst.*, B46:195–208.

Cipriani, F., Dauvergne, F., Gabriel, A., Wilkinson, C., & Lehmann, M.S., (1994). Image plate detectors for macromolecular neutron diffractometry. *Biophysical Chem.*, 53:5–13.

Harms, A., & Wyman, D.R., (1986). *Mathematics and Physics of Neutron Radiography,* Riedel.

Howard, J.A.K., Johnson, O., Schultz, A.J., & Stringer, A.M., (1987). Determination of the neutron absorption cross section for hydrogen as a function of wavelength with a pulsed neutron source. *J. Appl. Cryst.*, 20:120–122.

Niimura, N., Karasawa, Y., Tanaka, I., Miyahara, J., Kahashi, K., Saito, H., Koizumi, S., & Hidaka, M., (1994). An imaging plate neutron detector. *Nucl. Instr. Meth.*, A349:521–525.

Rausch, C., Bücherl, T., Gähler, R., von Seggern, H., & Winnacker, A., (1992). Recent developments in neutron detection. *Proc. SPIE,* 1737:255–263.

Roth, M., Lewit-Bentley, A., Michel, H., Diesenhofer, J., & Oesterhelt, D., (1989). Detergent structure in crystals of a bacterial photosynthetic reaction center. *Nature,* 340:659–662.

Schoenborn B.P., (1969). Neutron diffraction analysis of Myoglobin. *Nature,* 224:143–146.

von Seggern, H., Schwarzmichel, K., Bücherl, T., & Rausch, C., (1994). A new position-sensitive detector (PSD) for thermal neutrons. *Proc EPDIC-3* (in press).

Wilkinson, C., Gabriel, A., Lehman, M.S., Zemb, T., & Né, F., (1992). Image plate neutron detector. *Proc. SPIE,* 1737:324–329

Wlodawer A., Walter, J., Huber, R., & Sjölin, L., (1984). Structure of Trypsin inhibitor. *J. Mol. Biol.,* 180:301–329.

36

NEUTRON DIFFRACTOMETERS FOR STRUCTURAL BIOLOGY AT SPALLATION NEUTRON SOURCES

Benno P. Schoenborn and Eric Pitcher

Los Alamos National Laboratory
M888 Los Alamos, New Mexico 87545

ABSTRACT

Spallation neutron sources are ideal for diffraction studies of proteins and oriented molecular complexes (Schoenborn, 1992a). With spallation neutrons and their time dependent wavelength structure, it is easy to electronically select data with an optimal wavelength bandwidth and cover the whole Laue spectrum as time (wavelength) resolved snapshots. This optimizes data quality with best peak-to-background ratios and provides adequate spatial and energy resolution to eliminate peak overlaps. The application of this concept will use choppers to select the desired Laue wavelength spectrum and employ focusing optics and large cylindrical ^{3}He detectors to optimize data collection rates. Such a diffractometer will cover a Laue wavelength range from 1 to 5Å with a flight path length of 10m and an energy resolution of 0.25Å. Moderator concepts for maximal flux distribution within this energy range will be discussed using calculated flux profiles. Since the energy resolution required for such timed data collection in this super Laue techniques is not very high, the use of a linac only (LAMPF) spallation target is an exciting possibility with an order of magnitude increase in flux.

INTRODUCTION

From the outset it was recognised that neutron diffraction is intensity limited and that concerted strategies must be implemented to maximise the quantity and quality of data. Consequently a number of major projects in the development of instrumentation and experimental protocols were initiated. The early uses of neutron scattering and diffraction for the analysis of biological structures were paralleled by extensive development of techniques such as the unsuccessful Fourier chopper (Nunes *et al.*, 1971), the very useful multilayer monochromators (Schoenborn *et al.*, 1974), advanced detectors (Alberi, 1976; Cain

Neutrons in Biology, edited by Schoenborn and Knott
Plenum Press, New York, 1996

et al., 1976; Schoenborn *et al.*, 1978; Fischer *et al.*, 1983; Radeka *et al.*, 1995) and sophisticated data analysis techniques (Schoenborn, 1983a; 1983b).

Multilayer monochromators are extraordinarily efficient devices that can be designed for a specific application. High reflectivity (up to 98%), tuneable bandwidth and negligible harmonic contamination are unique features. In a field obsessed with more neutrons at the sample position, the multilayer monochromator system has been used with some considerable success (Saxena & Schoenborn, 1976; 1977; 1988; Lynn *et al.*, 1976; Saxena & Majkrzak, 1984; Knott, 1995). Multilayer concepts can also be used in focusing neutron optics and multi d-spaced multilayers are now being considered for toroidal 'mirrors' in spallation neutron instruments. Major improvements in both small angle scattering research and in protein crystallography were achieved by use of efficient position sensitive detectors with good resolution and positional stability (Convert & Forsyth, 1983; Glinka & Berk, 1983). After many years of demanding development, sophisticated detector technology is now more readily available and detectors of incredible performance can be designed for each application (Fischer *et al.*, 1983; Radeka *et al.*, 1995).

Collecting diffraction data on a position sensitive detector with advanced computer technology leads to the possibility of new and more efficient data collection protocols. Not only is it possible to collect a number of reflections simultaneously, it is possible to collect the data in multi-dimensional space and integrate the data and its background (Schoenborn, 1983a; 1983b; 1992b). This leads to a significant improvement in the counting statistics for each reflection for no greater investment in neutron beamtime. Further innovation in experimental techniques continues - crystal growing and deuteration techniques are improving, and structure as a function of temperature now offers the possibility of differentiating between disorder and thermal motion.

Many of the major developments highlighted above have occurred at the large neutron facilities. With time many of the benefits are adopted by the medium flux reactors (Knott, 1995; Niimura, 1995a; 1995b). Careful selection of problems can lead to a worthwhile contribution in the field of biological structure from instruments with limited intensity. As outlined in more detail below, the use of spallation neutron sources will add a new dimension to the field.

PULSED SPALLATION NEUTRON SOURCES

Pulsed neutron sources use high energy protons from an accelerator to bombard a heavy metal target and produce neutrons by spallation (Figures 1a and 1b). Proton pulses hit the target with a frequency of 10 to 120Hz and produce high energy neutrons in equivalent bursts. These neutrons are typically moderated by interaction with light elements like water, liquid hydrogen etc to reduce the energy to thermal values (1meV to 100meV, or 10Å to 1Å). The moderators at LANSCE contain a poison layer and are lined by decoupling materials to maintain the narrow pulse width for high resolution experiments but unfortunately at the expense of reduced flux (Figure 1b). The moderated neutrons appear in burst and travel along designated beam tubes. Since the velocity of neutrons is energy dependent, the neutrons travel at different speeds. By observation at a distance (eg 10m) from the primary source, the neutron arrival time is a function of energy with the short wavelength neutrons arriving first followed a few milliseconds later by the longer wavelength neutrons (Figure 2a). In this case, time acts as a monochromator.

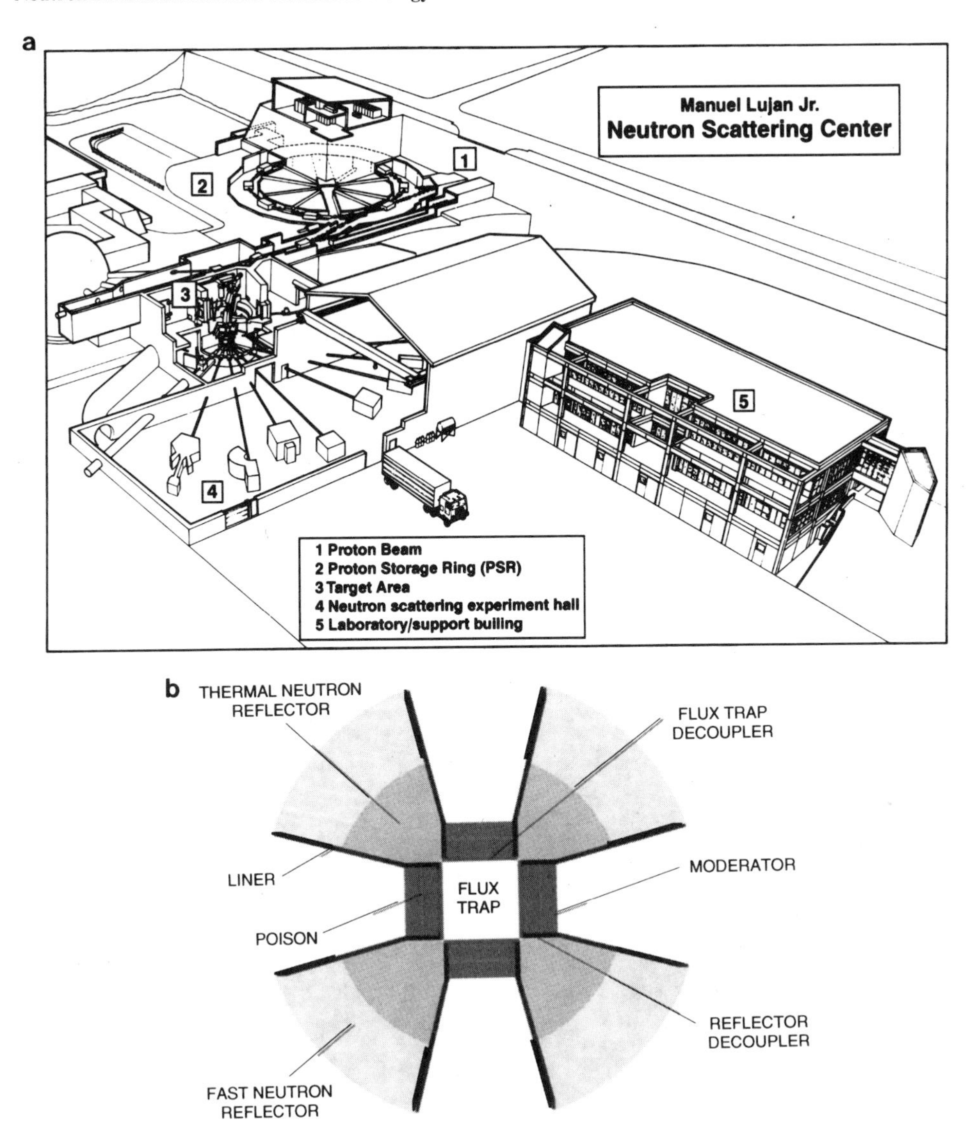

Figure 1. (a) Overview of the Los Alamos neutron scattering facilities. (b) Horizontal cut-away view of the target moderator area showing the liners (decouplers) and poison layers used to maintain the pulsed nature of the neutrons by eliminating 'stray neutrons' that would broaden the time/energy band width. The poison layer (Gd) in the moderator itself is used to limit the volume of moderator medium 'viewed' by the instrument. (A color version of this figure appears in the color insert following p. 214.)

Diffraction experiments are therefore carried out as a function of time and use a large part of the energy spectrum. This is particularly advantageous for protein crystallography. Data are collected in a stroboscopic fashion synchronised to the pulsed nature of the source. Each timed snapshot is a Laue pattern with gradually increasing wavelength but each with a small wavelength bandwidth of about 0.12Å. Since the inherent wavelength resolution of decoupled and poisoned moderators is much greater than required for protein crystallography, it would be advantageous to use a coupled unpoisoned system

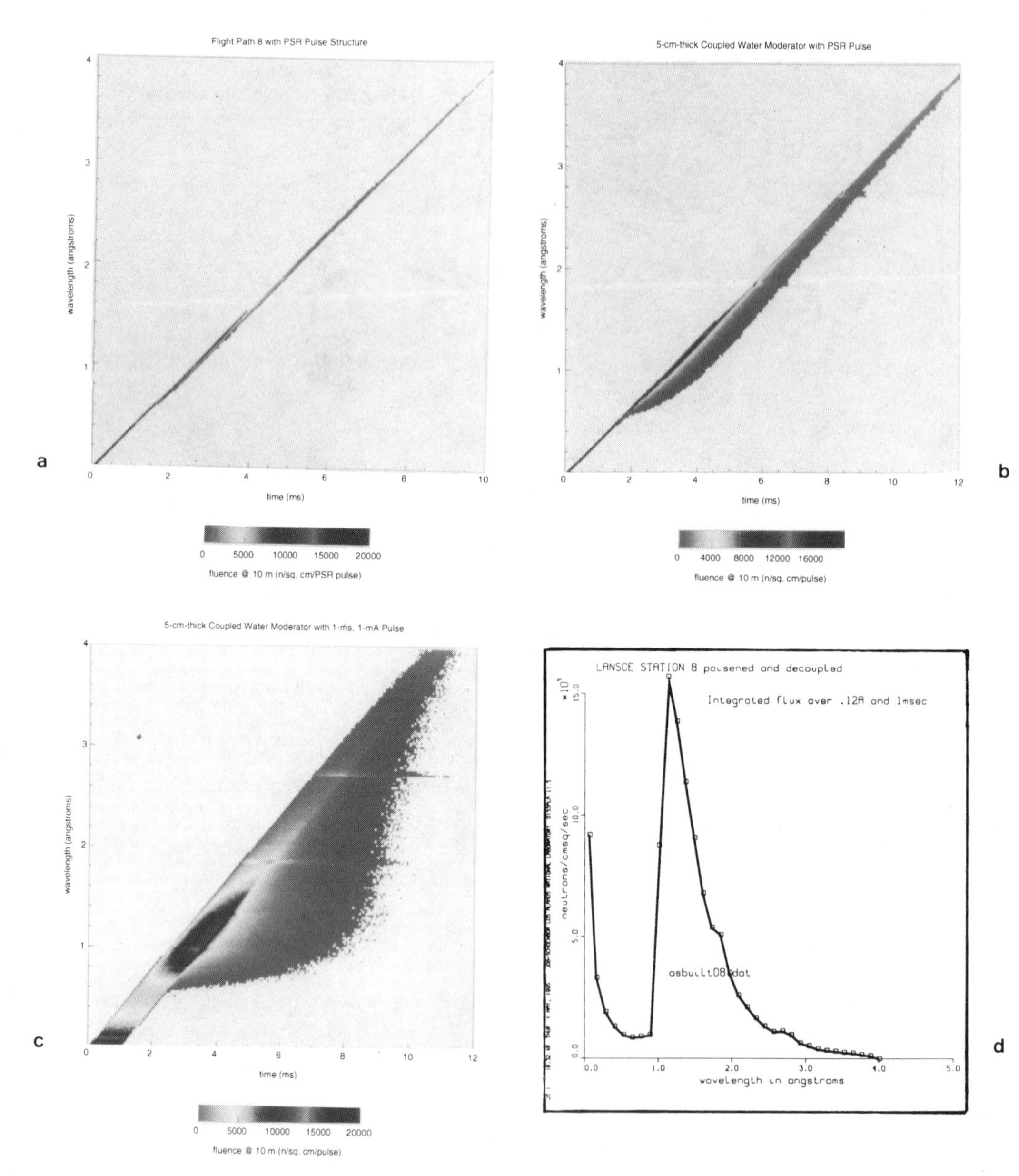

Figure 2. (a) Observed and calculated neutron flux at a sample position 10m from the neutron source of Station 8 at LANSCE using a 0.1mA pulse with a pulse duration less than 1μsec from the PSR (proton storage ring). Beamline 8 uses a water moderator. Use of a colder moderating medium, such as methane will shift the spectrum toward longer wavelength, as is desirable for protein and membrane crystallography where maximum flux between 1 and 5Å is most effective. (b) The flux distribution with the same pulse structure but using a fully coupled and unpoisoned water moderator. (c) A similar spectral representation but with a 1msec proton pulse (LPSS) with 1mA current at the end of the LAMPF Linac. Neutrons are moderated with a 50mm water moderator at ambient temperature. (d) Integrated flux over 1msec and 0.12Å as a function of wavelength for the case illustrated in Figure 1a; this is equivalent to the flux on the crystal for one snapshot of the timed Laue data collection. Departures from the smooth curve are thought to be artefact of the water scattering kernel used in the calculation, and are not expected in reality. Calculated flux profiles for the 'as built' conditions agree well with measured flux. (A color version of this figure appears in the color insert following p. 214.)

with its concomitant increase in flux. The performance of such coupled moderators depends on the specific details of the target system geometry. The full potential of a coupled moderator is realized only if all moderators in the target system are coupled. Coupled moderators have the additional benefit of lower γ backgrounds, as the decoupling material (Cd or Gd) generate γ radiation. Geometries to use such coupled and unpoisoned water moderators located above the present flux trap at LANSCE are being studied (Ferguson *et al.*, 1995).

It should be noted that the use of a time-of-flight Laue technique would also be of advantage at reactors and might be better than the quasi Laue approach which scans a reflection by a wavelength band that is just wide enough to 'cover' reflections (Schoenborn, 1992a). With the time-of-flight technique, the reactor neutron spectrum would be pulsed by a chopper producing bursts of neutrons similar to pulsed spallation neutron sources. The loss of flux by using only a time-fraction of the available neutrons is well compensated by using a wavelength band between 1 and 5Å in a super Laue type data collection technique. Optimal distribution of neutrons in this wavelength band will, however, require the development of new moderators that produce neutrons with energies between that produced by the conventional cold and liquid water moderators. In such a case, a 7m flight path and a 0.5msec pulse width would give adequate energy resolution with lowest achievable background and would allow a repetition rate of 100Hz. This would cut the monochromatic flux to 5% but allow the use of a 4Å wide wavelength band resulting in a significant gain as compared to the presently used monochromatic techniques. Fast efficient neutron detectors coupled to fast data acquisition systems are vital components in this enterprise. Essentially the data are collected in the Laue fashion but each diffraction peak has a wavelength associated with it thus eliminating physical and wavelength dependent spot overlaps - essentially the best of a monochromatic and a white radiation experiment are retained (Schoenborn, 1992a).

Long Pulse Spallation Neutron Sources (LPSS)

The present pulsed neutron sources rely on proton pulses of less than one microsecond duration yielding very high energy resolved neutrons (Figure 2a). Such resolution is, however, not required for protein crystallography as has been demonstrated by the quasi Laue technique (Schoenborn, 1992a) and, indeed, it is advantageous to use longer proton pulses. The use of such long pulse spallation sources (LPSS) for diffraction studies has been suggested by Bowman and described by Mezei (1994). These proposed devices would have a proton current increased by an order of magnitude with equivalent increases in spallation neutrons as shown in Figure 2c. The existing linear accelerator at Los Alamos (LAMPF) can produce proton pulses at 1mA current with a width of 1msec at 60Hz with a 10 fold increase in average neutron flux. In addition to increased initial proton current, the design of moderators tailored for protein crystallography can produce further increases in flux, enhancing the most useful wavelength range (1 to 5Å). The use of a LPSS with advanced large, efficient position sensitive detectors and focusing optics should enable the collection of high resolution protein data within days instead of months. A suitable device will be detailed later.

Spallation neutron sources are not the panacea for all neutron scattering but they are ideal for protein crystallography.

Flux Calculations Using a Monte Carlo Approach

In order to compare the neutron flux of the existing LANSCE source with different moderators and an LPSS, a Monte Carlo computation approach was used. The production of neutrons via spallation and their subsequent transport in flight tubes was simulated using the LAHET Code System (LCS) (Prael & Lichtenstein, 1989), which consists primarily of two Monte Carlo transport codes: LAHET and MCNP. LAHET, a descendent of HETC developed at Oak Ridge National Laboratory, is a model-based code that simulates the transport and interaction of high-energy nucleons, pions and muons with stationary nuclei. The user may select from a variety of physics models for simulating nuclear processes (Prael, 1994). The Bertini model is used to model the intranuclear cascade, wherein a high-energy subatomic particle interacts directly with individual nucleons in a target nucleus, imparting appreciable energy to an individual nucleon. This nucleon may be 'knocked' out of the nucleus with high energy, or it may rattle about within the nucleus, transferring its energy to the other nucleons and raising the nucleus in an excited state. In what is termed the 'evaporation phase', the nucleus then relaxes to a lower energy state by the emission of low-energy (a few MeV) neutrons, protons and/or photons. Low-energy neutrons created in this manner (about 15 per incident 800MeV proton) are then transported using MCNP, a data-based code that uses continuous-energy cross-section libraries to simulate the transport and nuclear interaction of low-energy neutrons. This code models the slowing down and thermalization of neutrons in the moderating medium, typically an hydrogenous material such as water, methane or liquid hydrogen. Thermalized neutrons then leak from the moderator, a small fraction of which happen to have a direction along designated beam tubes.

The user must provide to the code (LCS) a description of the geometry and a list of the materials appearing in the geometry. A detailed model of the as-built LANSCE target system geometry (Hughes, 1988) was used to calculate the neutron flux heading down Flight Tube 8 at a point 10m from the moderator leakage surface as a function of both time and wavelength. The moderator viewed by Flight Tube 8 is a 40mm thick high-intensity room-temperature water moderator, decoupled with 0.81mm of Cd and poisoned 25mm from its viewed surface with a 0.51mm thick foil of Gd. The viewed surface area is $130 \times 130\text{mm}^2$. Results are shown in Figure 2a for the proton pulse structure currently in operation at LANSCE: a 250nsec wide, 20Hz pulse and 0.1mA average beam current.

An increase in flux of about a factor of two with only a moderate line broadening can be obtained by using a coupled and unpoisoned moderator. Figure 2b shows the flux distribution with the above described pulse structure but using a coupled unpoisoned water moderator. Changes in reflector and moderator geometry and material can further increase flux of up to 8×.

A major increase in flux is achieved by using the LPSS concept and is depicted in Figure 2c. In this LPSS model, the flux profile is calculated 10m from a fully coupled unpoisoned water moderator in a target system of the same flux-trap configuration as the LANSCE target system but with a proton pulse structure of 60Hz, a 1msec square pulse, and an average beam current of 1mA. This is prototypical of the pulse characteristics proposed for the LANSCE LPSS. The neutron pulse is much broader as compared to the short pulse system depicted in Figures 2a and 2b. For protein crystallography a pulse width of 0.5msec would reduce the wavelength width to a nearly optimal quasi Laue width and would provide a best peak to background ratio. The existing LAMPF could be configured to such a pulse structure but with a slightly reduced current or an increased repetition rate.

A PROTEIN NEUTRON DIFFRACTOMETER FOR SPALLATION NEUTRONS

The proposed protein crystallography station at LANSCE, or at an LPSS, shown in Figure 3 will use focusing optics and choppers to produce the desired neutron pulse profile at the crystal. The concept of using such focusing devices based on super-mirror technology has already been tested for vertical focusing but has to be developed for toroidal geometry, so as to use all neutrons acceptable by the protein crystal. Measurements have shown that the large protein crystals used for neutron diffraction experiments have mosaic width of 0.2 to 0.3° and focusing within these limits will produce significant gains by a factor of approximately 7× or better (unpublished data Schoenborn *et al.*). A major component of any new data collection strategy is, however, the use of very large neutron detectors.

Detectors

At present a typical neutron detector has an active area of about 200 × 200mm^2 with a resolution of at best 1.3mm at fwhm. The size of these planar position sensitive detectors is largely given by the acceptable parallax and window thickness. The parallax is given by the gas thickness which is typically 12 to 15mm. The window is made of aluminium which has a low neutron absorption cross-section to minimize window scattering. To satisfy the safety requirements for pressure vessels as described in the 'Boiler code', a window thickness of at least 7mm is required for total gas pressures of close to 8atmos. The exact gas pressures depend on the wavelength range used, the resolution required and the choice of gases (Radeka *et al.*, 1995). To minimize parallax effects and reduce window

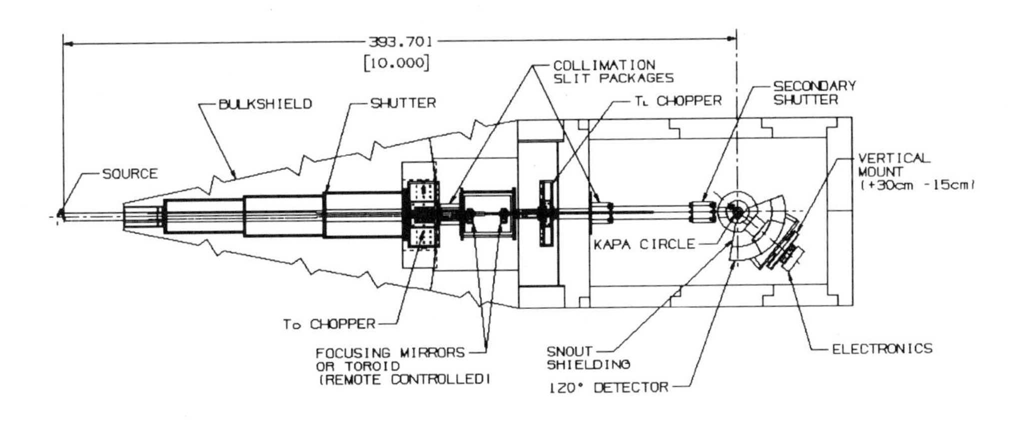

Figure 3. Conceptual design of a protein and membrane diffraction crystallographic station for a pulsed neutron source. This station will use choppers to define the wavelength band used and uses a toroid mirror to focus the beam. The first (T0) chopper opens the beam after the initial pulse with its high γ and fast neutron background is dissipated, thus protecting the sample and the detector from this initial radiation burst. The second chopper is used to prevent frame overlap and selects the long wavelength edge. The wavelength band desired for a particular experiment is easily chosen by tuning the speed of the choppers in synchronism with the accelerator's pulse structure. A large cylindrical position sensitive detector covering an arc of 120° with a 170mm height is planned with a resolution of 1.3mm and a counting rate exceeding 10^6 counts/sec (Alberi, 1976; Cain *et al.*, 1976).

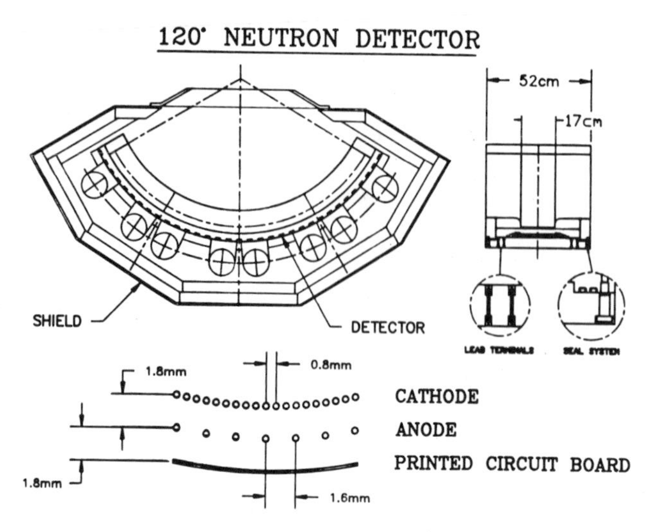

Figure 4. Orthogonal views of the ^3He detector.

thickness, cylindrical detectors are preferable. For crystallographic applications a detector with a radius of 700mm covering an arc of 120°, and a 170mm height, has been designed (Schoenborn, 1992b) (Figure 4).

An event counting detector system rather than a scintillation device has been chosen to take advantage of known technology and the large dynamic range these systems exhibit. Image plate type systems are not suitable in situations that use short time slices as encountered in inelastic experiments or with spallation sources. Phototube multiplier detectors are being developed for such application but have not yet demonstrated the large dynamic range required for protein crystallography. Solid state detectors are also under development but are not yet suitable for applications requiring large areas with resolutions approaching 1mm. Therefore, the readout system of a multiwire gas proportional detector will use the Radeka centroid finding filter (Fischer *et al.,* 1983) with a resolution approaching 1.3mm. The wire based system will allow single event counting with a counting rate approaching one million counts/sec. To achieve the high counting rates, the detector will be segmented into at least 4 subdivisions each with its own readout electronics and data acquisition system. The data acquisition system will utilize a commercial computer interface based on time slicing direct memory access processors (Motorola) with a reflective memory network system (VME Microsystems Inc). The detector will be constructed as a multiwire proportional chamber. Starting from the back, there will be the rear cathode, then an anode wire plane, then the front cathode; a window above the front cathode defines the drift region. The entire electrode structure will be contained in a stainless steel/aluminium pressure vessel and filled with a mixture of ^3He and C_3H_8. ^3He has an extremely high cross-section for thermal neutron absorption (for example, about 24,000 barns at 8Å), with the detection interaction resulting in the creation of a proton and a triton:

$$n + {}^3He \rightarrow p + {}^3H + 0.764\,MeV \tag{1}$$

The C_3H_8 performs two major functions: its high density limits the range of the proton and triton produced by the above nuclear reaction, thereby improving the spatial resolution of the detector relative to other commonly used gases, and it also acts as a quenching agent.

The two cathodes are constructed as position interpolating structures; they comprise an array of copper strips on a solid substrate for the rear cathode, and an array of wires for the upper cathode. The strips and wires are orthogonal to each other; therefore one cathode determines position in the X-axis, and the other determines position in the Y-axis. The sensitive region of the detector forms an arc which is nearly 1500mm along the detection-sensitive arc and 170mm wide. The containment structure is somewhat larger than these dimensions. The anode wires run in the planar direction, have a diameter of 15μm and a pitch of 1.6mm; therefore, just over 900 anode wires along the entire arc are needed. The upper cathode plane is constructed of 50μm wires running parallel to the anode wires, but with a pitch of 0.8mm. The lower cathode will be formed by a specially curved substrate of glass or printed circuit board, on which copper strips are deposited or etched, with the strips at a pitch of 1.4mm and running along the arc, ie at right angles to the anode wires. To simplify the fabrication of this electrode, it will be constructed in four sections. At the front of the detector, the gas volume is enclosed by a 6mm thick aluminium window, curved to provide a constant (radial) gas depth of 15mm between lower cathode and window. An operating gas mixture of 3atmos ^{3}He and 2.5atmos C_3H_8 provides high efficiency and good position resolution. Position resolution is almost inversely proportional to the pressure of C_3H_8.

Gas composition and purity is an important feature of the device. C_3H_8 is used to provide good stopping power for the proton and triton given off in the primary nuclear reaction. Propane is a superior quenching gas and has a relatively lower cross-section for γ ray absorption; this is crucial since there is nearly always a background of γ rays in neutron experiments. The helium-propane mixture must be kept very pure at all times. Because of cost considerations it is necessary to operate the detector with a closed gas, circulating system. Thus, a part of the circulating system contains a gas purifier, which is basically composed of two parts: a container of oxygen absorber, and a container filled with a molecular sieve material to absorb mainly water vapor. A small pump moves the gas at 1 to 2 liters per min, which is adequate for all normal operating conditions. With the counting gas of ^{3}He (3atmos) and propane (2.5atmos), the Raleigh resolution is designed to be in 1.2mm (vertical) and 1.8mm (horizontal) with a maximum parallax of 0.4mm.

Instrumentation for the Super Laue Method for Protein Crystallography

Pulsed sources with the wavelength resolved as a function of time provide an ideal use of the Laue method without the problem of increased background or the extreme number of spots, spot overlaps and spot multiplicity usually encountered in Laue crystallography - thus the name *super Laue*. The velocity of neutrons is energy (wavelength) dependent and allows the resolution of different wavelength as a function of their time of arrival at the detector. Experiments with 100Å unit cells require a flight path of 7 to 12m for LANSCE but longer for an LPSS. For the required energy resolution relatively few time slices are needed (15 slices using a source with a repetition rate of 40). The peak counting rate of the detector has, however, to be considered and the flight path length should be matched to the time interval of the wavelength bandwidth so that the detector is active for the whole time interval of the pulse structure. For such short flight path length, gravitational effects on the neutrons are negligible and amount to less than 0.2mm and therefore may be neglected. For such a short flight path, the actual length will be determined more by the shielding required than crystallographic considerations. For pulsed sources with its higher energy background produced by the initial pulse, the shielding required are much more demanding than in a reactor environment.

All time slices will be stored and archived. With a wavelength band from 1 to 5Å and a unit cell of 60Å, a single crystal setting will produce about 4800 reflections. An exact comparison of data collection efficiency between reactor and spallation based neutron sources (SNS, LPSS) is difficult since different data collection strategies are used. In the next section, an attempt is made to compare data collection strategies using measured flux and comparing different data collection modes.

Comparing Protein Data Collection Techniques

For these calculation, data efficiencies are calculated for (i) a reactor (HFBR) based on the presently used monochromatic technique, (ii) the quasi Laue and (iii) the time-of-flight super Laue for a spallation source (LANSCE) - all other conditions are equivalent. As noted previously, it is also possible to use a chopper at a reactor and then use a time-of-flight Laue technique. The true Laue at a reactor is not advisable since the background swamps the low intensity reflections. The background is proportional to the total flux incident on the crystal while a particular reflection's intensity is proportional to the intensity in the small wavelength band width that gives rise to the reflection. This was demonstrated with a Fourier chopper technique and a classical Laue approach using a wide wavelength band tailored by an MgO filter (Nunes & Schoenborn *unpublished*).

Spallation neutrons are ideal for protein crystallography and provide the best properties of the monochromatic and the Laue technique. Direct comparison of steady state and pulsed neutron sources for protein crystallographic measurements are difficult since the data collection strategies are different. The relative efficiencies of such sources can however be estimated by comparing the effective fluxes. The effective flux in this case is the integrated flux over the wavelength band $\Delta\lambda$ that gives rise to the diffraction intensity and the relative time it takes to measure reflections. The useful $\Delta\lambda$ is proportional to the mosaic spread of the crystal and for the large crystals used for neutron scattering, $\Delta\lambda$ is approximately 0.12Å. In the monochromatic case the $\Delta\lambda$ of the incident beam is small (~0.03Å) and the crystal is stepped through the reflecting condition by rotating the crystal (typically 10 steps at 0.1°). In the spallation Laue case no stepping is required since the $\Delta\lambda$ required to satisfy the diffraction condition is covered by the bandwidth of the incident beam and the timed structure of the beam. For the comparison made Table I, the flux is given as $n/cm^2/sec$ for the given $\Delta\lambda$.

To calculate these relative gain factors the following conditions were assumed:

- Measured flux for the HFBR with a Cu monochromator and average flux for LANSCE with an unpoisoned decoupled water moderator; (peak flux at 1.5Å is 3.3×10^6 $n/cm^2/sec$ with a $\Delta\lambda$ of 0.12Å and a 1msec time slice [integration time]).
- The same time interval of 10 for all 3 cases is used - it takes at least 10 steps to scan through reflections in the monochromatic case, 3 steps in quasi Laue (to catch reflections on the wavelength edge) and only one step in the Laue case since 'reflections' are completely 'illuminated' by the wavelength bands used.
- These calculations use the same detector size (45° acceptance angle).
- The number of reflections were calculated using the program Newlaue with a unit cell of $64 \times 30 \times 34$Å^3.
- No focusing, with optimal moderator view for a 0.2° converging beam. The use of focusing devices can further improve flux on target by a factor of 3 to 10.

Table I. Comparison of data collection strategy at a reactor and pulsed neutron source

	Monochromator HFBR	Quasi Laue HFBR	Laue LANSCE
Flux (neutrons/cm^2/sec)	6×10^6	24×10^6	0.8×10^6
Number of reflections	152	311	4835
Number of steps	10	3	1
Wavelength	1.5Å	1.5–1.65Å	1.0–5.0Å
Number of reflections × effective flux	9×10^9	24×10^9	38×10^9
Relative gain	1	2.7	4.2

The possible use of a coupled and unpoisoned water moderator increases the flux by a factor of 2 to 8 depending on the location of the moderator as described above (Figure 2b) and increases the gain relative to the conventional reactor model by at least 8×. These results strongly support the proposal that crystallography on a pulsed source has distinct advantage.

As outlined previously, further improvement can be achieved by targeting LAMPF directly onto a neutron target (LPSS). LAMPF can operate with a current of 1mA compared with the 0.1mA current of the existing compressor ring of the LANSCE facility. The resulting band width of the planned 1msec LPSS (Figure 2c) is however larger than can effectively be 'used' by a given reflection and the resulting gain is approximately 5. A 0.5msec pulse structure would be better and would result in a gain factor close to 10.

The development of the present LAMPF as an LPSS is straight forward and need basically only the addition of a neutron target complex and associated neutron beam facilities. Such a facility would cost about $US100 million, would take 2 years to construct and would provide a major new neutron source surpassing the capabilities of present sources for many experiments particularly protein, virus and membrane crystallography.

CONCLUSION

As describe above, spallation neutron sources provide an exciting new approach to protein, virus and membrane diffraction. Immediate improvements in data collection rates will be observed and it should be possible to collect high resolution data sets within weeks instead of months. As experience with spallation neutron crystallography increases and moderator, focusing, detector and data analysis designs improve, data collection rates should improve by order of magnitude.

ACKNOWLEDGMENTS

The described detector was designed in collaboration with V. Radeka, G.C. Smith and N.A. Schaknowski of the instrumentation division at BNL. The authors wish to thank R. Knott for editing this paper. This development was carried out under the auspices of the Office of Health and Environmental Research of the US Department of Energy.

REFERENCES

Alberi, J., (1976). Development of large-area, position-sensitive neutron detectors. In *Neutron Scattering for the Analysis of Biological Structures*. (B.P. Schoenborn, editor) ppVIII24-VIII42. (Springfield, VA: Nat. Technical Infor. Serv.; U.S. Dept. of Commerce).

Cain, J.E., Norvell, J.C., & Schoenborn, B.P., (1976). Linear position-sensitive counter system for protein crystallography. In *Neutron Scattering for the Analysis of Biological Structures*. (B.P. Schoenborn, editor) ppVIII43-VIII50. (Springfield, VA: Nat. Technical Inform. Serv.; U.S. Dept. of Commerce).

Convert, P.A., & Forsyth, J.B., (1983). *Position-Sensitive Detection of Thermal Neutrons*. (Academic Press, London).

Ferguson, P.D., Russell, G.J., & Pitcher, E.J., (1995). Reference moderator calculated performance for the LANSCE upgrade project. Proceedings of the Meetings ICANS-XIII and ESS-PM4, PSI Proceedings 95-02, 510.

Fischer, J., Radeka, V., & Boie, R.A., (1983). High position resolution and accuracy in ^{3}He two-dimensional thermal neutron PSDs. In *Position-Sensitive Detection of Thermal Neutrons*. (P. Convert and J.B. Forsyth, editors). pp129–140. (Academic Press, London).

Glinka, C.J., & Berk, N.F., (1983). The two-dimensional PSD at the National Bureau of Standards' Small Angle Neutron Scattering Facility. In *Position-Sensitive Detection of Thermal Neutrons*. (P. Convert and J.B. Forsyth, editors). pp141–148. (Academic Press, London).

Hughes III, H.G., (1988). Monte Carlo Simulation of the LANSCE Target Geometry. 97th ed. 1988; New York: Institute of Physics. 455p. Proceedings of the Tenth International Collaboration on Advanced Neutron Sources.

Knott, R., (1995). Neutron Scattering in Australia. In *Neutrons in Biology*. (B.P. Schoenborn and R. Knott, editors). (Plenum Publishing Corporation, New York).

Lynn, J.W., Kjems, J.K., Passell, L., Saxena, A.M., & Schoenborn, B.P., (1976). Iron germanium multilayer neutron polarizing monochromators. *J. Appl. Cryst.*, 9:454–459.

Mezei, F., (1994). On the comparison of continuous and pulsed sources. *Neutron News*, 5:2–3.

Niimura, N., (1995a). Neutron scattering and diffraction instrumentation for structural biology in Japan. In *Neutrons in Biology*. (B.P. Schoenborn and R. Knott, editors). (Plenum Publishing Corporation, New York).

Niimura, N., (1995b). Neutron diffractometer for bio-crystallography (BIX) with an imaging plate neutron detector. In *Neutrons in Biology*. (B.P. Schoenborn and R. Knott, editors). (Plenum Publishing Corporation, New York).

Nunes, A.C., Nathans, R., & Schoenborn, B.P., (1971). A neutron Fourier chopper for single crystal reflectivity measurements: some general design considerations. *Acta Cryst.*, A27:284–291.

Prael, R.E., (1994). A Review of the Physics Models in the LAHET Code. Los Alamos National Laboratory. LA-UR-94–1817.

Prael, R.E., & Lichtenstein, H., (1989). User Guide to LCS, the LAHET code system. CA-UR-89–3014.

Pynn, R., (1995). Reference Moderator Calculated Performance for the LANSCE Upgrade Project. 1995 Oct 11; ICANS-XIII.

Radeka, V., Schaknowski, N.A., Smith, G.C., & Yu, B., (1995). High precision thermal neutron detectors. In *Neutrons in Biology*. (B.P. Schoenborn and R. Knott, editors). (Plenum Publishing Corporation, New York).

Saxena, A.M., & Majkrzak, C.F., (1984). Neutron optics with multilayer monochromators. In *Neutrons in Biology*. (B.P. Schoenborn, editor). pp143–158. (Plenum Publishing Corporation, New York).

Saxena, A.M., & Schoenborn, B.P., (1976). Multilayer monochromators for neutron scattering. In *Neutron Scattering for the Analysis of Biological Structures*. (B.P. Schoenborn, editor). ppVIII30-VIII48. (Springfield, VA: Nat. Technical Infor. Serv.; U.S. Dept. of Commerce).

Saxena, A.M., & Schoenborn, B.P., (1977). Multilayer neutron monochromators. *Acta Cryst.*, 833:805–813.

Saxena, A.M., & Schoenborn, B.P., (1988). Multilayer monochromators for neutron spectrometers. *Material Science Forum*, 27/28:313–318.

Schoenborn, B.P., (1983a). Peak shape analysis for protein neutron crystallography with position-sensitive detectors. *Acta Cryst.*, A39:315–321.

Schoenborn, B.P., (1983b). Data processing in neutron protein crystallography using position-sensitive detectors. In *Position-Sensitive Detection of Thermal Neutrons*. (P. Convert and J.B. Forsyth, editors). pp321–331. (Academic Press, London).

Schoenborn, B.P., (1992a). Multilayer monochromators and super mirrors for neutron protein crystallography using a quasi Laue technique. *SPIE*, 1738:192–199.

Schoenborn, B.P., (1992b). Area detectors for neutron protein crystallography. *SPIE*, 1737:235–243.

Schoenborn, B.P., Alberi, J., Saxena, A.M., & Fischer, J., (1978). A low angle neutron data acquisition system for molecular biology. *J. Appl. Cryst.*, 11:455–460.

Schoenborn, B.P., Caspar, D.L.D., & Kammerer, P.F., (1974). A novel neutron monochromator. *J. Appl. Cryst.*, 7:508–510.

INDEX